AF750772

Vehicle–Track–Bridge System Subjected to Track Alignment

Xuyou Long · Jin Shi · Yingjie Wang

# Vehicle–Track–Bridge System Subjected to Track Alignment

## Dynamic Modeling and Applications

CHINA RAILWAY PUBLISHING HOUSE CO., LTD. Springer

Xuyou Long
China Railway Design Corporation (CRDC)
Tianjin, China

Jin Shi
School of Civil Engineering
Beijing Jiaotong University
Beijing, China

Yingjie Wang
School of Civil Engineering
Beijing Jiaotong University
Beijing, China

ISBN 978-981-95-1648-3 ISBN 978-981-95-1649-0 (eBook)
https://doi.org/10.1007/978-981-95-1649-0

Jointly published with China Railway Publishing House Co., Ltd.
The print edition is not for sale in China (Mainland). Customers from China (Mainland) please order the print book from: China Railway Publishing House Co., Ltd.

© China Railway Publishing House Co., Ltd. 2026

This work is subject to copyright. All rights are solely and exclusively licensed by the Publisher, whether the whole or part of the material is concerned, specifically the rights of reprinting, reuse of illustrations, recitation, broadcasting, reproduction on microfilms or in any other physical way, and transmission or information storage and retrieval, electronic adaptation, computer software, or by similar or dissimilar methodology now known or hereafter developed.
The use of general descriptive names, registered names, trademarks, service marks, etc. in this publication does not imply, even in the absence of a specific statement, that such names are exempt from the relevant protective laws and regulations and therefore free for general use.
The publishers, the authors, and the editors are safe to assume that the advice and information in this book are believed to be true and accurate at the date of publication. Neither the publishers nor the authors or the editors give a warranty, express or implied, with respect to the material contained herein or for any errors or omissions that may have been made. The publishers remain neutral with regard to jurisdictional claims in published maps and institutional affiliations.

This Springer imprint is published by the registered company Springer Nature Singapore Pte Ltd.
The registered company address is: 152 Beach Road, #21-01/04 Gateway East, Singapore 189721, Singapore

If disposing of this product, please recycle the paper.

# Contents

# Chapter 1
# Introduction

## 1.1 Overview of High-Speed Railway Development

The history of human transportation is fundamentally a continuous process of innovation aimed at increasing speed. Speed is the most important hallmark of modern transportation, often determining the rise and fall of specific transportation modes and tools. It not only reflects the level of social production technology but also directly promotes socio-economic development and scientific and technological progress.

The definition of high-speed railway (HSR) varies across countries and eras. According to the International Union of Railways (UIC), high-speed railways are defined as newly constructed rail lines designed for speeds of 250 km/h or higher, or upgraded rail lines designed for speeds of 200 km/h and potentially up to 250 km/h. In 1996, the European Union issued the "Directive 96/48/EC" which defined high-speed railways as newly constructed lines capable of a maximum operational speed of 250 km/h or above and upgraded lines capable of a maximum operational speed of 200 km/h. In Japan, the 1970 *Nationwide Shinkansen Railway Development Act* defined high-speed railways as railways where the maximum operational speed of trains in key sections reaches 200 km/h or above.

In 2013, the State Council of China released the *Railway Safety Management Regulations* (State Council Order No. 639), defining high-speed railways as passenger-dedicated lines designed for speeds exceeding 250 km/h and operational speeds of 200 km/h or more during the initial phase. According to the *Railway Major Technical Policy* (2013, Ministry of Railways Order No. 34), high-speed railways are defined as newly constructed passenger-dedicated railways designed for speeds of 250 km/h or above (including reserved speeds) and an initial operational speed of no less than 200 km/h. The *Railway Technical Management Regulations* (China Railway Corporation Technology [2014] No. 172) further specify that lines managed as high-speed railways include those with speeds of 200 km/h or above, as well as lines with speeds below 200 km/h on which only multiple unit (EMU) trains operate.

© China Railway Publishing House Co., Ltd. 2026

X. Long et al., *Vehicle–Track–Bridge System Subjected to Track Alignment*,
https://doi.org/10.1007/978-981-95-1649-0_1

Synthesizing the definitions from various countries and organizations, high-speed railways are generally considered to be railways where trains achieve a maximum operational speed of 200 km/h or above.

### 1.1.1 Overview of Global High-Speed Railway Development

High-speed railways represent a concentrated application of modern advanced technology in railway transportation. They reflect a country's technical progress in areas such as railway traction power, track structure, high-speed operation control, transportation organization, and management. High-speed railways are a response to socio-economic development and competition in the transportation market. They stimulate regional economic growth and urbanization, with particularly notable economic and social benefits in economically developed and densely populated regions.

Since the world's first railway was established in 1825, countries worldwide have made relentless efforts to increase train speeds. In 1903, Germany achieved a test speed of 210 km/h using an electric locomotive. In 1954, France reached a test speed of 243 km/h with an electric locomotive, and in 1962, Japan achieved a test speed of 256 km/h. On October 1, 1964, Japan inaugurated the world's first high-speed railway, the Tokaido Shinkansen, with a maximum operational speed of 210 km/h. In 1972, France achieved a test speed of 317 km/h using a diesel locomotive, and in 1981, the TGV train was officially put into operation in southeastern France with a speed of 260 km/h. Germany began high-speed train trials in 1985, reaching 345 km/h, and in 1988, the ICE train set a world record of 406.9 km/h. In 1990, France broke the record again with a test speed of 515.3 km/h. On April 3, 2007, the latest model of the French high-speed electric train, the TGV V-150, set a new world record with a speed of 574.8 km/h.

In 2008, China completed the Beijing-Tianjin Intercity Railway with a design speed of 350 km/h. On June 24, 2008, the domestically produced "Harmony" CRH3 EMU reached a test speed of 394.3 km/h on this line. On December 9, 2009, the Wuhan-Guangzhou High-Speed Railway achieved a world record test speed of 394.2 km/h in double-EMU operation. The Wuhan-Guangzhou High-Speed Railway officially opened for trial operation on December 26, 2009. On September 28, 2010, the Shanghai-Hangzhou High-Speed Railway set a test speed of 416.6 km/h using the domestically developed "Harmony" CRH380A next-generation high-speed EMU. On December 3, 2010, during the joint commissioning and comprehensive tests on the Zaozhuang-Bengbu section of the Beijing-Shanghai High-Speed Railway, the CRH380A achieved a test speed of 486.1 km/h, once again setting a world record for high-speed train operational testing. In July 2016, the China Standard EMU, independently designed and developed with full intellectual property rights, achieved speeds of 420 km/h in passing and double-EMU operation on the Zhengzhou-Xuzhou High-Speed Railway.

Currently, countries with high-speed railways include China, Japan, France, Italy, Sweden, Germany, Spain, the United Kingdom, Belgium, the Netherlands, South Korea, Turkey, Russia, Uzbekistan, and others. China has the largest high-speed railway network in the world. By the end of 2024, China's operational high-speed railway mileage exceeded 48,000 km, forming the largest and highest-standard high-speed railway network in the world. Based on technological characteristics, scope, and scale, the global development of high-speed railways can be divided into three stages: the exploratory and foundational stage, the expansion stage, and the rapid development stage.

1. **Exploratory and Foundational Stage**

From the 1960s to the late 1970s, marked by the inauguration of Japan's Tokaido Shinkansen in 1964, the first high-speed railway in the world, with a maximum operational speed of 210 km/h. Starting with the Tokaido Shinkansen, significant innovations were introduced in civil engineering, high-speed trains, traction power supply, and communication signals, transforming traditional railway systems. However, as high-speed railway development was still in its exploratory phase, with no prior experience to reference, progress was slow due to repeated testing and verification processes. It also took time to achieve consensus on accelerating high-speed railway development. Over nearly 20 years, Japan was the only country to build high-speed railways, completing the Tokaido Shinkansen in 1964 and the Sanyo Shinkansen in 1975, with a total mileage of 1069 km.

2. **Expansion Stage**

From the early 1980s to the end of the twentieth century, marked by the opening of France's first high-speed railway, the TGV Southeast Line, in 1981, with a maximum operational speed of 270 km/h. This represented a turning point, ushering in a new era where high-speed railways operated at speeds of 250–300 km/h. With continuous research, development, and application of high-speed railway technologies, comprehensive high-speed railway systems were established. In addition to the continued development of Japan's Shinkansen system, countries like France, Germany, and Italy created distinctive high-speed railway systems and series of products. They opened their first high-speed railway lines in 1981, 1991, and 1992, respectively, and initiated ambitious high-speed railway development plans.

Beginning in the 1990s, the successful operation of existing high-speed railways, coupled with the growing consensus on sustainable development, highlighted the critical role of high-speed railways in socio-economic sustainability. Other European countries, such as Spain and Belgium, began developing high-speed railways through technology transfer, opening their first lines in 1991 and 1997, respectively. Other nations, such as the Netherlands and Sweden, also drafted high-speed railway development plans. Over the next 20 years, Japan and Europe built more than 3000 km of new high-speed railways, more than triple the total built before the 1980s.

### 3. **Rapid Development Stage**

Since the early twenty-first century, marked by the rapid rise of China's high-speed railway, the world has entered a stage of accelerated development in this field. *China's Mid-to-Long-Term Railway Network Plan*, formulated in 2004 and revised in 2008, proposed the construction of a "Four Vertical and Four Horizontal" high-speed railway network. The 2016 version of the *Mid-to-Long-Term Railway Network Plan* expanded this to an "Eight Vertical and Eight Horizontal" blueprint, laying the foundation for a grand vision of high-speed railway development in China.

In just over a decade, China has become the country with the most comprehensive high-speed railway system, the strongest integration capabilities, the longest operational mileage, the highest running speeds, and the largest construction scale.

China's rapid high-speed railway development has injected strong momentum into global HSR growth and provided a powerful model for other nations, creating a worldwide "China HSR Effect." Countries such as the United States, Poland, Russia, and Turkey have accelerated their high-speed railway development plans, while developing countries in South America and Asia, such as Argentina, Brazil, Indonesia, and Thailand, have also joined the ranks of high-speed railway development.

By the end of 2024, the length and maximum speeds of high-speed railways in operation of the top ten countries with the longest high-speed rail operation lines in the world are in Table 1.1.

**Table 1.1** Overview of high-speed railway in operation of the top ten countries

| Serial number | Country | Operational mileage (km) | Maximum speed (km/h) | Serial number | Country | Operational mileage (km) | Maximum speed (km/h) |
|---|---|---|---|---|---|---|---|
| 1 | China | 48000 | 350 | 6 | Turkey | 1232 | 250 |
| 2 | Spain | 3993 | 310 | 7 | Finland | 1120 | 220 |
| 3 | Japan | 3146 | 320 | 8 | Italy | 921 | 300 |
| 4 | France | 2735 | 320 | 9 | Sweden | 895 | 250 |
| 5 | Germany | 1631 | 330 | 10 | South Korea | 873 | 300 |

*Notes* 1. The data of China's railways are as of the end of 2024 and do not include high-speed railway lines in Chinese Taiwan, which has a high-speed rail mileage of 349.5 km with a maximum operating speed of 300 km/h

2. The data of other countries are sourced from the International Union of Railways (UIC) and Wikipedia

### *1.1.2 Overview of High-Speed Railway Development in China*

Research on high-speed railways in China began in the 1990s and has undergone a long process of technological study, exploration, and development. Starting in 2004, China's high-speed railway (HSR) has progressed from nonexistence to existence, from small-scale to large-scale, from few to many, and from slow to fast, with continuous advancements at each stage. From the introduction of 200 km/h high-speed train technology to the independent development of the 350 and 380 km/h "Harmony" and "Fuxing" EMU (Electric Multiple Unit) trains, and from the opening of the Beijing-Tianjin Intercity Railway and the Wuhan-Guangzhou Passenger Dedicated Line to the Beijing-Shanghai High-Speed Railway, China has rapidly entered the "high-speed rail era".

Over the past decade, China's high-speed railway sector has experienced explosive growth, frequently breaking world records and achieving remarkable accomplishments. Although China was a latecomer to high-speed rail development, it started at a high level and developed rapidly. By introducing core technologies from abroad, assimilating and innovating upon them, and combining this with decades of accumulated domestic technological expertise, China has acquired the capability to construct and operate high-speed railways. The development of China's high-speed railway can be roughly divided into the following stages:

1. **Initial Exploration and Development Stage (1990–2004)**

Between the late 1980s and 1993, China began tracking the development of high-speed railway technologies globally while conducting foundational research on high-speed railway technologies domestically.

In 1993, the former State Science and Technology Commission, State Planning Commission, State Economic and Trade Commission, State Restructuring Commission, and Ministry of Railways jointly organized the "Preliminary Study on Major Technical and Economic Issues of the Beijing-Shanghai High-Speed Railway." This research provided a preliminary technical foundation for the development of high-speed railways. Through three major engineering practices—Guangzhou–Shenzhen Quasi-High-Speed Railway, the speed-up of existing lines, and the construction of the Qinhuangdao–Shenyang Passenger Dedicated Line—China began its initial exploration and accumulation of high-speed railway technology.

(1) Guangzhou–Shenzhen quasi-high-speed railway

To explore a development model for China's high-speed railways, the Guangzhou–Shenzhen Quasi-High-Speed Railway was completed and put into operation in 1994. With a total length of 147 km, passenger train speeds reached 160 km/h. By upgrading existing lines, constructing new segments, introducing Swedish tilting trains, and using the domestically produced Blue Arrow EMU, the railway achieved transit service for the first time, marking a qualitative leap in railway technology. This project pioneered large-scale speed increases in China's railways, yielding significant economic and social benefits.

(2) Five speed-up campaigns on existing lines

From 1997 to 2004, five rounds of speed upgrades were implemented on existing railway lines. The maximum train speeds on major trunk lines increased from 120 to 140, 160, and 200 km/h, with some sections reaching 250 km/h. By 2004, the total length of lines with speeds above 120 km/h had reached 22,000 km. These upgrades involved the extensive adoption of new technologies and equipment, as well as the introduction of overnight passenger trains.

- **First Speed-Up (April 1, 1997)**: Major trunk lines such as the Beijing-Shanghai, Beijing-Guangzhou, and Beijing-Harbin lines underwent system-wide upgrades. Forty pairs of express trains with a maximum speed of 140 km/h and an average speed of 90 km/h were introduced, along with 64 overnight trains.
- **Second Speed-Up (October 1, 1998)**: Speed limits on sections of the Beijing-Shanghai, Beijing-Guangzhou, and Beijing-Harbin trunk lines were increased to 140–160 km/h, while the Guangzhou–Shenzhen line adopted tilting trains with a maximum speed of 200 km/h. The average speed of passenger trains across the network reached 55.16 km/h.
- **Third Speed-Up (October 21, 2000)**: The focus was on lines such as the Eurasian Land Bridge (Lanzhou–Xinjiang Railway), the Beijing–Kowloon Railway, and the Zhejiang–Jiangxi Line. The nationwide railway network was reorganized, ticketing systems were integrated, and intercity ticketing services were introduced.
- **Fourth Speed-Up (October 21, 2001)**: The scope of speed upgrades expanded to cover major cities and regions nationwide, including the Wuhan–Chengdu, southern Beijing–Guangzhou, Beijing–Kowloon, Zhejiang–Jiangxi, Shanghai–Hangzhou, and Harbin–Dalian lines.
- **Fifth Speed-Up (April 18, 2004)**: The maximum speed on sections of major trunk lines such as the Beijing-Shanghai, Beijing-Guangzhou, and Beijing-Harbin lines was raised to 200 km/h.

(3) Construction of the Qinhuangdao–Shenyang passenger dedicated line

Between 1998 and 2002, China conducted in-depth research on high-speed railway technologies, focusing on major technical challenges. These research outcomes were prioritized for application in newly constructed passenger dedicated lines and speed-up projects for existing lines.

The Qinhuangdao–Shenyang Passenger Dedicated Line, with a total length of 404.6 km, was designed to accommodate speeds of up to 250 km/h. Construction began in 1999, and the line opened in 2003. Its construction marked a new milestone in China's railway engineering, achieving breakthroughs in design and construction standards, key technologies, and project management. It introduced a new model of passenger and freight segregation for trunk railways, creating high-capacity passenger corridors.

2. **Technology Introduction and Development Stage (2004–2008)**

During this stage, a significant amount of work was conducted on introducing foreign technologies and localizing and domesticating them for China's conditions.

In 2004, China began importing high-speed rail technologies, facing the challenge of adapting foreign technologies to China's unique conditions. For example, while foreign rail technology often overlooked the impact of subgrade conditions due to lower train speeds, China's vast territory spans diverse climatic and geological zones. These include soft soil foundations for the Beijing-Tianjin Intercity Railway, karst foundations for the Wuhan-Guangzhou High-Speed Railway, and collapsible loess foundations for the Zhengzhou-Xi'an High-Speed Railway. Building high-speed railways under such conditions required innovations in subgrade and embankment construction techniques, as foreign suppliers lacked experience with such complexities.

Domesticating imported technologies was equally critical. Beginning in 2004, China transitioned through phases of importing complete trains, importing components, assembling domestically, and achieving technological assimilation to increase localization rates. Without the foundational research and engineering practices accumulated over more than a decade, it would have been impossible to cost-effectively import advanced foreign technologies or innovate upon them.

The sixth speed-up campaign, completed in April 2007, marked a turning point. Through the strategy of introduction, digestion, absorption, and reinvention, the domestically produced Harmony EMU was put into large-scale practical use. China made breakthroughs in construction technology, high-speed train technology, operation and maintenance technology, and system integration technology, laying the foundation for a comprehensive high-speed railway technology system.

On August 1, 2008, the Beijing-Tianjin Intercity Railway, China's first high-speed railway capable of 350 km/h, opened for commercial operation. Through the introduction, absorption, and innovation of core technologies, China achieved a significant milestone, establishing a comprehensive set of technologies for 300–350 km/h high-speed railways, covering rolling stock, track infrastructure, communication and signaling, traction power supply, dispatching, and passenger services.

3. **Stage of Technological Re-innovation and Development (2009–2012)**

Through the successful implementation of the strategy of introduction, digestion, absorption, and re-innovation, China's high-speed railway industry has achieved leapfrog development. Remarkable achievements have been made in design and construction, product manufacturing, and operation management, propelling China from being a "follower" and "parallel runner" to becoming a "leader" in the field. Since 2008, new high-speed railways have been put into operation every year in China.

On December 26, 2009, the Wuhan-Guangzhou High-Speed Railway, the longest high-speed railway completed and put into operation at one time in the world, was officially opened. On June 30, 2011, the Beijing-Shanghai High-Speed Railway was completed and opened to traffic, marking the successful application of China's independently innovated high-speed railway technology. The Beijing-Shanghai High-Speed Railway, with a total length of 1318 km, was the largest and most technologically advanced project in the Mid-to-Long Term Railway Network Plan. Its construction adhered to the principles of independent innovation, starting from a high baseline and adhering to high standards while targeting the world's advanced levels. Through the process of introduction, digestion, absorption, and re-innovation, China established a high-speed railway technology system with independent intellectual property rights. China has preliminarily mastered the key technologies for the design and manufacturing of world-class high-speed passenger trains.

By leveraging major engineering projects, China created the CTCS-3 (China Train Control System Level 3) technical standards system with independent intellectual property rights. In addition, the CRH380 series, a new generation of high-speed trains, was successfully developed and produced. In 2010, a trial operation on the Beijing-Shanghai High-Speed Railway achieved a record-breaking speed of 486.1 km/h, setting a new milestone for the development of China's high-speed railway train technology. This achievement signified a leap in technological development that took foreign manufacturers decades to accomplish, ushering in a new era for China's high-speed railway development.

4. **Stage of Autonomous and Intelligent Development (2013–Present)**

Since the 18th National Congress of the Communist Party of China, with the development of the China Standard EMU (Electric Multiple Unit) as a focal point, China has comprehensively advanced the independent development of key high-speed rail technologies. A systematic mastery of core technologies in high-speed railway engineering, technical equipment, and operation management has been achieved, forming a comprehensive high-speed railway technology system.

Following a forward design concept, China developed the China Standard EMU, which possesses complete independent intellectual property rights and represents the world's advanced level. In 2016, the China Standard EMU achieved a successful trial run with a speed of 420 km/h and coupled operation on the Zhengzhou-Xuzhou High-Speed Railway. In June 2017, the China Standard EMU was officially named the "Fuxing" (Rejuvenation) train. In September 2017, it became the world's first high-speed train to commercially operate at a speed of 350 km/h on the Beijing-Shanghai High-Speed Railway. This marked a significant step forward from "manufacturing" to "innovating" in China's high-speed railway development.

In 2017, with the full-scale launch of the intelligent Beijing-Zhangjiakou High-Speed Railway project, China's high-speed railway began advancing toward greater intelligence, safety, comfort, and higher standards. Intelligent high-speed railways integrate cutting-edge technologies such as cloud computing, the Internet of Things (IoT), big data, BeiDou navigation, broadband mobile communication, and artificial intelligence (AI). These technologies enhance the levels of intelligent construction, intelligent equipment, and intelligent operation in high-speed railways, making passenger travel more convenient and efficient, railway operations safer and more effective, and railway equipment greener and more sustainable. With the successive

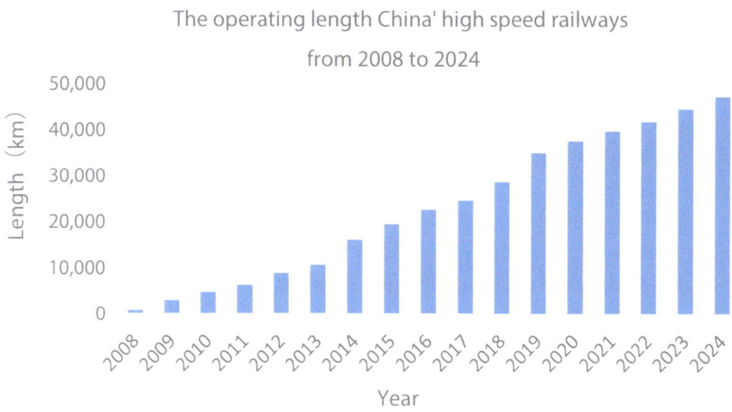


**Fig. 1.1** Statistics on China's high-speed railway operational lines

openings of the Beijing-Zhangjiakou High-Speed Railway at the end of 2019 and the Beijing-Xiong'an High-Speed Railway at the end of 2020, China's high-speed railways have made new strides in the path of independent innovation, entering the era of intelligent high-speed railways.

Following years of rapid development, by the end of 2021, China had established the world's largest high-speed railway network, with an operational length exceeding 40,000 km (see Fig. 1.1). Currently, China is implementing its "Going Global" strategy for high-speed railways. The complete technological system has been successfully applied to international projects such as Indonesia's Jakarta-Bandung High-Speed Railway. Additionally, Chinese Taiwan's high-speed railway network, which began construction in 1999 and was completed in 2007, spans a total length of 349.5 km, with a maximum operating speed of 300 km/h.

## 1.2 Overview of the Development of High-Speed Railway Track System Dynamics

### *1.2.1 Research Background*

Since the advent of railways, system dynamics has undergone a process from problem discovery to theoretical development and widespread application. The dynamics of vehicles, tracks, and bridge structures constitute one of the most fundamental theoretical supports for high-speed railways and have played a crucial role in the development of high-speed railway technology.

China has built the world's largest high-speed railway network, characterized by high operating speeds, long routes, wide regional coverage, and favorable track conditions. The continuous high-speed operation of trains over extended periods increases the complexity of operational safety and wheel-rail dynamic interactions. Moreover, track conditions have a more pronounced impact on operational safety and passenger comfort. High-speed railway track system dynamics focuses on the dynamic interaction between rolling stock and track under complex track conditions, providing the theoretical foundation for track parameter selection, alignment optimization, track smoothness management, and system design. This enables the coordinated design of high-speed railway trains and tracks, enhancing operational safety and comfort while minimizing wheel-rail dynamic interactions. Consequently, it supports the long-term safe and reliable service of high-speed railway infrastructure throughout its life cycle.

### *1.2.2 Development Overview*

1. **High-Speed Railway Track Design Standards**

Railway design is an extremely complex systematic engineering process, with track design playing a leading and coordinating role among various professional disciplines. It serves as the core component of railway design.

With the development of China's high-speed railways, relevant design codes and standards have been continuously refined. For example, to support the design of the Beijing-Shanghai High-Speed Railway, the "Interim Provisions on the Design of the Beijing-Shanghai High-Speed Railway" (Tie Jian She [2004] No. 157) were issued in 2004. Subsequently, incorporating the design experience of the Beijing-Tianjin and Beijing-Shanghai High-Speed Railways, the "Interim Provisions on the Design of Newly Constructed 300–350 km/h Passenger Dedicated Railways" (Tie Jian She [2007] No. 47) were released in 2007. Further integrating construction experience from the Wuhan-Guangzhou, Zhengzhou-Xi'an, Hefei-Wuhan, and Shijiazhuang-Taiyuan High-Speed Railways, the (*TB10621-2009*) "High-Speed Railway Design Specification (Trial)" was issued in 2009. In 2014, based on a systematic summary of China's experience in the construction and operation of 250–350 km/h high-speed railways, (*TB10621-2014*) "High-Speed Railway Design Specification" was officially released, refining regulations on key track parameters such as curve radius, transition curve length, intermediate straight length, vertical curve radius, maximum gradient, and minimum gradient segment length.

Internationally, track design standards mainly specify track parameters based on a target maximum speed of 350 km/h (300 km/h in Europe). The principles and computational methods in various national technical standards are generally consistent, differing only in certain boundary conditions. Major foreign track design standards include the International Union of Railways (UIC) "Track Alignment Design Characteristics for High-Speed Passenger Railways" (UIC 703), the European standard "Railway Applications—Track Alignment Design Parameters" (EN13803), Germany's "Railway Technical Management Regulations" (DS 800), as well as Japan's "Shinkansen Railway Structural Standards" and "Track Maintenance Principles".

Historically, China's railway track design standards, such as the "Railway Track Design Specification" (*TB10098-2017*) and the "High-Speed Railway Design Specification" (*TB10621-2014*), have been developed by integrating practical construction experience with simplified theoretical analysis. The selection of core alignment parameters is based on running dynamics, assuming the vehicle as a rigid body that fully follows the rigid track during operation, with passenger ride comfort mainly evaluated through superelevation deficiency and its rate of change. However, different national standards prescribe varying values for superelevation deficiency and its rate of change, leading to discrepancies in passengers' actual ride comfort. Furthermore, running dynamics methods neglect the impact of wheel-rail dynamic interactions, failing to adequately consider passenger discomfort caused by vehicle posture and internal component interactions. It is worth noting that running dynamics serves as the basis for determining horizontal and vertical alignment parameters in high-speed railway track design, and related studies have enriched and improved track standard formulation. However, the theoretical framework of running dynamics alone cannot fully reveal the variations in ride comfort as trains traverse track alignments. More importantly, for high-speed railways, excitations induced by track geometry can intensify train-track dynamic interactions, potentially affecting operational safety and structural performance. Therefore, the determination of track design standards must consider the impact of train-track dynamic interactions.

2. **Wheel–Rail Relationship**

The wheel–rail interaction is one of the most challenging aspects in the dynamics of railway systems. This complexity arises not only from the geometric characterization and solution of wheel–rail rolling contact problems, but also due to the randomness and uncertainty of the wheel–rail interaction under real operational conditions.

The contact region between the wheel and the rail is influenced by multiple factors. As the profile contours of the wheel and rail tread are curved, the contact positions of the left and right wheels affect each other, resulting in complex wheel–rail contact geometry. Wheel–rail contact mechanisms include single-point contact, multi-point contact, and conformal contact. Conventional geometric computations of wheel–rail contact typically neglect elastic deformation of the wheel and rail, and determine one or more contact points by identifying the locations with the minimum gap between the wheel and rail surfaces. After determining the contact points, the contact patch size and relevant contact parameters are calculated using Hertzian and non-Hertzian contact theories. The development of wheel–rail contact geometry has evolved from single-point contact to two-point, multi-point, and conformal contact types, while the calculation of contact forces has progressed from Hertzian to non-Hertzian models. In dynamic simulations, single-point contact remains the most mature and widely used model. In recent years, multi-point contact has seen some development, but conformal contact theory is still underdeveloped. Extensive empirical data from China's high-speed railways indicate a trend toward conformal contact due to tread wear, which significantly affects post-wear dynamic performance and poses challenges for dynamic simulation.

In 1926, Carter treated railway rails as elastic half-spaces and modeled the wheel as an elastic cylinder. He applied Hertzian theory and elastic half-space theory to solve two-dimensional elastic body rolling contact problems. In 1958, K. L. Johnson investigated the rolling contact of an elastic sphere on an elastic plane, extending Carter's work to three-dimensional rolling contact. He was the first to introduce the concept of spin creep into the study of wheel–rail rolling contact. In 1964, he further extended the model from circular to elliptical contact areas and proposed cubic asymptotic curves describing longitudinal and lateral creep force laws. Dutch scholar Kalker developed a linear theory for rolling contact, presenting a three-dimensional linear law for creep force, which remains widely applied today. This law has laid the foundation for subsequent simplified and approximate theories. Later, Kalker focused on simplified rolling contact theories and developed the FASTSIM algorithm, which is one of the most widely used and accurate creep theories in dynamic simulations today. Kalker also applied linear programming methods to solve elastic rolling contact problems, developing the comprehensive CONTACT program. This program addresses Hertzian and non-Hertzian normal contact problems, incremental slip contact problems, steady and unsteady rolling contact, and the elastic field within interacting bodies. It is currently the most complete theory for studying three-dimensional elastic body rolling contact. In 1983, Shen Zhiyun and others developed a fast algorithm by combining Kalker's linear theory with the Vermeulen–Johnson model. The algorithm simultaneously considered the effects of longitudinal and lateral creep rates, as well as spin creep, on longitudinal and lateral creep forces and spin moment. In 1999, Polach proposed a creep theory under large sliding conditions based on Kalker's linear theory, aiming to address the problem of adhesion degradation due to increased traction force. Both methods offer fast computation and meet engineering accuracy requirements, and have been widely applied in system dynamic simulations.

3. **Train-Track Dynamics**

With the rapid development of the railway industry, research on the dynamic interaction between trains and railway infrastructure has seen significant progress, encompassing vehicle–track and vehicle–bridge dynamic interaction issues.

In terms of vehicle–track dynamic interaction, numerous scholars have conducted extensive research incorporating issues such as track vibration, stability, and structural strength. Theoretical modeling has evolved from moving mass–track models and vehicle–track coupled models to more refined vehicle–track models that include substructure effects.

For example, Zhai Wanming et al. developed a comprehensive vehicle–track coupled dynamic theory, treating the vehicle and track systems as an integrated interacting system, and thoroughly considering key influencing factors of both vehicle and track structures. Lei Xiaoyan et al. established a finite element model of the high-speed train and track system, analyzing the dynamic behavior of the track under high-speed operation. Gao Liang et al. developed a refined dynamic model of high-speed railway vehicles and ballastless track, taking into account the interlayer interaction within the ballastless track structure. Wang Ping et al. proposed a coupled dynamic model of high-speed trains and turnouts, carefully accounting for the structural vibration characteristics of turnout components.

Regarding vehicle–bridge dynamic interaction, the emergence of high-speed railways has drawn increasing attention to the dynamic behavior of railway bridges under high-speed trains and its implications for safety during train passage. Countries such as France, Japan, Germany, Spain, South Korea, Sweden, as well as Taiwan, have developed their own vehicle–bridge coupled vibration analysis models during the advancement of high-speed rail. These models utilize computer simulation to systematically study the dynamic response of bridge structures under high-speed train loading, thereby providing essential dynamic parameters for high-speed railway bridge design. In Japan, Akio Matsuura was among the first to investigate the vertical dynamic impact of high-speed trains on bridges. In 1987, Tanabe developed a 31-degree-of-freedom vehicle model for four-axle passenger cars on the Shinkansen, solving it using finite element modal synthesis techniques. The spatial dynamic response of the vehicle–bridge system was analyzed, and the results have been directly applied to the design of high-speed railway bridges. L. Frýba studied analytical solutions for the vibration of beams and plates under moving loads. His work not only addressed theoretical analysis under various dynamic loads on railway bridges, but also included extensive experimental research to validate his theories. His findings have been adopted in regulations such as those by the International Union of Railways (UIC). Yang Yongbin, Yao Zhongda, and others proposed the VBI (vehicle–bridge interaction) element suitable for solving coupled vehicle–bridge problems. They systematically analyzed issues such as bridge and vehicle resonance using analytical methods and numerical finite element techniques. Since 1998, the former Ministry of Railways organized a research consortium composed of the China Academy of Railway Sciences, Southwest Jiaotong University, Beijing Jiaotong University, and Central South University. In conjunction with the construction of the Qinhuangdao–Shenyang Passenger Line and the Beijing–Shanghai High-Speed Railway, this team conducted systematic vehicle–bridge coupled vibration analyses on various types of bridges, including standard-span simply supported beam bridges, continuous beam bridges, continuous rigid-frame bridges, as well as simply supported and continuous steel–concrete composite bridges. The analysis results have been widely applied in engineering projects such as the Beijing–Tianjin Intercity Railway, Wuhan–Guangzhou, Shijiazhuang-Taiyuan, and Zhengzhou–Xi'an passenger lines, as well as the Beijing–Shanghai High-Speed Railway, providing essential support for the dynamic design of these bridges.

Regarding system dynamics problems induced by high-speed railway alignment geometry, increasing attention has been paid by both domestic and international research institutions and design organizations, and several studies have been carried out in this field. International research on high-speed railway dynamics has mainly focused on train ride quality and passenger comfort, examining how track alignment parameters influence these aspects and proposing corresponding evaluation frameworks. For example, K. Zboiński investigated the dynamic responses of vehicles traveling along curved tracks using a multi-body dynamic model, and separately analyzed the effects of train speed, superelevation, and curve radius. Ahmed A. Shabana explored wheel–rail interaction as a vehicle passes through curved track sections, based on an analysis of geometric characteristics of the track, and investigated the vehicle's dynamic behavior in such scenarios. L. T. Klauder Jr. assumed that ACELA trainsets experience lateral roll motion when passing through transition curves. By observing lateral wheel–rail forces and car body lateral acceleration, he proposed an improved cubic parabola for transition curve design. Mohamed Nejlaoui and colleagues used derailment coefficients and creep forces as safety indicators to study the dynamic performance of trains passing through small-radius curves and optimized curve parameters using a multi-objective approach. Chinese researchers have also conducted extensive investigations on key railway alignment parameters using vehicle–track dynamic modeling approaches. For example: Wang Kaiyun et al. evaluated the rational coordination of horizontal and vertical profiles for medium- and high-speed passenger-dedicated lines as well as for mixed passenger-freight lines from the perspective of dynamic operation safety and ride comfort. Focusing on high-speed railways, they investigated the dynamic interaction between wheel and rail on curved tracks, and analyzed the effects of track structure vibrations and key structural parameters on the wheel–rail dynamic performance in curves. Long Xuyou et al. studied the influence of superelevation configuration on transition curves, the minimum radius of circular curves, the minimum length of gradient segments, and the geometric parameters of overlapping horizontal and vertical curves on ride comfort. They also conducted dynamic evaluations of high-speed railway alignment schemes. Xuan Yan et al. established a finite element model of the coupled vehicle–track system for curved high-speed railway lines, and investigated the vibration responses of various components of the track structure under the action of high-speed EMUs, as well as the relationships among train operating speed, curve radius, and superelevation.

It is worth pointing out that although extensive research has been conducted in the field of train–track dynamics, with the continued development of high-speed railways and increasing engineering demands, future studies should further explore efficient dynamic simulation technologies suitable for complex spatial alignments, dynamic evaluation methods tailored for high-speed railway lines, and the correlation between alignment geometry and track smoothness.

## 1.3 Research Scope and Methodology of High-Speed Railway Track System Dynamics

### 1.3.1 Research Characteristics and Scope

The dynamics of high-speed railway track systems mainly focus on the dynamic response between vehicles and track structures under high-speed operation conditions. It aims to evaluate both the dynamic performance of train operation and the structural dynamics of the subgrade, thereby providing a basis for optimizing the design schemes of alignment geometry, track, and bridge structures, and ultimately enhancing the overall economic and social benefits of engineering projects. Compared with the more mature field of vehicle–track/bridge system dynamics, the dynamics of railway track systems exhibit the following key characteristics:

(1) It focuses on dynamic issues induced by track alignment, where the dynamic model must explicitly account for nonlinear wheel–rail interactions on curves as well as the spatial geometric conditions of the alignment, resulting in considerable theoretical complexity.
(2) The study of track alignment involves long railway segments, often stretching several kilometers or even hundreds of kilometers. New understandings are needed regarding how to select appropriate models, how to incorporate subgrade structures, and to what level of detail these should be refined, along with the development of novel solution techniques.
(3) Vibrations of the vehicle body caused by alignment irregularities mainly occur in the low-frequency range. Under high-speed conditions, such vibrations resemble those induced by long-wavelength track irregularities, thus requiring further study to distinguish between these effects.
(4) Commonly used comfort indices and evaluation methods for railway vehicles are primarily designed for vibrations above 0.4 Hz. These indicators and methods are not effective in distinguishing the effects of different alignment geometries.

The system dynamics of high-speed railway lines involves multiple engineering science disciplines, including structural dynamics, vehicle dynamics, vibration, and control. Current research mainly focuses on the following aspects:

(1) Study on the Influence of Design Parameters and Their Compatibility on the Dynamic Behavior of the Vehicle–Track System

Based on railway system dynamics theory, this research analyzes the effects of key horizontal and vertical alignment parameters on ride comfort and operational safety. It further explores the selection principles and coordination logic of critical parameters such as minimum curve radius, algebraic difference of gradients, minimum gradient segment length, and vertical curve radius, to evaluate the coherence and economic rationality of alignment design standards.

(2) Research on Evaluation Indicators and Methods for Operational Safety and Comfort at Higher Speeds

Through system dynamics analysis and full-scale train testing, this study aims to establish appropriate evaluation indices and methods for ensuring operational safety and ride comfort. These include dynamic performance indices for EMUs, structural dynamic performance indicators of the railway line, and safety evaluation criteria for wheel–rail interaction, thereby providing a scientific basis for the design of railway systems.

(3) Research on Dynamic Evaluation and Optimization of Alignment Schemes

For different combinations of alignment parameters, system dynamics methods are applied to predict EMU operation safety, passenger comfort, and wheel–rail wear under various alignment scenarios. By comparing the performance of different alignment schemes throughout their service life, this research provides guidance for optimizing system design.

(4) Study on the Influence of Track Condition on System Dynamics and Alignment Control

This research investigates how changes in track condition affect operational safety and comfort. It explores the impact of alignment geometry and track smoothness on dynamic indices during the service period. It also develops methods for reconstructing the horizontal and vertical alignment based on smoothness requirements, and techniques for precision adjustment of track geometry, providing technical support for fine-tuning and precision maintenance in high-speed railway systems.

(5) Study on the Dynamic Response and System Design of the Vehicle–Track–Bridge System under Complex Alignment Conditions

This research focuses on the dynamic responses of trains, tracks, and bridges under combined horizontal and vertical alignment conditions. It analyzes the conditions leading to resonance and vibration suppression in the vehicle–track–bridge system under complex line configurations, and investigates the influence of alignment parameters on system dynamic responses, thereby providing guidance for high-speed railway system design.

(6) Research on Smoothness Evaluation Methods and Alignment Parameter Adaptability for High-Speed Railways

Taking into full account the influence of vibrations in different frequency bands on passenger comfort, this study employs system dynamics methods to investigate the adaptability of horizontal and vertical alignment parameters to high-speed railway operations at various speeds, as well as the associated key comfort indicators. It further explores the principles for alignment design in accordance with ride quality management requirements.

### 1.3.2 Research Methodologies

Numerous scholars and research institutions, both domestic and international, have conducted extensive studies on the dynamic interaction between trains and railway lines, and have successively established and developed various theoretical models and methodologies. Many of these research outcomes have been successfully applied in engineering practice. Overall, the main research methods in track system dynamics include numerical simulation, experimental research, and comprehensive engineering analogy.

1. **Numerical Simulation**

The development of numerical integration techniques has made it possible to solve dynamic responses in complex systems and has promoted theoretical advancements in railway system dynamics. Given the high complexity and large degrees of freedom in train–track systems, increasing the dimensionality of theoretical models often results in significantly longer computation times and reduced accuracy. Moreover, due to limitations in computational resources and parameter conditions, numerical simulations often rely on approximations. Therefore, validating the rationality of the model becomes the primary concern. Researchers at home and abroad have developed various numerical integration methods tailored to the characteristics of train–track dynamic models, achieving significant improvements in both computational efficiency and solution accuracy. Today, numerical simulation has become the predominant method for analyzing railway system dynamics.

2. **Experimental Research**

Due to the complexity of train–track dynamic interactions, it is difficult to rely solely on numerical simulations to solve engineering problems. Therefore, field dynamic testing is essential. Targeted real-train experiments are conducted on test lines or operating railways to analyze variations in ride comfort, operational safety, and structural dynamic performance under different line conditions. Large amounts of experimental data are used to study the fundamental principles of track system dynamics and to validate the reliability of numerical simulation results.

3. **Comprehensive Engineering Analogy**

Engineering problems are influenced by a wide range of factors, making it difficult to resolve them using a single method. Thus, integrated studies that draw on past engineering practices, field experiments, and theoretical analysis are required. Additionally, the randomness of boundary conditions—such as track alignment, wheel–rail interfaces, and environmental factors—limits the accuracy of theoretical simulations. Comparative analysis of operational outcomes from similar projects at home and abroad can effectively support theoretical investigations.

# Chapter 2
# Vehicle System Dynamics Modeling

Railway vehicles can be viewed as complex multi-rigid-body systems composed of a body, frame, and wheelsets, with wheel-rail relationships exhibiting strong nonlinear characteristics. This chapter introduces the modeling methods for vehicle system dynamics under arbitrary geometric conditions. First, it briefly presents the multi-rigid-body dynamics methods suitable for vehicle modeling. Second, it discusses the geometrical relationships of wheel-rail contact and the theory of rolling contact between wheels and rails. Finally, it derives the motion equations of wheelsets and the complete vehicle motion equations under arbitrary geometric conditions in detail.

## 2.1 Dynamic Modeling Method of Multi-rigid-Body System

The first challenge in multi-rigid-body dynamics is to describe the relationship between the forces acting on the system and its motion through differential equations of motion. This section primarily introduces the dynamics equations of multi-rigid-body systems established using the Newton–Euler equations in Cartesian coordinates, with the inclusion of constraint equations via Lagrange multipliers.

### *2.1.1 Composition and Coordinates of Multi-rigid-Body System*

A multi-rigid-body system consists of multiple rigid bodies connected by various forms of connectors. Figure 2.1 illustrates a schematic of a multi-rigid-body system. The connection between two rigid bodies is referred to as a hinge. Hinge types include cylindrical hinges (which allow one degree of relative rotational freedom

© China Railway Publishing House Co., Ltd. 2026

X. Long et al., *Vehicle–Track–Bridge System Subjected to Track Alignment*,
https://doi.org/10.1007/978-981-95-1649-0_2

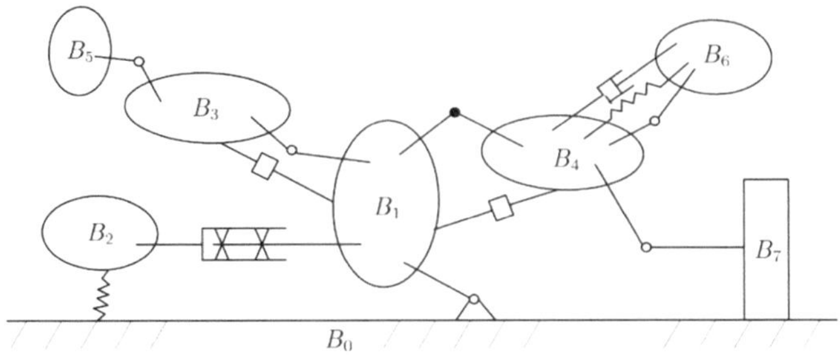


**Fig. 2.1** Composition of multi-rigid-body system

between adjacent bodies), universal joints (allowing two degrees of relative rotational freedom), and spherical hinges (allowing three degrees of relative rotational freedom). Other forms of motion constraints, such as relative sliding, may also be included, or there may be no kinematic constraints at all, with only forces acting (e.g., two adjacent rigid bodies connected by a spring). In summary, from the perspective of constraints and forces, hinges can encompass any nature of kinematic constraints (steady, unsteady, complete, or incomplete), and the corresponding reaction forces may or may not include kinematic constraints, existing solely as a coupling effect represented by certain forces. Due to the constraints imposed by hinges, the relative degrees of freedom of motion between two adjacent rigid bodies do not exceed six. It is stipulated that the interaction forces between two adjacent rigid bodies are concentrated at a single hinge. If there are more than two connectors between two adjacent rigid bodies, the principle of force translation can be applied. Thus, there exists only one hinge between each pair of adjacent rigid bodies, and conversely, a hinge connects only one pair of rigid bodies.

For a system consisting of $n$ rigid bodies, the configuration of each rigid body can be determined by the position vector of the center of mass of the rigid body and the generalized Euler angle describing the rotation of the rigid body. As shown in Fig. 2.2, the coordinate matrix of the position vector diameter $\boldsymbol{r}_{ri}$ of the centroid $C_i$ of the rigid body $B_i$ in the moving reference frame $\boldsymbol{e}^r$ is $\underline{\boldsymbol{r}}_{ri} = \left( r_{ri1} \ r_{ri2} \ r_{ri3} \right)^{\mathrm{T}}$; the three generalized Euler angles of rigid body $B_i$ on the body-fixed frame $\boldsymbol{e}^i$ are denoted by $\underline{\boldsymbol{\vartheta}}_i = \left( \psi_i \ \theta_i \ \varphi_i \right)^{\mathrm{T}}$. The three translational coordinates and three rotational coordinates of each rigid body are written into a $6 \times 1$ array, which is the Cartesian generalized coordinates of the generalized Euler angle of the rigid body.

$$\underline{\boldsymbol{q}}_i = \left( \underline{\boldsymbol{r}}_{ri}^{\mathrm{T}} \ \underline{\boldsymbol{\vartheta}}_i^{\mathrm{T}} \right)^{\mathrm{T}} = \left( r_{ri1} \ r_{ri2} \ r_{ri3} \ \psi_i \ \theta_i \ \varphi_i \right)^{\mathrm{T}} \tag{2.1.1}$$

The configuration of a multi-rigid-body system consisting of $n$ rigid bodies consists of $6n$ Cartesian generalized coordinates, which can be written as a $6n \times 1$ array, denoted by

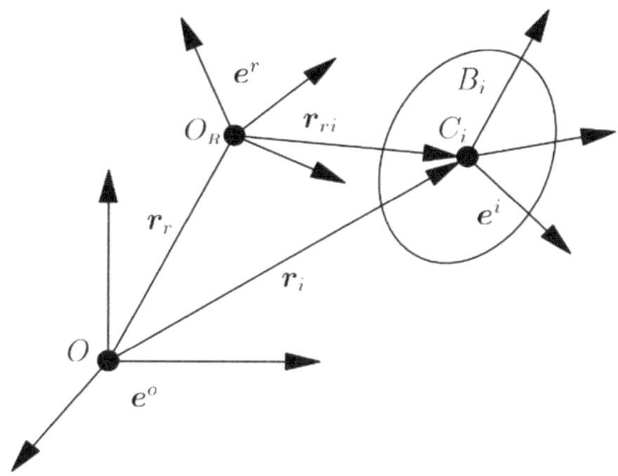


**Fig. 2.2** Body-fixed frame, moving reference frame, and inertial reference frame of rigid body

$$\underline{q} = \left( \underline{q}_1^{\mathrm{T}} \ \underline{q}_2^{\mathrm{T}} \ \cdots \ \underline{q}_n^{\mathrm{T}} \right)^{\mathrm{T}} = \left( q_{r1} \ q_{r2} \ \ldots \ q_{r6n} \right)^{\mathrm{T}} \tag{2.1.2}$$

## 2.1.2 *Introduction of the Kinematic Equations for the Moving Reference Frame*

In many practical problems, introducing a moving reference frame with a known motion is a convenient approach for analyzing the motion characteristics and dynamic behavior of a system. For example, when studying the dynamics of a railway vehicle–track system, the track, characterized by given track parameters, can be selected as the moving reference frame.

As shown in Fig. 2.2, let frame $\boldsymbol{e}^o$ be the inertial frame. Define $\boldsymbol{e}^r$ as the moving reference frame, which moves relative to the inertial frame, with its origin denoted as $O_R$. The frame $\boldsymbol{e}^i$ is co-moving frame of the rigid body $B_i$, with its origin located at the centroid $C_i$ of the rigid body. Denoting the position vectors of the centroid $C_i$ relative to the inertial frame and the moving reference frame as $\boldsymbol{r}_i$ and $\boldsymbol{r}_{ri}$, respectively, and the position vector of the reference point $O_R$ relative to the inertial frame as $\boldsymbol{r}_r$, the following relationship holds:

$$\boldsymbol{r}_i = \boldsymbol{r}_r + \boldsymbol{r}_{ri} \tag{2.1.3}$$

Let the angular velocity vectors of the rigid body $B_i$ relative to the inertial frame and the moving reference frame be $\boldsymbol{\omega}_i$ and $\boldsymbol{\omega}_{ri}$, respectively, and the angular velocity vector of the moving reference frame $\boldsymbol{e}^r$ relative to the inertial frame be $\boldsymbol{\omega}_r$. According to the principle of angular velocity superposition, the following relationship holds:

$$\boldsymbol{\omega}_i = \boldsymbol{\omega}_r + \boldsymbol{\omega}_{ri} \tag{2.1.4}$$

Taking the first and second time derivatives of Eq. (2.1.3) yields the expressions for the absolute velocity and acceleration of the centroid of each rigid body.

$$\boldsymbol{v}_i = \dot{\boldsymbol{r}}_i = \dot{\boldsymbol{r}}_r + \frac{{}^r\mathrm{d}}{\mathrm{d}t}\boldsymbol{r}_{ri} + \boldsymbol{\omega}_r \times \boldsymbol{r}_{ri} \tag{2.1.5}$$

$$\boldsymbol{a}_i = \ddot{\boldsymbol{r}}_i = \ddot{\boldsymbol{r}}_r + \frac{{}^r\mathrm{d}^2}{\mathrm{d}t^2}\boldsymbol{r}_{ri} + \dot{\boldsymbol{\omega}}_r \times \boldsymbol{r}_{ri} + \boldsymbol{\omega}_r \times (\boldsymbol{\omega}_r \times \boldsymbol{r}_{ri}) + 2\boldsymbol{\omega}_r \times \frac{{}^r\mathrm{d}}{\mathrm{d}t}\boldsymbol{r}_{ri} \tag{2.1.6}$$

In the formula, $(\dot{})$ denotes the absolute time derivative in the inertial frame, while $\frac{{}^r\mathrm{d}}{\mathrm{d}t}()$ represents the time derivative taken in the moving reference frame.

By taking the first absolute time derivative of Eq. (2.1.4), the expression for angular acceleration is obtained

$$\dot{\boldsymbol{\omega}}_i = \dot{\boldsymbol{\omega}}_r + \frac{{}^i\mathrm{d}}{\mathrm{d}t}\boldsymbol{\omega}_{ri} + \boldsymbol{\omega}_r \times \boldsymbol{\omega}_{ri} \tag{2.1.7}$$

$\frac{{}^i\mathrm{d}}{\mathrm{d}t}()$ denotes the time derivative taken in the co-moving frame.

### 2.1.3 Dynamic Equations of an Unconstrained Rigid Body

The Newton–Euler equation is used to establish the differential equation of motion of the multi-rigid-body system. Firstly, the system is disengaged and divided into several single rigid body particles. The constraint reaction force of each hinge will appear in the differential equation of motion, so it must be solved simultaneously with the constraint equation (algebraic equation). Although the number of the obtained equations is large and it is a differential–algebraic equation group, the derivation process is more intuitive and simple.

Each rigid body $B_i (i = 1, \ldots, n)$ is selected as the research object for force analysis. The active force and the constraint reaction force acting on the rigid body $B_i$ are simplified to the center of mass $C_i$ respectively, and the active force principal vector $\boldsymbol{F}_i^a$ and the principal moment $\boldsymbol{M}_i^a$, as well as the constraint reaction force principal vector $\boldsymbol{F}_i^c$ and the principal moment $\boldsymbol{M}_i^c$ are obtained. The Newton–Euler dynamic equation can be written for each rigid body $B_i$, namely

$$m_i\ddot{\boldsymbol{r}}_i = \boldsymbol{F}_i^a + \boldsymbol{F}_i^c \tag{2.1.8}$$

$$\boldsymbol{I}_i \cdot \dot{\boldsymbol{\omega}}_i + \boldsymbol{\omega}_i \times \boldsymbol{I}_i \cdot \boldsymbol{\omega}_i = \boldsymbol{M}_i^a + \boldsymbol{M}_i^c \tag{2.1.9}$$

Substituting Eqs. (2.1.6) and (2.1.7) into Eqs. (2.1.8) and (2.1.9), the dynamic equation of the rigid body relative to the dynamic reference base can be obtained.

$$m_i \frac{{}^r\mathrm{d}^2}{\mathrm{d}t^2} \boldsymbol{r}_{ri} = \boldsymbol{F}_i^a + \boldsymbol{F}_i^c + \boldsymbol{F}_i^I \tag{2.1.10}$$

$$\boldsymbol{I}_i \cdot \frac{{}^i\mathrm{d}}{\mathrm{d}t} \boldsymbol{\omega}_{ri} + \boldsymbol{\omega}_{ri} \times \boldsymbol{I}_i \cdot \boldsymbol{\omega}_{ri} = \boldsymbol{M}_i^a + \boldsymbol{M}_i^c + \boldsymbol{M}_i^I \tag{2.1.11}$$

In the formula, $\boldsymbol{F}_i^I$ and $\boldsymbol{M}_i^I$ are the inertia force and inertia moment generated by the traction motion of the moving base on the rigid body $B_i$, respectively, namely,

$$\boldsymbol{F}_i^I = -m_{ri} \left( \ddot{\boldsymbol{r}}_r + \dot{\boldsymbol{\omega}}_r \times \boldsymbol{r}_{ri} + \boldsymbol{\omega}_r \times (\boldsymbol{\omega}_r \times \boldsymbol{r}_{ri}) + 2\boldsymbol{\omega}_r \times \frac{{}^r\mathrm{d}}{\mathrm{d}t} \boldsymbol{r}_{ri} \right) \tag{2.1.12}$$

$$\boldsymbol{M}_i^I = -\boldsymbol{I}_i \cdot (\dot{\boldsymbol{\omega}}_r + \boldsymbol{\omega}_r \times \boldsymbol{\omega}_{ri}) - (\boldsymbol{\omega}_r + \boldsymbol{\omega}_{ri}) \times \boldsymbol{I}_i \cdot \boldsymbol{\omega}_r - \boldsymbol{\omega}_r \times I_i \cdot \boldsymbol{\omega}_{ri} \tag{2.1.13}$$

The coordinate matrix of the vector diameter $\boldsymbol{r}_{ri}$ on the moving reference base $\boldsymbol{e}^r$ is $\underline{r}_{ri}$, the coordinate matrix and the cross multiplication matrix of the angular velocity vector $\boldsymbol{\omega}_{ri}$ on the conjoined base $\boldsymbol{e}^i$ are $\underline{\omega}_{ri}'$ and $\underline{\tilde{\omega}}_{ri}'$, respectively, and the superscript '$\square'$' represents the coordinate matrix on the conjoined base $\boldsymbol{e}^i$ to distinguish the coordinate matrix on the moving reference base $\boldsymbol{e}^r$ without superscript.

Therefore, the coordinate formulas of the vector Eq. (2.1.12) on the dynamic reference basis and the vector Eq. (2.1.13) on the connected basis are

$$\underline{M}_i \underline{\ddot{r}}_{ri} = \underline{F}_i^a + \underline{F}_i^c + \underline{F}_i^I \tag{2.1.14}$$

$$\underline{I}_i' \underline{\dot{\omega}}_{ri}' + \underline{\tilde{\omega}}_{ri}' \underline{I}_i' \underline{\omega}_{ri}' = \underline{M}_i'^a + \underline{M}_i'^c + \underline{M}_i'^I \tag{2.1.15}$$

Among them, $\boldsymbol{F}_i^I$ and $\boldsymbol{M}_i'^I$ are the coordinate formulas of inertia force and inertia moment respectively, which are

$$\underline{F}_i^I = -\underline{M}_i \left( \underline{\ddot{r}} + \underline{\dot{\tilde{\omega}}}_r \underline{r}_{ri} + \underline{\tilde{\omega}}_r^2 \underline{r}_{ri} + 2\underline{\tilde{\omega}}_r \underline{\dot{r}}_{ri} \right) \tag{2.1.16}$$

$$\underline{M}_i'^I = -\underline{I}_i' \left( \underline{\dot{\omega}}_r' + \underline{\tilde{\omega}}_r' \underline{\omega}_{ri}' \right) - \left( \underline{\tilde{\omega}}_r' + \underline{\tilde{\omega}}_{ri}' \right) \underline{I}_i' \underline{\omega}_r' - \underline{\tilde{\omega}}_r' \underline{I}_i' \underline{\omega}_{ri}' \tag{2.1.17}$$

In the formula, $\underline{M}_i = m_i \underline{E}$ is the mass diagonal matrix of $3 \times 3$ of rigid body $B_i$; $\underline{I}_i'$ is the inertia matrix of rigid body $B_i$ in the body-fixed frame $3 \times 3$, which is a constant matrix.

The dynamic equation is a set of differential equations about velocity and angular velocity. When solving it, it must be combined with the attitude kinematics relationship to derive the second-order differential equations about attitude coordinates.

According to Euler's theorem, the rotation of a rigid body $B$ (conjoined base $\boldsymbol{e}^b$) rotating at a fixed point relative to the reference base $\boldsymbol{e}^r$ is decomposed into three consecutive fixed-axis rotations. According to the principle of angular velocity vector

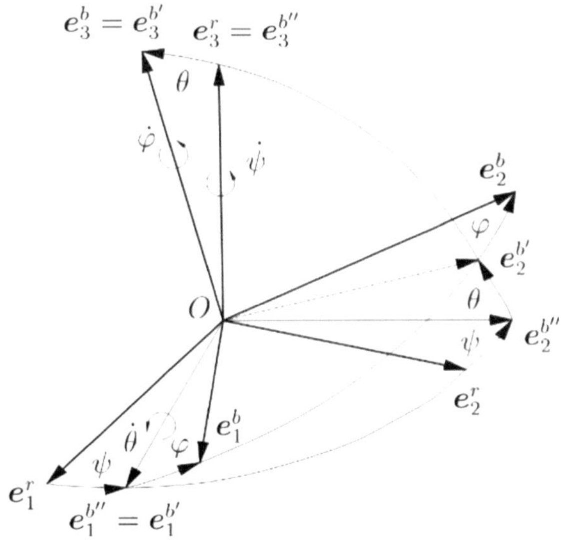


**Fig. 2.3** Euler angle coordinates

superposition, the Euler angle coordinate is used to represent the angular velocity. The angular velocity vector $\boldsymbol{\omega}$ can be decomposed into the sum of the three angular vectors of the three rotation axes along the Euler angle, namely

$$\boldsymbol{\omega} = \dot{\psi}\boldsymbol{e}_3^r + \dot{\theta}\boldsymbol{e}_1^{b'} + \dot{\varphi}\boldsymbol{e}_3^b \tag{2.1.18}$$

By decomposing the above equation on the connected basis $\boldsymbol{e}^b$, the Euler kinematic equation can be derived (Fig. 2.3).

$$\begin{bmatrix} \omega_1^b \\ \omega_2^b \\ \omega_3^b \end{bmatrix} = \begin{bmatrix} \sin\theta\sin\varphi & \cos\varphi & 0 \\ \sin\theta\cos\varphi & -\sin\varphi & 0 \\ \cos\theta & 0 & 1 \end{bmatrix} \begin{bmatrix} \dot{\psi} \\ \dot{\theta} \\ \dot{\varphi} \end{bmatrix} \tag{2.1.19}$$

The Cardan angle coordinate is used to represent the angular velocity, and the angular velocity vector $\boldsymbol{\omega}$ can be decomposed into the sum of three angular vectors along the three rotation axes of the Cardan angle, that is,

$$\boldsymbol{\omega} = \dot{\alpha}\boldsymbol{e}_1^r + \dot{\beta}\boldsymbol{e}_2^{b'} + \dot{\gamma}\boldsymbol{e}_3^b \tag{2.1.20}$$

By decomposing the above equation on the connected basis $\boldsymbol{e}^b$, the coordinate equation can be derived (Fig. 2.4).

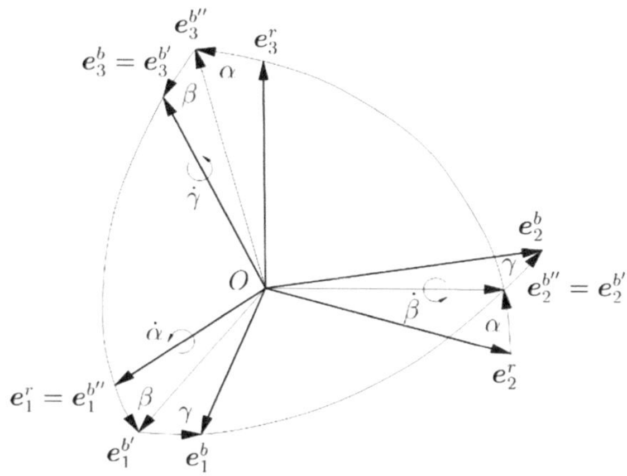


**Fig. 2.4** Cardan angle coordinate

$$\begin{bmatrix} \omega_1^b \\ \omega_2^b \\ \omega_3^b \end{bmatrix} = \begin{bmatrix} \cos\beta\cos\gamma & \sin\gamma & 0 \\ -\cos\beta\sin\gamma & \cos\gamma & 0 \\ \sin\beta & 0 & 1 \end{bmatrix} \begin{bmatrix} \dot{\alpha} \\ \dot{\beta} \\ \dot{\gamma} \end{bmatrix} \tag{2.1.21}$$

The angular velocity $\underline{\boldsymbol{\omega}}_{ri}{}'$ can be expressed as the derivative of the angular coordinate $\underline{\boldsymbol{\vartheta}}_i$ by Eqs. (2.1.19) or (2.1.21):

$$\underline{\boldsymbol{\omega}}_{ri}{}' = \underline{\boldsymbol{D}}_i \dot{\underline{\boldsymbol{\vartheta}}}_i \tag{2.1.22}$$

Substituting the Eq. (2.1.22) into the Eqs. (2.1.15) and (2.1.17), the equation expressed in angular coordinates can be obtained.

$$\underline{\boldsymbol{I}}_i{}'\underline{\boldsymbol{D}}_i\ddot{\underline{\boldsymbol{\vartheta}}}_i + (\underline{\boldsymbol{I}}_i{}'\dot{\underline{\boldsymbol{D}}}_i + \widetilde{\underline{\boldsymbol{D}}_i\dot{\underline{\boldsymbol{\vartheta}}}_i}\underline{\boldsymbol{I}}_i{}'\underline{\boldsymbol{D}}_i\dot{\underline{\boldsymbol{\vartheta}}}_i)\dot{\underline{\boldsymbol{\vartheta}}}_i = \underline{\boldsymbol{M}}_i'^a + \underline{\boldsymbol{M}}_i'^c + \underline{\boldsymbol{M}}_i'^I \tag{2.1.23}$$

$$\underline{\boldsymbol{M}}_i'^I = -\underline{\boldsymbol{I}}_i{}'\left(\dot{\underline{\boldsymbol{\omega}}}_r{}' + \tilde{\underline{\boldsymbol{\omega}}}_r{}'\underline{\boldsymbol{D}}_i\dot{\underline{\boldsymbol{\vartheta}}}_i\right) - \left(\underline{\boldsymbol{\omega}}_r{}' + \widetilde{\underline{\boldsymbol{D}}_i\dot{\underline{\boldsymbol{\vartheta}}}_i}\right)\underline{\boldsymbol{I}}_i{}'\underline{\boldsymbol{\omega}}_r{}' - \underline{\boldsymbol{\omega}}_r{}'\underline{\boldsymbol{I}}_i{}'\underline{\boldsymbol{D}}_i\dot{\underline{\boldsymbol{\vartheta}}}_i \tag{2.1.24}$$

The dynamic equation of the rigid body is composed of the motion equation of the center of mass (2.1.14) and the rotation equation around the center of mass (2.1.24).

$$\begin{bmatrix} m_i\underline{\boldsymbol{E}} & \underline{\boldsymbol{0}} \\ \underline{\boldsymbol{0}} & \underline{\boldsymbol{I}}_i{}'\underline{\boldsymbol{D}}_i \end{bmatrix} \begin{bmatrix} \ddot{\underline{\boldsymbol{r}}}_{ri} \\ \ddot{\underline{\boldsymbol{\vartheta}}}_i \end{bmatrix} = \begin{bmatrix} \underline{\boldsymbol{0}} \\ -(\underline{\boldsymbol{I}}_i{}'\dot{\underline{\boldsymbol{D}}}_i + \widetilde{\underline{\boldsymbol{D}}_i\dot{\underline{\boldsymbol{\vartheta}}}_i}\underline{\boldsymbol{I}}_i{}'\underline{\boldsymbol{D}}_i\dot{\underline{\boldsymbol{\vartheta}}}_i)\dot{\underline{\boldsymbol{\vartheta}}}_i \end{bmatrix} + \begin{bmatrix} \underline{\boldsymbol{F}}_i^a \\ \underline{\boldsymbol{M}}_i'^a \end{bmatrix} + \begin{bmatrix} \underline{\boldsymbol{F}}_i^c \\ \underline{\boldsymbol{M}}_i'^c \end{bmatrix} + \begin{bmatrix} \underline{\boldsymbol{F}}_i^I \\ \underline{\boldsymbol{M}}_i'^I \end{bmatrix} \tag{2.1.25}$$

The dynamic Eq. (2.1.25) of the 6-dimensional Cartesian coordinate matrix $\underline{\boldsymbol{q}}_i = \left(\underline{\boldsymbol{r}}_{ri}^{\mathrm{T}} \ \ \underline{\boldsymbol{\vartheta}}_i^{\mathrm{T}}\right)^{\mathrm{T}} (i = 1, 2, \ldots, n)$ and $B_i$ of the rigid body defined by Eq. (2.1.1) can be written as follows:

$$\underline{A}_i \ddot{\underline{q}}_i = \underline{b}_i \qquad (i = 1,2, \ldots, n) \tag{2.1.26}$$

Among them, $6 \times 6$ coefficient square matrix $\underline{A}_i$ and $6 \times 1$ array $\underline{b}_i$ are defined as

$$\underline{A}_i = \begin{bmatrix} m_i \underline{E} & \underline{0} \\ \underline{0} & \underline{I}_i' \underline{D}_i \end{bmatrix} \tag{2.1.27}$$

$$\underline{b}_i = \begin{bmatrix} \underline{F}_i^a + \underline{F}_i^c + \underline{F}_i^I \\ -(\underline{I}_i' \dot{\underline{D}}_i + \underline{D}_i \widetilde{\dot{\underline{\vartheta}}_i \underline{I}_i' \underline{D}_i \dot{\underline{\vartheta}}_i}) \dot{\underline{\vartheta}}_i + \underline{M}_i'^a + \underline{M}_i'^c + \underline{M}_i'^I \end{bmatrix} \tag{2.1.28}$$

Equation (2.1.26) of $n$ rigid bodies are written respectively, and the equations composed of $6n$ scalar equations are obtained, which are the dynamic equations of $n$ unconstrained rigid bodies.

$$\underline{A} \ddot{\underline{q}} = \underline{b} \tag{2.1.29}$$

where $\underline{q}$ is a $6n$-order array defined by (2.1.2), including all Cartesian coordinates. The $6n$-dimensional square matrix $\underline{A}$ and the array $\underline{b}$ are

$$\underline{A} = \mathtt{diag}(\underline{A}_1 \; \underline{A}_2 \; \ldots \underline{A}_n) \tag{2.1.30}$$

$$\underline{b} = \left( \underline{b}_1^{\mathrm{T}} \; \underline{b}_2^{\mathrm{T}} \ldots \underline{b}_n^{\mathrm{T}} \right)^{\mathrm{T}} \tag{2.1.31}$$

### 2.1.4 Constraints and Constraint Equations

In a multi-rigid body system, various types of hinges (or joints) exist between the rigid bodies, which means their movements are not independent and must satisfy given constraints. These conditions that restrict the relative motion between rigid bodies are referred to as kinematic constraints. The analytical expressions of these constraints are known as constraint equations. This section will introduce the general forms of the constraint equations for adjacent rigid bodies, as well as the basic constraint equations for points and vectors. Through combinations of these basic constraint equations, one can derive the constraint equations for general hinges.

1. **General Form of Constraint Equations for Adjacent Rigid Bodies**

For a system composed of $n$ rigid bodies, the generalized coordinate describing the configuration of the $i$ rigid body is $\underline{q}_i$, and the generalized coordinate can be Euler angle or direction cosine matrix. The generalized coordinate array of multi-rigid-body system can be expressed as $\underline{q} = \left( \underline{q}_1^{\mathrm{T}} \; \underline{q}_2^{\mathrm{T}} \ldots \underline{q}_n^{\mathrm{T}} \right)^{\mathrm{T}}$. The constraint equations describing the system can be generally expressed as

$$\underline{\boldsymbol{\Phi}}(\underline{\boldsymbol{q}}, \underline{\dot{\boldsymbol{q}}}, t) = 0 \tag{2.1.32}$$

$\underline{\boldsymbol{\Phi}} = \left(\Phi_1 \ldots, \Phi_{n_c}\right)^{\mathrm{T}}$, $n_c$ are the number of constraint equations. When the constraint equation contains the velocity $\underline{\dot{\boldsymbol{q}}}$ of the coordinate and the equation is not integrable, this constraint is called a nonholonomic constraint. When the equation contains only coordinate $\underline{\boldsymbol{q}}$ and time $t$, or the equation containing velocity $\underline{\dot{\boldsymbol{q}}}$ can be integrated, it is called a holonomic constraint. Most of the constraints in the project are complete constraints, and their constraint equations can be expressed as generalized coordinates.

$$\underline{\boldsymbol{\Phi}}(\underline{\boldsymbol{q}}, t) = \underline{\boldsymbol{0}} \tag{2.1.33}$$

If the equation does not contain time $t$, it is called a steady constraint, otherwise it is called an unsteady constraint. The constraint equation is usually a nonlinear equation of system coordinates.

Deriving the constraint equation (2.1.33) to time, the velocity constraint equation of the system is obtained, that is,

$$\underline{\boldsymbol{\Phi}}_q \underline{\dot{\boldsymbol{q}}} + \underline{\boldsymbol{\Phi}}_t = \underline{\boldsymbol{0}} \tag{2.1.34}$$

Among them, $\underline{\boldsymbol{\Phi}}_q = \left(\frac{\partial \Phi_i}{\partial q_i}\right)_{n_c \times n_q}$ is a $n_c \times n_q$ order matrix, and $n_q$ is the number of generalized coordinates; $\underline{\boldsymbol{\Phi}}_t = \left(\frac{\partial \Phi_i}{\partial t}\right)_{n_c \times 1}$ is a matrix of order $n_c$, and $\Phi_q$ is called the Jacobian matrix of the constraint equation.

The acceleration constraint equation of the system is obtained by taking the derivative of the velocity constraint equation (2.1.34) to the time.

$$\underline{\boldsymbol{\Phi}}_q \underline{\ddot{\boldsymbol{q}}} - \underline{\boldsymbol{\gamma}} = \underline{\boldsymbol{0}} \tag{2.1.35}$$

$$\underline{\boldsymbol{\gamma}} = -(\underline{\boldsymbol{\Phi}}_q \underline{\dot{\boldsymbol{q}}})_q \underline{\dot{\boldsymbol{q}}} - 2\underline{\boldsymbol{\Phi}}_{qt} \underline{\dot{\boldsymbol{q}}} - \underline{\boldsymbol{\Phi}}_{tt} \tag{2.1.36}$$

2. **Basic Constraint Equations for Points and Vectors**

The kinematic constraints of rigid bodies can be divided into two categories: relative translational constraints and relative rotational constraints. The relative translational constraint can be described by the relative distance between the points on the two rigid bodies, and the relative rotation can be described by the relative relationship of the vectors fixed on the two rigid bodies respectively. The mathematical expression describing these vector relations is called the basic constraint equation. Using these basic constraint equations, the constraint equations of various hinges can be constructed.

As shown in Fig. 2.5, the rigid body $B_i$ and the rigid body $B_j$ are connected by hinges, and the installation positions of the hinges on the two rigid bodies are point $P$ and point $Q$, respectively. Let base $\underline{\boldsymbol{e}}^o$ be the global reference base. Connected bases

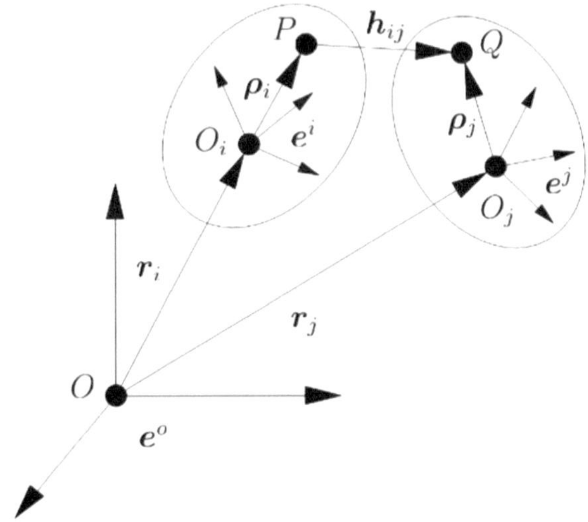


**Fig. 2.5** Kinematic relationship of adjacent rigid body

$\underline{e}^i$ and $\underline{e}^j$ are established on the points $O_i$ and $O_j$ of rigid body $B_i$ and rigid body $B_j$, respectively. The vector diameters of connected bases and inertial bases are $\boldsymbol{r}_i$ and $\boldsymbol{r}_j$, respectively. The vectors $\boldsymbol{\rho}_i$ and $\boldsymbol{\rho}_j$ are defined as the consolidated rigid body $B_i$ and the rigid body $B_j$, respectively. These two vectors will be used to define the rotation between rigid bodies, so they can be defined at any position of the rigid body. In addition, for the convenience of calculation, they are often defined as unit vectors.

(1) The basic constraint equation of point

The relative translational constraint between two rigid bodies can be determined by limiting the size and direction of the vector $\boldsymbol{h}_{ij} = \overrightarrow{PQ}$, and the vector $\boldsymbol{h}_{ij}$ can be expressed as

$$\boldsymbol{h}_{ij} = \boldsymbol{r}_j + \boldsymbol{\rho}_j - \boldsymbol{r}_i - \boldsymbol{\rho}_i \tag{2.1.37}$$

The constraint equation of the point $P$ and the point $Q$ always coincide can be written as

$$\boldsymbol{\Phi}(\boldsymbol{h}_{ij}) = \boldsymbol{h}_{ij} = \boldsymbol{0} \tag{2.1.38}$$

The global coordinate matrix constraint equation is

$$\underline{\boldsymbol{\Phi}}(\underline{\boldsymbol{h}}_{ij}) = \underline{\boldsymbol{h}}_{ij} = \underline{\boldsymbol{r}}_j + \underline{\boldsymbol{A}}^j \underline{\boldsymbol{\rho}}_j^j - \underline{\boldsymbol{r}}_i - \underline{\boldsymbol{A}}^i \underline{\boldsymbol{\rho}}_i^i = \underline{\boldsymbol{0}} \tag{2.1.39}$$

In the formula, $\underline{\boldsymbol{A}}^i$ and $\underline{\boldsymbol{A}}^j$ are the coordinate matrix of the rigid body base respectively, and $\underline{\boldsymbol{\rho}}_i^i$ and $\underline{\boldsymbol{\rho}}_j^j$ are the coordinate array of the two vectors on the base of the rigid body base respectively.

The constraint equation contains three scalar equations.

The distance between point $P$ and point $Q$ is always given, and the constraint equation can be expressed as

$$\boldsymbol{\Phi}(\boldsymbol{h}_{ij}, t) = \boldsymbol{h}_{ij} \cdot \boldsymbol{h}_{ij} - \boldsymbol{c}(t)^2 = \boldsymbol{0} \tag{2.1.40}$$

The global coordinate matrix constraint equation is

$$\underline{\boldsymbol{\Phi}}(\underline{\boldsymbol{h}}_{ij}, t) = \underline{\boldsymbol{h}}_{ij}^{\mathrm{T}} \underline{\boldsymbol{h}}_{ij} - \underline{\boldsymbol{c}}(t)^2 = \underline{\boldsymbol{0}} \tag{2.1.41}$$

(2) The basic constraint equation of vector

The vertical constraint equations of vectors $\boldsymbol{\rho}_i$ and $\boldsymbol{\rho}_j$ are

$$\boldsymbol{\Phi}(\boldsymbol{\rho}_i, \boldsymbol{\rho}_j) = \boldsymbol{\rho}_i \cdot \boldsymbol{\rho}_j = \boldsymbol{0} \tag{2.1.42}$$

The global coordinate matrix constraint equation is

$$\underline{\boldsymbol{\Phi}}(\underline{\boldsymbol{\rho}}_i, \underline{\boldsymbol{\rho}}_j) = \underline{\boldsymbol{\rho}}_i^{i\mathrm{T}} \underline{\boldsymbol{A}}^{i\mathrm{T}} \underline{\boldsymbol{A}}^j \underline{\boldsymbol{\rho}}_j^j = \boldsymbol{0} \tag{2.1.43}$$

The constraint equation contains a scalar equation.

Another commonly used vector constraint relation $\boldsymbol{\rho}_j$ and $\boldsymbol{h}_{ij}$ is vertical, and the constraint equation can be written as

$$\boldsymbol{\Phi}(\boldsymbol{h}_{ij}, \boldsymbol{\rho}_j) = \boldsymbol{h}_{ij} \boldsymbol{\rho}_j = \boldsymbol{0} \tag{2.1.44}$$

The global coordinate matrix constraint equation is

$$\underline{\boldsymbol{\Phi}}(\underline{\boldsymbol{h}}_{ij}, \underline{\boldsymbol{\rho}}_j) = \underline{\boldsymbol{h}}_{ij}^{\mathrm{T}} \underline{\boldsymbol{A}}^j \underline{\boldsymbol{\rho}}_j^j = \underline{\boldsymbol{0}} \tag{2.1.45}$$

The velocity and acceleration constraint equations can be obtained by solving the first and second derivatives of the constraint equation with respect to time.

### *2.1.5 Dynamics Equations of Constrained Multi-rigid Body Systems*

For a system consisting of $n$ rigid bodies, the total degree of freedom of the system is $n_c$ when the system is subject to $n_f = 6n - n_c$ independent complete constraints. Obviously, the generalized coordinate array $\underline{\boldsymbol{q}}$ of the multi-rigid-body system contains non-independent coordinates, which can be processed by the Lagrange multiplier method. Firstly, the terms of the unconstrained system dynamics equation (2.1.29) are multiplied by the variation $\delta \underline{\boldsymbol{q}}$ of the absolute coordinates to obtain

$$(\underline{A}\ddot{\underline{q}} - \underline{b})^{\mathrm{T}}\delta\underline{q} = \underline{0} \tag{2.1.46}$$

Among them, Variation $\delta\underline{q}$ is not an independent variable and must satisfy the restriction of $n_c$ constraint equations (2.1.32). Consideration formula (2.1.34), the coordinate of the time-varying points have

$$\underline{\Phi}_q\delta\underline{q} = \underline{0} \tag{2.1.47}$$

$n_c$ Lagrangian multiplier $\lambda_k(k = 1, 2, \ldots, n_c)$ is introduced to form the array $\underline{\lambda}$

$$\underline{\lambda} = \begin{pmatrix} \lambda_1 & \lambda_2 & \ldots & \lambda_{n_c} \end{pmatrix}^{\mathrm{T}} \tag{2.1.48}$$

By multiplying equation (2.1.47) with the corresponding Lagrange multipliers and adding them to the dynamic equation (2.1.46), we obtain

$$\left(\underline{A}\ddot{\underline{q}} - \underline{b} + \underline{\Phi}_q^{\mathrm{T}}\underline{\lambda}\right)^{\mathrm{T}}\delta\underline{q} = \underline{0} \tag{2.1.49}$$

Let the coefficients of $n_c$ non-independent variations in $\delta\underline{q}$ be zero, and only the sum related to $6n - n_c$ independent variations is left in the above formula. The sufficient necessity of Eq. (2.1.49) requires independent variables to be zero. The result is that all the elements in the bracket are zero, and considering that the sum of the element work done by the ideal constraint reaction force at any virtual displacement is zero, the dynamic equation with Lagrange multiplier is derived:

$$\underline{A}\ddot{\underline{q}} = \underline{b}' - \underline{\Phi}_q^{\mathrm{T}}\underline{\lambda} \tag{2.1.50}$$

In the formula, $\underline{b}'$ is the corresponding term of the array $\underline{b}$ minus the ideal constraint reaction force and torque $\underline{F}_i^c$ and $\underline{M}_i'^c$. It can be seen that $-\underline{\Phi}_q^{\mathrm{T}}\underline{\lambda}$ is equivalent to the ideal constraint reaction force and torque missing in the array $\underline{b}$.

Equation (2.1.50) has $6n + n_c$ unknown variables, which must be solved simultaneously with the acceleration constraint equation (2.1.35).

$$\begin{bmatrix} \underline{A} & \underline{\Phi}_q^{\mathrm{T}} \\ \underline{\Phi}_q & \underline{0} \end{bmatrix}\begin{bmatrix} \ddot{\underline{q}} \\ \underline{\lambda} \end{bmatrix} = \begin{bmatrix} \underline{b}' \\ \underline{\gamma} \end{bmatrix} \tag{2.1.51}$$

This equation represents the general form of the multibody system dynamic equations in Cartesian absolute coordinates. The equation contains $6n$ absolute coordinates ($j = 1,2, \ldots, 6n$) and $n_c$ Lagrange multipliers $\lambda_k(k = 1,2, \ldots, n_c)$.

### 2.1.6 Numerical Analysis Methods for Dynamics Equations

The dynamic equation of multibody system (2.1.51) is a differential–algebraic mixed equations. There are two kinds of numerical methods for solving these equations: augmented formulation and embedded formulation. The augmented formulation regards generalized coordinates and Lagrange multipliers as unknowns, and solves $6n + n_c$ equations simultaneously. The embedded formulation is to divide the coordinates into independent coordinates and non-independent coordinates, find the relationship between the two, and convert the dynamic equation into a pure differential equation of independent coordinates for solution.

3. **Augmented Formulation**

The characteristic of the augmented formulation is to directly solve the differential–algebraic mixed equations. The numerical method for solving differential–algebraic equations is far less mature than the initial value problem of pure differential equations. The inherent ill-condition of this kind of equations will cause many difficulties in numerical calculation. There are two methods to solve the problem: direct method and default stability method. For $6n + n_c$ unknown variables $\ddot{\underline{q}}$ and $\underline{\boldsymbol{\lambda}}$, the equation is a closed algebra. It can be proved that the coefficient matrix of the equation is nonsingular, so if the right side of the equation is nonzero, there is a unique solution.

The basic idea of the direct method is to use the numerical method of solving algebraic equations to obtain $\ddot{\underline{q}}$ and $\underline{\boldsymbol{\lambda}}$, and then use the numerical integration method to solve $\dot{\underline{q}}$ and $\underline{q}$ by $\ddot{\underline{q}}$. The direct method uses the acceleration constraint equation. Therefore, the numerical solutions $\dot{\underline{q}}^{(i)}$ and $\underline{q}^{(i)}$ do not necessarily satisfy the constraint equation (2.1.33) and the velocity constraint equation (2.1.34).

$$\underline{\boldsymbol{\Phi}}(\underline{q}, t) = \underline{\boldsymbol{\varepsilon}} \neq \underline{\mathbf{0}} \tag{2.1.52}$$

$$\underline{\boldsymbol{\Phi}}_q \dot{\underline{q}} + \underline{\boldsymbol{\Phi}}_t = \underline{\boldsymbol{\varepsilon}}_1 \neq \underline{\mathbf{0}} \tag{2.1.53}$$

As the number of numerical calculation steps increases, the error accumulation will cause the numerical solution to diverge, and the default stability method is an improvement of the direct method. The idea of default stability method comes from the principle of control feedback. Constants $\alpha$ and $\beta$ are introduced. The equation of the default stability method in the $\underline{\boldsymbol{\gamma}} - 2\alpha\underline{\boldsymbol{\varepsilon}}_1 - \beta^2\underline{\boldsymbol{\varepsilon}}$ substitution (2.1.51) is

$$\begin{bmatrix} \underline{\boldsymbol{A}} & \underline{\boldsymbol{\Phi}}_q^{\mathrm{T}} \\ \underline{\boldsymbol{\Phi}}_q & \underline{\mathbf{0}} \end{bmatrix} \begin{bmatrix} \ddot{\underline{q}} \\ \underline{\boldsymbol{\lambda}} \end{bmatrix} = \begin{bmatrix} \underline{b}' \\ \underline{\boldsymbol{\gamma}} - 2\alpha\underline{\boldsymbol{\varepsilon}}_1 - \beta^2\underline{\boldsymbol{\varepsilon}} \end{bmatrix} \tag{2.1.54}$$

It can be seen that when $\underline{\boldsymbol{\varepsilon}}$ and $\underline{\boldsymbol{\varepsilon}}_1$ are zero, Eq. (2.1.54) is consistent with Eq. (2.1.51). Once a breach occurs, the correction term automatically works. The empirical values of $\alpha$ and $\beta$ are between 5 and 50, and the stable response is faster when $\alpha = \beta$ is used. Due to the introduction of the correction term, the original dynamic equation of the system is destroyed, so the above-mentioned default stabilization method is only suitable for the case of small default. The calculation of the default stability method is not much different from the direct method. As long as the calculation of $\underline{\boldsymbol{\varepsilon}}$ and $\underline{\boldsymbol{\varepsilon}}_1$ values is increased, the $\underline{\boldsymbol{\gamma}}$ in Eq. (2.1.51) can be replaced by the function $\underline{\boldsymbol{\gamma}} - 2\alpha\underline{\boldsymbol{\varepsilon}}_1 - \beta^2\underline{\boldsymbol{\varepsilon}}$.

4. **Embedded Formulation**

The essence of the embedded formulation is to select the appropriate algorithm to deal with the constraint equation. Select independent generalized coordinates, find the relationship between independent and non-independent coordinates, and then express the dynamic equation as a pure differential equation of generalized coordinates for numerical solution.

Due to the existence of constraints, the generalized coordinates describing the multibody system are not independent. By matrix decomposition (such as LU decomposition, QR decomposition, etc.), the Jacobian matrix $\underline{\boldsymbol{\Phi}}_q$ of the constraint equation can be decomposed into the following form.

$$(\underline{\boldsymbol{\Phi}}_q)_{s\times n} = \left[(\underline{\boldsymbol{\Phi}}_u)_{s\times s},\ (\underline{\boldsymbol{\Phi}}_v)_{s\times(n-s)}\right] \tag{2.1.55}$$

The velocity constraint equation (2.1.34) can be written as

$$\underline{\boldsymbol{\Phi}}_u\underline{\dot{\boldsymbol{u}}} + \underline{\boldsymbol{\Phi}}_v\underline{\dot{\boldsymbol{v}}} = -\underline{\boldsymbol{\Phi}}_t \tag{2.1.56}$$

In the formula, $\underline{\boldsymbol{u}}$ and $\underline{\boldsymbol{v}}$ are the non-independent and independent generalized arrays of the system, respectively, namely

$$\underline{\boldsymbol{q}} = \left[\underline{\boldsymbol{u}}^{\mathrm{T}},\ \underline{\boldsymbol{v}}^{\mathrm{T}}\right]^{\mathrm{T}} \tag{2.1.57}$$

From Eqs. (2.1.56) and (2.1.57), the relationship between the generalized coordinates and the derivative of the independent coordinates of the system can be obtained.

$$\underline{\dot{\boldsymbol{q}}} = \underline{\boldsymbol{H}}\,\underline{\dot{\boldsymbol{v}}} + \underline{\boldsymbol{c}} \tag{2.1.58}$$

In the formula, the expressions of velocity transformation matrices $\underline{\boldsymbol{H}}$ and $\underline{\boldsymbol{c}}$ are

$$\underline{\boldsymbol{H}} = \begin{bmatrix} -\underline{\boldsymbol{\Phi}}_u^{-1}\underline{\boldsymbol{\Phi}}_v \\ \underline{\boldsymbol{E}} \end{bmatrix} \tag{2.1.59}$$

$$\underline{\boldsymbol{c}} = \begin{bmatrix} -\underline{\boldsymbol{\Phi}}_u^{-1}\underline{\boldsymbol{\Phi}}_t \\ \underline{\boldsymbol{0}} \end{bmatrix} \tag{2.1.60}$$

The generalized coordinate acceleration of the system is

$$\ddot{\underline{q}} = \underline{H}\,\ddot{\underline{v}} + \dot{\underline{H}}\,\dot{\underline{v}} + \dot{\underline{c}} = \underline{H}\,\ddot{\underline{v}} + \underline{d} \tag{2.1.61}$$

In the formula

$$\underline{d} = \begin{bmatrix} -\underline{\Phi}_u^{-1}\left[\left(\underline{\Phi}_q\dot{\underline{q}}\right)_q\dot{\underline{q}} + 2\underline{\Phi}_{qt}\dot{\underline{q}} + \underline{\Phi}_{tt}\right] \\ \underline{0} \end{bmatrix} \tag{2.1.62}$$

The velocity transformation matrix $\underline{H}$ can eliminate the system constraint reaction term $-\underline{\Phi}_q^{\mathrm{T}}\underline{\lambda}$, and there is

$$\underline{H}^{\mathrm{T}}\underline{\Phi}_q^{\mathrm{T}} = \left[-\underline{\Phi}_v^{\mathrm{T}}\left(\underline{\Phi}_u^{-1}\right)^{\mathrm{T}}\underline{E}\right]\begin{bmatrix}\underline{\Phi}_u^{\mathrm{T}} \\ \underline{\Phi}_v^{\mathrm{T}}\end{bmatrix} = -\underline{\Phi}_v^{\mathrm{T}}\left(\underline{\Phi}_u^{-1}\right)^{\mathrm{T}}\underline{\Phi}_u^{\mathrm{T}} + \underline{\Phi}_v^{\mathrm{T}} = \underline{0} \tag{2.1.63}$$

Therefore, by multiplying the two ends of the equal sign of Eq. (2.1.61) by $\underline{H}^{\mathrm{T}}$ and substituting it into Eq. (2.1.50), the dynamic differential equations with equal number of equations and degrees of freedom and independent generalized coordinates as variables can be obtained.

$$\overline{\underline{A}}\,\ddot{\underline{v}} = \overline{\underline{b}} \tag{2.1.64}$$

In the formula, $\overline{\underline{A}} = \underline{H}^{\mathrm{T}}\underline{A}\,\underline{H}$, $\overline{\underline{b}} = \underline{H}^{\mathrm{T}}\left(\underline{b}' - \underline{A}\,\underline{d}\right)$.

## 2.2 Wheel-Rail Contact Geometric Relationship

The connection between railway vehicles and tracks is realized through wheel-rail contact. This relationship links the vehicle and the track, allowing for interaction and mutual influence, which to some extent determines the operational characteristics of the vehicle. The wheel-rail contact relationship consists of the geometric relationship of wheel-rail contact and the theory of rolling contact. This section primarily discusses the characteristics of wheel-rail profiles and methods for calculating geometric contact parameters.

### 2.2.1 *Basic Characteristics and Parameters of Wheel-Rail*

The railway vehicle wheelset consists of an axle and two wheels fixed on either side, with no allowance for any looseness during motion. Therefore, the wheelset can be considered as a whole, focusing on the contact between the left and right wheels and the rail. The contact surface between the wheel and the rail head is referred to as the tread. The characteristics of the tread and the profile of the rail head, as well as the

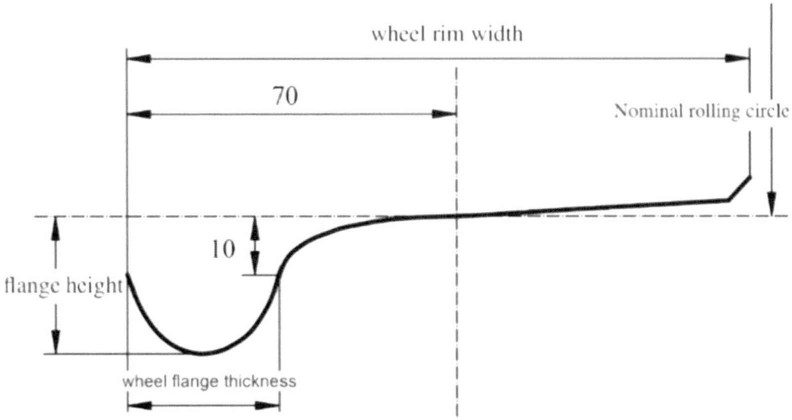


**Fig. 2.6** Wheel tread shape

nominal rolling diameter and contact position, significantly influence the dynamic performance of the vehicle.

1. **Wheel Shape and Dimensions**

Figure 2.6 shows the outer contour line of the tread's cross-section, known as the tread profile. Important geometric dimensions include the flange, nominal wheel diameter, wheelset gauge, and tread slope.

(1) Flange

The flange is a crucial component that keeps the vehicle running along the rail and prevents the wheel from derailing. The flange height is the radial distance from the top of the flange to the point on the tread where the nominal rolling circle is located. Vertical wear on the tread increases the flange height. The flange thickness is defined as the distance from the intersection of a horizontal line 10 mm above the nominal rolling circle to the inner side of the flange. Both the height and thickness of the flange are related to the safety of the wheelset when passing through turnouts.

(2) Nominal diameter of wheels

The nominal diameter of a wheel is defined as the diameter of the rolling circle located 70 mm from the inner side of the wheel flange, due to the wheel tread being composed of multiple arc segments and having varying diameters. A smaller wheel diameter can lower the vehicle's center of gravity, reduce the unsprung mass, shorten the fixed wheelbase of the bogie, decrease the construction boundary, and lower engineering costs. However, smaller diameter wheels can increase rolling resistance, raise the contact stress between the wheel and rail, accelerate tread wear, and amplify vibrations at track depressions and joints.

(3) Wheelset inner distance

The wheelset inner distance refers to the distance between the inner sides of the wheel flanges. During operation, this distance affects the safety and stability of the vehicle. It is essential to maintain a certain clearance between the wheel flange and the rail to ensure the automatic centering effect of the wheelset, preventing excessive sliding and uneven wear between the wheel and rail. Additionally, it must limit the amplitude of the wheelset's hunting motion to prevent excessive gaps from degrading ride comfort.

In China, the inner distance of the rail wheelset is 1353 mm, while in Europe and Japan, it is 1360 mm. If the center of the wheelset is aligned with the center of the track, the clearance between the wheels and rails in China is 9 mm, whereas in Europe and Japan, it is 5.5 mm. These two clearance configurations have their respective advantages and disadvantages. China's larger inner distance can reduce the difficulties associated with track laying, maintenance, and repairs, while also minimizing wheel-rail wear. However, the downside is that it allows for a greater lateral movement during train operation, which can affect ride comfort. Conversely, the smaller inner distance in Europe and Japan requires higher precision in track laying, maintenance, and repairs, which may accelerate wheel-rail wear but is beneficial for improving ride comfort (Fig. 2.7).

(4) Wheel tread slope

When navigating curves, the length of the outer rail is greater than that of the inner rail. To ensure that the distance traveled by the outer wheel is greater than that of the inner wheel within the same time interval, the rolling radius of the outer wheel on the same axle must be greater than that of the inner wheel. Therefore, the wheel tread must have a slope to reduce operational resistance, minimize wheel-rail wear, and prevent derailment.

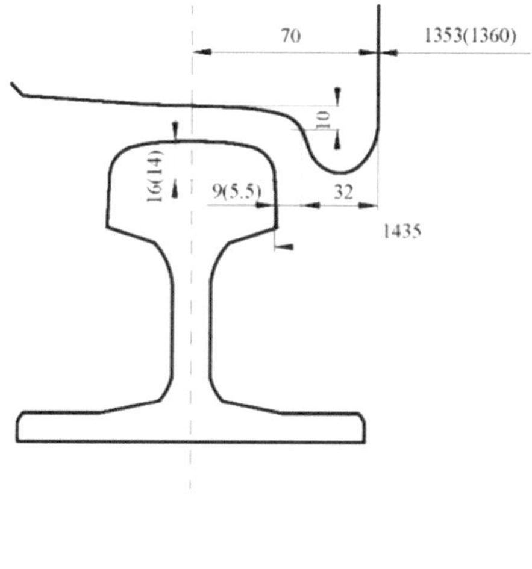


**Fig. 2.7** Wheel rail gap schematic

2. **Type of Wheel Tread**

Currently, the wheel treads used in various countries can be classified into three types: cylindrical tread, conical tread, and wear-type tread.

(1) Cylindrical tread

Cylindrical treads have no slope in their outer profile. This type of tread is found only on a few track inspection vehicles in Japan, designed to eliminate the adverse effects of tread slope on the test results for track elevation and horizontal irregularities during inspections.

(2) Conical tread

The conical tread was commonly used in China's railways in the past. The flange's inner side features a slope of 1:20 over a distance of 48 mm to 100 mm, followed by a 1:10 slope that facilitates navigation through small-radius curves and switches. The wheel flange consists of four arc segments, with an R = 18 mm arc connecting the flange to the straight segment. The position of this arc's center is critical, as its shape affects whether two-point contact occurs between the wheel and rail, as well as the vertical distance between these contact points, significantly influencing flange wear.

(3) Wear-type tread

The wear-type tread was developed based on research and improvements to the conical tread. Operational data from various countries indicate that conical treads wear quickly during use but stabilize after reaching a certain level of wear. Practical experience has shown that designing the wheel tread to resemble the shape of a worn tread from the outset—i.e., a wear-type tread—can significantly reduce wear and contact stress between the wheel and rail, ensuring both operational stability and curve navigation capability. The main wear-type treads used in China's railway vehicles include LM and LMA types. The LMA type was developed in the 1990s for high-speed trains.

3. **Rail Head Profile**

Rails consist of the rail head, web, and base, with the rail head profile being the primary factor influencing the geometric relationship between the wheel and rail contact. The rail head profile comprises multiple arc segments, and to ensure that the normal force between the wheel and rail passes through the center line of the rail, a slope is set at the rail base. The standard rail types widely used in China include 50, 60, and 75 kg/m.

The cross-section of the 60 kg/m standard rail is designed based on the wear-type principle and is used in conjunction with LM tread wheels. The rail head part of the 60 kg/m standard rail is tangent to two $R = 13$ mm arcs, two $R = 80$ mm arcs, and one $R = 300$ mm arc. The rail head width is 73 mm, and the side surfaces of the rail head form a 1:20 slope with the center line of the rail, with a height of 176 mm.

In addition to the rail head profile, the track gauge between the two rails and the rail base slope also significantly impact the contact geometry. The track gauge is measured at 16 mm below the common tangent plane of the rail head, with China's standard gauge being 1435 mm. To ensure that the normal forces between the wheel and rail pass through the center line of the rail, the rails must be inclined inward during installation, creating a lateral slope between the rail base and the track surface, known as the rail base slope. There are two types of rail base slopes: 1:20 and 1:40 (Fig. 2.8).

Table 2.1 Shows the basic parameters of rails and wheels used in China railway, Japan Shinkansen and European high-speed railways.

Figures 2.9 and 2.10 show the profiles of $LM_A$ tread and CHN60 rail in China, as well as the profiles of S1002 tread and UIC60 rail in European railway. It can be seen that there is little difference between the two profiles.

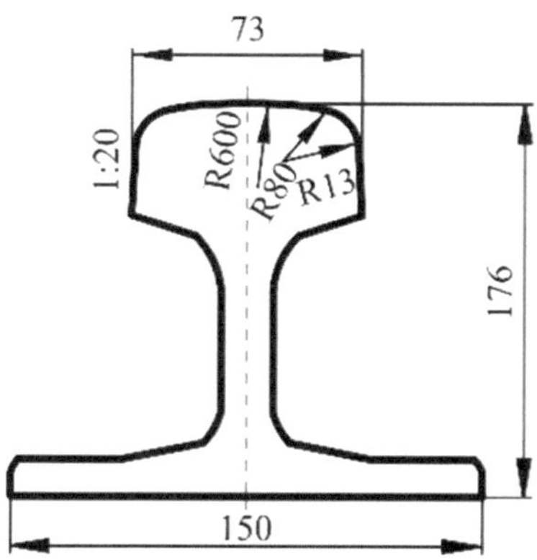


**Fig. 2.8** Shape of rail head

**Table 2.1** Basic parameters of rail and wheel

| | Rail profile | Standard gauge/mm | Rail cant | Wheel tread | The inner distance of the wheelset / mm | Rolling radius/ mm | Wheel flange thickness/ mm |
|---|---|---|---|---|---|---|---|
| Chinese railway | CHN60 | 1435 | 1:40 | $LM_A$ | 1353 | 915/860 | 32.0 |
| Japanese Shinkansen | JIS60 | 1435 | 1:40 | JP-ARC | 1360 | 860/790 | 32.5 |
| European railway | UIC60 | 1435 | 1:40(20) | S1002 | 1360 | 920/850 | 32.5 |

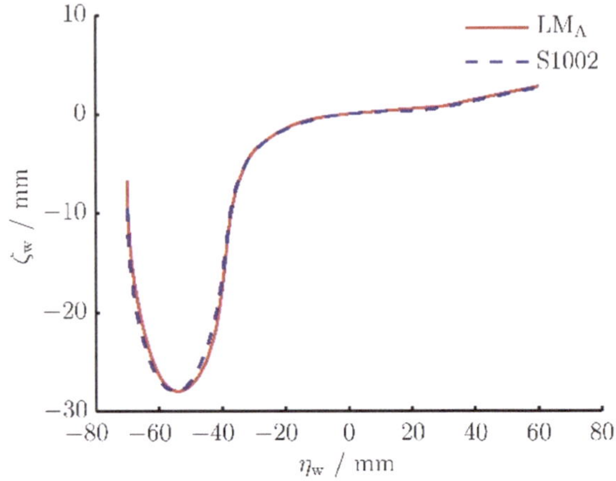


**Fig. 2.9** LM$_A$ and S1002 tread shape

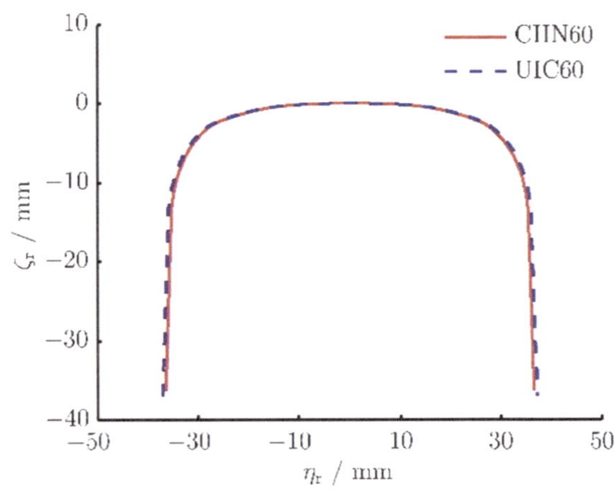


**Fig. 2.10** CHN60 and UIC60 rail head shape

### *2.2.2 Calculation of Wheel-Rail Geometric Contact Relations*

The geometric contact relations and parameters between the wheel and rail are fundamental for studying the dynamic interaction between vehicles and tracks. A free wheelset in space has six degrees of freedom, comprising three translational and three rotational freedoms. Due to the relatively small contact area between the wheel and rail, the wheel-rail contact is typically treated as a single point, with both the wheel and rail considered as rigid bodies. Thus, when the wheelset moves along two rigid tracks, two constraints are generated, leaving four degrees of freedom. In vehicle dynamics analysis, the lateral movement and roll motion of the wheelset are generally chosen as independent degrees of freedom, while vertical movement and yawing motion are constrained. Under normal conditions, the wheel and rail maintain constant contact, and the geometric parameters of this contact are independent of the longitudinal position and rotation of the wheelset on the rail. Therefore, the geometric parameters are essentially functions of the wheelset's lateral displacement and roll angle relative to the track.

1. **Wheel-Rail Contact Parameters and Influencing Factors**

The position of the wheel-rail contact point changes continuously with the lateral displacement and roll of the wheelset. Additionally, several factors impact the location of the contact point, including the shape of the wheel tread and rail head, track gauge, cant, and rail base slope. Geometric parameters at the contact point, such as radius of curvature, equivalent conicity, and contact angle, are critical factors that determine the size of the contact patch between the wheel and rail (see Fig. 2.11). The geometric analysis of wheel-rail contact involves determining the position of the contact point on the track and the associated geometric parameters. This analysis is essential for understanding the load distribution, wear characteristics, and overall performance of the wheel-rail system.

The wheel-rail contact geometric parameters that have a greater impact on the dynamic performance of rolling stock during operation are mainly (see Fig. 2.12):

(1) The actual rolling circle radius $r_L$ and $r_R$ of the left and right wheels at the wheel-rail contact point
(2) The radius of curvature $\rho_{wL}$ and $\rho_{wR}$ of the left and right wheels at the wheel-rail contact point
(3) The radius of curvature $\rho_{rL}$ and $\rho_{rR}$ of the rail head section of the left and right wheels at the wheel-rail contact point
(4) The contact angles $\delta_{wL}$ and $\delta_{wR}$ of the left and right wheels and the left and right rails at the wheel-rail contact point, that is, the angle between the wheel-rail common section and the center line of the wheelset

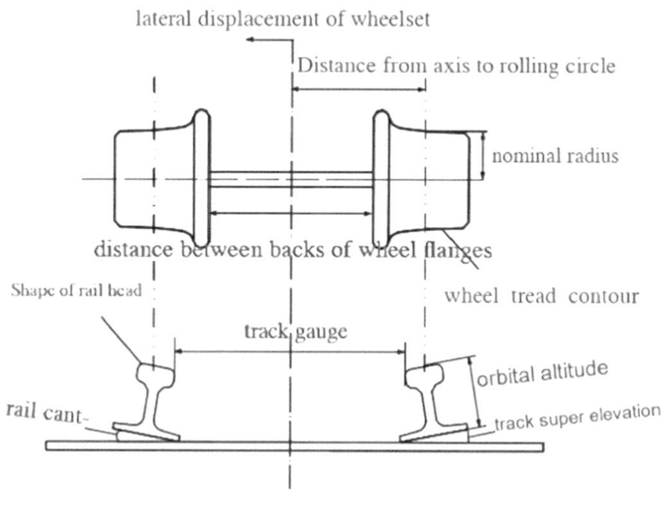


Fig. 2.11 The main factors affecting wheel/rail contact

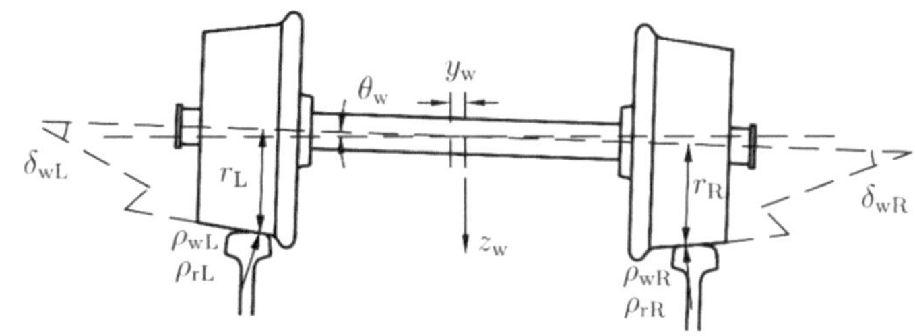


**Fig. 2.12** Wheel-rail contact geometric parameter

(5) The side roll angle $\theta_w$ of the wheelset, the side roll angle of the wheelset will cause the side of the bogie and the carbody
(6) The vertical displacement $z_w$ of the wheelset center will cause the vertical displacement of the bogie and the carbody.

## 2. Wheel-Rail Profile Fitting

Newly manufactured wheels and rails typically have relatively regular geometric shapes. However, the profiles of wheel treads and rail heads after operation often become irregular and cannot be described using standard analytical expressions. Therefore, the worn profiles of the wheel and rail need to be represented by fitting a curve to the surface discrete points obtained through measurement. This process involves measuring the surface points of the worn wheel and rail profiles to obtain a set of discrete data points. Even if the wheel-rail profiles are initially expressed using piecewise functions, they can be converted into discrete data points for analysis. Subsequently, mathematical methods are employed to find a fitting curve that closely conforms to the actual outline of the profiles. The fitting curve should also maintain continuity in its first and second derivatives across the entire curve to ensure a smooth transition and realistic representation of the profile changes. The resulting fitted curve accurately captures the worn shapes of the wheel and rail, which is essential for further analysis of their performance and interaction.

There are many methods of curve fitting, and the result of fitting wheel-rail profile curve by cubic spline function curve is satisfactory. A set of data is given on interval $[a, b]$:

$$\begin{aligned} a = x_0 < x_1 < \cdots < x_n = b \\ y_1 < y_2 < \cdots < y_n \end{aligned} \tag{2.2.1}$$

The cubic spline curve $S(x)$ is a piecewise defined function with a continuous bivalent derivative on the interval $[a, b]$. By selecting the second-order derivative value $S''(x_i) = M_i (i = 0, 1, \ldots, n)$ on the node as the parameter constructor $S(x)$, the following conditions are satisfied:

(1) $S(x)$ has continuous first and second derivatives on $[a, b]$.
(2) On each subinterval$\left[x_i, x_{i+1}\right]$, $S(x)$ is a polynomial of no more than three times;

(3) $S(x_i) = y_i$ $(i = 0, 1, \ldots, n)$, means that this function represents all points on the curve.

The $n$ cubic polynomial segments can be expressed as:

$$S(x_i) = a_i + b_i(x - x_i) + c_i(x - x_i)^2 + d_i(x - x_i)^3 \tag{2.2.2}$$

In the formula, $a_i$, $b_i$, $c_i$ and $d_i$ are $4n$ unknown coefficients. These coefficients can be obtained according to the above three conditions and two end conditions.

3. **Calculation of Wheel-Rail Geometric Contact Relations in Plane Problems**

In the absence of yaw angle, the degrees of freedom reduce to one, leading to what is known as a plane problem. The various geometric contact parameters between the wheel and rail are functions solely of the lateral displacement of the wheelset's center. To analyze this, an upright plane (the main cross-section) is created by cutting through the wheel. The intersection of this plane with the wheel's profile defines the main contour line of the wheel tread. In the context of the plane problem, the contact point between the wheel and rail lies along this main contour line.

To compute the wheel-rail contact point under rigid wheel-rail constraints, the method involves identifying points along the main contour lines of the left and right rails where the distance between the two curves is zero. This point must also be a tangent point of the two curves. In practical calculations, the left and right rails are vertically offset downward by a certain distance. After the wheelset undergoes lateral displacement, the side roll angles of the wheel-rail system are adjusted iteratively to minimize the vertical distance between the two main contour lines of the rails. The process continues until the distances on the left and right sides approach equality. These two points represent the contact points for the left and right rails. Once the wheel-rail contact points are determined, other geometric contact parameters, such as contact angle and rolling radius, can be derived analytically. This systematic approach allows for an accurate understanding of the wheel-rail interaction under the specified conditions.

A fixed coordinate system $(O; y, z)$ is made in the vertical plane of the center line of the fixed wheelset, and the origin of the coordinate is in the vertical plane passing through the center line of the track, as shown in Fig. 2.13. For the wheel-rail tread shape, the plane coordinate systems $(O_w; \eta_w, \zeta_w)$ and $(O_r; \eta_r, \zeta_r)$ are established respectively on the wheel-rail and rail treads with the wheel-rail contact point at the center of the track as the origin, as shown in Figs. 2.9 and 2.10. Wheelset and rail profiles can be described by fitted functions, respectively.

$$\zeta_w = f(\eta_w), \quad \zeta_r = g(\eta_r) \tag{2.2.3}$$

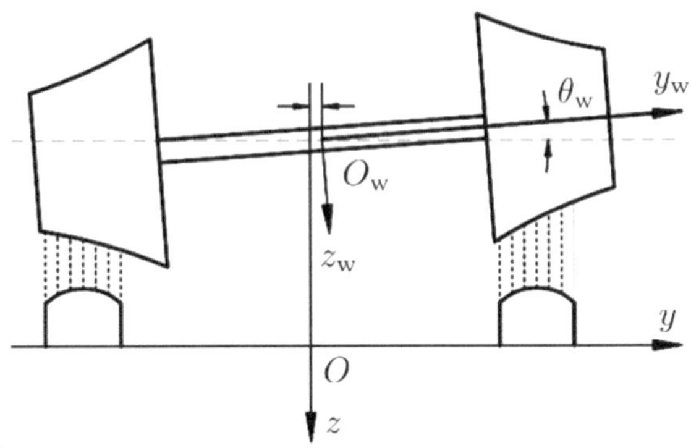


**Fig. 2.13** Plane problem calculation of wheel-rail contact geometric parameters

When the wheelset lateral displacement $y_w$ occurs, the vertical distance between the wheel and rail is calculated for each small horizontal distance between the left and right wheel and rail, and the minimum vertical distance between the left and right wheel and rail is found to be $\Delta z_{Lmin}$ and $\Delta z_{Rmin}$, respectively, and the corresponding lateral position is $y_{Lmin}$ and $y_{Rmin}$. If $\Delta z_{Lmin}=\Delta z_{Rmin}$, when the wheelset moves downward, the left and right wheels contact with the left and right rails at the same time, and $y_{Lmin}$ and $y_{Rmin}$ are the left and right wheel-rail contact points. When the two are not equal, it is necessary to adjust the side roll angle of the wheelset to make it conform to the contact relationship. If $\Delta z_{Rmin}>\Delta z_{Lmin}$, the wheelset can be rotated counterclockwise by an angle $\gamma$.

$$\gamma = \frac{z_{Rmin} - z_{Lmin}}{y_{Rmin} - y_{Lmin}} \tag{2.2.4}$$

After the wheelset rotates, the above scanning process is repeated, that is, the minimum distance between the left and right wheels is checked again at the new position of the wheelset. When the absolute value of the difference between the two minimum distances is less than a given small positive integer, the iteration can be ended. At this time, the side roll angle of the wheelset is

$$\theta_w = \theta_{w0} + \sum_{1}^{k} \gamma_i (i = 1, 2, \ldots, k) \tag{2.2.5}$$

In the formula, $\theta_{w0}$ is the initial roll angle of the wheelset, $\gamma_i$ is the roll angle adjusted by each iteration, and $k$ is the number of iterations.

According to the position $y_{Lmin}$ and $y_{Rmin}$ of the wheel-rail contact point in the fixed coordinate system, the abscissa $\eta_{wL}$, $\eta_{wR}$, $\eta_{rL}$ and $\eta_{rR}$ of the left and right wheelsets and rail contact points in the relative body coordinate system can be obtained. Substituting them into Eqs. (2.2.6) and (2.2.7) respectively, the contact angle and radius of curvature of the wheel-rail contact point, wheel-rail tread can be obtained.

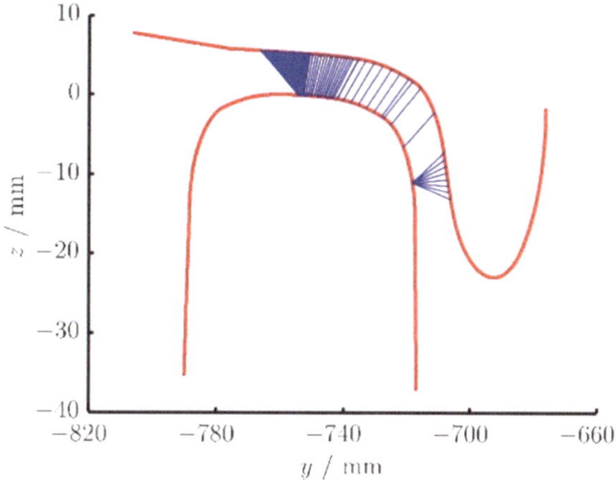


**Fig. 2.14** The distribution of contact points between $LM_A$ tread and CHN60 rail

$$\delta_w = \frac{d\zeta_w}{d\eta_w} = f'(\eta_w), \quad \delta_r = \frac{d\zeta_r}{d\eta_r} = g'(\eta_r) \tag{2.2.6}$$

$$\rho_w = \left| \frac{\left(1 + f'^2(\eta_w)\right)^{\frac{3}{2}}}{f''(\eta_w)} \right|, \quad \rho_r = \left| \frac{\left(1 + g'^2(\eta_r)\right)^{\frac{3}{2}}}{g''(\eta_r)} \right| \tag{2.2.7}$$

The vertical displacement of the wheelset can be obtained according to the geometric relationship determined by the side roll angle and the vertical coordinate change at the left or right wheel-rail contact point:

$$z_w = -a_0\theta_w - \zeta_{wR} + \zeta_{rR}, \quad z_w = -a_0\theta_w - \zeta_{wL} + \zeta_{rL} \tag{2.2.8}$$

The actual rolling circle half A and B of the left and right wheels at the wheel-rail contact point can be obtained by the nominal rolling circle radius of the wheel and the vertical coordinate of the wheelset tread at the contact point:

$$r_L = r_0 - \zeta_{wL}, \quad r_R = r_0 - \zeta_{wR} \tag{2.2.9}$$

In the approximate calculation, the rolling circle radius $r_L$ and $r_R$ are expressed as the function of the lateral displacement $y_w$ of the wheelset.

$$r_L = r_0 - \lambda_0 y_w, \quad r_R = r_0 + \lambda_0 y_w \tag{2.2.10}$$

In the formula, $\lambda_0$ is called the equivalent conicity, which has

$$\lambda_0 = \frac{\delta_0 \rho_w (1 + \rho_r \delta_0 / a_0)}{(\rho_w - \rho_r)(1 - r_0 \delta_0 / a_0)} \tag{2.2.11}$$

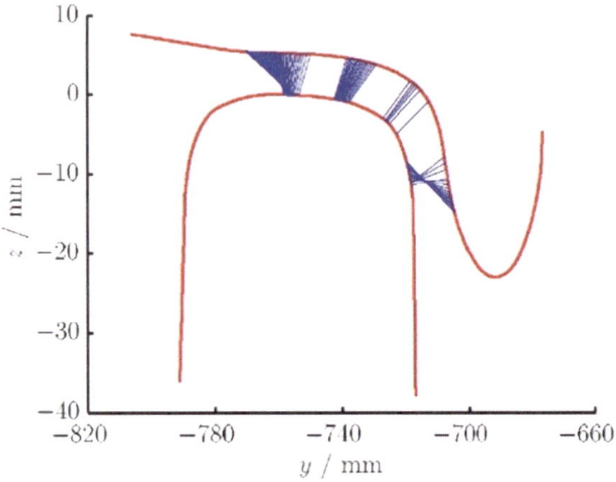


**Fig. 2.15** Distribution of contact points between S1002 rail and UIC60 rail

The contact point distribution maps of $LM_A$ type tread and CHN60 type rail and S1002 type tread and UIC60 type rail are given in Figs. 2.14 and 2.15, respectively.

4. **Calculation of Wheel-Rail Geometric Relations in Space Problems**

In the spatial problem of wheel-rail geometric relations, the presence of yaw angle means that the contact points are generally not located on the main contour line of the wheel; instead, they are distributed on the surface of the wheel profile near the main contour line. The method for determining the wheel-rail contact points is fundamentally similar to that used in plane problems, but it requires an expanded approach to find the minimum vertical distances on both the left and right wheel treads and the top surfaces of the rails.

One commonly used method for calculating wheel-rail spatial contact geometry is the trace method. The basic principle of this method is that when the yaw angle and side roll angle of the wheelset are fixed, the set of spatial contact points between the wheel and rail forms a spatial curve on the wheel tread. This curve represents the trajectory of the contact points on the wheel tread. As a result, the calculation range for determining the wheel-rail contact points is reduced from a surface area to a curve, allowing the computational methods and complexity to remain similar to those in plane problems. The key difference is that in the plane problem, the contact points are determined along a planar curve (the main contour line of the wheel), while in this method, they are determined along a spatial curve (the trace).

The trace method for calculating wheel-rail spatial contact geometry is based on the following fundamental assumptions: (1) Both the wheel and rail are treated as rigid bodies, with no elastic deformation affecting the contact relations. In other words, no point on the wheel surface can penetrate the rail. (2) The contact area between the wheel and rail is considered as a point or a patch, meaning that there are no coplanar or collinear contact situations between the wheel tread and the rail. (3) The left wheel and left rail, as well as the right wheel and right rail of the same wheelset, are assumed to be in contact simultaneously, with no occurrence of one side losing contact.

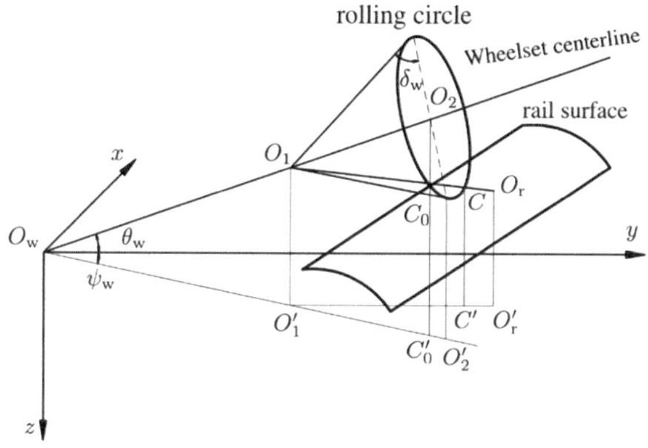


**Fig. 2.16** Calculation of wheel-rail contact geometric parameters by trace method

Taking the right wheel-rail shown in Fig. 2.16 as an example, the wheelset coordinate system ($O_w$; $x$, $y$, $z$) is established with the centroid $O_w$ of the wheelset as the origin, and the lateral movement, yaw movement and roll movement of the wheelset are expressed as $y_w$, $\psi_w$ and $\theta_w$, respectively. The wheel-rail contact point is $C$, and the $C_0$ point is the lowest point of the rolling circle. Due to the yaw angle $\psi_w$, the wheel-rail contact point is not at the lowest point of the rolling circle. $O_1O_r$ is the common normal line passing through the point $C$ tread and the rail cylinder. Points $O_1$ and $O_r$ are located on the wheel-rail center line and the rail center line respectively.

According to Fig. 2.16, the coordinates of the wheel-rail contact point can be obtained as

$$\begin{cases} x_C = x_{O_2} + l_x r_w \tan\delta_w \\ y_C = y_{O_2} - \frac{r_w}{1-l_x^2}\left(l_x^2 l_y \tan\delta_w + l_z m\right) + y_w \\ z_C = z_{O_2} - \frac{r_w}{1-l_x^2}\left(l_x^2 l_z \tan\delta_w - l_y m\right) \end{cases} \tag{2.2.12}$$

$r_w$ and $\delta_w$ are the wheel rolling circle radius and the right wheel-rail contact angle, respectively. $l_x$, $l_y$ and $l_z$ are the cosine of the axis direction of the wheelset; $x_{O_2}$, $y_{O_2}$ and $z_{O_2}$ are the coordinates of the rolling center of the wheel. There is the following formula

$$\begin{cases} l_x = -\cos\theta_w \sin\psi_w \\ l_y = \cos\theta_w \cos\psi_w \\ l_z = \sin\theta_w \end{cases} \tag{2.2.13}$$

$$\begin{cases} x_{O_2} = b_w l_x \\ y_{O_2} = b_w l_y \\ z_{O_2} = b_w l_z \end{cases} \tag{2.2.14}$$

$b_w$ is the lateral distance from the center of the wheelset to the discrete point of the wheelset tread. Parameter $m$ has

$$m = \sqrt{l - l_x^2\left(1 + \tan^2 \delta_w\right)} \tag{2.2.15}$$

The trace of the left wheel-rail contact point can be obtained by the same method. Since the side roll angle $\theta_w$ is an unknown quantity, it is necessary to first specify an initial value for iterative solution. When the wheelset traverses, shakes and rolls are fixed, the track of wheel-rail spatial contact can be obtained by changing the position of discrete points on the wheel tread. Then, similar to the plane problem, by looking for the point where the vertical distance between the wheel and rail is the smallest and the left and right distances are equal as the contact point, if the distance between the left and right minimum points is not equal, the side roll angle is adjusted to repeat the above steps, and the solution process is shown in Fig. 2.17.

It can be seen that when the track line method is used to find the wheel-rail contact point, the set of wheel tread contact points is obtained by temporarily abandoning the rail profile. In the iterative solution process, in addition to the lateral displacement of the wheelset and the shaking head as the input, if the rail vibration is considered, the vertical, lateral displacement and torsion angle of the left and right rails need to be taken as the input. When considering the excitation of track irregularity, the influence of track gauge and horizontal irregularity should also be considered.

Figure 2.18 shows the space curve of wheel-rail contact trace when the wheelset yaw angles are 0°, 1°, 3° and 5° respectively. It can be seen that when the wheelset has no yaw angle, the trace of the wheel-rail contact point is a plane curve, which is completely consistent with the main contour line of the wheel tread. When the wheelset has a yaw angle, the trace of the contact point will become a three-dimensional space curve, and the slope of the trace point at the root of the wheelset is more obvious than that of other parts.

Figure 2.19 shows the calculated curves of the rolling circle radius difference, contact angle difference and equivalent conicity of the left and right wheels with the lateral displacement of the wheelset. It can be seen that when the lateral displacement of the wheelset is small, the rolling circle radius difference and the contact angle radius difference change approximately linearly. The equivalent conicity changes approximately linearly with the lateral displacement of the wheelset. When the wheelset yaw angle is small, it has little influence on the wheel-rail contact geometric parameters.

In the dynamic simulation analysis of vehicle systems, there are two main methods for calculating wheel-rail contact parameters. The first method uses a lookup table, where the wheel-rail contact parameters are initially stored as a function of the wheelset's lateral displacement and yaw angle prior to the simulation. During the calculation, the corresponding wheel-rail contact parameters are obtained through interpolation based on real-time values of lateral displacement and yaw angle. This method is fast but has certain limitations in accuracy. The second method involves real-time online calculations, which provide higher precision but may impact computational efficiency.

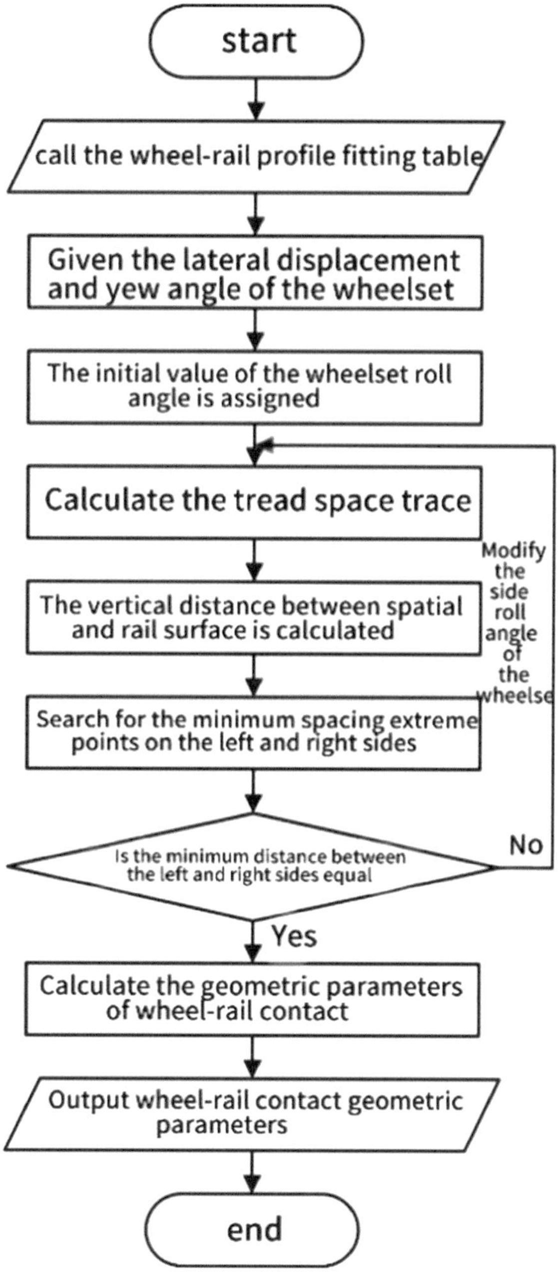


**Fig. 2.17** The calculation process of wheel-rail contact geometric parameters

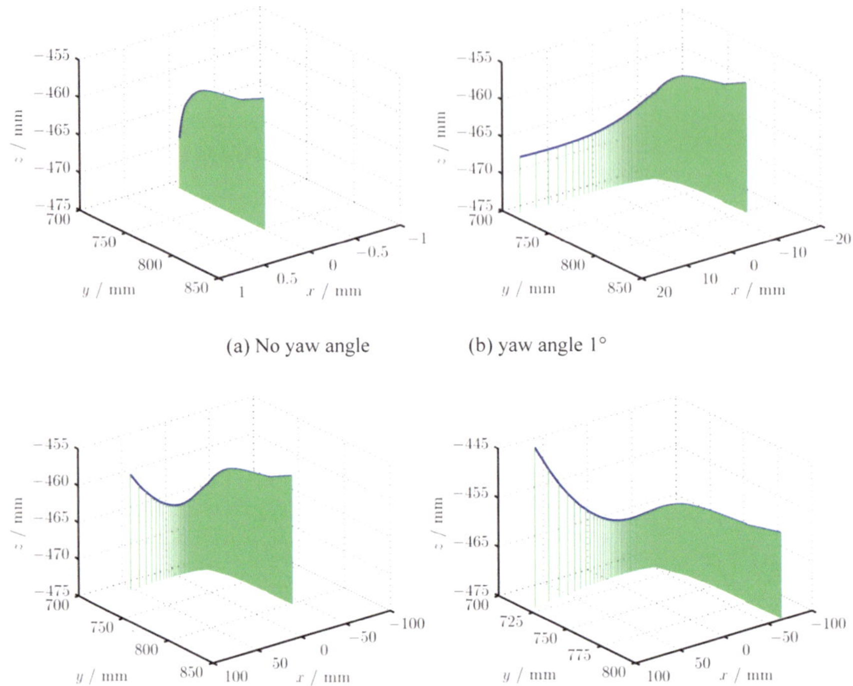


(a) No yaw angle (b) yaw angle 1°

(c) yaw angle 3° (d) yaw angle 5°

**Fig. 2.18** Wheel-rail contact trace

## 2.3 Wheel-Rail Rolling Contact Theory

Wheel-rail rolling contact is a key feature that distinguishes rail vehicles from other modes of transportation in their interaction with surrounding media. The theory of wheel-rail rolling contact is also a core research topic in the dynamics of rail vehicles. The interaction between the wheel and rail significantly influences the dynamic performance of the vehicle system. This section first introduces the application of Hertz contact theory to wheel-rail contact, which describes the shape and size of contact between two elastic bodies. It then discusses the phenomenon of creep between two rolling objects in the contact zone, along with methods for calculating the wheel-rail creep ratio and creep forces. The focus is on Kalker's linear creep theory, which is widely adopted, as well as methods for approximating and correcting nonlinear creep forces.

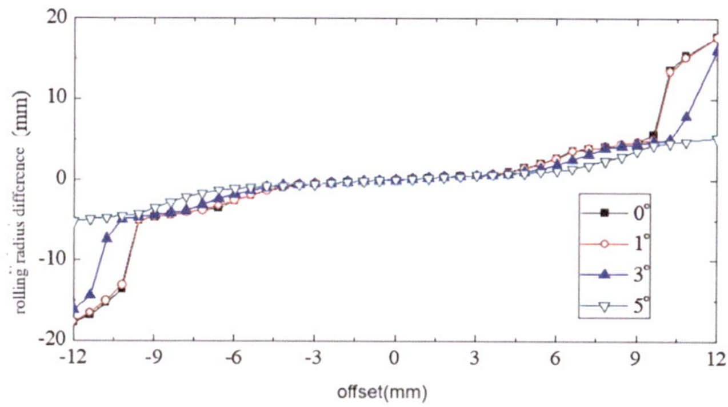


(a) Rolling radius difference

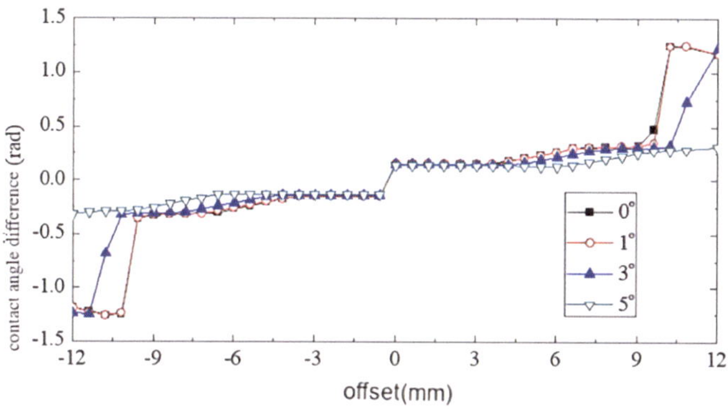


(b) Contact angle difference

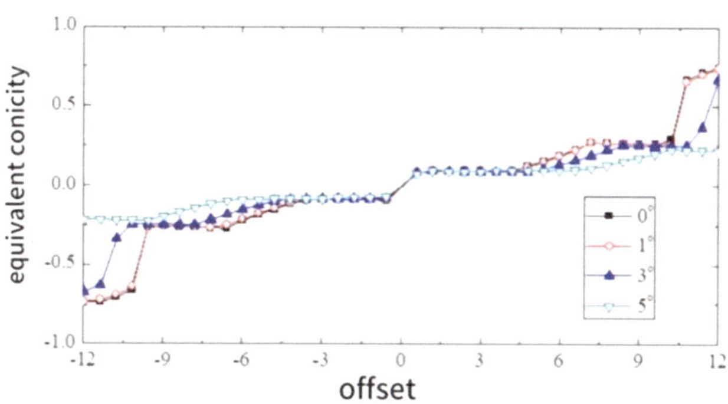


(c) Equivalent conicity

**Fig. 2.19** Curves of wheel-rail contact parameters changing with wheelset lateral displacement and yaw angle

### 2.3.1 Hertz Contact Theory

As early as in the nineteenth century, Hertz studied the contact problem of two elastic bodies by using the theory of elastic mechanics. It is considered that the shape of the contact patch between two elastic bodies is an ellipse, and the calculation formulas of the long and short semi-axis a and b of the contact patch ellipse are derived, as shown in Fig. 2.20.

Two constants $A$ and $B$ are defined, which can be expressed as functions of the angle between the principal curvature planes. Let $\rho_1$ and $\rho_2$ be the principal radius of curvature of one of the elastomers at its contact point, $\rho_{t1}$ and $\rho_{t2}$ be the principal radius of curvature of the other elastomer in contact with it at the same contact point, and the angle $\eta$ be the angle between the plane containing $1/\rho_1$ and the plane containing $1/\rho_{t1}$. Then the constants $A$ and $B$ can be determined by the following formula:

$$\begin{cases} A+B=\frac{1}{2}\left(\frac{1}{\rho_1}+\frac{1}{\rho_2}+\frac{1}{\rho_{t1}}+\frac{1}{\rho_{t2}}\right) \\ B-A=\frac{1}{2}\left[\left(\frac{1}{\rho_1}-\frac{1}{\rho_2}\right)^2+\left(\frac{1}{\rho_{t1}}-\frac{1}{\rho_{t2}}\right)^2+2\left(\frac{1}{\rho_1}-\frac{1}{\rho_2}\right)\cos 2\eta\right]^{\frac{1}{2}} \end{cases} \tag{2.3.1}$$

Considering the specific situation of the contact between the railway vehicle and the rail, the main curvature radius is the direction along the line and the direction perpendicular to the line, and the angle $\eta$ is equal to zero. Thus, each principal radius of curvature can be defined as follows:

$\rho_1 = R_w$—The radius of the wheel rolling circle is related to the nominal radius of the wheel;

$\rho_2 = r_w$—Cross-sectional shape radius of wheel tread;

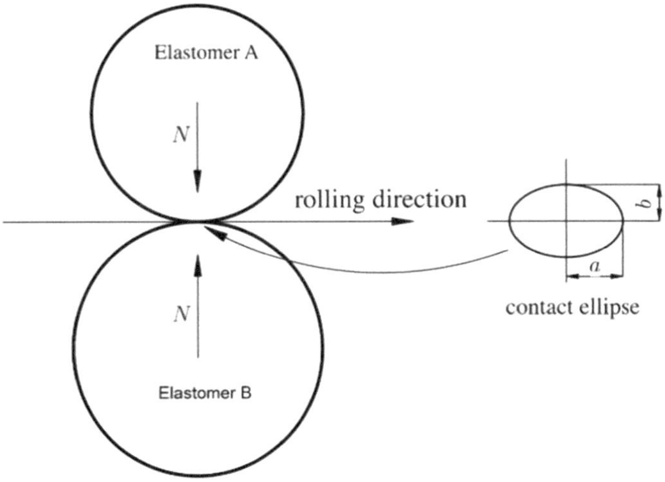


**Fig. 2.20** Elastomer contact and contact ellipse

$\rho_{t1} = \infty$—The rail head is straight along the length of the track, so its radius of curvature is infinite.

$\rho_t = r_r$—Radius of rail head cross section.

For the sign of the radius of curvature, Hertz stipulates that if the center of curvature is located inside the elastic body, it takes a positive sign, otherwise it takes a negative sign. Thus Eq. (2.3.1) can be further simplified as

$$\begin{cases} A + B = \frac{1}{2}(\frac{1}{R_w} + \frac{1}{r_r} + \frac{1}{r_w}) \\ B - A = \frac{1}{2}(\frac{1}{R_w} - \frac{1}{r_r} - \frac{1}{r_w}) \end{cases} \tag{2.3.2}$$

Hertz gave two other constants $k_1$ and $k_2$:

$$k_1 = \frac{1 - \sigma_1^2}{\pi E_1}, \quad k_2 = \frac{1 - \sigma_2^2}{\pi E_2} \tag{2.3.3}$$

In Eq. (2.3.3), it is assumed that two elastomers have different elastic modulus $E$ and Poisson's ratio $\sigma$. For wheels and rails, the elastic modulus $E$ and Poisson's ratio $\boldsymbol{\sigma}$ of the material can be considered to be the same, so there are:

$$k_1 + k_2 = \frac{2(1 - \sigma^2)}{\pi E} \tag{2.3.4}$$

Hertz gave the formulas for calculating the long and short radius $a$ and $b$ of the contact ellipse:

$$\begin{cases} a = m\left[\frac{3\pi N(k_1+k_2)}{4(A+B)}\right]^{\frac{1}{3}} \\ b = n\left[\frac{3\pi N(k_1+k_2)}{4(A+B)}\right]^{\frac{1}{3}} \end{cases} \tag{2.3.5}$$

In the formula, $N$ is the normal load on the contact elliptical spot; $m$ and $n$ are coefficients related to $A$ and $B$. Denote $\beta = \arccos|(B - A)/(A + B)|$, corresponding to different $\beta$ values, $m$ and $n$ can be determined by interpolation in Table 2.2.

## 2.3.2 *Wheel-Rail Creep Theory*

### 1. **Wheel-Rail Creep Phenomenon**

When a wheelset rolls on the track, relative displacement and velocity differences occur between the wheel and rail. The contact patch between the two not only experiences slight elastic deformation but also has a small velocity difference (micro-creep). This minor elastic sliding of the metallic material is termed "creep" and represents an intermediate state between pure rolling and pure sliding. Due to the sliding that occurs between the wheel and rail, a velocity difference is generated,

**Table 2.2** The relationship between $\beta$ and hertz coefficients $m$ and $n$

| $\beta$ (deg) | $m$ | $n$ | $\beta$ (deg) | $m$ | $n$ | $\beta$ (deg) | $m$ | $n$ |
|---|---|---|---|---|---|---|---|---|
| 0.5 | 61.4 | 0.1018 | 10 | 6.604 | 0.3112 | 60 | 1.486 | 0.717 |
| 1 | 36.89 | 0.1314 | 20 | 3.813 | 0.4125 | 65 | 1.378 | 0.759 |
| 1.5 | 27.48 | 0.1522 | 30 | 2.731 | 0.493 | 70 | 1.284 | 0.802 |
| 2 | 22.26 | 0.1691 | 35 | 2.397 | 0.530 | 75 | 1.202 | 0.846 |
| 3 | 16.5 | 0.1964 | 40 | 2.136 | 0.567 | 80 | 1.128 | 0.893 |
| 4 | 13.31 | 0.2188 | 45 | 1.926 | 0.604 | 85 | 1.061 | 0.944 |
| 6 | 9.79 | 0.2552 | 50 | 1.754 | 0.641 | 90 | 1.0 | 1.0 |
| 8 | 7.86 | 0.285 | 55 | 1.611 | 0.678 | | | |

forming a contact patch between the two pressed elastic bodies. The presence of friction results in tangential forces at the contact patch, known as creep forces. The study of how the tangential force at the contact patch varies with creep has led to the development of wheel-rail creep theory.

2. **Definition of Creep Rate**

In general, the circumferential velocity of the contact surface of two rolling objects is not equal. This velocity difference can be used to define the dimensionless creep rate when two objects are pure rolling. As early as the 1920s, Carter found the creep phenomenon between wheel and rail, and gave the definition of longitudinal creep rate $\xi_x$ and lateral creep rate $\xi_y$:

$$\begin{cases} \xi_x = \dfrac{\text{The actual forward speed of the wheel} - \text{The forward speed of pure rolling}}{\text{The forward speed generated by the pure rolling of the wheel}} \\ \xi_y = \dfrac{\text{The actual lateral speed of the wheel} - \text{Transverse velocity in pure rolling}}{\text{The forward speed generated by the pure rolling of the wheel}} \end{cases} \tag{2.3.6}$$

In addition, due to the difference between the rotational speeds of the wheelset and the rail rotating around the normal line perpendicular to the plane of the contact surface, spin creep occurs. The corresponding spin creep rate $\xi_{sp}$ is defined as:

$$\xi_{sp} = \frac{\begin{array}{c}\text{The angular velocity of the wheel around the normal direction of the contact spot} \\ -\text{The angular velocity of the rail around the normal direction of the contact spot}\end{array}}{\text{The forward speed generated by the pure rolling of the wheel}} \tag{2.3.7}$$

In the early 1970s, the International Union of Railways (UIC) made a more precise definition of creep rate considering the movement characteristics of wheels on rails in the case of large creep. Refer to Fig. 2.21, the $\{O; 1, 2, 3\}$ coordinate system is established with the center of the wheel-rail contact ellipse as the origin. The $O1$

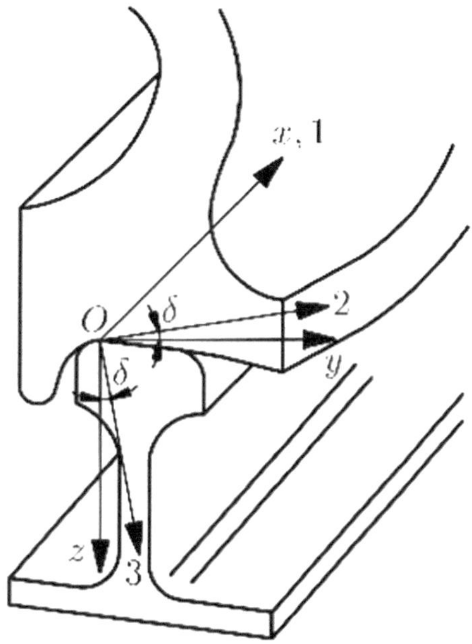


**Fig. 2.21** Two coordinate systems on the wheel-rail contact ellipse

axis is the forward direction of the wheel, which coincides with the $Ox$ axis of the wheelset coordinate system. In the wheel-rail contact plane, the $O2$-axis is roughly parallel to the axis of the axle (with a yaw angle $\psi$ in the direction of the axle), and in the $yz$-plane, the angle between the $O2$-axis and the $Oy$-axis is the contact angle $\delta$; the $O3$ axis is the normal direction of the contact ellipse. In fact, as long as the $\{O; x, y, z\}$ coordinate system rotates a contact angle $\delta$ around the $Ox$ axis, the $\{O; 1, 2, 3\}$ coordinate system is obtained.

Assuming that the speed of the wheel in the contact spot ellipse along the $O1$-axis, the $O2$-axis and the rotation speed around the $O3$-axis are $v_{w1}$, $v_{w2}$ and $\Omega_{w3}$, respectively, and the speed of the contact spot on the rail is $v_{r1}$, $v_{r1}$ and $\Omega_{r3}$, respectively, the definition of the creep rate is as follows:

$$\begin{cases} \xi_1 = \dfrac{v_{r1} - v_{w1}}{(v_{r1} + v_{w1})/2} \\ \xi_2 = \dfrac{v_{r2} - v_{w2}}{(v_{r1} + v_{w1})/2} \\ \xi_3 = \dfrac{\Omega_{r3} - \Omega_{w3}}{(v_{r1} + v_{w1})/2} \end{cases} \tag{2.3.8}$$

In the formula, $(v_{r1} + v_{w1})/2$ denotes the average speed of the wheelset moving along the rail.

3. **Introduction of Wheel-Rail Creep Theory**

The relationship between creep rate and creep force is quite complex. Solving the relationship between creep coefficient, creep force and creep rate has always been the main problem solved by wheel-rail creep theory. In practical application, it is always simplified according to some theories.

In 1926, Carter published a one-dimensional wheel-rail rolling contact model, which systematically analyzed the relationship between creep rate and creep force. He used the theory of elastic rolling contact to divide the wheel-rail contact spot into sliding zone and adhesion zone. The rail is simulated as an elastic half-space, and the wheel is simulated as an elastic cylinder. The two-dimensional elastic problem of two half-space contact is solved, and the calculation method of longitudinal creep rate and transverse creep rate is proposed.

In 1964, Johnson and Vermeulen developed the longitudinal and lateral wheel-rail creep rate/force model in the process of wheel-rail rolling contact according to Carter's research idea. The relationship and calculation formula of longitudinal creep, lateral creep and tangential force are given, and the experimental demonstration is also made. The model is two-dimensional nonlinear, ignoring the influence of wheel-rail spin creep rate.

In 1967, Kalker developed a linear wheel-rail creep theoretical model considering the spin effect by means of series theory in his doctoral thesis. This model established the linear relationship law of wheel-rail creep rate/force and wheel-rail creep force couple/spin. Subsequently, the 'simplified theory' model was developed on the basis of the linear model.

In 1983, with the help of Johnson-Vermeulen's two-dimensional nonlinear theoretical model and Kalker's three-dimensional linear theoretical model, Shen et al. developed a three-dimensional nonlinear wheel-rail creep rate model considering the influence of small spin.

These models adopt the Hertz contact theory assumption when considering the normal force. In the 1970s, Kalker developed a three-dimensional non-Hertz contact theoretical model based on the variational principle and developed the corresponding numerical program 'CONTACT'. The program can obtain the wheel-rail related information that cannot be obtained by the above five classical theories. But its calculation speed is slow.

Jin Xuesong et al. compared and analyzed the numerical analysis results of the above models (except Carter model). The results showed that the results of the other models were close to each other except Kalker's linear theoretical model [33]. Considering the efficiency requirements of the dynamic simulation analysis of the vehicle-line system, the wheel-rail creep rate/force model of Shen et al. Not only considers the factors comprehensively, but also has high computational efficiency, which is widely used at present. In addition, since the 1980s, the finite element method has been applied to the study of wheel-rail rolling contact problems, and good results have been achieved in the study of wheel-rail unsteady rolling, high-frequency vibration and vibration caused by rail corrugation in high-speed railways.

4. **Kalker's Linear Creep Theory**

In practical engineering applications, the calculation of creep force is more based on Kalker linear theory, and then nonlinear correction is carried out.

In Kalker's linear creep theory, it is assumed that the contact area is all adhesive area, and the distribution of tangential force is symmetrical, so the longitudinal creep force is independent of the transverse creep rate. The linear relationship between creep force and creep rate is given as follows:

$$\begin{cases} F'_{cx} = -f_{11}\xi_x \\ F'_{cy} = -f_{22}\xi_y - f_{23}\xi_{sp} \\ M'_{cz} = f_{32}\xi_y - f_{33}\xi_{sp} \end{cases} \tag{2.3.9}$$

where $f_{ij}$ is the creep coefficient (where $f_{23} = f_{32}$), determined by the following formula:

$$\begin{cases} f_{11} = E(ab)C_{11} \\ f_{22} = E(ab)C_{22} \\ f_{23} = E(ab)^{3/2}C_{23} \\ f_{33} = E(ab)^2C_{33} \end{cases} \tag{2.3.10}$$

In the formula, $a$ and $b$ are the long and short axes of the contact ellipse, which can be obtained by the Hertz theoretical formula. $E$ is the Young's elastic modulus, and $C_{ij}$ is the dimensionless Kalker coefficient.

After determining the major and minor axes of the contact ellipse, $C_{ij}$ can be obtained, which is related to $a/b$ and Poisson's ratio $\sigma$. For steel wheels and rails, the values of $\sigma = 0.3$ and $C_{ij}$ can be obtained from Table 2.3 according to the value of $a/b$.

**Table 2.3** Numerical calculation table of Kalker coefficient $C_{ij}(\sigma = 0.3)$

| $a/b$ | $C_{11}$ | $C_{22}$ | $C_{23}$ | $C_{33}$ | $b/a$ | $C_{11}$ | $C_{22}$ | $C_{23}$ | $C_{33}$ |
|---|---|---|---|---|---|---|---|---|---|
| 0.1 | 1.35 | 0.98 | 0.195 | 3.34 | 0.9 | 1.7 | 1.49 | 0.628 | 0.425 |
| 0.2 | 1.37 | 1.01 | 0.242 | 1.74 | 0.8 | 1.75 | 1.56 | 0.689 | 0.396 |
| 0.3 | 1.40 | 1.06 | 0.288 | 1.18 | 0.7 | 1.81 | 1.65 | 0.768 | 0.366 |
| 0.4 | 1.44 | 1.11 | 0.328 | 0.925 | 0.6 | 1.90 | 1.76 | 0.875 | 0.336 |
| 0.5 | 1.47 | 1.18 | 0.368 | 0.766 | 0.5 | 2.03 | 1.93 | 1.04 | 0.304 |
| 0.6 | 1.50 | 1.22 | 0.410 | 0.661 | 0.4 | 2.21 | 2.15 | 1.27 | 0.275 |
| 0.7 | 1.54 | 1.28 | 0.451 | 0.588 | 0.3 | 2.51 | 2.54 | 1.71 | 0.246 |
| 0.8 | 1.57 | 1.32 | 0.493 | 0.533 | 0.2 | 3.08 | 3.26 | 2.64 | 0.215 |
| 0.9 | 1.60 | 1.39 | 0.535 | 0.492 | 0.1 | 4.60 | 5.15 | 5.81 | 0.183 |
| 1.0 | 1.65 | 1.43 | 0.579 | 0.458 | | | | | |

5. **Nonlinear Correction of Creep Force**

Using the method provided by Kalker linear theory to calculate the creep coefficient is simple and practical, and has a certain accuracy. However, Kalker's linear creep theory is only applicable to the case of small creep rate and small spin. In the case of large creep or even complete creep, the creep force is saturated, and the relationship between creep force and creep rate is nonlinear. Therefore, when studying the vehicle curve passing under the condition of large creep, using the linear theory to calculate the creep force will produce a large error. Johnson-Vermeulen theory gives a nonlinear correction method under arbitrary creep rate. Shen et al. considered the influence of spin on this basis.

Based on the longitudinal creep force $F'_{cx}$ and transverse creep force $F'_{cy}$ obtained by Kalker linear theory, the synthetic creep force is calculated.

$$F'_{\mathrm{c}} = \sqrt{F'^2_{cx} + F'^2_{cy}} \tag{2.3.11}$$

According to the Coulomb friction law, the synthetic creep force does not exceed the pure sliding friction force $\mu N$. According to the Johnson-Vermeulen theory, the creep force $F'_{\mathrm{c}}$ is limited by the nonlinear quantity $F_{\mathrm{c}}$, and the relationship is as follows

$$F_{\mathrm{c}} = \begin{cases} \mu N\left[\dfrac{F'_{\mathrm{c}}}{\mu N} - \dfrac{1}{3}\left(\dfrac{F'_{\mathrm{c}}}{\mu N}\right)^2 + \dfrac{1}{27}\left(\dfrac{F'_{\mathrm{c}}}{\mu N}\right)^3\right], & \text{if } F'_{\mathrm{c}} \le 3\mu N \\ \mu N, & \text{if } F'_{\mathrm{c}} > 3\mu N \end{cases} \tag{2.3.12}$$

In the formula, $\mu$ is the friction coefficient between wheel and rail, and $N$ is the normal force at the contact point.

The creep force correction coefficient is introduced

$$\varepsilon = \frac{F_{\mathrm{c}}}{F'_{\mathrm{c}}} \tag{2.3.13}$$

Therefore, the corrected creep force and torque are:

$$\begin{cases} F_{cx} = \varepsilon F'_{cx} \\ F_{cy} = \varepsilon F'_{cy} \\ M_{cz} = \varepsilon M'_{cz} \end{cases} \tag{2.3.14}$$

## 2.4 Wheelset Motion Equation

### 2.4.1 Wheelset Coordinate System

The inertial coordinate system $\boldsymbol{e}^{\mathrm{I}}$, the track coordinate system $\boldsymbol{e}^{\mathrm{T}}$ and the wheelset coordinate system $\boldsymbol{e}^{\mathrm{W}}$ are established, as shown in Fig. 2.22. The absolute motion of the wheelset can be obtained by the combination of the motion of the wheelset relative to the track coordinate system and the motion of the track coordinate system relative to the inertial coordinate system.

1. **Track Coordinate System**

The origin $O_{\mathrm{T}}$ of the track coordinate system $\boldsymbol{e}^{\mathrm{T}}$ is located at the center line of the track and moves forward with the vehicle speed $v$ along the tangent direction of the track, which reflects the geometric alignment of the line and the track irregularity.

The track coordinate system $\boldsymbol{e}^{\mathrm{T}}$ is obtained by three continuous rotations of the inertial coordinate system $\boldsymbol{e}^{\mathrm{I}}$. Two auxiliary coordinate systems $\boldsymbol{e}^{\mathrm{T}'}$ and $\boldsymbol{e}^{\mathrm{T}''}$ are introduced to reflect the position after the first two rotations.

(1) By rotating the $\psi_{\mathrm{T}}$ angle around the $z_{\mathrm{I}}$ axis of the inertial coordinate system $\boldsymbol{e}^{\mathrm{I}}$, the $\boldsymbol{e}^{\mathrm{T}'}$ is obtained, which reflects the plane alignment.
(2) By rotating the $\varphi_{\mathrm{T}}$ angle around the $y_{\mathrm{T}'}$ axis of the coordinate system $\boldsymbol{e}^{\mathrm{T}'}$, the $\boldsymbol{e}^{\mathrm{T}''}$ is obtained, which reflects the vertical alignment.
(3) By rotating the $\theta_{\mathrm{T}}$ angle around the $\boldsymbol{x}_{\mathrm{T}''}$ axis of the coordinate system $\boldsymbol{e}^{\mathrm{T}''}$, the $\boldsymbol{e}^{\mathrm{T}}$ is obtained, which reflects the superelevation.

The transformation matrices $\underline{\boldsymbol{A}}_z^{\psi}$, $\underline{\boldsymbol{A}}_x^{\theta}$ and $\underline{\boldsymbol{A}}_y^{\varphi}$ between adjacent coordinate systems are

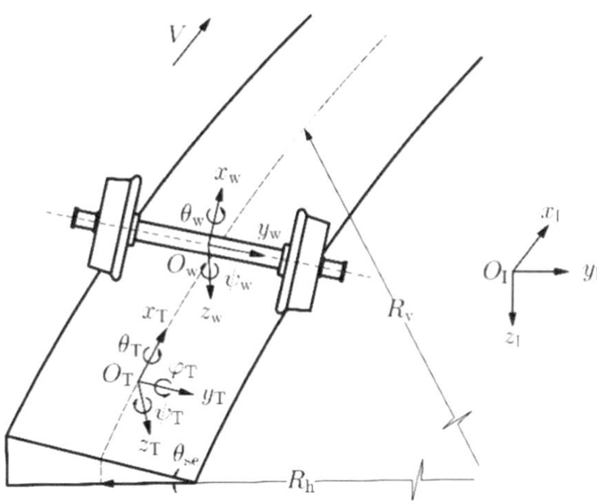


Fig. 2.22 Wheelset motion coordinate system

$$\underline{\boldsymbol{A}}_z^{\psi} = \begin{bmatrix} \cos\psi & \sin\psi & 0 \\ -\sin\psi & \cos\psi & 0 \\ 0 & 0 & 1 \end{bmatrix} \tag{2.4.1}$$

$$\underline{\boldsymbol{A}}_x^{\theta} = \begin{bmatrix} 1 & 0 & 0 \\ 0 & \cos\theta & \sin\theta \\ 0 & -\sin\theta & \cos\theta \end{bmatrix}$$

$$\underline{\boldsymbol{A}}_y^{\varphi} = \begin{bmatrix} \cos\varphi & 0 & -\sin\varphi \\ 0 & 1 & 0 \\ \sin\varphi & 0 & \cos\varphi \end{bmatrix}$$

The transformation matrix $\underline{\boldsymbol{G}}^{\mathrm{TI}}, \underline{\boldsymbol{G}}^{\mathrm{TT}'}, \underline{\boldsymbol{G}}^{\mathrm{TT}''}$ from coordinate system $\boldsymbol{e}^{\mathrm{I}}, \boldsymbol{e}^{\mathrm{T}'}, \boldsymbol{e}^{\mathrm{T}''}$ to track coordinate system $\boldsymbol{e}^{\mathrm{T}}$ can be written as

$$\underline{\boldsymbol{G}}^{\mathrm{TI}} = \underline{\boldsymbol{A}}_x^{\theta}\underline{\boldsymbol{A}}_y^{\varphi}\underline{\boldsymbol{A}}_z^{\psi}, \quad \underline{\boldsymbol{G}}^{\mathrm{TT}'} = \underline{\boldsymbol{A}}_x^{\theta}\underline{\boldsymbol{A}}_y^{\varphi}, \quad \underline{\boldsymbol{G}}^{\mathrm{TT}''} = \underline{\boldsymbol{A}}_x^{\theta} \tag{2.4.2}$$

There will be

$$\underline{\boldsymbol{G}}^{\mathrm{TI}} = \begin{bmatrix} \cos\varphi_{\mathrm{T}}\cos\psi_{\mathrm{T}} & \cos\varphi_{\mathrm{T}}\sin\psi_{\mathrm{T}} & -\sin\varphi_{\mathrm{T}} \\ \sin\theta_{\mathrm{T}}\sin\varphi_{\mathrm{T}}\cos\psi_{\mathrm{T}} - \cos\theta_{\mathrm{T}}\sin\psi_{\mathrm{T}} & \sin\theta_{\mathrm{T}}\sin\varphi_{\mathrm{T}}\sin\psi_{\mathrm{T}} + \cos\theta_{\mathrm{T}}\cos\psi_{\mathrm{T}} & \sin\theta_{\mathrm{T}}\cos\varphi_{\mathrm{T}} \\ \cos\theta_{\mathrm{T}}\sin\varphi_{\mathrm{T}}\cos\psi_{\mathrm{T}} + \sin\theta_{\mathrm{T}}\sin\psi_{\mathrm{T}} & \cos\theta_{\mathrm{T}}\sin\varphi_{\mathrm{T}}\sin\psi_{\mathrm{T}} - \sin\theta_{\mathrm{T}}\cos\psi_{\mathrm{T}} & \cos\theta_{\mathrm{T}}\cos\varphi_{\mathrm{T}} \end{bmatrix} \tag{2.4.3}$$

According to the properties of the transformation matrix, the transformation matrix $\underline{\boldsymbol{G}}^{\mathrm{IT}}$ from the track coordinate system $\boldsymbol{e}^{\mathrm{T}}$ to the inertial coordinate system $\boldsymbol{e}^{\mathrm{I}}$ is obtained.

$$\underline{\boldsymbol{G}}^{\mathrm{IT}} = (\underline{\boldsymbol{G}}^{\mathrm{TI}})^{-1} = (\underline{\boldsymbol{G}}^{\mathrm{TI}})^{\mathrm{T}} \tag{2.4.4}$$

The three rotation angles $\psi_{\mathrm{T}}$, $\varphi_{\mathrm{T}}$ and $\theta_{\mathrm{T}}$ of the track coordinate system are obtained by the parameters of the horizontal and vertical sections of the line, which are functions of the line length $s$. The rotation angles $\psi_{\mathrm{T}}$ and $\varphi_{\mathrm{T}}$ of the plane and the longitudinal section can be represented by the curvature radii $R_h$ and $R_v$ of the plane and the longitudinal section respectively.

$$\psi_{\mathrm{T}} = \psi_{\mathrm{T}0} + \int_{s_0}^{s} \frac{\mathrm{d}s}{R_{\mathrm{h}}(s)}, \quad \mathrm{d}\psi_{\mathrm{T}} = \frac{\mathrm{d}s}{R_{\mathrm{h}}(s)} \tag{2.4.5}$$

$$\varphi_{\mathrm{T}} = \varphi_{\mathrm{T}0} + \int_{s_0}^{s} \frac{\mathrm{d}s}{R_{\mathrm{v}}(s)}, \quad \mathrm{d}\varphi_{\mathrm{T}} = \frac{\mathrm{d}s}{R_{\mathrm{v}}(s)} \tag{2.4.6}$$

The angle $\theta_{\mathrm{T}}$ of the cross section can be expressed by the super high angle $\theta_{\mathtt{se}}$ from the super high angle $h_{\mathtt{se}}$.

$$\theta_{\mathrm{T}} = \theta_{\mathtt{se}}(s) = h_{\mathtt{se}}(s)/l_{\mathrm{s}} \tag{2.4.7}$$

In the formula, $l_{\mathrm{s}}$ is the distance between the center lines of the two rails, which is 1500 mm.

When the plane is a right-turning curve, $R_{\mathrm{h}}$ and $\theta_{\mathtt{se}}$ take a positive sign, and the left-turning curve is negative; the vertical section is a concave vertical curve with a positive sign, and the convex vertical curve $R_{\mathrm{v}}$ takes a negative sign.

Then the track coordinate system moves along the $O_{\mathrm{T}}x_{\mathrm{T}}$ speed $v = \mathrm{d}s/\mathrm{d}t$, so that the motion of the track coordinate system $\boldsymbol{e}^{\mathrm{T}}$ can be completely determined. From Eqs. (2.4.5) and (2.4.6), the angular velocity of the track coordinate system is

$$\frac{\mathrm{d}\psi_{\mathrm{T}}}{\mathrm{d}t} = \frac{v}{R_{\mathrm{h}}(s)} \tag{2.4.8}$$

$$\frac{\mathrm{d}\varphi_{\mathrm{T}}}{\mathrm{d}t} = \frac{v}{R_{\mathrm{v}}(s)} \tag{2.4.9}$$

According to the definition of track coordinate system, there are

$$\dot{\boldsymbol{r}}_{\mathrm{T}}^{\mathrm{I}} = \underline{\dot{\boldsymbol{r}}}_{\mathrm{T}}^{\mathrm{I}}\boldsymbol{e}^{\mathrm{T}}, \quad \underline{\dot{\boldsymbol{r}}}_{\mathrm{T}}^{\mathrm{I}} = \begin{bmatrix} v & 0 & -a_0\dot{\theta}_{\mathrm{T}} \end{bmatrix}^{\mathrm{T}} \tag{2.4.10}$$

$$\boldsymbol{\omega}_{\mathrm{TI}} = \dot{\psi}_{\mathrm{T}}\boldsymbol{e}_z^{\mathrm{I}} + \dot{\varphi}_{\mathrm{T}}\boldsymbol{e}_y^{\mathrm{T}'} + \dot{\theta}_{\mathrm{T}}\boldsymbol{e}_x^{\mathrm{T}''} = \underline{\boldsymbol{\omega}}_{\mathrm{TI}}\boldsymbol{e}^{\mathrm{T}}$$

$$\underline{\boldsymbol{\omega}}_{\mathrm{TI}} = \begin{bmatrix} -\dot{\psi}_{\mathrm{T}}\sin\varphi_{\mathrm{T}} + \dot{\theta}_{\mathrm{T}} \\ \dot{\psi}_{\mathrm{T}}\sin\theta_{\mathrm{T}}\cos\varphi_{\mathrm{T}} + \dot{\varphi}_{\mathrm{T}}\cos\theta_{\mathrm{T}} \\ \dot{\psi}_{\mathrm{T}}\cos\theta_{\mathrm{T}}\cos\varphi_{\mathrm{T}} - \dot{\varphi}_{\mathrm{T}}\sin\theta_{\mathrm{T}} \end{bmatrix} \approx \begin{bmatrix} \dot{\theta}_{\mathrm{T}} \\ \dot{\varphi}_{\mathrm{T}} + \dot{\psi}_{\mathrm{T}}\theta_{\mathrm{T}} \\ \dot{\psi}_{\mathrm{T}} - \dot{\varphi}_{\mathrm{T}}\theta_{\mathrm{T}} \end{bmatrix} \tag{2.4.11}$$

2. **Wheelset Coordinate System**

The spatial position of the wheelset relative to the track coordinate system $\boldsymbol{e}^{\mathrm{T}}$ can be determined by successive three consecutive rotations of the yaw $\psi_{\mathrm{w}}$, roll $\theta_{\mathrm{w}}$ and rotation $\varphi_{\mathrm{w}}$. Since the rotation angle $\varphi_{\mathrm{w}}$ of the wheelset around the axis $O_{\mathrm{w}}y_{\mathrm{w}}$ does not affect the position of the axis $O_{\mathrm{w}}x_{\mathrm{w}}$ and the axis $O_{\mathrm{w}}z_{\mathrm{w}}$ relative to the track coordinate system and the geometry and mass distribution of the wheelset, the wheelset coordinate system $\boldsymbol{e}^{\mathrm{w}}$ defines that the origin $O_{\mathrm{w}}$ is located at the center of mass of the wheelset, including only the first two yaw and roll rotation motions.

(1) The auxiliary coordinate system $\boldsymbol{e}^{\mathrm{w}'}$ of the wheelset is obtained by rotating the yaw angle $\psi_{\mathrm{w}}$ around the $z_{\mathrm{w}}$ axis of the inertial coordinate system $\boldsymbol{e}^{\mathrm{T}}$, and the auxiliary coordinate system D of the wheelset is obtained.
(2) By rotating the side roll angle $\theta_{\mathrm{w}}$ around the $x_{\mathrm{w}'}$ axis of the coordinate system $\boldsymbol{e}^{\mathrm{w}'}$, the wheel set coordinate system $\boldsymbol{e}^{\mathrm{w}}$ is obtained.

The transformation matrix $\underline{G}^{\mathrm{wT}}$ from the track coordinate system $e^{\mathrm{T}}$ to the wheelset coordinate system $e^{\mathrm{w}}$ can be written as

$$\underline{G}^{\mathrm{wT}} = \underline{A}^{\theta}_{x_{\mathrm{w}}}\,\underline{A}^{\psi}_{z_{\mathrm{w}}} = \begin{bmatrix} \cos\psi_{\mathrm{w}} & \sin\psi_{\mathrm{w}} & 0 \\ -\cos\theta_{\mathrm{w}}\sin\psi_{\mathrm{w}} & \cos\theta_{\mathrm{w}}\cos\psi_{\mathrm{w}} & \sin\theta_{\mathrm{w}} \\ \sin\theta_{\mathrm{w}}\sin\psi_{\mathrm{w}} & -\sin\theta_{\mathrm{w}}\cos\psi_{\mathrm{w}} & \cos\theta_{\mathrm{w}} \end{bmatrix} \tag{2.4.12}$$

The transformation matrix $\underline{G}^{\mathrm{Tw}}$ from the wheelset coordinate system $e^{\mathrm{w}}$ to the track coordinate system $e^{\mathrm{T}}$ can be obtained by the following formula:

$$\underline{G}^{\mathrm{Tw}} = (\underline{G}^{\mathrm{wT}})^{-1} = (\underline{G}^{\mathrm{wT}})^{\mathrm{T}} \tag{2.4.13}$$

3. **Contact Spot Coordinate System**

The coordinate system $e^{\mathrm{cL}}$ and $e^{\mathrm{cR}}$ of the left and right contact spot of the wheelset are established with the wheel-rail contact point as the origin, as shown in Fig. 2.23. In the $x_{\mathrm{w}}$ axis of the wheelset coordinate system $e^{\mathrm{w}}$, the contact angle $\delta_{\mathrm{L}}$ and $\delta_{\mathrm{R}}$ are rotated to obtain the contact spot coordinate system $e^{\mathrm{cL}}$ and $e^{\mathrm{cR}}$ of the wheelset coordinate system. The change matrix $\underline{G}^{\mathrm{cw}}$ from the wheelset coordinate system to the contact spot coordinate system is

$$\underline{G}^{\mathrm{cw}} = \begin{bmatrix} 1 & 0 & 0 \\ 0 & \cos\delta & \sin\delta \\ 0 & -\sin\delta & \cos\delta \end{bmatrix} \tag{2.4.14}$$

The change matrix $\underline{G}^{\mathrm{cT}}$ from the track coordinate system to the contact spot coordinate system is

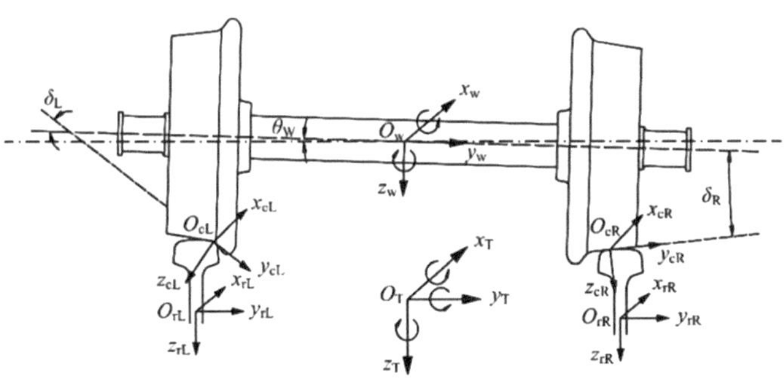


**Fig. 2.23** Wheel set contact spot coordinate system diagram

$$\underline{G}^{\mathrm{cT}} = \begin{bmatrix} \cos\psi_{\mathrm{w}} & \sin\psi_{\mathrm{w}} & 0 \\ -\cos(\theta_{\mathrm{w}}+\delta)\sin\psi_{\mathrm{w}} & \cos(\theta_{\mathrm{w}}+\delta)\cos\psi_{\mathrm{w}} & \sin(\theta_{\mathrm{w}}+\delta) \\ \sin(\theta_{\mathrm{w}}+\delta)\sin\psi_{\mathrm{w}} & -\sin(\theta_{\mathrm{w}}+\delta)\cos\psi_{\mathrm{w}} & \cos(\theta_{\mathrm{w}}+\delta) \end{bmatrix} \tag{2.4.15}$$

In the formula, the contact angle $\delta_{\mathrm{L}}$ of the left wheelset is positive, and the contact angle $\delta_{\mathrm{R}}$ of the right wheelset is negative.

### 2.4.2 Acceleration of Mass Center of Wheelset

The motion of the mass center of the wheelset can be expressed as the sum of the motion of the wheelset relative to the track coordinate system and the motion of the track coordinate system relative to the inertial coordinate system.

$$\boldsymbol{r}_{\mathrm{W}} = \boldsymbol{r}_{\mathrm{T}} + \boldsymbol{\rho}_{\mathrm{W}} \tag{2.4.16}$$

In the formula, $\boldsymbol{r}_{\mathrm{T}}$ is the radius vector of the origin of the track coordinate system. According to the definition, it is located at the centroid position of the wheelset at the center of the track and moves forward at speed $\boldsymbol{v}$. $\boldsymbol{\rho}_{\mathrm{W}}$ is the vector diameter of the wheelset centroid relative to the origin of the track coordinate system.

The absolute velocity of the mass center of the wheelset can be obtained by the derivation of the above formula to time.

$$\boldsymbol{v}_{\mathrm{W}}^{\mathrm{I}} = \dot{\boldsymbol{r}}_{\mathrm{W}}^{\mathrm{I}} = \dot{\boldsymbol{r}}_{\mathrm{T}}^{\mathrm{I}} + \dot{\boldsymbol{\rho}}_{\mathrm{W}}^{\mathrm{T}} + \boldsymbol{\omega}_{\mathrm{TI}} \times \boldsymbol{\rho}_{\mathrm{W}} \tag{2.4.17}$$

In the formula, the superscripts 'I' and 'T' are relative to the inertial coordinate system and the track coordinate system, respectively. $\dot{\boldsymbol{\rho}}_{\mathrm{W}}^{\mathrm{T}}$ represents the derivative of vector $\boldsymbol{\rho}_{\mathrm{W}}$ in the track coordinate system to time, and $\boldsymbol{\omega}_{\mathrm{TI}}$ represents the angular velocity vector of the track coordinate system relative to the inertial coordinate system.

The absolute acceleration of the centroid of the wheelset can be expressed as:

$$\boldsymbol{a}_{\mathrm{W}}^{\mathrm{I}} = \ddot{\boldsymbol{r}}_{\mathrm{W}}^{\mathrm{I}} = \ddot{\boldsymbol{r}}_{\mathrm{T}}^{\mathrm{I}} + \ddot{\boldsymbol{\rho}}_{\mathrm{W}}^{\mathrm{T}} + \dot{\boldsymbol{\omega}}_{\mathrm{TI}}^{\mathrm{I}} \times \boldsymbol{\rho}_{\mathrm{W}} + \boldsymbol{\omega}_{\mathrm{TI}} \times (\boldsymbol{\omega}_{\mathrm{TI}} \times \boldsymbol{\rho}_{\mathrm{W}}) + 2\boldsymbol{\omega}_{\mathrm{TI}} \times \dot{\boldsymbol{\rho}}_{\mathrm{W}}^{\mathrm{T}} \tag{2.4.18}$$

The displacement $\boldsymbol{\rho}_{\mathrm{W}}$ of the center of mass of the wheelset in the track coordinate system can be expressed as

$$\boldsymbol{\rho}_{\mathrm{W}} = \underline{\boldsymbol{\rho}}_{\mathrm{W}}^{\mathrm{T}} \boldsymbol{e}^{\mathrm{T}}, \quad \underline{\boldsymbol{\rho}}_{\mathrm{W}} = \begin{bmatrix} x_{\mathrm{W}} & y_{\mathrm{W}} & z_{\mathrm{W}} - r_0 \end{bmatrix}^{\mathrm{T}} \tag{2.4.19}$$

The relative track coordinate system velocity and acceleration can be expressed as

$$\dot{\boldsymbol{\rho}}_{\mathrm{W}}^{\mathrm{T}} = \dot{\underline{\boldsymbol{\rho}}}_{\mathrm{W}}^{\mathrm{TT}} \boldsymbol{e}^{\mathrm{T}}, \quad \ddot{\boldsymbol{\rho}}_{\mathrm{W}}^{\mathrm{T}} = \ddot{\underline{\boldsymbol{\rho}}}_{\mathrm{W}}^{\mathrm{TT}} \boldsymbol{e}^{\mathrm{T}} \tag{2.4.20}$$

Substituting (2.4.11), (2.4.19) and (2.4.20) into (2.4.18), and considering the small angle and ignoring the high order term, the absolute acceleration coordinate matrix of the wheelset centroid in the track coordinate system can be obtained.

$$\underline{\boldsymbol{a}}_{\mathrm{w}}^{\mathrm{I}} \approx \begin{bmatrix} \ddot{x}_{\mathrm{w}} + \dot{v} - r_0\ddot{\varphi}_{\mathrm{T}} \\ \ddot{y}_{\mathrm{w}} + v\dot{\psi}_{\mathrm{T}} - v\dot{\varphi}_{\mathrm{T}}\theta_{\mathrm{T}} + r_0\ddot{\theta}_{\mathrm{T}} \\ \ddot{z}_{\mathrm{w}} - v\dot{\varphi}_{\mathrm{T}} - v\dot{\psi}_{\mathrm{T}}\theta_{\mathrm{T}} - a_0\ddot{\theta}_{T} \end{bmatrix} \tag{2.4.21}$$

## 2.4.3 Wheelset Angular Velocity and Moment of Momentum

### 1. Wheelset Angular Velocity

The wheelset coordinate system does not rotate around the wheel shaft. The absolute angular velocity of the wheelset coordinate system is the sum of the angular velocity of the track coordinate system relative to the inertial coordinate system and the angular velocity of the wheelset coordinate system relative to the track coordinate system:

$$\boldsymbol{\omega}_{\mathrm{W_bI}} = \boldsymbol{\omega}_{\mathrm{TI}} + \boldsymbol{\omega}_{\mathrm{W_bT}} \tag{2.4.22}$$

The absolute angular velocity of the wheelset is based on the motion of the wheelset coordinate system, and the rotational angular velocity $\varphi_{\mathrm{w}}$ of the wheelset needs to be considered, including:

$$\boldsymbol{\omega}_{\mathrm{wI}} = \boldsymbol{\omega}_{\mathrm{TI}} + \boldsymbol{\omega}_{\mathrm{W_bT}} + \varphi_{\mathrm{w}} = \boldsymbol{\omega}_{\mathrm{TI}} + \boldsymbol{\omega}_{\mathrm{wT}} \tag{2.4.23}$$

The angular velocity $\boldsymbol{\omega}_{\mathrm{W_bT}}$ of the wheelset coordinate system relative to the track coordinate system can be obtained in the coordinate matrix of the wheelset coordinate system.

$$\boldsymbol{\omega}_{\mathrm{W_bT}} = \dot{\psi}_{\mathrm{w}}\boldsymbol{e}_z^{\mathrm{T}} + \dot{\theta}_{\mathrm{w}}\boldsymbol{e}_x^{\mathrm{w}'} = \underline{\boldsymbol{\omega}}_{\mathrm{W_bT}}^{\mathrm{T}}\boldsymbol{e}^{\mathrm{w}} \tag{2.4.24}$$

$$\underline{\boldsymbol{\omega}}_{\mathrm{W_bT}} = \begin{bmatrix} \dot{\theta}_{\mathrm{w}} \\ \dot{\psi}_{\mathrm{w}}\sin\theta_{\mathrm{w}} \\ \dot{\psi}_{\mathrm{w}}\cos\theta_{\mathrm{w}} \end{bmatrix}$$

The rotation angular velocity $\varphi_{\mathrm{w}}$ of the wheelset can be written as

$$\varphi_{\mathrm{w}} = \dot{\varphi}_{\mathrm{w}}\boldsymbol{e}_y^{\mathrm{w}}, \quad \dot{\varphi}_{\mathrm{w}} = \Omega + \dot{\chi} \tag{2.4.25}$$

In the formula, $\Omega = -v/r_0$ is the nominal rolling speed of the wheelset, $\dot{\chi}$ is the disturbance of the rotational speed of the wheelset, and $r_0$ is the nominal rolling circle radius of the wheelset.

The angular velocity $\boldsymbol{\omega}_{\mathrm{wT}}$ of the wheelset coordinate system relative to the track coordinate system can be obtained in the coordinate matrix of the wheelset coordinate system.

$$\boldsymbol{\omega}_{\mathrm{wT}} = \boldsymbol{\omega}_{\mathrm{w_bT}} + \boldsymbol{\varphi}_{\mathrm{w}} = \underline{\boldsymbol{\omega}}_{\mathrm{wT}}^{\mathrm{T}} \boldsymbol{e}^{\mathrm{w}},$$

$$\underline{\boldsymbol{\omega}}_{\mathrm{wT}} = \begin{bmatrix} \dot{\theta}_{\mathrm{w}} \\ \dot{\psi}_{\mathrm{w}} \sin\theta_{\mathrm{w}} + \Omega + \dot{\chi} \\ \dot{\psi}_{\mathrm{w}} \cos\theta_{\mathrm{w}} \end{bmatrix} \tag{2.4.26}$$

Considering the small angle and ignoring the high price, the coordinate matrix of the wheel set coordinate system and the absolute angular velocity of the wheel set in the wheel set coordinate system can be written as

$$\underline{\boldsymbol{\omega}}_{\mathrm{w_bI}} \approx \begin{bmatrix} \dot{\theta}_{\mathrm{T}} + \dot{\theta}_{\mathrm{w}} \\ \dot{\varphi}_{\mathrm{T}} + \dot{\psi}_{\mathrm{T}}\theta_{\mathrm{T}} + \dot{\psi}_{\mathrm{w}} \sin\theta_{\mathrm{w}} \\ \dot{\psi}_{\mathrm{T}} - \dot{\varphi}_{\mathrm{T}}\theta_{\mathrm{T}} + \dot{\psi}_{\mathrm{w}} \cos\theta_{\mathrm{w}} \end{bmatrix} \tag{2.4.27}$$

$$\underline{\boldsymbol{\omega}}_{\mathrm{wI}} \approx \begin{bmatrix} \dot{\theta}_{\mathrm{T}} + \dot{\theta}_{\mathrm{w}} \\ \dot{\varphi}_{\mathrm{T}} + \dot{\psi}_{\mathrm{T}}\theta_{\mathrm{T}} + \dot{\psi}_{\mathrm{w}} \sin\theta_{\mathrm{w}} + \Omega + \dot{\chi} \\ \dot{\psi}_{\mathrm{T}} - \dot{\varphi}_{\mathrm{T}}\theta_{\mathrm{T}} + \dot{\psi}_{\mathrm{w}} \cos\theta_{\mathrm{w}} \end{bmatrix} \tag{2.4.28}$$

3. **Wheelset Moment of Momentum**

The absolute moment of momentum of the wheelset relative to its centroid is

$$\boldsymbol{L}_{\mathrm{w}} = \boldsymbol{I}_{\mathrm{w}} \cdot \boldsymbol{\omega}_{\mathrm{wI}} \tag{2.4.29}$$

In the formula, $\boldsymbol{I}_{\mathrm{w}}$ is the inertia tensor of the mass center of the wheelset, and there is

$$\boldsymbol{I}_{\mathrm{w}} = \boldsymbol{e}^{\mathrm{wT}} \begin{bmatrix} I_{\mathrm{w}x} & 0 & 0 \\ 0 & I_{\mathrm{w}y} & 0 \\ 0 & 0 & I_{\mathrm{w}z} \end{bmatrix} \boldsymbol{e}^{\mathrm{w}} \tag{2.4.30}$$

The derivative of the angular momentum of the wheelset to time can be written as:

$$\dot{\boldsymbol{L}}_{\mathrm{w}}^{\mathrm{I}} = \dot{\boldsymbol{L}}_{\mathrm{w}}^{\mathrm{w_b}} + \boldsymbol{\omega}_{\mathrm{w_bI}} \times \boldsymbol{L}_{\mathrm{w}} = \boldsymbol{I}_{\mathrm{w}} \cdot \dot{\boldsymbol{\omega}}_{\mathrm{wI}} + \boldsymbol{\omega}_{\mathrm{w_bI}} \times \boldsymbol{I}_{\mathrm{w}} \cdot \boldsymbol{\omega}_{\mathrm{wI}} \tag{2.4.31}$$

Considering the small angle and ignoring the high price, the coordinate matrix of the derivative of the angular momentum of the wheelset to time in the wheelset coordinate system can be written as

$$\dot{\underline{\boldsymbol{L}}}_{\mathrm{w}}^{\mathrm{I}} \approx \begin{bmatrix} I_{\mathrm{w}x}(\ddot{\theta}_{\mathrm{w}} + \ddot{\theta}_{\mathrm{T}}) - I_{\mathrm{w}y}(\dot{\varphi}_{\mathrm{w}} + \dot{\varphi}_{\mathrm{T}})(\dot{\psi}_{\mathrm{w}} + \dot{\psi}_{\mathrm{T}}) + I_{\mathrm{w}z}(\dot{\psi}_{\mathrm{w}} + \dot{\psi}_{\mathrm{T}})\varphi_{\mathrm{T}} \\ I_{\mathrm{w}y}(\ddot{\varphi}_{\mathrm{w}} + \ddot{\varphi}_{\mathrm{T}}) + I_{\mathrm{w}x}(\dot{\theta}_{\mathrm{w}} + \dot{\theta}_{\mathrm{T}})(\dot{\psi}_{\mathrm{w}} + \dot{\psi}_{\mathrm{T}}) - I_{\mathrm{w}z}(\dot{\psi}_{\mathrm{w}} + \dot{\psi}_{\mathrm{T}})(\dot{\theta}_{\mathrm{w}} + \dot{\theta}_{\mathrm{T}}) \\ I_{\mathrm{w}z}(\ddot{\psi}_{\mathrm{w}} + \ddot{\psi}_{\mathrm{T}}) - I_{\mathrm{w}x}(\dot{\theta}_{\mathrm{w}} + \dot{\theta}_{\mathrm{T}})\varphi_{\mathrm{T}} + I_{\mathrm{w}y}(\dot{\varphi}_{\mathrm{w}} + \dot{\varphi}_{\mathrm{T}})(\dot{\theta}_{\mathrm{w}} + \dot{\theta}_{\mathrm{T}}) \end{bmatrix} \tag{2.4.32}$$

## 2.4.4 Equation of Motion of Free Wheelset

The motion equation of free wheelset can be obtained from Newton–Euler equation.

$$m_{\mathrm{w}}\boldsymbol{a}_{\mathrm{w}}^{\mathrm{I}} = \boldsymbol{F}_{\mathrm{w}} \tag{2.4.33}$$

$$\dot{\boldsymbol{L}}_{\mathrm{w}}^{\mathrm{I}} = \boldsymbol{M}_{\mathrm{w}} \tag{2.4.34}$$

$\boldsymbol{F}_{\mathrm{w}}$ is the external force vector of the wheelset, which is composed of left and right wheel-rail creep force ($\boldsymbol{F}_{\mathrm{L}}, \boldsymbol{F}_{\mathrm{R}}$), wheel-rail normal force ($\boldsymbol{N}_{\mathrm{L}}, \boldsymbol{N}_{\mathrm{R}}$), suspension force $\boldsymbol{F}_{\mathrm{susp}_{\mathrm{w}}}$ and gravity $\boldsymbol{F}_{\mathrm{g}}$, as shown in Fig. 2.24, so

$$\boldsymbol{F}_{\mathrm{w}} = \boldsymbol{F}_{\mathrm{L}} + \boldsymbol{F}_{\mathrm{R}} + \boldsymbol{N}_{\mathrm{L}} + \boldsymbol{N}_{\mathrm{R}} + \boldsymbol{F}_{\mathrm{susp}_{\mathrm{w}}} + \boldsymbol{F}_{\mathrm{g}} \tag{2.4.35}$$

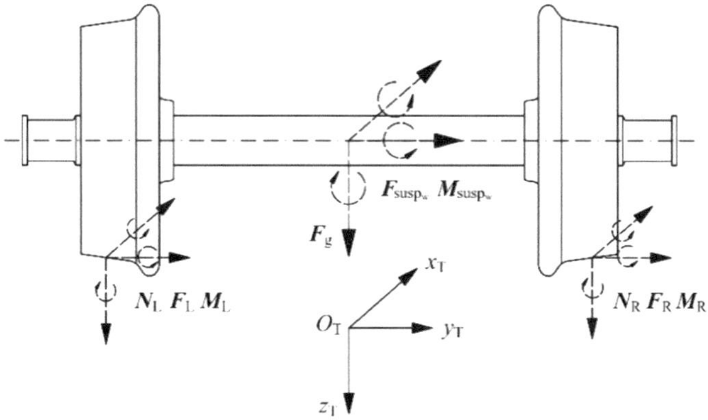


**Fig. 2.24** Wheel set force diagram

$\boldsymbol{M}_{\mathrm{w}}$ is the wheel external torque vector, which is composed of wheel-rail tangential creep and spin creep torque ($\boldsymbol{M}_{\mathrm{L}}$, $\boldsymbol{M}_{\mathrm{R}}$), wheel-rail normal torque and suspension torque $\boldsymbol{M}_{\mathrm{susp}_{\mathrm{w}}}$, as shown in Fig. 2.24, so

$$\boldsymbol{M}_{\mathrm{w}} = \boldsymbol{r}_{\mathrm{cwL}} \times (\boldsymbol{F}_{\mathrm{L}} + \boldsymbol{N}_{\mathrm{L}}) + \boldsymbol{r}_{\mathrm{cwR}} \times (\boldsymbol{F}_{\mathrm{R}} + \boldsymbol{N}_{\mathrm{R}}) + \boldsymbol{M}_{\mathrm{L}} + \boldsymbol{M}_{\mathrm{R}} + \boldsymbol{M}_{\mathrm{susp}_{\mathrm{w}}} \tag{2.4.36}$$

In the formula, $\boldsymbol{r}_{\mathrm{cwL}}$ and $\boldsymbol{r}_{\mathrm{cwR}}$ are the position vector diameters from the left and right wheel-rail contact points to the center of mass of the wheelset respectively, which are projected into the wheelset coordinate system.

$$\underline{r}_{\mathrm{cwL}} = \begin{bmatrix} 0 & -a_0 & r_{\mathrm{L}} \end{bmatrix}^{\mathrm{T}}, \quad \underline{r}_{\mathrm{cwR}} = \begin{bmatrix} 0 & a_0 & r_{\mathrm{R}} \end{bmatrix}^{\mathrm{T}} \tag{2.4.37}$$

The wheel-rail normal force, creep force and torque are defined in the contact spot coordinate system. The left wheel-rail force is defined as follows, and the right subscript is changed accordingly.

$$\boldsymbol{N}_{\mathrm{L}} = \underline{\boldsymbol{N}}_{\mathrm{L}}{}^{\mathrm{T}} \boldsymbol{e}^{\mathrm{c}}, \quad \underline{\boldsymbol{N}}_{\mathrm{L}} = \begin{bmatrix} 0 & 0 & N_{\mathrm{cL}} \end{bmatrix}^{\mathrm{T}} \tag{2.4.38}$$

$$\boldsymbol{F}_{\mathrm{L}} = \underline{\boldsymbol{F}}_{\mathrm{L}}{}^{\mathrm{T}} \boldsymbol{e}^{\mathrm{c}}, \quad \underline{\boldsymbol{F}}_{\mathrm{L}} = \begin{bmatrix} F_{\mathrm{cL}x} & F_{\mathrm{cL}y} & 0 \end{bmatrix}^{\mathrm{T}} \tag{2.4.39}$$

$$\boldsymbol{M}_{\mathrm{L}} = \underline{\boldsymbol{M}}_{\mathrm{L}}{}^{\mathrm{T}} \boldsymbol{e}^{\mathrm{c}}, \quad \underline{\boldsymbol{M}}_{\mathrm{L}} = \begin{bmatrix} 0 & 0 & M_{\mathrm{cL}} \end{bmatrix}^{\mathrm{T}} \tag{2.4.40}$$

In general, the translational motion of the wheelset is projected to the track coordinate system, the rotation of the wheelset is projected to the wheelset coordinate system, and the wheel-rail force is also projected accordingly, as shown in Fig. 2.25.

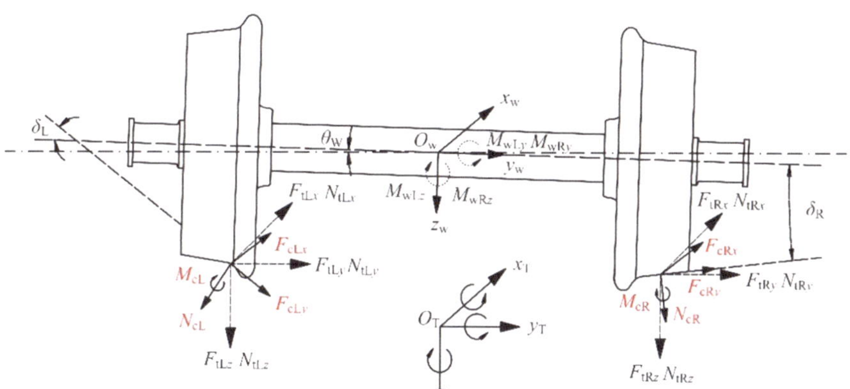


**Fig. 2.25** Schematic diagram of wheel-rail force and moment decomposed into track and wheelset coordinate system

The coordinate matrix of the external force vector $\boldsymbol{F}_{\mathrm{w}}$ of the wheelset in the track coordinate system is

$$\underline{\boldsymbol{F}}_{\mathrm{w}} = \underline{\boldsymbol{G}}^{\mathrm{TcL}}(\underline{\boldsymbol{F}}_{\mathrm{L}} + \underline{\boldsymbol{N}}_{\mathrm{L}}) + \underline{\boldsymbol{G}}^{\mathrm{TcR}}(\underline{\boldsymbol{F}}_{\mathrm{R}} + \underline{\boldsymbol{N}}_{\mathrm{R}}) + \underline{\boldsymbol{F}}_{\mathrm{susp_w}} + \underline{\boldsymbol{F}}_{\mathrm{g}}$$
$$= \begin{bmatrix} F_{\mathrm{tL}x} + F_{\mathrm{tR}x} + N_{\mathrm{tL}x} + N_{\mathrm{tR}x} + F_{\mathrm{susp}_{\mathrm{w}x}} \\ F_{\mathrm{tL}y} + F_{\mathrm{tR}y} + N_{\mathrm{tL}y} + N_{\mathrm{tR}y} + F_{\mathrm{susp}_{\mathrm{w}y}} + m_{\mathrm{w}}g\theta_{\mathrm{T}} \\ F_{\mathrm{tL}z} + F_{\mathrm{tR}z} + N_{\mathrm{tL}z} + N_{\mathrm{tR}z} + F_{\mathrm{susp}_{\mathrm{w}z}} + m_{\mathrm{w}}g \end{bmatrix} \tag{2.4.41}$$

Among them, the left and right wheel-rail normal force and creep force in the track coordinate system can be expressed as:

$$\begin{cases} N_{\mathrm{tL}x} = N_{\mathrm{cL}}\sin(\delta_{\mathrm{L}} + \theta_{\mathrm{w}})\psi_{\mathrm{w}} \\ N_{\mathrm{tL}y} = -N_{\mathrm{cL}}\sin(\delta_{\mathrm{L}} + \theta_{\mathrm{w}}) \\ N_{\mathrm{tL}z} = N_{\mathrm{cL}}\cos(\delta_{\mathrm{L}} + \theta_{\mathrm{w}}) \\ F_{\mathrm{tL}x} = F_{\mathrm{cL}x} - F_{\mathrm{cL}y}\cos(\delta_{\mathrm{L}} + \theta_{\mathrm{w}})\psi_{\mathrm{w}} \\ F_{\mathrm{tL}y} = F_{\mathrm{cL}x}\psi_{\mathrm{w}} + F_{\mathrm{cL}y}\cos(\delta_{\mathrm{L}} + \theta_{\mathrm{w}}) \\ F_{\mathrm{tL}z} = F_{\mathrm{cL}y}\sin(\delta_{\mathrm{L}} + \theta_{\mathrm{w}}) \end{cases} \tag{2.4.42}$$

$$\begin{cases} N_{\mathrm{tR}x} = -N_{\mathrm{cR}}\sin(\delta_{\mathrm{R}} - \theta_{\mathrm{w}})\psi_{\mathrm{w}} \\ N_{\mathrm{tR}y} = N_{\mathrm{cR}}\sin(\delta_{\mathrm{R}} - \theta_{\mathrm{w}}) \\ N_{\mathrm{tR}z} = N_{\mathrm{cR}}\cos(\delta_{\mathrm{R}} - \theta_{\mathrm{w}}) \\ F_{\mathrm{tR}x} = F_{\mathrm{cR}x} - F_{\mathrm{cR}y}\cos(\delta_{\mathrm{R}} - \theta_{\mathrm{w}})\psi_{\mathrm{w}} \\ F_{\mathrm{tR}y} = F_{\mathrm{cR}x}\psi_{\mathrm{w}} + F_{\mathrm{cR}y}\cos(\delta_{\mathrm{R}} - \theta_{\mathrm{w}}) \\ F_{\mathrm{tR}z} = -F_{\mathrm{cR}y}\sin(\delta_{\mathrm{R}} - \theta_{\mathrm{w}}) \end{cases} \tag{2.4.43}$$

The coordinate matrix of the wheel external torque vector $\boldsymbol{M}_{\mathrm{w}}$ in the wheelset coordinate system is

$$\underline{\boldsymbol{M}}_{\mathrm{w}} = \underline{\tilde{\boldsymbol{r}}}_{\mathrm{cwL}}\underline{\boldsymbol{G}}^{\mathrm{wcL}}(\underline{\boldsymbol{F}}_{\mathrm{L}} + \underline{\boldsymbol{N}}_{\mathrm{L}}) + \underline{\tilde{\boldsymbol{r}}}_{\mathrm{cwR}}\underline{\boldsymbol{G}}^{\mathrm{wcR}}(\underline{\boldsymbol{F}}_{\mathrm{R}} + \underline{\boldsymbol{N}}_{\mathrm{R}}) + \underline{\boldsymbol{G}}^{\mathrm{wcL}}\underline{\boldsymbol{M}}_{\mathrm{L}} + \underline{\boldsymbol{G}}^{\mathrm{wcR}}\underline{\boldsymbol{M}}_{\mathrm{R}}$$
$$+ \underline{\boldsymbol{G}}^{\mathrm{wT}}\underline{\boldsymbol{M}}^{\mathrm{T}}_{\mathrm{susp_w}} \tag{2.4.44}$$

Substituting into the wheel-rail normal force, creep force and torque, it can be obtained that

$$\underline{\boldsymbol{M}}_{\mathrm{w}} = \begin{bmatrix} -a_0(N_{\mathrm{cL}}\cos\delta_{\mathrm{L}} + F_{\mathrm{cL}y}\sin\delta_{\mathrm{L}}) - r_{\mathrm{L}}(F_{\mathrm{cL}y}\cos\delta_{\mathrm{L}} + N_{\mathrm{cL}}\sin\delta_{\mathrm{L}})\cdots \\ \cdots + a_0(N_{\mathrm{cR}}\cos\delta_{\mathrm{R}} - F_{\mathrm{cR}y}\sin\delta_{\mathrm{R}}) - r_{\mathrm{R}}(F_{\mathrm{cR}y}\cos\delta_{\mathrm{R}} - N_{\mathrm{cR}}\sin\delta_{\mathrm{R}}) \\ r_{\mathrm{L}}F_{\mathrm{cL}x} + r_{\mathrm{R}}F_{\mathrm{cR}x} - M_{\mathrm{cL}}\sin\delta_{\mathrm{L}} + M_{\mathrm{cR}}\sin\delta_{\mathrm{R}} \\ a_0(F_{\mathrm{cL}x} - F_{\mathrm{cR}x}) + M_{\mathrm{cL}}\cos\delta_{\mathrm{L}} + M_{\mathrm{cR}}\cos\delta_{\mathrm{R}} \end{bmatrix}$$
$$+ \underline{\boldsymbol{G}}^{\mathrm{wT}}\underline{\boldsymbol{M}}_{\mathrm{susp_w}} \tag{2.4.45}$$

From the Eqs. (2.4.21) and (2.4.41), the coordinate formula of the vector Eq. (2.4.33) of the wheelset motion in the track coordinate system can be obtained.

Equations (2.4.32) and (2.4.44), the coordinate formula of vector Eq. (2.4.34) in wheelset coordinate system can be obtained:

$$m_{\mathrm{w}}\underline{\boldsymbol{a}}_{\mathrm{w}}^{\mathrm{I}} = \underline{\boldsymbol{F}}_{\mathrm{w}} \tag{2.4.46}$$

$$\dot{\underline{\boldsymbol{L}}}_{\mathrm{w}}^{\mathrm{I}} = \underline{\boldsymbol{M}}_{\mathrm{w}} \tag{2.4.47}$$

The actual movement of the wheelset is subject to certain constraints. Generally, the longitudinal velocity of the wheelset is assumed to be $v$, that is, $\dot{x}_{\mathrm{w}} = v$, $x_{\mathrm{w}} = 0$. At the same time, the wheelset is constrained by the contact points of the left and right rails. Therefore, the above equations are not necessarily independent, and the consideration of constraint conditions and the solution method of the equations determine the degree of freedom of the wheelset motion.

### 2.4.5 *Solution of Wheel-Rail Force*

The determination of the wheel-rail tangential force and the spin creep force first needs to obtain the wheel-rail normal force and the wheel-rail contact geometric parameters. Therefore, the method of solving the wheelset motion equation mainly depends on the method of solving the wheel-rail normal force. The main difference between different normal forces is the rigid or elastic assumption of the constraint conditions at the contact point of the wheelset.

1. **Wheel-Rail Normal Force**

There are usually two methods to solve the wheel-rail normal force, namely the wheel-rail elastic contact method and the wheel-rail rigid constraint method. The solution method of wheel-rail normal force determines the independent degree of freedom of wheelset motion.

(1) Wheel-rail elastic contact method

The wheel-rail elastic contact method calculates the elastic contact compression between the wheelset and the rail on the contact surface, and then calculates according to the Hertz nonlinear elastic contact theory.

There are many factors affecting the normal compression of the wheel-rail contact surface, such as the vertical displacement of the wheelset, the lateral displacement, the yaw angle, the roll angle, the vibration displacement of the rail and the track irregularity. The compression vector at the contact point can be expressed as

$$\boldsymbol{r}_{\mathrm{p}} = \boldsymbol{r}_{\mathrm{cr}} - \boldsymbol{\rho}_{\mathrm{w}} - \boldsymbol{r}_{\mathrm{cw}} \tag{2.4.48}$$

As shown in Fig. 2.26, $\boldsymbol{r}_{\mathrm{cr}}$ is the radius of the wheel-rail contact point on the rail relative to the origin of the track coordinate system, $\boldsymbol{\rho}_{\mathrm{w}}$ is the radius of the

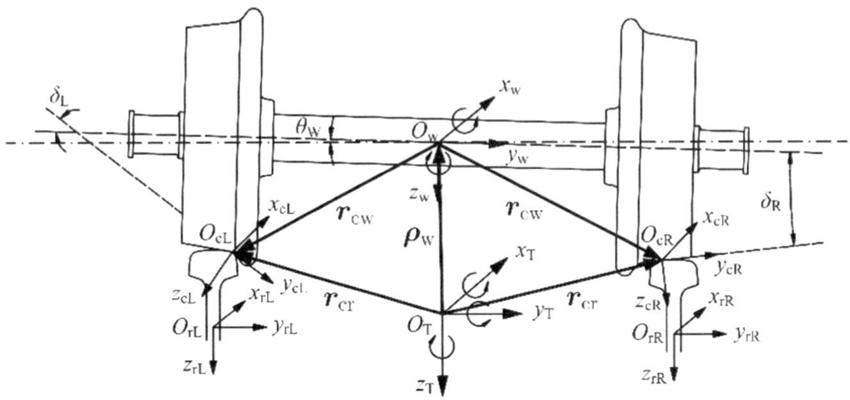


Fig. 2.26 The schematic diagram of wheel-rail normal compression solution

centroid of the wheelset relative to the origin of the track coordinate system, and $\boldsymbol{r}_{\mathrm{cw}}$ is the radius of the wheel-rail contact point on the wheelset relative to the centroid of the wheelset. The vector diameter $\boldsymbol{r}_{\mathrm{cr}}$ and $\boldsymbol{\rho}_{\mathrm{w}}$ are defined in the track coordinate system with $\boldsymbol{r}_{\mathrm{cr}} = \underline{\boldsymbol{r}}_{\mathrm{cr}}{}^{\mathrm{T}}\boldsymbol{e}^{\mathrm{T}}$ and $\boldsymbol{\rho}_{\mathrm{w}} = \underline{\boldsymbol{\rho}}_{\mathrm{w}}{}^{\mathrm{T}}\boldsymbol{e}^{\mathrm{T}}$, $\boldsymbol{r}_{\mathrm{cw}}$ is defined in the wheel set coordinate system $\boldsymbol{r}_{\mathrm{cw}} = \underline{\boldsymbol{r}}_{\mathrm{cw}}{}^{\mathrm{T}}\boldsymbol{e}^{\mathrm{W}}$, and the contact point coordinate system compression $\boldsymbol{r}_{\mathrm{p}}$ can be expressed as

$$\underline{\boldsymbol{r}}_{\mathrm{p}} = \underline{\boldsymbol{G}}^{\mathrm{cw}}(\underline{\boldsymbol{G}}^{\mathrm{wT}}(\underline{\boldsymbol{r}}_{\mathrm{cr}} - \underline{\boldsymbol{\rho}}_{\mathrm{w}}) - \underline{\boldsymbol{r}}_{\mathrm{cw}}) \tag{2.4.49}$$

The normal elastic compression $\delta q(t)$ on the wheel-rail contact surface takes the normal component of $\underline{\boldsymbol{r}}_{\mathrm{p}}$. According to the Hertz nonlinear elastic contact theory, the normal force of the wheelset is

$$N(t) = [\frac{1}{G}\delta q(t)]^{\frac{3}{2}} \tag{2.4.50}$$

In the formula, $G$ is the wheel-rail contact constant (m/N$^{2/3}$). For the tapered tread wheel, there is $G = 4.57 r_{\mathrm{w}}{}^{-0.149} \times 10^{-8}$, for the worn tread wheel, there is $G = 3.86 r_{\mathrm{w}}{}^{-0.115} \times 10^{-8}$, $r_{\mathrm{w}}$ is the wheel rolling circle radius.

When using this method, the rail has no constraint condition for the wheelset motion. The vertical motion and roll motion of the wheelset are independent degrees of freedom, and the wheelset has 6 degrees of freedom.

The advantage of this method is that the mutual compression and separation of wheel and rail can be considered, so it is suitable for the study of derailment, flange contact and other issues. However, due to the large Hertz stiffness, about 109 N/m, a smaller time step is required, which has a certain impact on the calculation efficiency.

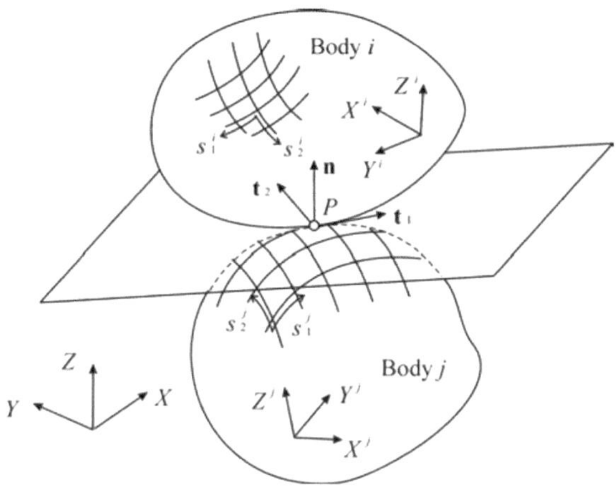


**Fig. 2.27** Contact sketch of two rigid bodies

(2) **Wheel-rail rigid constraint method**

According to the surface theory, any point on the surface of the rigid body can be determined by two coordinate parameters under this system. Let the contact point of rigid body $i$ and rigid body $j$ be $P$, and its position coordinates in the two rigid body coordinate systems can be expressed by two parameters respectively. See Fig. 2.27, that is, there are

$$\underline{\boldsymbol{u}}_P^i(s_1^i, s_2^i) = \begin{bmatrix} x^i(s_1^i, s_2^i) \\ y^i(s_1^i, s_2^i) \\ z^i(s_1^i, s_2^i) \end{bmatrix}, \quad \underline{\boldsymbol{u}}_P^j(s_1^j, s_2^j) = \begin{bmatrix} x^j(s_1^j, s_2^j) \\ y^j(s_1^j, s_2^j) \\ z^j(s_1^j, s_2^j) \end{bmatrix} \tag{2.4.51}$$

It is determined that the four parameters of the contact points on the two rigid bodies can be written as array $s$.

$$\underline{s} = \begin{bmatrix} s_1^i & s_2^i & s_1^j & s_2^j \end{bmatrix}^{\mathrm{T}} \tag{2.4.52}$$

The orthogonal tangent vectors and normal vectors at the contact points on rigid body $i$ can be expressed as

$$\underline{\boldsymbol{t}_1}^i = \frac{\partial \underline{\boldsymbol{u}}_P^i}{\partial s_1^i}, \quad \underline{\boldsymbol{t}_2}^i = \frac{\partial \underline{\boldsymbol{u}}_P^i}{\partial s_2^i} \tag{2.4.53}$$

$$\underline{\boldsymbol{n}}^i = \underline{\boldsymbol{t}_1}^i \times \underline{\boldsymbol{t}_2}^i \tag{2.4.54}$$

The parameter coordinates on the rail and the wheel are shown in Fig. 2.28.

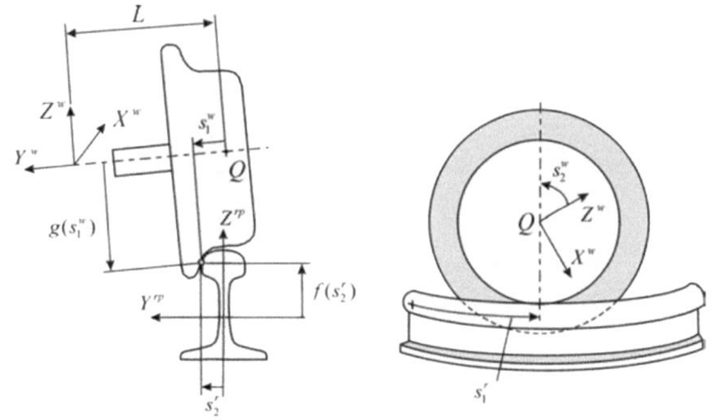

**Fig. 2.28** Rail and wheel surface parameters

Two rigid bodies contact with a smooth surface, it can be proved that the following two conditions need to be satisfied: (1) The contact points on the two rigid bodies are the same point in space. (2) The contact points on the two rigid bodies have the same common section, that is, the normal force is parallel.

These two conditions can be written as the following constraint equations:

$$\underline{\boldsymbol{\Phi}}^{k}(\underline{\boldsymbol{q}}^{\mathrm{w}}, \underline{\boldsymbol{q}}^{\mathrm{r}}, \underline{\boldsymbol{s}}^{\mathrm{w}k}, \underline{\boldsymbol{s}}^{\mathrm{r}k}) = \begin{bmatrix} \underline{\boldsymbol{r}}_P^{\mathrm{w}} - \underline{\boldsymbol{r}}_P^{\mathrm{r}} \\ \underline{\boldsymbol{t}}_1^{\mathrm{w}} \cdot \underline{\boldsymbol{n}}^{\mathrm{r}} \\ \underline{\boldsymbol{t}}_2^{\mathrm{w}} \cdot \underline{\boldsymbol{n}}^{\mathrm{r}} \end{bmatrix}^{k} = \underline{\boldsymbol{0}} \tag{2.4.55}$$

The first row of vector coordinates is expanded and can be written as

$$\underline{\boldsymbol{\Phi}}^{k}(\underline{\boldsymbol{q}}^{\mathrm{w}}, \underline{\boldsymbol{q}}^{\mathrm{r}}, \underline{\boldsymbol{s}}^{\mathrm{w}k}, \underline{\boldsymbol{s}}^{\mathrm{r}k}) = \begin{bmatrix} \underline{\boldsymbol{t}}_1^{\mathrm{r}} \cdot (\underline{\boldsymbol{r}}_P^{\mathrm{w}} - \underline{\boldsymbol{r}}_P^{\mathrm{r}}) \\ \underline{\boldsymbol{t}}_2^{\mathrm{r}} \cdot (\underline{\boldsymbol{r}}_P^{\mathrm{w}} - \underline{\boldsymbol{r}}_P^{\mathrm{r}}) \\ \underline{\boldsymbol{n}}^{\mathrm{r}} \cdot (\underline{\boldsymbol{r}}_P^{\mathrm{w}} - \underline{\boldsymbol{r}}_P^{\mathrm{r}}) \\ \underline{\boldsymbol{t}}_1^{\mathrm{w}} \cdot \underline{\boldsymbol{n}}^{\mathrm{r}} \\ \underline{\boldsymbol{t}}_2^{\mathrm{w}} \cdot \underline{\boldsymbol{n}}^{\mathrm{r}} \end{bmatrix}^{k} = \underline{\boldsymbol{0}} \tag{2.4.56}$$

In the formula, the superscript k represents the number of wheel-rail contact points. There are eight contact points in four sets of wheelsets of a vehicle. The superscripts w and r represent the wheels and rails respectively, and qw and qr represent the wheel-rail generalized coordinates respectively.

It can be seen that the above equation has five nonlinear algebraic equations. Due to the introduction of four position variables s, a wheel-rail contact will limit one degree of freedom. The wheel-rail contact constraint equations of the vehicle are written as the following form

$$\underline{\boldsymbol{\Phi}}(\underline{\boldsymbol{q}}, \underline{\boldsymbol{s}}) = \underline{\boldsymbol{0}} \tag{2.4.57}$$

The second derivative of the constraint equation is solved to obtain the wheel-rail acceleration constraint equation.

$$\underline{\Phi}_q \underline{\ddot{q}} + \underline{\Phi}_s \underline{\ddot{s}} = \underline{Q}_d \tag{2.4.58}$$

where $\underline{\Phi}_q$ and $\underline{\Phi}_s$ are the Jacobian matrices of the constraint equations.

The augmented vehicle dynamics equation can be obtained by Lagrange multiplier method, which can solve the wheel-rail geometric relationship online.

$$\begin{bmatrix} \underline{M} & \underline{0} & \underline{\Phi}_q^{\mathrm{T}} \\ \underline{0} & \underline{0} & \underline{\Phi}_s^{\mathrm{T}} \\ \underline{\Phi}_q & \underline{\Phi}_s & \underline{0} \end{bmatrix} \begin{bmatrix} \underline{\ddot{q}} \\ \underline{\ddot{s}} \\ \underline{\lambda} \end{bmatrix} = \begin{bmatrix} \underline{Q} \\ \underline{0} \\ \underline{Q}_d \end{bmatrix} \tag{2.4.59}$$

Another way is to use independent generalized coordinates, that is, the contraction method. Since the wheel and rail always keep contact, the wheel movement is constrained by the left and right rails, that is, the movement of a wheel has two independent constraints and four independent generalized coordinates. Generally, the vertical motion and roll motion of the wheelset are selected as non-independent motion, and the wheel-rail normal force can be obtained by the vertical motion and roll motion equations. Because the wheel set motion equation contains both the wheel-rail normal force and the creep force component, it needs to be solved iteratively. The solution process of the wheel-rail force is shown in Fig. 2.29.

First of all, the wheel-rail normal force can be solved by the wheel-set floating and rolling motion equations. The wheel-rail normal force and creep force are interrelated, so it is necessary to carry out iterative calculation.

By Eqs. (2.4.46) and (2.4.47), the equations of the wheel set's sinking, floating and rolling can be written as follows:

$$m_{\mathrm{w}}(\ddot{z}_{\mathrm{w}} - v\dot{\varphi}_{\mathrm{T}} - v\dot{\psi}_{\mathrm{T}}\theta_{\mathrm{T}} - a_0\ddot{\theta}_{\mathrm{T}}) = F_{\mathrm{tL}z} + F_{\mathrm{tR}z} + N_{\mathrm{tL}z} + N_{\mathrm{tR}z} + F_{\mathrm{susp}_{\mathrm{w}z}} + m_{\mathrm{w}}g \tag{2.4.60}$$

$$\begin{aligned} I_{\mathrm{w}x}\big(\ddot{\theta}_{\mathrm{w}} + \ddot{\theta}_{\mathrm{T}}\big) \quad & - I_{\mathrm{w}y}(\dot{\varphi}_{\mathrm{w}} + \dot{\varphi}_{\mathrm{T}})\big(\dot{\psi}_{\mathrm{w}} + \dot{\psi}_{\mathrm{T}}\big) + I_{\mathrm{w}z}\big(\dot{\psi}_{\mathrm{w}} + \dot{\psi}_{\mathrm{T}}\big)\varphi_{\mathrm{T}} \\ &= -a_0\big(N_{\mathrm{cL}}\cos\delta_{\mathrm{L}} + F_{\mathrm{cL}y}\sin\delta_{\mathrm{L}}\big) - r_{\mathrm{L}}\big(F_{\mathrm{cL}y}\cos\delta_{\mathrm{L}} + N_{\mathrm{cL}}\sin\delta_{\mathrm{L}}\big) \\ &\quad + a_0\big(N_{\mathrm{cR}}\cos\delta_{\mathrm{R}} - F_{\mathrm{cR}y}\sin\delta_{\mathrm{R}}\big) - r_{\mathrm{R}}\big(F_{\mathrm{cR}y}\cos\delta_{\mathrm{R}} - N_{\mathrm{cR}}\sin\delta_{\mathrm{R}}\big) + M_{\mathrm{susp}_{\mathrm{w}x}} \end{aligned} \tag{2.4.61}$$

The terms involved in the wheel-rail normal force in the equation are independent, and there is

$$\begin{cases} F_{\mathrm{w}z} = N_{\mathrm{cL}}\cos(\theta_{\mathrm{w}} + \delta_{\mathrm{L}}) + N_{\mathrm{cR}}\cos(\theta_{\mathrm{w}} - \delta_{\mathrm{R}}) \\ M_{\mathrm{w}\theta} = -N_{\mathrm{cL}}(a_0\cos\delta_{\mathrm{L}} + r_{\mathrm{L}}\sin\delta_{\mathrm{L}}) + N_{\mathrm{cR}}(a_0\cos\delta_{\mathrm{R}} + r_{\mathrm{R}}\sin\delta_{\mathrm{R}}) \end{cases} \tag{2.4.62}$$

Let

$$\begin{cases} a_1 = \cos(\theta_{\mathrm{w}} + \delta_{\mathrm{L}}) \\ a_2 = \cos(\theta_{\mathrm{w}} - \delta_{\mathrm{R}}) \\ b_1 = -(a_0\cos\delta_{\mathrm{L}} + r_{\mathrm{L}}\sin\delta_{\mathrm{L}}) \\ b_1 = (a_0\cos\delta_{\mathrm{R}} + r_{\mathrm{R}}\sin\delta_{\mathrm{R}}) \end{cases} \tag{2.4.63}$$

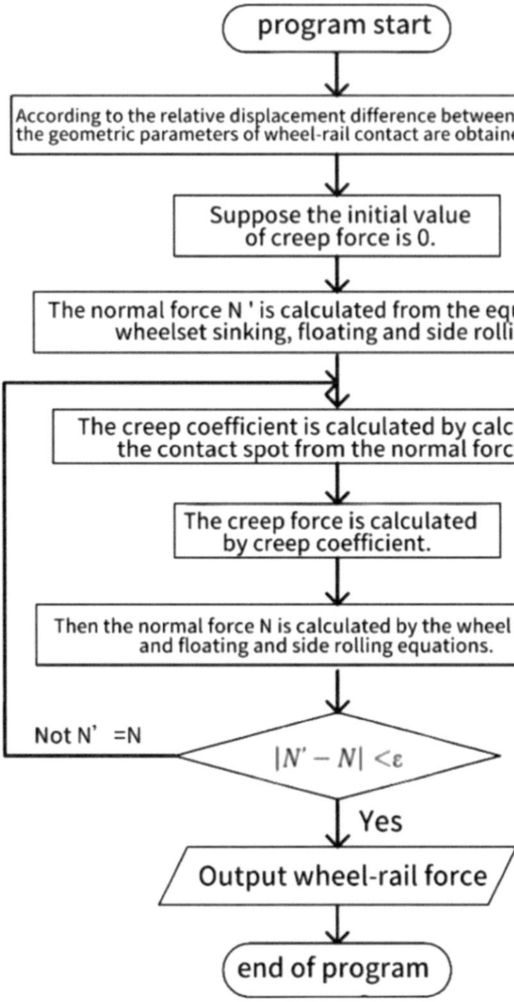


**Fig. 2.29** Iterative solution process of wheel-rail force

Thus, the wheel-rail normal forces $N_{\mathrm{cL}}$ and $N_{\mathrm{cR}}$ can be obtained.

$$\begin{cases} N_{\mathrm{cL}} = \dfrac{b_2 F_{\mathrm{wz}} - a_2 M_{\mathrm{w}\theta}}{a_1 b_2 - a_2 b_1} \\ N_{\mathrm{cR}} = \dfrac{-b_1 F_{\mathrm{wz}} + a_1 M_{\mathrm{w}\theta}}{a_1 b_2 - a_2 b_1} \end{cases} \tag{2.4.64}$$

In the iterative solution, the displacement and velocity of the wheel set are determined by the wheel-rail contact geometry.

## 2. Creep Rate and Creep Force

When the rail motion is not considered, the velocity and angular velocity at the contact point of the wheelset can be obtained from the velocity and angular velocity of the centroid of the wheelset.

$$\underline{v}_{\mathrm{c}}^{\mathrm{I}} = \underline{v}_{\mathrm{w}}^{\mathrm{I}} + \underline{\omega}_{\mathrm{wI}} \times \underline{r}_{\mathrm{cw}} \tag{2.4.65}$$

$$\omega_{\mathrm{cI}} = \underline{G}^{\mathrm{cW}} \omega_{\mathrm{wI}} \tag{2.4.66}$$

$r_{\mathrm{cw}}$ is the position vector diameter from the wheel-rail contact point to the centroid of the wheelset, which is defined in the wheelset coordinate system.

$$r_{\mathrm{cw}} = \underline{r}_{\mathrm{cw}}^{\mathrm{T}} e^{\mathrm{w}}, \quad \underline{r}_{\mathrm{cw}} = \begin{bmatrix} 0 & r_{\mathrm{cw}y} & r_{\mathrm{cw}z} \end{bmatrix}^{\mathrm{T}} \tag{2.4.67}$$

In the formula, the left wheelset $r_{\mathrm{cw}z}$ takes a negative value, and the right wheelset $r_{\mathrm{cw}y}$ takes a positive value.

It can be seen from the formula (2.4.17) that the coordinate matrix of the absolute speed of the wheelset in the track coordinate system is

$$\underline{v}_{\mathrm{w}}^{\mathrm{I}} \approx \begin{bmatrix} v + \dot{x}_{\mathrm{w}} - r_{\mathrm{cw}z}\dot{\varphi}_{\mathrm{T}} \\ \dot{y}_{\mathrm{w}} + r_{\mathrm{cw}z}\dot{\theta}_{\mathrm{T}} \\ \dot{z}_{\mathrm{w}} - r_{\mathrm{cw}y}\dot{\theta}_{\mathrm{T}} \end{bmatrix} \tag{2.4.68}$$

The relative velocity in the contact spot coordinate system is

$$\underline{v}_{\mathrm{c}}^{\mathrm{I}} = \underline{G}^{\mathrm{cW}} (\underline{G}^{\mathrm{WT}} \underline{v}_{\mathrm{w}}^{\mathrm{I}} + \underline{\tilde{\omega}}_{\mathrm{wI}} \underline{r}_{\mathrm{cw}}) \tag{2.4.69}$$

Substituting Eqs. (2.4.12), (2.4.14), (2.4.28), (2.4.67) and (2.4.69), the wheel-rail relative velocity in the contact spot coordinate system can be obtained as follows:

$$\underline{v}_{\mathrm{c}}^{\mathrm{I}} \approx \underline{G}^{\mathrm{cW}} \begin{bmatrix} v + \dot{x}_{\mathrm{w}} - r_{\mathrm{cw}y}(\dot{\psi}_{\mathrm{w}} + \dot{\psi}_{\mathrm{T}}) + r_{\mathrm{cw}z}\dot{\varphi}_{\mathrm{w}} + \psi_{\mathrm{w}}(\dot{y}_{\mathrm{w}} + r_{\mathrm{cw}z}\dot{\theta}_{\mathrm{T}}) \\ \dot{y}_{\mathrm{w}} - r_{\mathrm{cw}z}\dot{\theta}_{\mathrm{w}} - \psi_{\mathrm{w}}(v + \dot{x}_{\mathrm{w}} - r_{\mathrm{cw}z}\varphi_{\mathrm{T}}) + \theta_{\mathrm{w}}\dot{z}_{\mathrm{w}} \\ \dot{z}_{\mathrm{w}} - \theta_{\mathrm{w}}(\dot{y}_{\mathrm{w}} + r_{\mathrm{cw}z}\theta_{\mathrm{T}}) + r_{\mathrm{cw}y}(\theta_{\mathrm{w}} + \theta_{\mathrm{T}}) - a_0\dot{\theta}_{\mathrm{T}} \end{bmatrix} \tag{2.4.70}$$

Assuming that the normal relative velocity between the wheel and rail in the contact spot coordinate system is zero, that is, $v_{\mathrm{c}}^{\mathrm{I}}|_z = 0$, the relative velocity of the wheelset is

$$\underline{v}_{\mathrm{c}}^{\mathrm{I}} \approx \begin{bmatrix} v + \dot{x}_{\mathrm{w}} - r_{\mathrm{cw}y}(\dot{\psi}_{\mathrm{w}} + \dot{\psi}_{\mathrm{T}}) + r_{\mathrm{cw}z}\dot{\varphi}_{\mathrm{w}} + \psi_{\mathrm{w}}(\dot{y}_{\mathrm{w}} + r_{\mathrm{cw}z}\dot{\theta}_{\mathrm{T}}) \\ (\dot{y}_{\mathrm{w}} - r_{\mathrm{cw}z}\dot{\theta}_{\mathrm{w}} - \psi_{\mathrm{w}}(v + \dot{x}_{\mathrm{w}} - r_{\mathrm{cw}z}\varphi_{\mathrm{T}}) + \theta_{\mathrm{w}}\dot{z}_{\mathrm{w}})/\cos\delta \\ 0 \end{bmatrix} \tag{2.4.71}$$

$$\underline{\boldsymbol{\omega}}_{\mathrm{cI}} \approx \begin{bmatrix} \dot{\theta}_{\mathrm{T}} + \dot{\theta}_{\mathrm{w}} \\ (\dot{\varphi}_{\mathrm{T}} + \dot{\varphi}_{\mathrm{w}}) \cos\delta + (\dot{\psi}_{\mathrm{T}} + \dot{\psi}_{\mathrm{w}}) \sin\delta \\ (\dot{\psi}_{\mathrm{T}} + \dot{\psi}_{\mathrm{w}}) \cos\delta - (\dot{\varphi}_{\mathrm{T}} + \dot{\varphi}_{\mathrm{w}}) \sin\delta \end{bmatrix} \tag{2.4.72}$$

According to the definition of creep coefficient (2.3.8), the creep coefficient at the left and right wheels and rails can be obtained:

$$\xi_x = v_{\mathrm{c}}^{\mathrm{I}}\big|_x / v = [v + \dot{x}_{\mathrm{w}} - r_{\mathrm{cwy}}(\dot{\psi}_{\mathrm{w}} + \dot{\psi}_{\mathrm{T}}) + r_{\mathrm{cwz}}\dot{\varphi}_{\mathrm{w}} + \psi_{\mathrm{w}}(\dot{y}_{\mathrm{w}} + r_{\mathrm{cwz}}\dot{\theta}_{\mathrm{T}})]/v \tag{2.4.73}$$

$$\xi_y = v_{\mathrm{c}}^{\mathrm{I}}\big|_y / v = [(\dot{y}_{\mathrm{w}} - r_{\mathrm{cwz}}\dot{\theta}_{\mathrm{w}} - \psi_{\mathrm{w}}(v + \dot{x}_{\mathrm{w}} - r_{\mathrm{cwz}}\varphi_{\mathrm{T}}) + \theta_{\mathrm{w}}\dot{z}_{\mathrm{w}}) / \cos\delta]/v \tag{2.4.74}$$

$$\xi_{\mathrm{sp}} = \omega_{\mathrm{cI}}\big|_z / v = [(\dot{\psi}_{\mathrm{T}} + \dot{\psi}_{\mathrm{w}}) \cos\delta - (\dot{\varphi}_{\mathrm{T}} + \dot{\varphi}_{\mathrm{w}}) \sin\delta]/v \tag{2.4.75}$$

Let the coordinates of the contact points of the left and right wheelsets be $\underline{\boldsymbol{r}}_{\mathrm{cwL}} = \begin{bmatrix} 0 & -a_0 & r_{\mathrm{L}} \end{bmatrix}^{\mathrm{T}}$ and $\underline{\boldsymbol{r}}_{\mathrm{cwR}} = \begin{bmatrix} 0 & -a_0 & r_{\mathrm{R}} \end{bmatrix}^{\mathrm{T}}$ respectively, and take $\dot{\varphi}_{\mathrm{w}} \approx \Omega = -v/r_0$ and $r_0 = (r_{\mathrm{L}} + r_{\mathrm{R}})/2$. Ignoring the higher order terms, the creep force at the contact spots of the left and right wheels can be simplified as

$$\begin{cases} \xi_{x\mathrm{L}} = 1 - \frac{r_{\mathrm{L}}}{r_0} + \frac{a_0}{v}\dot{\psi}_{\mathrm{w}} + \frac{a_0}{R_{\mathrm{h}}} \\ \xi_{y\mathrm{L}} = \frac{1}{v}(\dot{y}_{\mathrm{w}} - r_{\mathrm{L}}\dot{\theta}_{\mathrm{w}} - v\psi_{\mathrm{w}}) \\ \xi_{\mathrm{spL}} = \frac{\delta_{\mathrm{L}}}{r_0} + \frac{\dot{\psi}_{\mathrm{w}}}{v} + \frac{1}{R_{\mathrm{h}}} \end{cases} \tag{2.4.76}$$

$$\begin{cases} \xi_{x\mathrm{R}} = 1 - \dfrac{r_{\mathrm{R}}}{r_0} - \dfrac{a_0}{v}\dot{\psi}_{\mathrm{w}} - \dfrac{a_0}{R_{\mathrm{h}}} \\ \xi_{y\mathrm{R}} = \dfrac{1}{v}(\dot{y}_{\mathrm{w}} - r_{\mathrm{R}}\dot{\theta}_{\mathrm{w}} - v\psi_{\mathrm{w}}) \\ \xi_{\mathrm{spR}} = -\dfrac{\delta_{\mathrm{R}}}{r_0} + \dfrac{\dot{\psi}_{\mathrm{w}}}{v} + \dfrac{1}{R_{\mathrm{h}}} \end{cases} \tag{2.4.77}$$

After obtaining the wheel-rail creep rate, the wheel-rail creep force can be obtained by using the Kalker linear theory formula (2.3.9) and then using the formula (Eq. 2.3.14) for nonlinear correction.

## 2.5 Equations of Motion for Vehicles

The dynamic model forms the foundation of dynamic simulation and analysis. The model should not be overly simplistic, as it would fail to capture the primary aspects and influencing factors of the system's dynamic response. At the same time, the model should not be overly complex, as this could lead to cumbersome analysis, prolonged computation time, and even obscure the effects of key factors in the system. Therefore, reasonable assumptions should be made based on the issue under investigation, emphasizing the main factors while neglecting secondary ones, to establish a rational and efficient model.

### 2.5.1 Vehicle Model and Degrees of Freedom

When constructing a dynamic model of a vehicle, the axle box suspension and the central suspension system are typically used as dividing interfaces. The vehicle mass is divided into three main components: the car body, the bogie frame, and the wheelsets. The suspension system is treated as a massless spring-damper system, while the car body, bogie frame, and wheelsets are considered rigid bodies. A rigid body has six degrees of freedom, which include three translational displacements along the coordinate axes and three rotational displacements about the coordinate axes. Consequently, the spatial position of each component is fully described by six basic motion forms: heave, lateral displacement, longitudinal displacement, yaw, pitch, and roll, as illustrated in Fig. 2.30.

A high-speed passenger car can generally be simplified as a multi-rigid-body system composed of the car body, front and rear bogies, and four wheelsets. The longitudinal motion along the direction of train operation has minimal influence on the vertical and lateral vibrations of the vehicle system. Therefore, in the vehicle model, the vibrations of components along the train's running direction are typically not considered. As a result, each rigid body has five degrees of freedom, which are bounce, lateral displacement, roll, pitch, and yaw. The bounce and roll motions of the wheelsets are determined based on the method used to solve the normal wheel-rail forces, and whether they are treated as independent degrees of freedom depends on this method (see Table 2.4). The vehicle dynamics model is shown in Fig. 2.31.

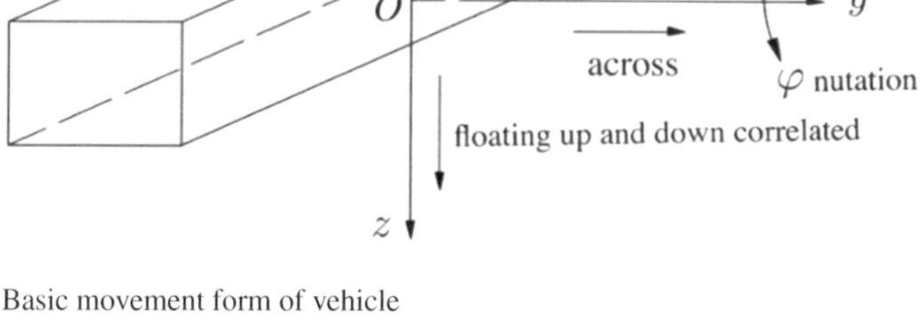


**Fig. 2.30** Basic movement form of vehicle

**Table 2.4** Degree of freedom of vehicle model

| Components | Lateral displacement | Bounce | Rolling | Yawing | Pitch |
|---|---|---|---|---|---|
| Carbody structure | $y_c$ | $z_c$ | $\theta_c$ | $\psi_c$ | $\varphi_c$ |
| Framework ($i = 1{,}2$) | $y_{ti}$ | $z_{ti}$ | $\theta_{ti}$ | $\psi_{ti}$ | $\varphi_{ti}$ |
| Wheel pair ($j = 1 - 4$) | $y_{wj}$ | $(z_{wj})$ | $(\theta_{wj})$ | $\psi_{wj}$ | $\varphi_{wj}$ |

*Note* The bounce and roll motions of the wheelsets need to be determined as independent degrees of freedom based on the method used to solve the normal wheel-rail forces

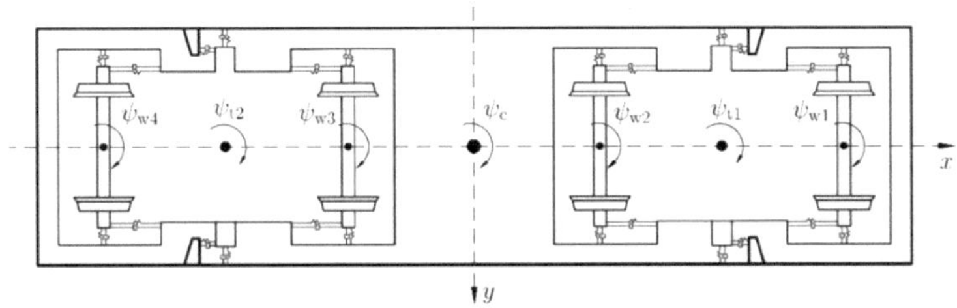


(a) Top view of vehicle model

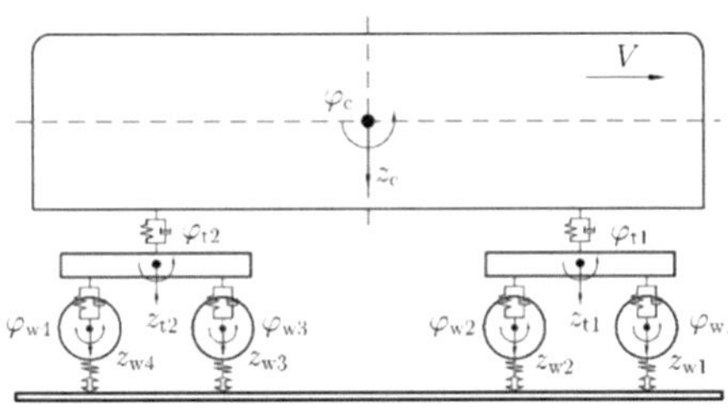


(b) Side view of vehicle model

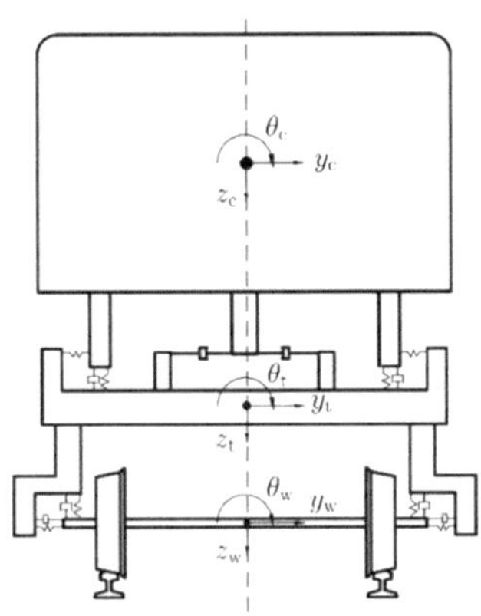


(c) Vehicle model front view

**Fig. 2.31** Vehicle dynamics model illustration

## 2.5.2 Vehicle Motion Equation

The derivation method of the motion equation of the frame and the carbody is basically consistent with the wheel set motion equation. The main difference is that the frame and the carbody do not rotate. Therefore, the body coordinate system moves completely with the frame and the carbody, that is, $\boldsymbol{\omega}_{\mathrm{t_bI}} = \boldsymbol{\omega}_{\mathrm{tI}}$, $\boldsymbol{\omega}_{\mathrm{c_bI}} = \boldsymbol{\omega}_{\mathrm{cI}}$.

The transformation matrix $\underline{\boldsymbol{G}}^{\mathrm{bT}}$ from the track coordinate system $\boldsymbol{e}^{\mathrm{T}}$ to the body coordinate system $\boldsymbol{e}^{\mathrm{b}}$ can be written as

$$\underline{\boldsymbol{G}}^{\mathrm{bT}} = \underline{\boldsymbol{A}}_y^{\varphi}\underline{\boldsymbol{A}}_x^{\theta}\underline{\boldsymbol{A}}_z^{\psi} \tag{2.5.1}$$

According to Eq. (2.4.1), there is

$$\underline{\boldsymbol{G}}^{\mathrm{bT}} = \begin{bmatrix} \cos\psi\cos\varphi - \sin\psi\sin\theta\sin\varphi & \cos\varphi\sin\psi + \cos\psi\sin\theta\sin\varphi & -\cos\theta\sin\varphi \\ -\cos\theta\sin\psi & \cos\psi\cos\theta & \sin\theta \\ \cos\psi\sin\varphi + \cos\varphi\sin\psi\sin\theta & \sin\psi\sin\varphi - \cos\psi\cos\varphi\sin\theta & \cos\theta\cos\varphi \end{bmatrix} \tag{2.5.2}$$

1. **Wheelset Equation**

The wheelset equation has been derived in detail in Sect. 2.4.

2. **Framework Equation**

In the track coordinate system, the displacement of the mass center of the frame can be expressed as

$$\boldsymbol{\rho}_{\mathrm{t}} = \underline{\boldsymbol{\rho}}_{\mathrm{t}}^{\mathrm{T}}\boldsymbol{e}^{\mathrm{T}}, \quad \underline{\boldsymbol{\rho}}_{\mathrm{t}} = [x_{\mathrm{t}} \quad y_{\mathrm{t}} \quad z_{\mathrm{t}} - r_0 - h_3]^{\mathrm{T}} \tag{2.5.3}$$

Considering small displacement and small angle, ignoring the high-order term, the absolute acceleration coordinate matrix of the center of mass of the frame in the track coordinate system and the angular velocity coordinate matrix of the frame in the body coordinate system can be obtained.

$$\underline{\boldsymbol{a}}_{\mathrm{t}}^{\mathrm{I}} \approx \begin{bmatrix} \ddot{x}_{\mathrm{t}} + \dot{v} - (r_0 + h_3)\ddot{\varphi}_{\mathrm{T}} \\ \ddot{y}_{\mathrm{t}} + v\dot{\psi}_{\mathrm{T}} - v\dot{\varphi}_{\mathrm{T}}\theta_{\mathrm{T}} + (r_0 + h_3)\ddot{\theta}_{\mathrm{T}} \\ \ddot{z}_{\mathrm{t}} - v\dot{\varphi}_{\mathrm{T}} - v\dot{\psi}_{\mathrm{T}}\theta_{\mathrm{T}} - a_0\ddot{\theta}_{\mathrm{T}} \end{bmatrix} \tag{2.5.4}$$

$$\underline{\boldsymbol{\omega}}_{\mathrm{tI}} \approx \begin{bmatrix} \dot{\theta}_{\mathrm{T}} + \dot{\theta}_{\mathrm{t}} \\ \dot{\varphi}_{\mathrm{T}} + \dot{\varphi}_{\mathrm{t}} + \dot{\psi}_{\mathrm{T}}\theta_{\mathrm{T}} + \dot{\psi}_{\mathrm{T}}\sin\theta_{\mathrm{t}} \\ \dot{\psi}_{\mathrm{T}} - \dot{\varphi}_{\mathrm{T}}\theta_{\mathrm{T}} + \dot{\psi}_{\mathrm{T}}\cos\theta_{\mathrm{t}} \end{bmatrix} \tag{2.5.5}$$

From the formula (2.1.9), the coordinate matrix of the time derivative of the angular momentum of the frame in the frame body coordinate system can be obtained.

$$\dot{\boldsymbol{L}}_{\mathrm{t}}^{\mathrm{I}} = \boldsymbol{I}_{\mathrm{t}} \cdot \dot{\boldsymbol{\omega}}_{\mathrm{tI}} + \boldsymbol{\omega}_{\mathrm{tI}} \times \boldsymbol{I}_{\mathrm{t}} \cdot \boldsymbol{\omega}_{\mathrm{tI}} \tag{2.5.6}$$

$$\dot{\underline{\boldsymbol{L}}}_{\mathrm{t}}^{\mathrm{I}} \approx \begin{bmatrix} I_{\mathrm{t}x}(\ddot{\theta}_{\mathrm{t}} + \ddot{\theta}_{\mathrm{T}}) - I_{\mathrm{t}y}(\dot{\varphi}_{\mathrm{t}} + \dot{\varphi}_{\mathrm{T}})(\dot{\psi}_{\mathrm{t}} + \dot{\psi}_{\mathrm{T}}) + I_{\mathrm{t}z}(\dot{\psi}_{\mathrm{t}} + \dot{\psi}_{\mathrm{T}})(\varphi_{\mathrm{t}} + \varphi_{\mathrm{T}}) \\ I_{\mathrm{t}y}(\ddot{\varphi}_{\mathrm{t}} + \ddot{\varphi}_{\mathrm{T}}) + I_{\mathrm{t}x}(\dot{\theta}_{\mathrm{t}} + \dot{\theta}_{\mathrm{T}})(\dot{\psi}_{\mathrm{t}} + \dot{\psi}_{\mathrm{T}}) - I_{\mathrm{t}z}(\dot{\psi}_{\mathrm{t}} + \dot{\psi}_{\mathrm{T}})(\dot{\theta}_{\mathrm{t}} + \dot{\theta}_{\mathrm{T}}) \\ I_{\mathrm{t}z}(\ddot{\psi}_{\mathrm{t}} + \ddot{\psi}_{\mathrm{T}}) - I_{\mathrm{t}x}(\dot{\theta}_{\mathrm{t}} + \dot{\theta}_{\mathrm{T}})(\varphi_{\mathrm{t}} + \varphi_{\mathrm{T}}) + I_{\mathrm{t}y}(\dot{\varphi}_{\mathrm{t}} + \dot{\varphi}_{\mathrm{T}})(\dot{\theta}_{\mathrm{t}} + \dot{\theta}_{\mathrm{T}}) \end{bmatrix} \tag{2.5.7}$$

The motion equation of the frame can be obtained from the Newton–Euler equation.

$$m_{\mathrm{t}}\boldsymbol{a}_{\mathrm{t}}^{\mathrm{I}} = \boldsymbol{F}_{\mathrm{t}} \tag{2.5.8}$$

$$\dot{\boldsymbol{L}}_{\mathrm{t}}^{\mathrm{I}} = \boldsymbol{M}_{\mathrm{t}} \tag{2.5.9}$$

In the formula, $\boldsymbol{F}_{\mathrm{t}}$ is the external force vector of the frame, which is composed of suspension force and gravity. In the track coordinate system, it can be expressed as

$$\boldsymbol{F}_{\mathrm{t}} = \underline{\boldsymbol{F}}_{\mathrm{t}}^{\mathrm{T}}\boldsymbol{e}^{\mathrm{T}}, \quad \underline{\boldsymbol{F}}_{\mathrm{t}} = [F_{\mathrm{susp}_{\mathrm{t}x}} \quad F_{\mathrm{susp}_{\mathrm{t}y}} + m_{\mathrm{t}}g\theta_{\mathrm{T}} \quad F_{\mathrm{susp}_{\mathrm{t}z}} + m_{\mathrm{t}}g]^{\mathrm{T}} \tag{2.5.10}$$

$\boldsymbol{M}_{\mathrm{t}}$ is the external torque vector of the frame, which is composed of suspension torque. In the body coordinate system, it can be expressed as

$$\boldsymbol{M}_{\mathrm{t}} = \underline{\boldsymbol{M}}_{\mathrm{t}}^{\mathrm{T}}\boldsymbol{e}^{\mathrm{t}}, \quad \underline{\boldsymbol{M}}_{\mathrm{t}} = [M_{\mathrm{susp}_{\mathrm{t}x}} \quad M_{\mathrm{susp}_{\mathrm{t}y}} \quad M_{\mathrm{susp}_{\mathrm{t}z}}]^{\mathrm{T}} \tag{2.5.11}$$

3. **Carbody Equation**

In the track coordinate system, the displacement of the center of mass of the carbody can be expressed as

$$\boldsymbol{\rho}_{\mathrm{c}} = \underline{\boldsymbol{\rho}}_{\mathrm{c}}^{\mathrm{T}}\boldsymbol{e}^{\mathrm{T}}, \quad \underline{\boldsymbol{\rho}}_{\mathrm{c}} = [x_{\mathrm{c}} \quad y_{\mathrm{c}} \quad z_{\mathrm{c}} - r_0 - h_3 - h_2 - h_1]^{\mathrm{T}} \tag{2.5.12}$$

Considering the small displacement and small angle, ignoring the high order term, the absolute acceleration coordinate matrix of the carbody mass center in the track coordinate system and the angular velocity coordinate matrix of the carbody in the body coordinate system can be obtained.

$$\underline{\boldsymbol{a}}_{\mathrm{c}}^{\mathrm{I}} \approx \begin{bmatrix} \ddot{x}_{\mathrm{c}} + \dot{v} - (r_0 + h_3 + h_2 + h_1)\ddot{\varphi}_{\mathrm{T}} \\ \ddot{y}_{\mathrm{c}} + v\dot{\psi}_{\mathrm{T}} - v\dot{\varphi}_{\mathrm{T}}\theta_{\mathrm{T}} + (r_0 + h_3 + h_3 + h_2 + h_1)\ddot{\theta}_{\mathrm{T}} \\ \ddot{z}_{\mathrm{c}} - v\dot{\varphi}_{\mathrm{T}} - v\dot{\psi}_{\mathrm{T}}\theta_{\mathrm{T}} - a_0\ddot{\theta}_{\mathrm{T}} \end{bmatrix} \tag{2.5.13}$$

$$\underline{\boldsymbol{\omega}}_{\mathrm{cI}} \approx \begin{bmatrix} \dot{\theta}_{\mathrm{T}} + \dot{\theta}_{\mathrm{c}} \\ \dot{\varphi}_{\mathrm{T}} + \dot{\varphi}_{\mathrm{c}} + \dot{\psi}_{\mathrm{T}}\theta_{\mathrm{T}} + \dot{\psi}_{\mathrm{c}}\sin\theta_{\mathrm{c}} \\ \dot{\psi}_{\mathrm{T}} - \dot{\varphi}_{\mathrm{T}}\theta_{\mathrm{T}} + \dot{\psi}_{\mathrm{c}}\cos\theta_{\mathrm{c}} \end{bmatrix} \tag{2.5.14}$$

From the formula (2.1.9), the coordinate matrix of the time derivative of the angular momentum of the carbody in the frame body coordinate system can be obtained.

$$\dot{\boldsymbol{L}}_{\mathrm{c}}^{\mathrm{I}} = \boldsymbol{I}_{\mathrm{c}} \cdot \dot{\boldsymbol{\omega}}_{\mathrm{cI}} + \boldsymbol{\omega}_{\mathrm{cI}} \times \boldsymbol{I}_{\mathrm{c}} \cdot \boldsymbol{\omega}_{\mathrm{cI}} \tag{2.5.15}$$

$$\underline{\dot{\boldsymbol{L}}}_{\mathrm{c}}^{\mathrm{I}} \approx \begin{bmatrix} I_{\mathrm{c}x}(\ddot{\theta}_{\mathrm{c}} + \ddot{\theta}_{\mathrm{T}}) - I_{\mathrm{c}y}(\dot{\varphi}_{\mathrm{c}} + \dot{\varphi}_{\mathrm{T}})(\dot{\psi}_{\mathrm{c}} + \dot{\psi}_{\mathrm{T}}) + I_{\mathrm{c}z}(\dot{\psi}_{\mathrm{c}} + \dot{\psi}_{\mathrm{T}})(\varphi_{\mathrm{c}} + \varphi_{\mathrm{T}}) \\ I_{\mathrm{c}y}(\ddot{\varphi}_{\mathrm{c}} + \ddot{\varphi}_{\mathrm{T}}) + I_{\mathrm{c}x}(\dot{\theta}_{\mathrm{t}} + \dot{\theta}_{\mathrm{T}})(\dot{\psi}_{\mathrm{c}} + \dot{\psi}_{\mathrm{T}}) - I_{\mathrm{c}z}(\dot{\psi}_{\mathrm{c}} + \dot{\psi}_{\mathrm{T}})(\dot{\theta}_{\mathrm{c}} + \dot{\theta}_{\mathrm{T}}) \\ I_{\mathrm{c}z}(\ddot{\psi}_{\mathrm{c}} + \ddot{\psi}_{\mathrm{T}}) - I_{\mathrm{c}x}(\dot{\theta}_{\mathrm{c}} + \dot{\theta}_{\mathrm{T}})(\varphi_{\mathrm{c}} + \varphi_{\mathrm{T}}) + I_{\mathrm{c}y}(\dot{\varphi}_{\mathrm{c}} + \dot{\varphi}_{\mathrm{T}})(\dot{\theta}_{\mathrm{c}} + \dot{\theta}_{\mathrm{T}}) \end{bmatrix} \tag{2.5.16}$$

The motion equation of the carbody can be obtained from the Newton–Euler equation.

$$m_{\mathrm{c}}\boldsymbol{a}_{\mathrm{c}}^{\mathrm{I}} = \boldsymbol{F}_{\mathrm{c}} \tag{2.5.17}$$

$$\dot{\boldsymbol{L}}_{\mathrm{c}}^{\mathrm{I}} = \boldsymbol{M}_{\mathrm{c}} \tag{2.5.18}$$

In the formula, $\boldsymbol{F}_{\mathrm{c}}$ is the external force vector of the carbody, which is composed of suspension force and gravity. In the track coordinate system, it can be expressed as

$$\boldsymbol{F}_{\mathrm{c}} = \underline{\boldsymbol{F}}_{\mathrm{c}}^{\mathrm{T}}\boldsymbol{e}^{\mathrm{T}} \quad \underline{\boldsymbol{F}}_{\mathrm{c}} = [F_{\mathrm{susp}_{\mathrm{c}x}} \quad F_{\mathrm{susp}_{\mathrm{c}y}} + m_{\mathrm{c}}\mathrm{g}\theta_{\mathrm{T}} \quad F_{\mathrm{susp}_{\mathrm{c}z}} + m_{\mathrm{c}}\mathrm{g}]^{\mathrm{T}} \tag{2.5.19}$$

$\boldsymbol{M}_{\mathrm{c}}$ is the external torque vector of the carbody, which is composed of suspension torque. In the body coordinate system, it can be expressed as

$$\boldsymbol{M}_{\mathrm{c}} = \underline{\boldsymbol{M}}_{\mathrm{c}}^{\mathrm{T}}\boldsymbol{e}^{\mathrm{t}}, \quad \underline{\boldsymbol{M}}_{\mathrm{c}} = [M_{\mathrm{susp}_{\mathrm{c}x}} \quad M_{\mathrm{susp}_{\mathrm{c}y}} \quad M_{\mathrm{susp}_{\mathrm{c}z}}]^{\mathrm{T}} \tag{2.5.20}$$

### 2.5.3 Suspension Force

The suspension force can be obtained by the relative displacement and relative velocity between the rigid bodies. As shown in Fig. 2.32, the coordinate matrix of the center of mass of rigid body $i$ in the corresponding track coordinate system $\boldsymbol{e}^{\mathrm{T}i}$ is

$$\underline{\boldsymbol{r}}_{\mathrm{b}i} = [x_{\mathrm{b}i} \quad y_{\mathrm{b}i} + y_{\mathrm{b}i0} \quad z_{\mathrm{b}i} + z_{\mathrm{b}i0}]^{\mathrm{T}} \tag{2.5.21}$$

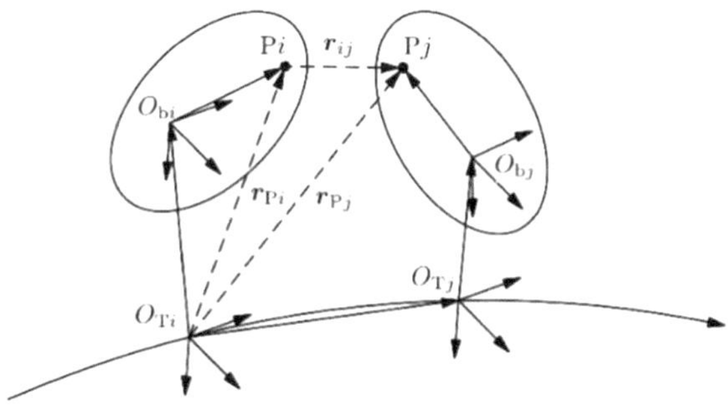


**Fig. 2.32** The relative displacement between the points on two rigid bodies

In the formula, $[0 \quad y_{bi0} \quad z_{bi0}]^{T}$ is the equilibrium position of the center of mass. Points $P_i$ and $P_j$ are fixed on rigid bodies $i$ and $j$, respectively. The relative positions of points $P_i$ and $P_j$ can be expressed as

$$\boldsymbol{r}_{ij} = \boldsymbol{r}_{Pj} - \boldsymbol{r}_{Pi} \tag{2.5.22}$$

The coordinates of the two points are converted to the corresponding track coordinate system of the center of mass of the rigid body $i$, which is $\boldsymbol{e}^{Ti}$. See Fig. 2.33, there are

$$\underline{\boldsymbol{r}}_{Pi} = \begin{bmatrix} x_{bi} \\ y_{bi} + y_{bi0} \\ z_{bi} + z_{bi0} \end{bmatrix} + \underline{\boldsymbol{G}}^{Tibi} \begin{bmatrix} x_{Pi} \\ y_{Pi} \\ z_{Pi} \end{bmatrix} \tag{2.5.23}$$

$$\underline{\boldsymbol{r}}_{Pj} = \underline{\boldsymbol{G}}^{TiTj} \left( \begin{bmatrix} x_{bj} \\ y_{bj} + y_{bj0} \\ z_{bj} + z_{bj0} \end{bmatrix} + \underline{\boldsymbol{G}}^{Tibj} \begin{bmatrix} x_{Pj} \\ y_{Pj} \\ z_{Pj} \end{bmatrix} \right) + \begin{bmatrix} x_{ij} \\ x_{ij}(\psi_{Tj} - \psi_{Ti})/2 \\ x_{ij}(\varphi_{Tj} - \varphi_{Ti})/2 \end{bmatrix} \tag{2.5.24}$$

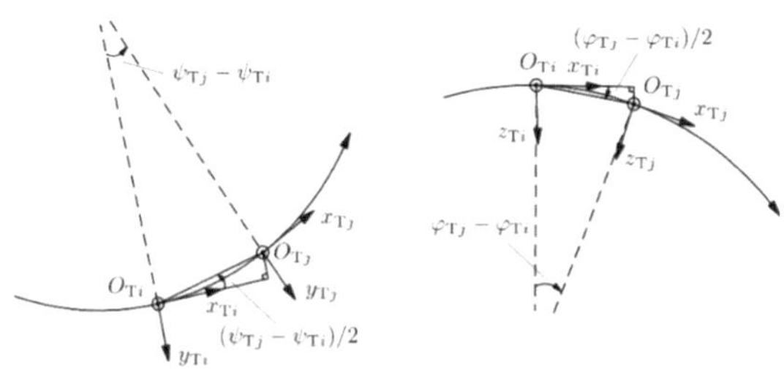


**Fig. 2.33** The relative displacement of two points on the curve

(a) The relative displacement of two points on the plane curve (b) The relative displacement of two points on the vertical curve

The relative displacement between points $P_i$ and $P_j$ can be expressed as

$$\Delta \boldsymbol{r}_{ij} = \boldsymbol{r}_{Pj} - \boldsymbol{r}_{Pi} - \boldsymbol{r}_{ij0} \tag{2.5.25}$$

$\boldsymbol{r}_{ij0}$ is the nominal displacement under equilibrium condition.

$$\underline{r}_{ij0} = \begin{bmatrix} x_{Pj} + x_{ij} - x_{Pi} \\ y_{b0j} + y_{Pj} - y_{b0i} + y_{Pi} \\ z_{b0j} + z_{Pj} - z_{b0i} + z_{Pi} \end{bmatrix} \tag{2.5.26}$$

Taking into account small displacements and rotations, and ignoring higher-order quantities, it is possible to obtain

$$\Delta \underline{r}_{ij} \approx \begin{bmatrix} [x_{b^*} - \psi_{b^*} y_{P^*} + \varphi_{b^*} z_{P^*}]_i^j - (y_{b0j} + y_{Pj})(\psi_{Tj} - \psi_{Ti}) + (z_{b0j} + z_{Pj})(\varphi_{Tj} - \varphi_{Ti}) \\ [y_{b^*} + \psi_{b^*} x_{P^*} - \theta_{b^*} z_{P^*}]_i^j + x_{Pj}(\psi_{Tj} - \psi_{Ti}) + x_{ij}(\psi_{Tj} - \psi_{Ti})/2 \\ [z_{b^*} - \varphi_{b^*} x_{P^*} + \theta_{b^*} y_{P^*}]_i^j - x_{Pj}(\varphi_{Tj} - \varphi_{Ti}) - x_{ij}(\varphi_{Tj} - \varphi_{Ti})/2 \end{bmatrix} \tag{2.5.27}$$

In the formula, the relationship between the two rotation angles is

$$\psi_{Tj} - \psi_{Ti} = x_{ij}/R_h \tag{2.5.28}$$

$$\varphi_{Tj} - \varphi_{Ti} = x_{ij}/R_v \tag{2.5.29}$$

By substituting the corresponding motion of the carbody, frame and wheelset into the formula (2.5.27), the suspension forces in three directions of the primary and secondary suspensions of the vehicle can be obtained. The suspension force and moment acting on the wheelset, frame and carbody can be synthesized by the primary and secondary suspension forces, as shown in Fig. 2.34. The force and moment are the components of the track coordinate system. Before being substituted into the vehicle motion equation, the suspension moment also needs to be used (2.5.2), which is transferred from the track coordinate system to the body coordinate system.

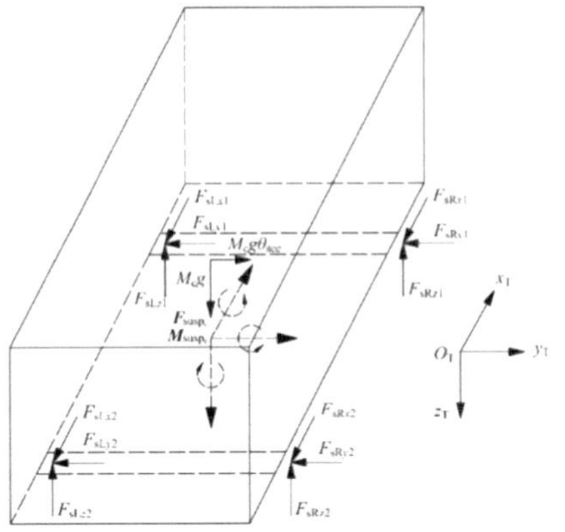

(a) carbody suspension force diagram

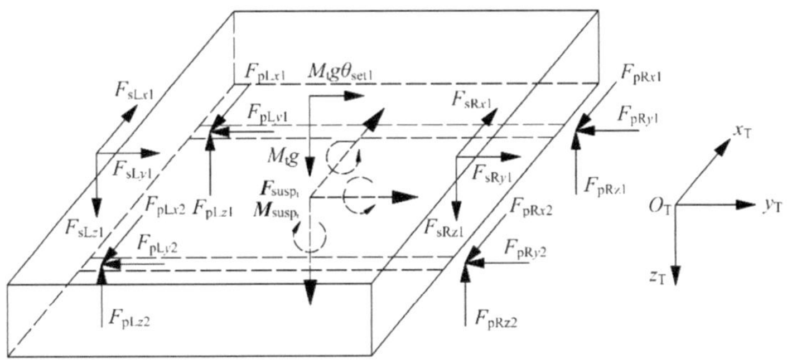


(b) Frame suspension force diagram

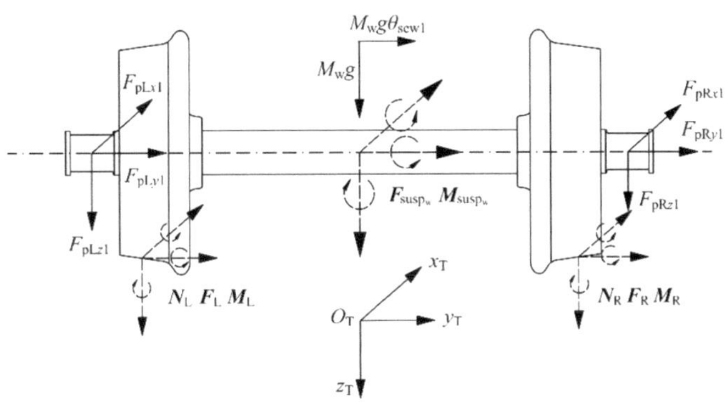


(c) The schematic diagram of wheelset suspension force

**Fig. 2.34** Schematic diagram of vehicle suspension force

# Chapter 3
# Vehicle-Track System Dynamics Modeling

The system dynamics analysis includes the basic steps of modeling, solving, and validating corrections. This chapter first introduces the dynamic modeling methods of track and bridge structures in the vertical and spatial models of the vehicle-track system, including the hypothetical modal method, the finite element method, and the comprehensive modal method. Secondly, the numerical solution of the system equations is introduced. Finally, the development and validation of the dynamics software are introduced.

## 3.1 System Dynamics Analysis Methods

### *3.1.1 Basic Steps of Dynamics Analysis*

The main objective of structural dynamic analysis is to develop both the mechanical (physical) model and the mathematical model of the system. Mechanical models in structural dynamics can be divided into two primary categories: discrete models (also known as lumped parameter models) and continuous models. For discrete systems, the mathematical model is typically represented by an ordinary differential equation for single-degree-of-freedom systems, or by a set of ordinary differential equations for multi-degree-of-freedom systems. In contrast, continuous systems are generally described by partial differential equations.

To facilitate computational processing, the time variable is usually discretized, which allows all these mathematical models to be reformulated as algebraic equations. By solving the equations of motion, we can determine the dynamic response of the system. To validate the analysis, it is crucial to conduct experimental verification of the dynamic model, compare it with experimental data, and make any necessary adjustments. The fundamental steps for analyzing engineering structural dynamics problems are illustrated in Fig. 3.1.

© China Railway Publishing House Co., Ltd. 2026

X. Long et al., *Vehicle–Track–Bridge System Subjected to Track Alignment*,
https://doi.org/10.1007/978-981-95-1649-0_3

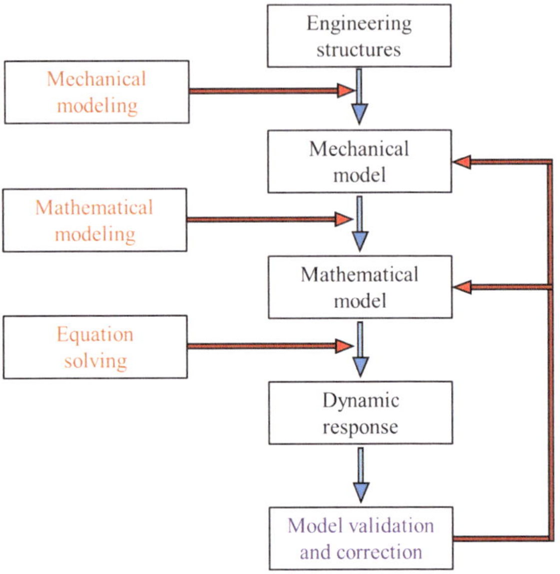


**Fig. 3.1** The basic steps in the study of vibration problems in engineering structures

### *3.1.2 Discretization Methods for Continuous Systems*

In the dynamic analysis of the vehicle–track system, the various components of the vehicle are generally assumed to be rigid bodies, forming a multi-body system; the modeling methods for this have been detailed in the previous chapter. This chapter focuses on the modeling methods for subgrade structures such as tracks and bridges. Most subgrade structures are continuous elastic systems, such as rails and bridges, and the exact solutions for continuous systems are only applicable to components with simple geometries and boundary conditions. When component geometries or boundary conditions are complex, approximate methods must be employed. For the convenience of problem analysis and computer processing, these systems are typically simplified into discrete systems. Common approximate methods include the lumped mass method, the assumed modes method (generalized displacement method), the finite element method, and the combined modal method. The common characteristic of these approximation methods is that an infinite-degree-of-freedom system is approximated by a system with a finite number of degrees of freedom.

1. **Lumped Mass Method**

The lumped mass method involves concentrating the mass of the continuous system into a finite number of points or sections. Components with relatively high inertia and stiffness can be treated as concentrated mass points and rigid bodies (e.g., sleepers). Components with low inertia and high elasticity can be modeled as massless springs, where their mass may be neglected or incorporated into the lumped mass. Systems

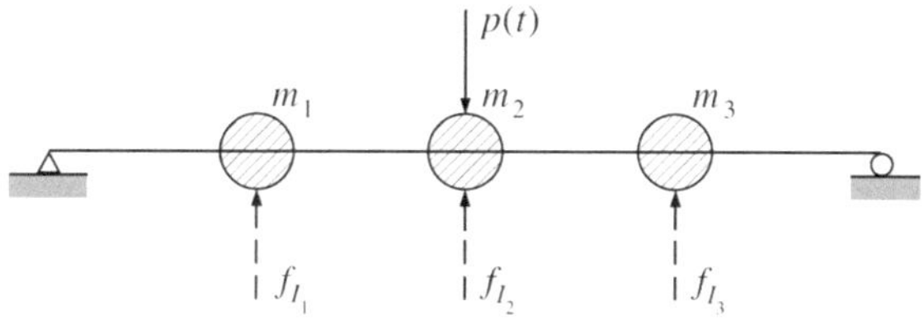


**Fig. 3.2** Schematic representation of the lumped mass method for simply supported beams

with uniformly distributed physical parameters can also be approximately decomposed into a finite number of lumped masses. The number of lumped masses depends on the required calculation accuracy. Once the continuous system is discretized into a finite-degree-of-freedom system, analysis methods for multi-degree-of-freedom systems can be applied.

Figure 3.2 shows a simply supported beam discretized into three discrete mass points or blocks. If only the vertical motion of these three mass points is considered, the system is a three-degree-of-freedom (3DOF) system. If these masses are not concentrated at points but possess finite rotational inertia, the rotational displacements of these points must also be taken into account, resulting in a system with six degrees of freedom.

2. **Assumed Modes Method (Generalized Displacement Method)**

The assumed modes method is assumed to determine the laws of motion of a system using a linear combination of a finite number of functions. It is assumed that the mass of the system is distributed fairly uniformly everywhere, and then, in order to limit the degrees of freedom, the shape of the deflection curve of the structure can be assumed to be represented by the sum of a series of prescribed displacement curves. For example, the deflection curve of a simply supported beam can be expressed in terms of a triangular series, e.g., the Fig. 3.3, which can be expressed as follows

$$u(x) = \sum_{i=1}^{\infty} b_i \sin \frac{i\pi x}{L} \tag{3.1.1}$$

The advantage of this method is that the shape of the actual beam can be well approximated by the magnitude of the finite term sine wave components. In general, Any function $\psi(x)$ that is compatible with the described geometric boundary conditions and maintains the required continuity of internal displacements can be used, and thus, the system's vibrational displacement is expressed as the sum of the first $\boldsymbol{n}$ modes of vibration

$$u(x, t) = \sum_{i=1}^{n} \psi_i(x) q_i(t) \tag{3.1.2}$$

$\psi_i(x)$, which is the modal function of the system, and the shape of the structure it forms depends on the magnitude term $q_i(t)$, i.e., the generalized coordinates. The

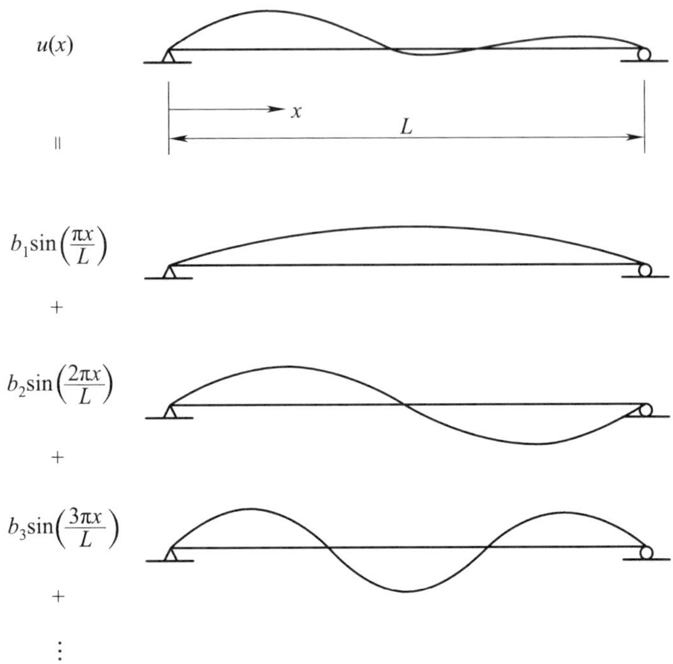


**Fig. 3.3** Schematic representation of assumed modes method for simply supported beams

number of shape curves assumed, the $n$ represents the number of degrees of freedom considered. In general, the assumed modal method is more accurate than the lumped mass method, but will require more computational effort for each degree of freedom.

3. **Finite Element Method**

The finite element method (FEM) is a method developed in the 1950s and 1960s, drawing on the advantages of the lumped mass method and the assumed modal method. The finite element method divides a complex structure into a finite number of units, the endpoints of which are called nodes, and the displacements of the nodes are taken as generalized coordinates, and the mass and stiffness of the units are concentrated on the nodes. Each unit as an elastomer, the displacement of each point in the unit with the node displacement of the interpolation function (unit of the assumed mode), because only the unit, rather than the entire structure to take the assumed mode, so the modal function can be obtained very simple, and can make the modal of each unit is the same, see Fig. 3.4. Finally, a finite degree of freedom problem is formed with all the node displacements as the variables to be solved. The finite element method utilizes the ideas of spatial discretization and assumption of modes at the same time, which is suitable for most of the complex structures, and it is

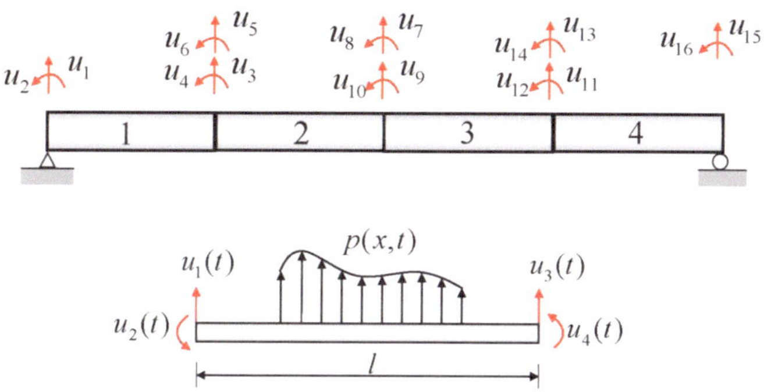


**Fig. 3.4** Schematic of the finite element method for simply supported beams

the most widely used approximation method for the analysis of structural mechanical properties.

4. **Modal Synthesis Method**

The basic idea of modal synthesis method is to "turn the whole into zero, accumulate zero into whole", which is a mature method formed with the development of finite element and substructure methods. For a complex system composed of many components, it is difficult to find the assumed mode suitable for the whole system. The complex structure is decomposed into a number of simpler substructures, the assumed modes are selected for each substructure, and then the assumed modes of the substructures are synthesized into the modal function of the overall system according to the coordinated conditions of displacements and forces on the docking surfaces.

Since the 1970s, the modal synthesis method has been widely used in engineering. It is not only used in the dynamic analysis of large complex structures, but also used in random vibration, nonlinear vibration and other kinds of complex problems. Low-order modes account for the main components of the actual engineering problems, so only a small number of low-order modes need to be calculated for each substructure and then synthesized. Substructures should be divided so that they are easy to analyze, and the interface surface should be as small as possible to reduce the coupling between the substructures. The interconnections between substructures are called interfaces. According to the method of interface processing, the modal synthesis method is divided into fixed interface method and free interface method. The fixed interface method fixes the degrees of freedom on the interface completely; the free interface method does not constrain the degrees of freedom on the interface.

### *3.1.3 Methods for Establishing the Equations of Motion*

When a trainset operates on bridges or tracks, if the bridge and track are considered as rigid bodies and only the vertical vibration of the vehicle is considered, the trainset can

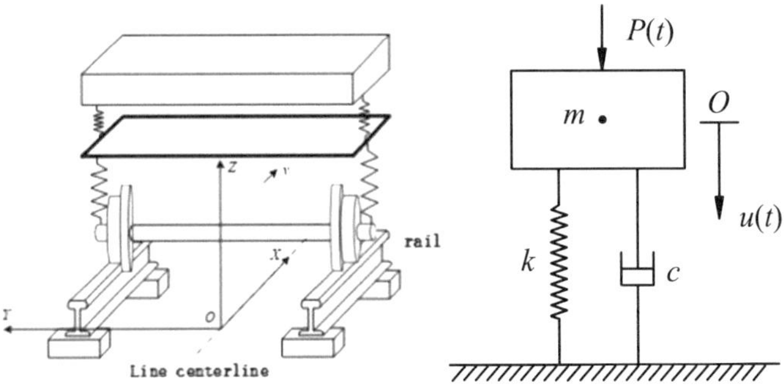


**Fig. 3.5** A simplified single-degree-of-freedom system

be simplified to a single-degree-of-freedom system. As shown in Fig. 3.5, a railway vehicle is supported by a suspension system; when establishing the single-degree-of-freedom system, the vehicle is simplified to a mass block, and the suspension system is simplified to a spring-damper element.

As shown in Fig. 3.5, the vehicle mass is denoted as m, and the stiffness and damping of the suspension system are represented by $k$ and $c$, respectively. The point $O$, where the vehicle is in static equilibrium, is selected as the initial coordinate. Assume that at time $t$, the displacement of the vehicle is $u(t)$, and the external load acting on the vehicle is $P(t)$. The vehicle mass block is isolated for force analysis, as shown in Fig. 3.6.

Such as Fig. 3.6 as shown, assuming that the vehicle mass block m is in equilibrium at time t, the load acting on the mass block m has an inertial force $f_I = m\ddot{u}$ and spring return force $f_s = k(u + \Delta_{st})$, the damping force $f_D$ and the external load $f_D = c\dot{u}$.

The establishment of the equations of motion of a system is the most important and sometimes the most difficult in the whole process of dynamic analysis. Commonly used methods to establish the equations of motion are mainly three types, the first type is the use of Newton's second law and D'Alembert (D'Alembert) principle of the direct balance method, the method belongs to the vector mechanics method, has the advantage of strong geometric intuition, but in dealing with the constrained system, due to the binding force of the emergence of the appearance of the very cumbersome. The second category is the D'Alembert's principle and the principle

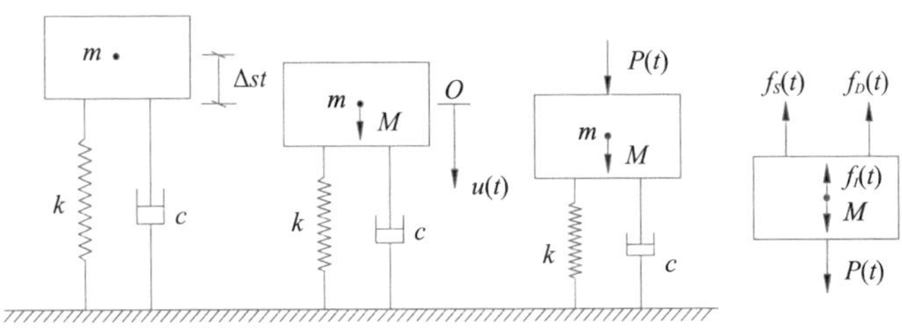


**Fig. 3.6** Force analysis of single-degree-of-freedom systems

of virtual displacement of the virtual work of the dynamic form of the universal equations, the method can avoid solving the constraint forces between the mass or rigid body, the formation of the equation for the variable equations, but the calculation of the virtual work of the force and the displacement is still a vector. The third category for Hamilton (Hamilton) principle, the second type of Lagrange (Lagrange) equations and other methods of analytical mechanics, these methods to establish the equation no longer contains inertial and elastic forces, etc., but based on kinetic and potential energy, has the advantage of dealing with pure scalars only. Appropriate methods should be chosen according to the characteristics of the dynamical system and the problem under study.

(1) D'Alembert's principle establishes the equation of motion

D'Alembert's principle can be described as follows: at any instant of motion, in addition to the total load actually acting on the system, plus the imaginary inertia force, then the system will be in the imaginary equilibrium state at that instant, which is called the dynamic equilibrium state. Remembering $F_i$ and $f_{Ii}$ as the total load and inertia force on the mass point $m_i$, respectively, D'Alembert's principle can be described as follows

$$F_i - f_{Ii} = 0 \quad (i = 1, 2, \ldots, N) \tag{3.1.3}$$

Usually the total load $F_i$ is the sum of the loads on the system, including the external load $P_i$, the damping force $f_D$ and the spring restoring force $f_S$. The method of establishing the equation of motion of the system by D'Alembert's principle is called dynamic-static method or inertial force method, which is also the most direct and convenient method of establishing the equation of motion.

The equilibrium equation for the mass block is obtained from D'Alembert's principle as

$$f_I + f_D + f_S = P(t) + M \tag{3.1.4}$$

will $f_I, f_D, f_S$ Substituting the above equation, and taking into account $\Delta st = M/k$, the equations of motion of the gravity-influenced mass-block single-degree-of-freedom structural system are obtained as

$$m\ddot{u} + c\dot{u} + ku = P(t) \tag{3.1.5}$$

In the study of the dynamic response of the structure, the effect of gravity can be completely disregarded to establish the dynamic equations of the system, and the dynamic solution of the structural system can be obtained by directly solving the equations of motion under the dynamic loads. When the influence of gravity is considered, the total displacement (deformation) is equal to the static solution plus the dynamic solution, i.e. the principle of superposition is established. Therefore, in the structural reaction problem, applying the principle of superposition, the static problem (generally gravity problem) and the dynamic problem can be calculated separately, and then superpose the results to get the real structural reaction.

(2) The universal equations of dynamics establish the equations of motion that

When the structural system is complex and contains many interconnected masses or masses of finite size, it may be difficult to write the equilibrium equations directly for the forces acting on the system. In this case, it is possible to use the principle of virtual work in combination with D'Alembert's principle to establish the equations of motion. The principle of virtual work refers to the ideal constraints of the mass system motion, in any instant, the total load and inertia force in any virtual displacement on the virtual work done by the sum is equal to zero.

Let the total load on the ith mass of the system be $F_i$, the inertia force be $f_{Ii}$, and the virtual displacement be $\delta_{ui}$ Don't write the virtual work equation from the virtual work principle as

$$\sum_{i=1}^{N}(F_i - f_{Ii})\delta u_i = 0 \tag{3.1.6}$$

Therefore, in establishing the equations of motion using the principle of virtual work, it is first necessary to specify all the loads acting on the mass of the system, including inertial forces. The main advantage of using the virtual work principle is that the virtual work is scalar and can be added algebraically, while the forces acting on the structure are vectorial and need to be superimposed according to the vector.

(3) Hamilton's principle, Lagrange's equations to establish the equations of motion, the

When the energy of the system is expressed in a variational form, the Hamiltonian principle can be used to establish the equation of motion, which can be expressed as follows

$$\int_{t_1}^{t_2} \delta L dt + \int_{t_1}^{t_2} \delta W_{nc} dt = 0 \tag{3.1.7}$$

where, $L$ is the Lagrangian function of the system has $L(q, \dot{q}, t) = T - V$, which are system-generalized coordinates $q$ and generalized speed $\dot{q}$. $T$ is the total kinetic energy of the system, $V$ is the potential energy of the system, including strain energy and any conservative force (due to the action of the force, the object moves from the starting point to the end point, if the work done by the force does not change because of the path of the different, the force is called conservative force); $W_{nc}$ is the work done by the non-conservative forces acting on the system (any work related to the path of the force is called non-conservative forces, including damping and any accidental loads); $\delta$ is the variation in the specified time. The work done by the non-conservative forces acting on the system (any force whose work is related to the path is called a non-conservative force, including damping forces and any accidental loads); $\delta$ is the variational component in the specified time.

The ideal, complete system of Lagrangian equations of the second type is that the

$$\frac{d}{dt}\frac{\partial L}{\partial \dot{q}_\alpha} - \frac{\partial L}{\partial q_\alpha} = Q_\alpha, \quad \alpha = 1, 2 \ldots, s \tag{3.1.8}$$

In the formula, the $Q_\alpha$ are generalized forces, including non-conservative active forces, and the conservative force part has been reflected in the Lagrangian function $L$.

The difference between Hamilton's principle, Lagrange's equation and the principle of virtual work lies in the fact that the inertial and elastic forces are not obviously used, but are replaced by kinetic and potential energy respectively. The advantage of this method of building equations is therefore that it relates only to a purely scalar quantity, energy, whereas in the principle of virtual work, although the work itself is a scalar quantity, the forces and displacements used to calculate the work are vector quantities.

## 3.2 Vertical Dynamic Analysis Model of the Vehicle-Track System

The vertical dynamic analysis model of the vehicle-track-bridge system consists of the train, track, bridge, and subgrade. It is primarily used for analyzing the vertical dynamics of track structures, including: dynamic design and optimization of track structural parameters, vehicle-bridge resonance analysis, dynamic analysis of track transition sections, dynamic analysis of road-bridge transition zones. A schematic representation is shown in Fig. 3.7.

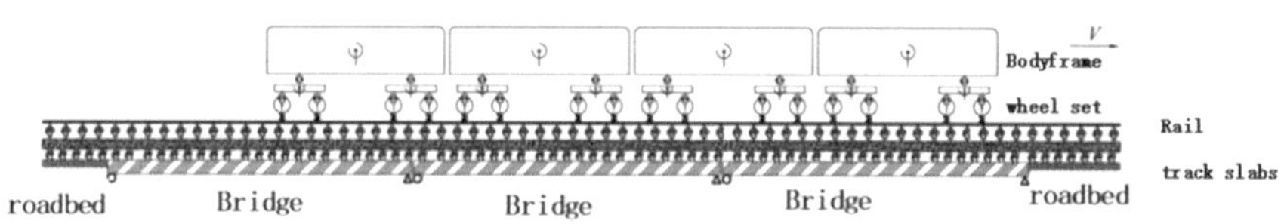


**Fig. 3.7** Vehicle-rail-bridge structural dynamic analysis model

### 3.2.1 Vehicle Modeling

A train consists of multiple identical vehicles. For passenger cars or high-speed trainsets, a two-stage suspension system is considered. Each vehicle has a total of 10 degrees of freedom, including: the pitch and heave motions of the car body, the pitch and heave motions of the front and rear bogie frames, the heave motions of the four wheelsets. A schematic representation is shown in Fig. 3.8.

The equations of the vehicle equations of motion can be derived from D'Alembert's principle, the

$$\boldsymbol{M}_{\mathrm{v}}\ddot{\boldsymbol{u}}_{\mathrm{v}} + \boldsymbol{C}_{\mathrm{v}}\dot{\boldsymbol{u}}_{\mathrm{v}} + \boldsymbol{K}_{\mathrm{v}}\boldsymbol{u}_{\mathrm{v}} = \boldsymbol{F}_{\mathrm{v}} \tag{3.2.1}$$

where, displacement, force vector $\boldsymbol{u}_{\mathrm{v}}, \boldsymbol{F}_{\mathrm{v}}$ and mass and stiffness matrices $\boldsymbol{M}_{\mathrm{v}}, \boldsymbol{K}_{\mathrm{v}}$. Can be see respectively in Eq. (3.2.2)–(3.2.5). The damping matrix $C_{\mathrm{v}}$ can be obtained by replacing the letter "$K$" in the stiffness matrix $K_{\mathrm{v}}$ with "$C$".

$$\boldsymbol{u}_{\mathrm{v}} = \begin{bmatrix} Z_{\mathrm{c}} & \varphi_{\mathrm{c}} & Z_{\mathrm{t1}} & \varphi_{\mathrm{t1}} & Z_{\mathrm{t2}} & \varphi_{\mathrm{t2}} & Z_{\mathrm{w1}} & Z_{\mathrm{w2}} & Z_{\mathrm{w3}} & Z_{\mathrm{w4}} \end{bmatrix}^{\mathrm{T}} \tag{3.2.2}$$

$$\boldsymbol{F}_{\mathrm{v}} = \begin{bmatrix} M_{\mathrm{c}}g & 0 & M_{\mathrm{t}}g & 0 & M_{\mathrm{t}}g & 0 & M_{\mathrm{w}}g - P_{\mathrm{w1}} & M_{\mathrm{w}}g - P_{\mathrm{w2}} & M_{\mathrm{w}}g - P_{\mathrm{w3}} & M_{\mathrm{w}}g - P_{\mathrm{w4}} \end{bmatrix}^{\mathrm{T}} \tag{3.2.3}$$

$$\boldsymbol{M}_{\mathrm{v}} = \begin{pmatrix} M_{\mathrm{c}} & I_{\mathrm{c}} & M_{\mathrm{t}} & I_{\mathrm{t}} & M_{\mathrm{t}} & I_{\mathrm{t}} & M_{\mathrm{w}} & M_{\mathrm{w}} & M_{\mathrm{w}} & M_{\mathrm{w}} \end{pmatrix}^{\mathrm{T}} \tag{3.2.4}$$

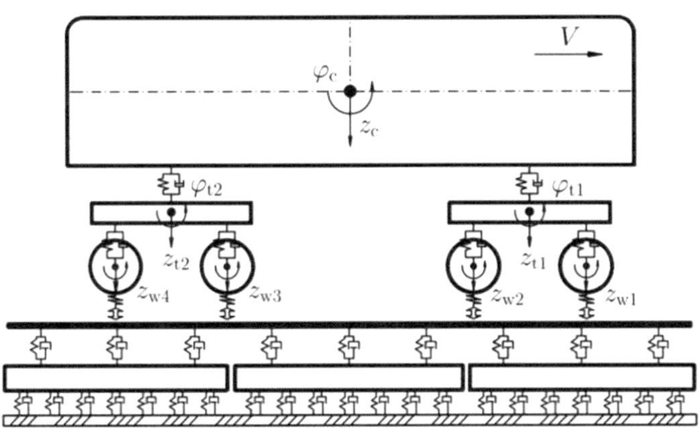


**Fig. 3.8** Vehicle-slab track vertical dynamic analysis model

$$
\boldsymbol{K}_{\mathrm{v}} = \begin{bmatrix}
2K_{2z} & 0 & -K_{2z} & 0 & -K_{2z} & 0 & 0 & 0 & 0 & 0 \\
 & 2K_{2z}l_{\mathrm{c}}^2 & -K_{2z}l_{\mathrm{c}} & 0 & K_{2z}l_{\mathrm{c}} & 0 & 0 & 0 & 0 & 0 \\
 & & K_{2z}+2K_{1z} & 0 & 0 & 0 & -K_{1z} & -K_{1z} & 0 & 0 \\
 & & & 2K_{1z}l_{\mathrm{t}}^2 & 0 & 0 & -K_{1z}l_{\mathrm{t}} & K_{1z}l_{\mathrm{t}} & 0 & 0 \\
 & & & & K_{2z}+2K_{1z} & 0 & 0 & 0 & -K_{1z} & -K_{1z} \\
 & & & & & 2K_{1z}l_{\mathrm{t}}^2 & 0 & 0 & -K_{1z}l_{\mathrm{t}} & K_{1z}l_{\mathrm{t}} \\
 & & & & & & K_{1z} & 0 & 0 & 0 \\
 & & \text{Sym} & & & & & K_{1z} & 0 & 0 \\
 & & & & & & & & K_{1z} & 0 \\
 & & & & & & & & & K_{1z}
\end{bmatrix} \tag{3.2.5}
$$

In the formula, the $Z_{\mathrm{c}}, Z_{\mathrm{t}i}, Z_{\mathrm{w}j}$ are the vertical line displacements of the vehicle body, the ith bogie, and the jth wheelset, respectively; the $\varphi_{\mathrm{c}}, \varphi_{\mathrm{t}i}$ Nodal angular displacement of the car body, ith bogie. $M_{\mathrm{c}}, M_{\mathrm{t}}, M_{\mathrm{w}}$ are the masses of the body, bogie, and wheelset, respectively. $I_{\mathrm{c}}, I_{\mathrm{t}}$ Nodal moment of inertia of the body and bogie, respectively. $K_{1z}, K_{2z}$ The vertical spring stiffnesses of the first and second series suspensions, respectively. $l_{\mathrm{c}}, l_{\mathrm{t}}$ Separate half of the vehicle's fixed distance and half of the bogie's wheelbase.

## 3.2.2 *Track Modeling*

There are two structural types of track: ballastless track and ballasted track. The track structure assumption modal method and finite element method **modeling** methods are introduced here respectively.

1. **Vertical Track Modal Analysis Method**

(1) Slab track modeling

The slab track model consists of rails and track slabs, the rails can be regarded as Euler or Timoshenko beams on continuous elastic discrete point supports, supported on the track slabs; the track slabs are regarded as free beams on elastic supports at both ends, see Fig. 3.8.

1) The rail equation

$$
E_{\mathrm{r}}I_{\mathrm{ry}}\frac{\partial^4 Z_{\mathrm{r}}(x,t)}{\partial x^4} + m_{\mathrm{r}}\frac{\partial^2 Z_{\mathrm{r}}(x,t)}{\partial t^2} = -\sum_{i=1}^{N_{\mathrm{s}}} F_{\mathrm{rs}i}(t)\delta(x - x_{\mathrm{s}i}) + \sum_{j=1}^{4} P_{\mathrm{w}j}(t)\delta\big(x - x_{\mathrm{w}j}\big) \tag{3.2.6}
$$

In the formula, the $E_{\mathrm{r}}$ is the modulus of elasticity of the rail, the $I_{xy}$ is the pair of rail sections $y$ Moment of inertia of the shaft, and $N_s$ is the number of sleepers within the calculated length of the rail. $P_{\mathrm{w}j}$ For the first $j$ the vertical force acting on the rail by the individual wheels. $F_{rsi}$ For rail No $i$. The vertical reaction force of the pivot points, there are

$$F_{rsi}(t) = K_p[Z_r(x_{si}, t) - Z_s(x_{si}, t)] + C_p\left[\dot{Z}_r(x_{si}, t) - \dot{Z}_s(x_{si}, t)\right] \tag{3.2.7}$$

In the formula, the $Z_s(x_{si}, t)$ represents the displacement of the track slab or tie under the support of the rail.

$x_{si}$ represents the x-coordinate of the i-th support point of the rail.

$$x_{si} = il_s \tag{3.2.8}$$

In the formula, the $l_s$ is the spacing of the rails.

$x_{wi}$ represents the $x$-coordinate of the $i$-th support point of the rail and the x-coordinate of the $j$-th wheelset; the motion coordinates $x_{wj} = (j = 1 - 4)$ are as follows:

$$\begin{cases} x_{w1}(t) = x_0 + 2(l_c + l_t) + vt \\ x_{w2}(t) = x_0 + 2l_c + l_t + vt \\ x_{w3}(t) = x_0 + 2l_t + vt \\ x_{w4}(t) = x_0 + vt \end{cases} \tag{3.2.9}$$

In the formula, the $x_0$ are the coordinates of the 4th wheel at the starting moment.

For numerical analysis, the fourth order partial differential equations are transformed into second order ordinary differential equations by assuming modal method. Introducing the rail canonical vibration coordinates, the $q_{zk}(t)$, Applying the rail vertical vibration function.

$$Z_k(t) = \sqrt{\frac{2}{m_r l}} \sin \frac{k\pi x}{l} \tag{3.2.10}$$

The displacement of the rail can be expressed as

$$Z_r(x, t) = \sum_{k=1}^{NMR} Z_k(x) q_{zk}(x) \tag{3.2.11}$$

In the formula, the $l$ is the length of the rail, the *NMR* For the rail vertical vibration mode of the cut-off mode order, generally require the cut-off frequency in the analyzed rail effective frequency of more than two times.

Will be the formula (3.2.11) with carry-over (3.2.6) This results in a system of second-order ordinary differential equations for the rail vibration coordinates of the following form.

$$\ddot{q}_{zk}(t) + \frac{c_r}{m_r}\dot{q}_{zk}(t) + \frac{E_r I_{ry}}{m_r}\left(\frac{k\pi}{l}\right)^4 q_{zk}(t) = -\sum_{i=1}^{N_s} F_{rsi}(t) Z_k(x_{Fi})$$

$$+\sum_{j=1}^{4} P_{wj}(t) Z_k(x_{wj}) \qquad (k = 1 \sim NMR) \tag{3.2.12}$$

In the derivation of the equations, the effects of shear deformation and moment of inertia of the beam in the process of motion are not considered. For the case where the deflection of the beam is much smaller than its length, i.e., when the bending radius of the beam is much larger compared with the height of the beam, the bending deformation is the main one, and the effects of shear deformation and moment of inertia are so small that they can be neglected, and the above equation has sufficient accuracy. According to this assumption, only consider the bending deformation of the beam is called Euler beam.

A more accurate differential equation of the beam bending motion will be obtained by considering the effects of rotational inertia and shear deformation

$$\left[m\frac{\partial^2 u}{\partial t^2} + EI\frac{\partial^4 u}{\partial x^4} - P\right] - \rho I\frac{\partial^2 u}{\partial x^2 \partial t^2} + \frac{EI}{k'AG}\frac{\partial^2}{\partial x^2}\left(P - m\frac{\partial^2 u}{\partial t^2}\right)$$

$$-\frac{\rho I}{k'AG}\frac{\partial^2}{\partial t^2}\left(P - m\frac{\partial^2 u}{\partial t^2}\right) = 0 \tag{3.2.13}$$

In the formula, the $k'$ is a constant factor determined by the shape of the cross-section, which is called the effective shear coefficient of the cross-section; $A$ is the area of the cross-section; and $G$ is the shear modulus of elasticity.

This is the partial differential equation for the transverse bending motion of the beam when the moment of inertia and shear deformation are considered. In this equation, the first term is the basic case without considering the moment of inertia and shear deformation, the second term considers the effect of moment of inertia, the third term the effect of shear deformation, and the fourth term considers the coupling of the effects of moment of inertia and shear deformation. The beam created by considering the moment of inertia and shear deformation is called Timoshenko beam. Generally for the length ratio is relatively small (such as less than 10), or higher order of the intrinsic frequency calculation, should take into account the rotational inertia and shear deformation of the impact, such as high-frequency vibration of the wheel and rail problems.

2) The track slab equation

The track slab is regarded as an elastic continuous support on the two ends of a free Euler beam (Fig. 3.10), the equation of motion of the track slab is given by

$$E_s I_s \frac{\partial^4 Z_s(x,t)}{\partial x^4} + \frac{M_s}{L_s}\frac{\partial^2 Z_s(x,t)}{\partial t^2} = \sum_{i=1}^{N_s} F_{rsi}(t)\delta(x - x_{si}) - \sum_{j=1}^{N_b} F_{sbj}(t)\delta\left(x - x_{bj}\right) \tag{3.2.14}$$

In the formula, the $E_s$ is the Young's modulus of elasticity of the bridge, the $I_s$ is the moment of inertia of the bridge. $Z_s(x, t)$ is the vibration displacement of the track slab. $M_s$ is the mass of the track slab. $F_{rsi}$ for the rail and the track slab between the force, see formula (3.2.7). $F_{rbj}$ for the subgrade or bridge support discrete after the first *j* the vertical reaction force of the pivot point, the $N_b$ is the number of supporting nodes, with In this equation, $E_s$ is the Young's modulus of the bridge, $I_s$ is the moment of inertia of the bridge, $Z_s(x, t)$ is the vibration displacement of the track slab; $M_s$ is the mass of the track slab, and $F_{rsi}$ is the interaction force between the rail and the track slab, as shown in Eq. (3.2.7). $F_{rbj}$ is the vertical reaction force at the j-th support point after discretization of the subgrade or bridge support, and $N$ is the number of support nodes, as follows:

$$F_{sbj}(t) = K_s\left[Z_s(x_{bj}, t) - Z_b(x_{bj}, t)\right] + C_s\left[\dot{Z}_s(x_{bj}, t) - \dot{Z}_b(x_{bj}, t)\right] \tag{3.2.15}$$

The assumed modal method is used to convert this fourth order partial differential equation into a second order system of ordinary differential equations. Selecting *NMS* generalized coordinates, the $T_n t$, introducing a system of free beam orthogonal functions $\{X_n\}$ (n = 1 ~ *NMS*), where

$$\begin{cases} X_1 = 1 \\ X_2 = \sqrt{3}(1 - 2x/L_s) \\ X_m = (\cosh\beta_m x + \cos\beta_m x) - C_m(\sinh\beta_m x + \sin\beta_m x) \quad (m > 2) \end{cases} \tag{3.2.16}$$

In the formula, the $C_m$, $\beta_m$ is a constant whose value is given in Table 3.1, $L_s$ is the length of the track slab.

The vertical displacement $Z_s(x, t)$ of the track slab can be expressed by separating variables as follows:

$$Z_s(x,t) = \sum_{n=1}^{NMS} X_n(x) T_n(t) \tag{3.2.17}$$

**Table 3.1** Free beam factor

| $m$ | 1 | 2 | 3 | 4 | 5 | $\geq 6$ |
|---|---|---|---|---|---|---|
| $C_m$ | – | – | 0.982,502 | 1.000,777 | 0.999,966 | 1.000,000 |
| $\beta_m L_s$ | 0 | 0 | 4.72,004 | 7.8,5320 | 10.9956 | (2m−3) π/2 |

By substituting Eq. (3.2.17) into Eq. (3.2.14), the second-order differential equation system for the modal coordinates of the track slab is obtained as follows:

$$\ddot{T}_n(t) + \frac{E_s I_s L_s \beta_n^4}{M_s} T_n(t) = \sum_{i=1}^{N_s} \frac{F_{rsi}(t)}{M_s} X_n(x_{si}) - \sum_{j=1}^{N_b} \frac{F_{sbj}(t)}{M_s} X_n(x_{bj}) \quad (3.2.18)$$

(2) Ballasted track modeling

The ballasted track model consists of rails, sleepers, and beds, and adopts a three-layer elastic discrete point-supported Euler beam model. The model takes into account the vibration participation of rails, sleepers and roadbed, as well as the elasticity of the fastener, subgrade and subgrade (bridge) structure, see Fig. 3.9.

The rail equations of motion of the ballasted tack model are the same as the slab track model (Fig. 3.8).

① the equations of motion of the rails

The sleeper can be regarded as a mass block, and the force analysis of the sleeper $i$ can obtain its equation of motion.

$$\begin{aligned} & M_{si}\ddot{Z}_{si}(t) + K_{pi}[Z_{si}(t) - Z_r(x_i, t)] + C_{pi}\left[\dot{Z}_{si}(t) - \dot{Z}_r(x_i, t)\right] \\ & + K_{bi}[Z_{si}(t) - Z_{di}(t)] + C_{bi}\left[\dot{Z}_{si}(t) - \dot{Z}_{di}(t)\right] = 0 \end{aligned} \quad (3.2.19)$$

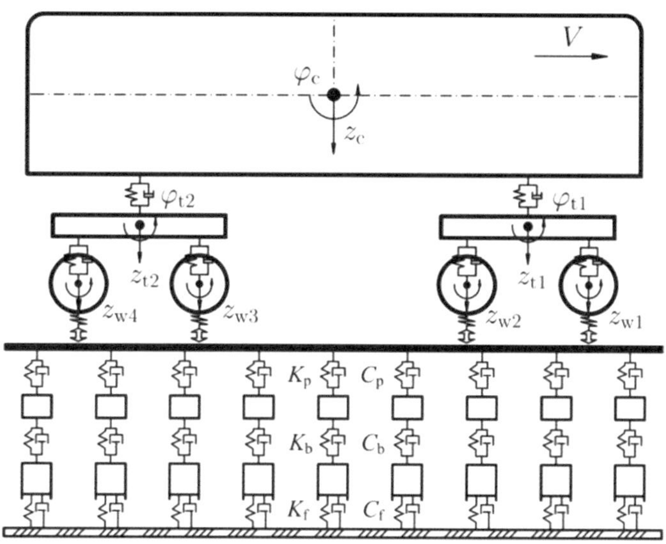


**Fig. 3.9** Vehicle-ballasted track vertical dynamic analysis model

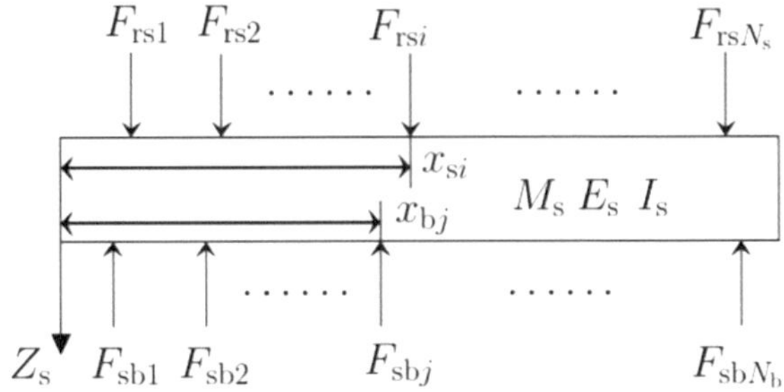


**Fig. 3.10** Schematic diagram of a model of a track slab

② the equation of motion of the roadbed

The subgrade is discretized into mass blocks, and the force analysis of the subgrade block $i$ yields the following equations of motion.

$$\begin{aligned} & M_{di}\ddot{Z}_{di}(t) + K_{bi}[Z_{di}(t) - Z_s(x_i,t)] + C_{bi}\left[\dot{Z}_{di}(t) - \dot{Z}_s(x_i,t)\right] \\ & + K_{fi}[Z_{di}(t) - Z_{bi}(t)] + C_{fi}\left[\dot{Z}_{di}(t) - \dot{Z}_{bi}(t)\right] = 0 \end{aligned} \tag{3.2.20}$$

2. **Vertical Track Finite Element Modeling**

(1) The general idea of finite element modeling

After the complex structure is discretized into a finite number of units, a reasonable unit displacement pattern is selected, which needs to meet the requirements of completeness and compatibility. The displacement of any point in the unit after discretization in the spatial domain can be interpolated as follows

$$\boldsymbol{u} = \boldsymbol{N}\boldsymbol{u}^e \tag{3.2.21}$$

Among them

$$\boldsymbol{u} = \begin{Bmatrix} u(x,y,z) \\ v(x,y,z) \\ w(x,y,z) \end{Bmatrix}, \quad \boldsymbol{N} = \left[\boldsymbol{N}_1\ \boldsymbol{N}_2 \ldots \boldsymbol{N}_n\right],$$
$$\boldsymbol{N}_i = N_i\boldsymbol{I}_{3\times3} \quad (i = 1,2,\ldots,n)$$

$\boldsymbol{N}$ is the matrix of form functions, the $\boldsymbol{u}^e$ is the unit node displacement vector, the $\boldsymbol{I}$ is the unit matrix.

The unitary strain component can be obtained from Eq. (3.2.21) asked for

$$\boldsymbol{\varepsilon} = \boldsymbol{\Delta u} = \boldsymbol{\Delta N u}^e = \boldsymbol{B u}^e \tag{3.2.22}$$

$\boldsymbol{\Delta}$ is the matrix of differential operators for calculating the unit strain, the $\boldsymbol{B}$ is the unit strain matrix, the $\boldsymbol{B} = \boldsymbol{\Delta N}$.

The unit stress components are

$$\boldsymbol{\sigma} = \boldsymbol{D\varepsilon} = \boldsymbol{DBu}^e = \boldsymbol{Su}^e \tag{3.2.23}$$

$\boldsymbol{D}$ is the elasticity matrix, the $\boldsymbol{S}$ is the stress matrix.

The unit strain energy is

$$\boldsymbol{V}^e = \frac{1}{2}\int_V \boldsymbol{\varepsilon}^{\mathrm{T}}\boldsymbol{\sigma}\,dv = \frac{1}{2}\int_V \boldsymbol{u}^{e\mathrm{T}}\boldsymbol{B}^{\mathrm{T}}\boldsymbol{DBu}^e\,dv = \frac{1}{2}\boldsymbol{u}^{e\mathrm{T}}\left(\int_V \boldsymbol{B}^{\mathrm{T}}\boldsymbol{DB}dv\right)\boldsymbol{u}^e = \frac{1}{2}\boldsymbol{u}^{e\mathrm{T}}\boldsymbol{K}^e\boldsymbol{u}^e \tag{3.2.24}$$

$\boldsymbol{K}^e = \int_V \boldsymbol{B}^{\mathrm{T}}\boldsymbol{DB}dv$ is the cell stiffness matrix.

Element Strain Energy is

$$\boldsymbol{T}^e = \frac{1}{2}\int_V \rho\dot{\boldsymbol{u}}^{\mathrm{T}}\dot{\boldsymbol{u}}dv = \frac{1}{2}\dot{\boldsymbol{u}}^{e\mathrm{T}}\left(\int_V \rho\boldsymbol{N}^{\mathrm{T}}\boldsymbol{N}dv\right)\dot{\boldsymbol{u}}^e = \frac{1}{2}\dot{\boldsymbol{u}}^{e\mathrm{T}}\boldsymbol{m}^e\dot{\boldsymbol{u}}^e \tag{3.2.25}$$

$\boldsymbol{m}^e = \int_V \rho\boldsymbol{N}^{\mathrm{T}}\boldsymbol{N}dv$ is the unit consistent quality matrix.

If there is volumetric force $\boldsymbol{q}^e$, whose equivalent nodal force $\boldsymbol{R}^e$. It can be obtained from the principle of virtual work

$$\boldsymbol{R}^e = \int_V \boldsymbol{N}^{\mathrm{T}}\boldsymbol{q}^e dv \tag{3.2.26}$$

If there is a viscous damping force $\boldsymbol{q}_f^e$, which can be expressed as follows

$$\boldsymbol{q}_f^e = -\gamma\dot{\boldsymbol{u}} = -\gamma\boldsymbol{N}\dot{\boldsymbol{u}}^e \tag{3.2.27}$$

Its equivalent nodal force $\boldsymbol{R}_f^e$ For

$$\boldsymbol{R}_f^e = -\int_V \gamma\boldsymbol{N}^{\mathrm{T}}\boldsymbol{N}\dot{\boldsymbol{u}}^e dv = -\left(\int_V \gamma\boldsymbol{N}^{\mathrm{T}}\boldsymbol{N}dv\right)\dot{\boldsymbol{u}}^e = -\boldsymbol{C}\dot{\boldsymbol{u}}^e \tag{3.2.28}$$

Equation $\gamma$ is the damping factor, the $\boldsymbol{C} = \int_v \gamma\boldsymbol{N}^{\mathrm{T}}\boldsymbol{N}dv$ is the damping matrix.

Nodal displacement and force vectors in the local coordinate system are converted to the global coordinate system by the coordinate transformation matrix, and then the total strain energy can be specifically obtained $\boldsymbol{T}$, total kinetic energy $\boldsymbol{V}$ and the sum of the virtual work of the external and damping forces, the $\delta\boldsymbol{W}$, as follows

$$V = \sum_e V^e = \frac{1}{2}\sum_e \boldsymbol{u}^{e\mathrm{T}}\boldsymbol{K}^e\boldsymbol{u}^e = \frac{1}{2}\boldsymbol{u}^{\mathrm{T}}\left(\sum_e \boldsymbol{K}^e\right)\boldsymbol{u} = \frac{1}{2}\boldsymbol{u}^{\mathrm{T}}\boldsymbol{K}\boldsymbol{u} \tag{3.2.29}$$

$$T = \sum_e T^e = \frac{1}{2}\sum_e \dot{\boldsymbol{u}}^{e\mathrm{T}}\boldsymbol{m}^e\dot{\boldsymbol{u}}^e = \frac{1}{2}\dot{\boldsymbol{u}}^{\mathrm{T}}\left(\sum_e \boldsymbol{m}^e\right)\dot{\boldsymbol{u}} = \frac{1}{2}\dot{\boldsymbol{u}}^{\mathrm{T}}\boldsymbol{m}\dot{\boldsymbol{u}} \tag{3.2.30}$$

$$\delta \boldsymbol{W} = \sum_e \delta \boldsymbol{W}^e = \sum_e \delta \boldsymbol{u}^{e^{\mathrm{T}}}\left(\boldsymbol{R}^e + \boldsymbol{R}_f^e\right)^e = \delta \boldsymbol{u}^{\mathrm{T}}\left(\sum_e \boldsymbol{R}^e + \sum_e \boldsymbol{R}_f^e\right) = \delta \boldsymbol{u}^{\mathrm{T}}(\boldsymbol{R} + \boldsymbol{R}_f) \tag{3.2.31}$$

utilizing the formula (3.2.28) gain

$$\boldsymbol{R}_f = \sum \boldsymbol{R}_f^e = -\sum_e \boldsymbol{C}\dot{\boldsymbol{u}}^e = -\left(\sum_e \boldsymbol{C}^e\right)\dot{\boldsymbol{u}} = -\boldsymbol{C}\dot{\boldsymbol{u}} \tag{3.2.32}$$

And so the Eq. (3.2.31) can be changed to

$$\delta \boldsymbol{W} = \delta \boldsymbol{u}^{\mathrm{T}}\boldsymbol{R}' \tag{3.2.33}$$

where $\boldsymbol{R}' = \boldsymbol{R} + \boldsymbol{R}_f = \boldsymbol{R} - \boldsymbol{C}\dot{\boldsymbol{u}}$

Bringing the formula (3.2.29), (3.2.30) and (3.2.33) in the Lagrange equation yields

$$\boldsymbol{M}\ddot{\boldsymbol{u}} + \boldsymbol{C}\dot{\boldsymbol{u}} + \boldsymbol{K}\boldsymbol{u} = \boldsymbol{R} \tag{3.2.34}$$

(2) Ballasted track finite element model

Since the cross-section of the rail is symmetric about the vertical axis, the lateral and vertical deformations of the rail can be considered independent of each other under the external load, and they can be considered as beam units. Each node has 4 degrees of freedom, and each unit has 8 degrees of freedom. The unit displacement is expressed as follows.

The rail can be regarded as a beam unit, and between two adjacent sleepers can be regarded as a unit with 2 freedoms at each end, i.e., vertical displacement and rotation (zri, φri). In order to reduce the scale of the overall stiffness matrix of the equations after finite-element combination, a centralized beam unit model is used, and the vertical displacements of the sleeper and the subgrade mass block under the node can be attached to the node. Therefore, each node possesses 4 degrees of freedom (DOF) and each element has 8 DOF, as illustrated in Fig. 3.11.

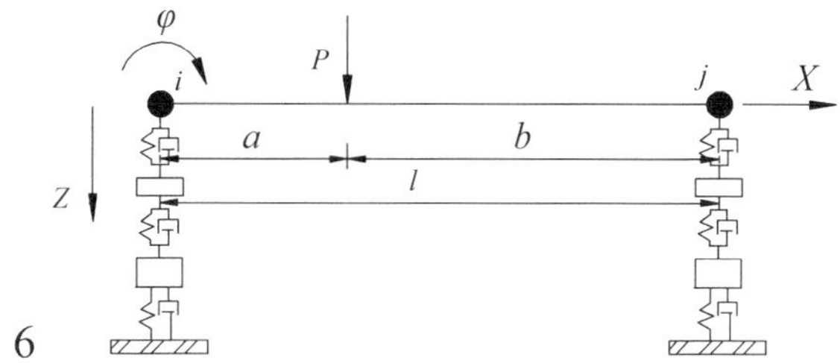


**Fig. 3.11** Schematic diagram of a centralized girder unit for ballasted track

Therefore, the displacement vector of the centralized beam unit can be expressed as follows

$$\boldsymbol{x}_l^e = \left\{ z_{ri} \ \varphi_{ri} \ z_{si} \ z_{bj} \ z_{rj} \ \varphi_{rj} \ z_{sj} \ z_{bj} \right\}^T \tag{3.2.35}$$

The unit stiffness matrix is given by

$$\boldsymbol{K}_l^e = \boldsymbol{K}_0^e + \boldsymbol{K}_1^e = \begin{bmatrix} \frac{12EI}{l^3} & \frac{6EI}{l^2} & 0 & 0 & \frac{-12EI}{l^3} & \frac{6EI}{l^2} & 0 & 0 \\ & \frac{4EI}{l} & 0 & 0 & -\frac{6EI}{l^2} & \frac{2EI}{l} & 0 & 0 \\ & & 0 & 0 & 0 & 0 & 0 & 0 \\ & & & 0 & 0 & 0 & 0 & 0 \\ & & & & \frac{12EI}{l^3} & -\frac{6EI}{l^2} & 0 & 0 \\ & & \text{Sym} & & & \frac{4EI}{l} & 0 & 0 \\ & & & & & & 0 & 0 \\ & & & & & & & 0 \end{bmatrix}$$

$$+ \frac{1}{2} \begin{bmatrix} K_p & 0 & -K_p & 0 & 0 & 0 & 0 & 0 \\ & 0 & 0 & 0 & 0 & 0 & 0 & 0 \\ & & K_p + K_b & -K_b & 0 & 0 & 0 & 0 \\ & & & K_b + K_f & 0 & 0 & 0 & 0 \\ & & & & K_p & 0 & -K_p & 0 \\ & & \text{Sym} & & & 0 & 0 & 0 \\ & & & & & & K_p + K_b & -K_b \\ & & & & & & & K_b + K_f \end{bmatrix} \tag{3.2.36}$$

In the formula, the $\boldsymbol{K}_0^e$, $\boldsymbol{K}_1^e$ Respectively for the beam unit stiffness matrix and elastic support stiffness matrix, $EI$ for the rail bending stiffness, l for the spacing of the rail sleeper, $m$ for the mass of rail per linear meter, $K_p$, $K_b$, $K_f$ for the fastener, subgrade and foundation support stiffness.

The beam unit mass matrix can be written as, the

$$[M]_l^e = [M]_0^e + [M]_1^e = \frac{ml}{420}\begin{bmatrix} 156 & 22l & 0 & 0 & 54 & -13l & 0 & 0 \\ & 4l^2 & 0 & 0 & 13l & -3l^2 & 0 & 0 \\ & & 0 & 0 & 0 & 0 & 0 & 0 \\ & & & 0 & 0 & 0 & 0 & 0 \\ & & & & 156 & -22l & 0 & 0 \\ & & \text{Sym} & & & 4l^2 & 0 & 0 \\ & & & & & & 0 & 0 \\ & & & & & & & 0 \end{bmatrix} + \frac{1}{2}\begin{bmatrix} 0 & 0 & 0 & 0 & 0 & 0 & 0 & 0 \\ & 0 & 0 & 0 & 0 & 0 & 0 & 0 \\ & & M_s & 0 & 0 & 0 & 0 & 0 \\ & & & M_b & 0 & 0 & 0 & 0 \\ & & & & 0 & 0 & 0 & 0 \\ & \text{Sym} & & & & 0 & 0 & 0 \\ & & & & & & M_s & 0 \\ & & & & & & & M_b \end{bmatrix} \quad (3.2.37)$$

$\boldsymbol{M}_0^e$, $\boldsymbol{M}_1^e$ is the beam unit mass matrix and the sleeper and subgrade unit mass matrix, respectively, $M_s$, $M_b$ are the half of the mass of the sleeper and subgrade mass block.

The unit damping matrix can be Rayleigh damping, which is related to the damping ratio and the intrinsic frequency of the system, and can be calculated from the unit stiffness and mass matrices as well as from the fastener and subgrade damping

$$\boldsymbol{C}_l^e = \boldsymbol{C}_0^e + \boldsymbol{C}_1^e = \alpha \boldsymbol{K}_0^e + \beta \boldsymbol{C}_0^e + \boldsymbol{C}_1^e \quad (3.2.38)$$

where $\alpha$ and $\beta$ are the damping coefficients. $\boldsymbol{C}_1^e$ is the under-rail support damping matrix, which can be obtained from the under-rail stiffness matrix, the $\boldsymbol{K}_1^e$ in "$K$" is obtained by replacing "$C$".

The unit stage force can be expressed as that

$$\boldsymbol{Q}_l^e = \left( \frac{b^2(l+2a)P}{l^3} \; \frac{ab^2P}{l^2} \; 0 \; 0 \; \frac{a^2(l+2b)}{l^3} \; -\frac{a^2b}{l^2} \; 0 \; 0 \right)^{\mathrm{T}} \quad (3.2.39)$$

Finally, the finite element dynamic equations of the ballasted track structure can be obtained by combining the unit matrices

$$\boldsymbol{M}_l \ddot{\boldsymbol{x}}_l + \boldsymbol{C}_l \dot{\boldsymbol{x}}_l + \boldsymbol{K}_l \boldsymbol{x}_l = \boldsymbol{Q}_l \quad (3.2.40)$$

where $\boldsymbol{M}_l = \sum_e \boldsymbol{M}_l^e,\ \boldsymbol{K}_l = \sum_e \boldsymbol{K}_l^e,\ \boldsymbol{C}_l = \sum_e \boldsymbol{C}_l^e,\ \boldsymbol{Q}_l = \sum_e \boldsymbol{Q}_l^e$

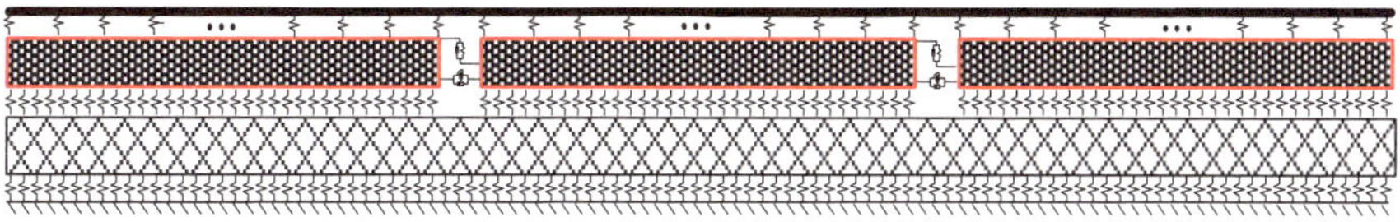

**Fig. 3.12** Schematic diagram of the finite element track structure model

(3) Finite element modeling of slab track

Ballastless track of high-speed railroad in China mainly adopts the types of slab track and double-block track. Slab track is mainly composed of rails, fasteners, track slabs, CA mortar (or self-compacting concrete), concrete support layer and under-rail subgrade or bridge structure. As illustrated in Fig. 3.12, the rail, track slab, concrete bearing layer are two nodes, four degrees of freedom beam unit simulation, the use of springs, damper simulation fasteners, CA mortar, subgrade or bridge support; which fasteners for the spacing spring damper, CA mortar (or self-compacting concrete), subgrade or bridge are uniformly supported spring damper for the longitudinal characteristics of the II slab track, the track slab joints can be increased in the position of the For the longitudinal characteristics of type II slab track, spring dampers can be added at the joint position of the track slab for shear and bending resistance.

The direct stiffness method is used to establish the dynamic equations of ballast track as follows

$$\boldsymbol{M}_t\ddot{\boldsymbol{U}}_t + \boldsymbol{C}_t\dot{\boldsymbol{U}}_t + \boldsymbol{K}_t\boldsymbol{U}_t = \boldsymbol{F}_t \tag{3.2.41}$$

$$\boldsymbol{M}_t = \begin{bmatrix} \boldsymbol{M}_r & & \\ & \boldsymbol{M}_s & \\ & & \boldsymbol{M}_c \end{bmatrix} \tag{3.2.42}$$

$$\boldsymbol{K}_t = \begin{bmatrix} \boldsymbol{K}_r + \boldsymbol{K}_{rs} & -\boldsymbol{K}_{rs} & \\ -\boldsymbol{K}_{rs} & \boldsymbol{K}_s + \boldsymbol{K}_{rs} + \boldsymbol{K}_{sc} & -\boldsymbol{K}_{sc} \\ & -\boldsymbol{K}_{sc} & \boldsymbol{K}_c + \boldsymbol{K}_{sc} + \boldsymbol{K}_{cb} \end{bmatrix} \tag{3.2.43}$$

$$\boldsymbol{C}_t = \begin{bmatrix} \boldsymbol{C}_r + \boldsymbol{C}_{rs} & -\boldsymbol{C}_{rs} & \\ -\boldsymbol{C}_{rs} & \boldsymbol{C}_s + \boldsymbol{C}_{rs} + \boldsymbol{C}_{sc} & -\boldsymbol{C}_{sc} \\ & -\boldsymbol{C}_{sc} & \boldsymbol{C}_c + \boldsymbol{C}_{sc} + \boldsymbol{C}_{cb} \end{bmatrix} \tag{3.2.44}$$

In the formula, the $\boldsymbol{M}_t$, $\boldsymbol{C}_t$, $\boldsymbol{K}_t$ representing the mass matrix, damping matrix, stiffness matrix, and displacement vector of the ballast track, respectively. $\boldsymbol{F}_t$ is the external force vector of the ballast track. $\boldsymbol{K}_{rs}$ is the stiffness matrix formed by the fastener; the $\boldsymbol{K}_{sc}$ is the stiffness matrix formed by the CA mortar; the $\boldsymbol{K}_{cb}$ is the stiffness matrix formed by the roadbed. $\boldsymbol{C}_{rs}$ is the damping matrix formed by the fasteners; the $\boldsymbol{C}_{sc}$ is the damping matrix formed by the CA mortar; the $\boldsymbol{C}_{cb}$ is the damping matrix formed by the roadbed. $\boldsymbol{M}_r$, $\boldsymbol{C}_r$, $\boldsymbol{K}_r$ represent the overall stiffness matrix, mass matrix, and damping matrix of the rail layer, respectively. $\boldsymbol{M}_s$, $\boldsymbol{C}_s$, $\boldsymbol{K}_s$ represent the overall stiffness matrix, mass matrix, and damping matrix of the track slab layer, respectively, taking into account the inter-slab discontinuities and shear hinges. $\boldsymbol{M}_c$, $\boldsymbol{C}_c$, $\boldsymbol{K}_c$ represent the overall stiffness matrix, mass matrix, and damping matrix of the concrete support layer, respectively.

The overall matrix of the rail is assembled from the unit matrix, which can be expressed as

$$\boldsymbol{K}_r = \sum_{i=1}^{n} \boldsymbol{k}_r \quad \boldsymbol{M}_r = \sum_{i=1}^{n} \boldsymbol{m}_r \quad \boldsymbol{C}_r = \sum_{i=1}^{n} \boldsymbol{c}_r \tag{3.2.45}$$

The unit matrix for the rail unit is as follows

$$\boldsymbol{k}_r = \begin{bmatrix} 12E_rI_r/l^3 & -6E_rI_r/l^2 & -12E_rI_r/l^3 & -6E_rI_r/l^2 \\ -6E_rI_r/l^2 & 4E_rI_r/l & 6E_rI_r/l^2 & 2E_rI_r/l \\ -12E_rI_r/l^3 & 6E_rI_r/l^2 & 12E_rI_r/l^3 & 6E_rI_r/l^2 \\ -6E_rI_r/l^2 & 2E_rI_r/l & 6E_rI_r/l^2 & 4E_rI_r/l \end{bmatrix} \tag{3.2.46}$$

$$\boldsymbol{m}_r = (m_1 l/420) \begin{bmatrix} 156 & -22l & 54 & 13l \\ -22l & 4l^2 & -13l & -3l^3 \\ 54 & -13l & 156 & 22l \\ 13l & -3l^2 & 22l & 4l^2 \end{bmatrix} \tag{3.2.47}$$

The damping matrix of the rail element adopts proportional damping, which is related to the damping ratio and the system's natural frequencies

$$\boldsymbol{c}_r = \alpha_r \boldsymbol{m}_r + \beta_r \boldsymbol{k}_r \tag{3.2.48}$$

In the above equation, the $l$ is the length of the cell. $m_1$ is the mass of rail per unit length. $E_r$ is the Young's modulus of the rail; the $I_r$ is the cross sectional inertia distance of the rails; $A_r$ are the cross-sectional areas of the rails, respectively. $\alpha_r$, $\beta_r$ is the damping scale factor.

The overall matrix of the track slab is assembled from a matrix of cells, which can be expressed as follows:

$$\boldsymbol{K}_s = \sum_{i=1}^{n} \boldsymbol{k}_s \quad \boldsymbol{M}_s = \sum_{i=1}^{n} \boldsymbol{m}_s \quad \boldsymbol{C}_s = \sum_{i=1}^{n} \boldsymbol{c}_s \tag{3.2.49}$$

Similarly, the overall matrix of the concrete support layer is assembled from a matrix of cells as follows

$$\boldsymbol{K}_c = \sum_{i=1}^{n} \boldsymbol{k}_c \quad \boldsymbol{M}_c = \sum_{i=1}^{n} \boldsymbol{m}_c \quad \boldsymbol{C}_c = \sum_{i=1}^{n} \boldsymbol{c}_c \tag{3.2.50}$$

**Fig. 3.13** Schematic diagram of discrete point supports for rails

As the track slab, concrete support layer using the unit matrix and rail unit matrix consistent, only need to replace the corresponding parameters for the track slab, concrete support layer parameters can be, here will not repeat.

The ballasted track overall support matrix is assembled from the unit support matrix, which is expressed as follows

$$
\begin{aligned}
&\boldsymbol{K}_{rs} = \sum_{i=1}^{n} \boldsymbol{k}_{rs},\ \boldsymbol{C}_{rs} = \sum_{i=1}^{n} \boldsymbol{c}_{rs},\ \boldsymbol{K}_{sc} = \sum_{i=1}^{n} \boldsymbol{k}_{sc},\ \boldsymbol{C}_{sc} = \sum_{i=1}^{n} \boldsymbol{c}_{sc},\\
&\boldsymbol{K}_{cb} = \sum_{i=1}^{n} \boldsymbol{k}_{cb},\ \boldsymbol{C}_{cb} = \sum_{i=1}^{n} \boldsymbol{c}_{cb}
\end{aligned}
\tag{3.2.51}
$$

The rails are supported at discrete points, such as Fig. 3.13 shown.

The matrix of support units is as follows

$$
\boldsymbol{k}_{rs} = \begin{bmatrix} k_{z1} & 0 & 0 & 0 \\ 0 & 0 & 0 & 0 \\ 0 & 0 & k_{z1} & 0 \\ 0 & 0 & 0 & 0 \end{bmatrix} \tag{3.2.52}
$$

Track slabs and concrete slabs are uniformly supported, e.g., the Fig. 3.14 shown.

The matrix of support units is as follows

$$
\boldsymbol{k}_{sc} = k_{z2}\frac{l}{420}\begin{bmatrix} 156 & -22l & 54 & 13l \\ -22l & 4l^2 & -13l & -3l^2 \\ 54 & -13l & 156 & 22l \\ 13l & -3l^2 & 22l & 4l^2 \end{bmatrix} \tag{3.2.53}
$$

In the above equation, $k_{z1}$ represents the support stiffness of the fasteners, and $k_{z2}$ represents the equivalent support stiffness of the CA mortar. Since the support condition of the concrete slab is identical to that of the track slab, the support stiffness of the CA mortar in the track slab's support matrix can simply be replaced with the subgrade support stiffness. The damping elements are arranged in the same manner

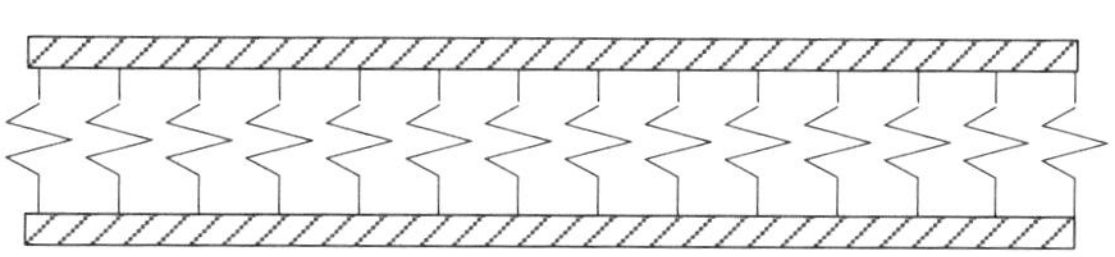

**Fig. 3.14** Schematic diagram of uniform support of track slab and concrete slab

as the elastic support elements; Therefore, the damping matrix of the support unit can be obtained by replacing $k$ with $c$.

### 3.2.3 Bridge Modeling

Simply supported bridges are considered as Euler beams per span. The equation of motion of the bridge in the vertical direction is given by

$$E_{\mathrm{b}}I_{\mathrm{b}}\frac{\partial^4 Z_{\mathrm{b}}(x,t)}{\partial x^4}+m_{\mathrm{b}}\frac{\partial^2 Z_{\mathrm{b}}(x,t)}{\partial t^2}+c_{\mathrm{b}}\frac{\partial Z_{\mathrm{b}}(x,t)}{\partial t} = \sum_{i=1}^{N_{\mathrm{b}}} F_{\mathrm{sb}i}(t)\delta(x-x_{\mathrm{b}i}) \tag{3.2.54}$$

In the formula, the $E_b$ is the Young's modulus of elasticity of the bridge, the $I_b$ is the bridge moment of inertia, and $m_b$ is the distributed mass per linear meter of the bridge.$c_b$ is the distributed damping per linear meter for the bridge.

Similar to the treatment of the rail motion equation, a second-order ordinary differential equation system for the bridge modal coordinates $q_{bk}(t)$ can be derived as follows:

$$\ddot{q}_{\mathrm{b}k}(t)+2\xi_{\mathrm{b}k}\omega_{\mathrm{b}k}\dot{q}_{\mathrm{b}k}(t)+\omega_{\mathrm{b}k}^2 q_k(t)=\sum_{i=1}^{N_{\mathrm{b}}} F_{\mathrm{sb}i}(t)R_k(x_{\mathrm{b}i}) \tag{3.2.55}$$

Among them

$$Z_{\mathrm{b}}(x,t)=\sum_{k=1}^{\mathrm{NMB}} R_k(x)q_{\mathrm{b}k}(t) \tag{3.2.56}$$

$$R_k(x)=\sqrt{\frac{2}{m_{\mathrm{b}}L_{\mathrm{b}}}}\sin\left(\frac{k\pi x}{L_{\mathrm{b}}}\right) \tag{3.2.57}$$

$$\omega_{\mathrm{b}k}=(k^2\pi^2/L_{\mathrm{b}}^2)\sqrt{E_{\mathrm{b}}I_{\mathrm{b}}/m_{\mathrm{b}}} \tag{3.2.58}$$

$$c_{\mathrm{b}}=2\xi_{\mathrm{b}k}m_{\mathrm{b}}\omega_{\mathrm{b}k} \tag{3.2.59}$$

In the formula, the $q_{bk}(t)$ are the coordinates of the regular vibration pattern of the beam, $R_k(x)$ is the vertical mode function of the beam, $L_b$ is the length of the beam, $\omega_{bk}$ is the frequency of each vibration mode, $\xi_{bk}$ is the damping coefficient for each vibration mode.

### 3.2.4 Wheel-Rail Forces

The wheel-rail forces can be obtained by linearizing the Hertzian elastic contact theory.

$$P_{wj} = \begin{cases} K_c\left[Z_{wj}(t) - Z_r(x_{wj}, t)\right] & \text{if } \left(Z_{wj}(t) - Z_r(x_{wj}, t)\right) > 0 \\ 0 & \text{if } \left(Z_{wj}(t) - Z_r(x_{wj}, t)\right) \leq 0 \end{cases} \tag{3.2.60}$$

In the formula, the

$$K_c = \frac{3}{2G}\left(\frac{W}{2}\right)^{(1/3)} \tag{3.2.61}$$

where $W$ is the axle weight of the vehicle and $G$ is the wheel-rail contact constant.

For tapered tread wheels

$$G = 4.57R^{-0.149} \times 10^{-8} \tag{3.2.62}$$

For abrasive tread wheels

$$G = 3.86R^{-0.115} \times 10^{-8} \tag{3.2.63}$$

where $R$ is the wheel radius (m).

### 3.2.5 Model Validation

The rail, track slab and bridge equations can be transformed into bivalent ordinary differential equations by the mode superposition method. Finally, the equations can be combined to obtain the time-varying linear second order ordinary differential equations. The central difference method or Newmark—β and other direct integration methods can be used to solve the equations.

1. **Functional Validation of Dynamic Analysis of Track Structures**

The dynamic analysis of the track structure is verified by comparison with the example in the paper "Vibration Characteristics of Floating Slab Track" [70] by Chen-Ming Kuo et al. The model is schematically shown in Fig. 3.15.The computational parameters were taken from the literature, see Table 3.2.

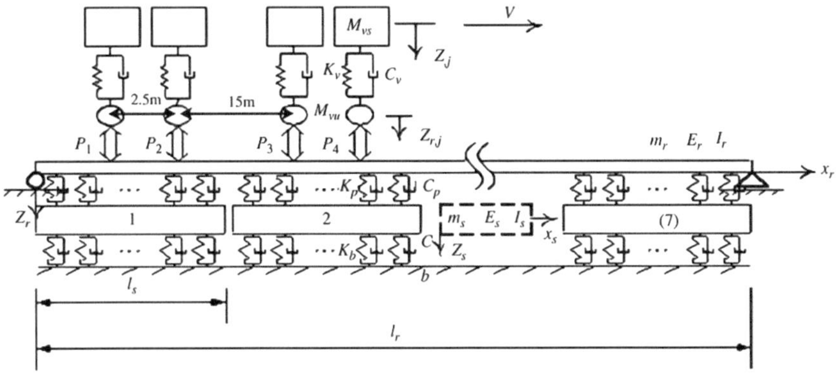


**Fig. 3.15** Simplified model for vehicle-track calculations [70]

## Bogie

**Table 3.2** Track and vehicle parameters [70]

| Project | | Unit | Parameter values |
|---|---|---|---|
| Rails | Mass of rail $m_r$ | kg/m | 60.34 |
| | Young's modulus of elasticity of rails $E_r$ | N/m$^2$ | $2.0 \times 10^{11}$ |
| | Rail moment of inertia $I_r$ | m4 | $3.05 \times 10^{-5}$ |
| | Sleeper spacing $H_p$ | m | 0.65 |
| | Fastener stiffness $K_p$ | MN/m | 100 |
| | Fastener damping $C_p$ | Ns/m | 30,000 |
| Track slab | Track slab mass $M_s$ | kg | 1640 |
| | Young's modulus of elasticity $E_r$ for track slabs | N/m$^2$ | $3.5 \times 10^{10}$ |
| | Track slab moment of inertia $I_s$ | m4 | $1.89 \times 10^{-4}$ |
| | Length of track slab $l_s$ | m | 1.25 |
| | Track slab support spacing $H_b$ | m | 0.65 |
| | Track slab support stiffness $K_b$ | MN/m | 5.8 |
| | Damping for track slab support $C_b$ | Ns/m | 35,000 |
| Vehicles | Body mass $W_{car}$ | kg | 41,750 |
| | Secondary suspension stiffness $K_2$ | N/m | $2.65 \times 10^5$ |
| | Secondary suspension Damping $C_2$ | Ns/m | $4.51 \times 10^4$ |
| | Bogie mass $M_b$ | kg | 3040 |
| | Primary suspension stiffness $K_1$ | N/m | $1.18 \times 10^6$ |
| | Primary suspension Damping $C_1$ | Ns/m | $3.92 \times 10^4$ |
| | Mass of wheelset $m_w$ | kg | 890 |

Figures 3.16 and 3.17 present the comparison of displacement and acceleration calculation results for the rail and track plate. In the calculations of this program, the running speed is taken as 22 m/s, consistent with the value in reference [70], and the track length is 80 m. Figure 3.16a shows the displacement time history curves for the rail and track plate at the midpoint of the track calculated by this program, with maximum values of 3.5 mm and 3.1 mm, respectively. Figure 3.16b presents the displacement time history curves for the rail and track plate from reference [70]. Figure 3.16c shows the maximum displacement and acceleration of the rail and track plate at the midpoint of the track for different track lengths as given in reference [70]. It can be observed that the track vibration displacement stabilizes when the track length reaches 60 track slabs (75 m), with the maximum displacement of the rail and track plate converging between 3.2 mm and 3.7 mm. The results calculated by this program are in good agreement with the curves and peak values presented in reference [70].

Figure 3.17a shows the acceleration time history curves for the rail and track plate calculated by this program, with maximum values of 0.66 $m/s^2$ and 0.57 $m/s^2$, respectively. Figure 3.17b presents the maximum acceleration values for the rail and track plate at different track lengths as given in reference [70]. It can be observed that the maximum acceleration for the rail fluctuates around 0.6 $m/s^2$, and for the track plate, it fluctuates around 0.55 $m/s^2$. Similarly, the comparison of the track structure vibration acceleration further confirms the consistency of the two calculation structures.

From the comparison of the track structure vibrations, it is evident that both the vibration displacement and acceleration for the rail and track plate calculated by this program match well with the results from reference [70].

2. **Functional Validation of Dynamic Analysis of Bridge Structures**

The structural dynamic analysis of the bridge is verified by comparison with the examples in the paper "Vehicle-Bridge Interaction Element for Dynamic Analysis" [71] by Yeong-Bin Yang and Jong-Dar Yau. An example of vibration analysis of a simply supported beam with suspended mass acting on it is given in the literature [71]. The model is shown in Fig. 3.18, for calculation parameters, see Table 3.3.

Figures 3.19 and 3.20 present the comparison of the beam mid-span vibration displacement and acceleration. In the calculation model of this program, the suspended mass is simplified as a moving force, and the bridge vibration modes are taken as the 1st, 5th, and 10th modes. From the comparison between the calculation results of this program and those in reference [71], it can be observed that the vibration displacement and acceleration curves, as well as their peak values, match well.

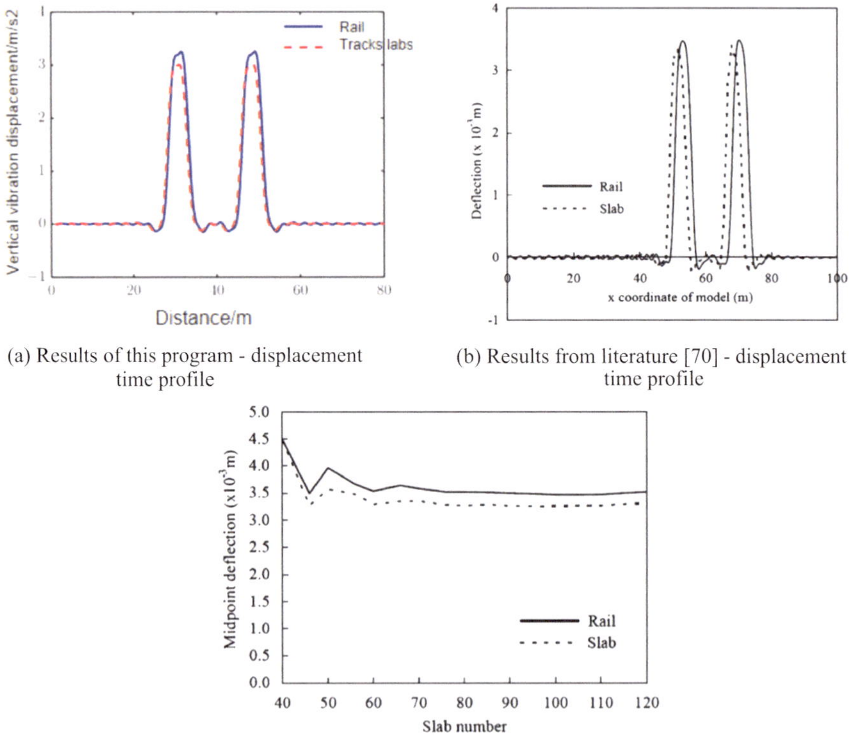


(a) Results of this program - displacement time profile

(b) Results from literature [70] - displacement time profile

(c) Results from the literature [70] - displacement versus track length curves

**Fig. 3.16** Comparison of vibration displacements of rails and track slabs

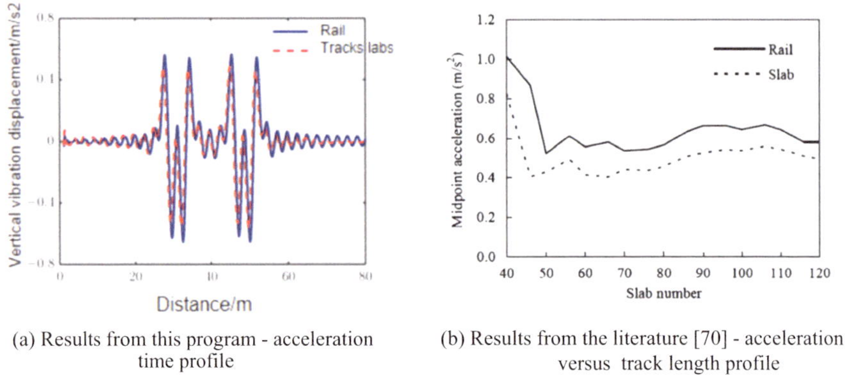


(a) Results from this program - acceleration time profile

(b) Results from the literature [70] - acceleration versus track length profile

**Fig. 3.17** Comparison of vibration acceleration of rails and track slabs

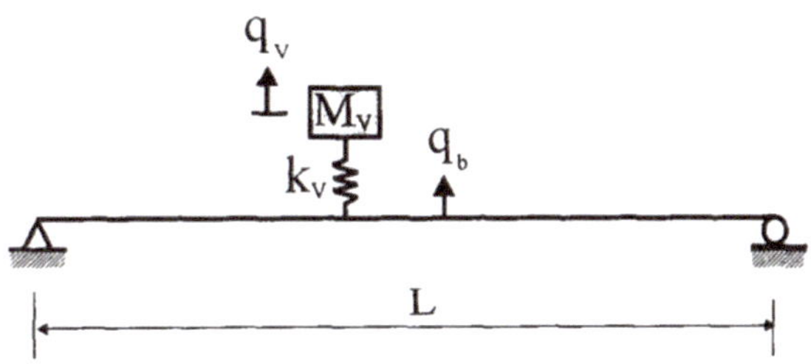


**Fig. 3.18** Simplified model for vehicle-bridge calculations [71]

**Table 3.3** Simply supported beams and vehicle parameters [71]

| Project | Unit | Parameter values |
|---|---|---|
| Beam length L | m | 25 |
| Young's modulus of elasticity E | N/m$^2$ | $2.89 \times 10^9$ |
| Poisson's ratio σ | | 0.2 |
| Moment of inertia I | m4 | 2.9 |
| Mass m per linear meter | kg/m | 2303 |
| Suspension stiffness k v | kN/m | 1595 |
| Suspended mass M v | kg | 5750 |
| Travel speed V | km/h | 100 |

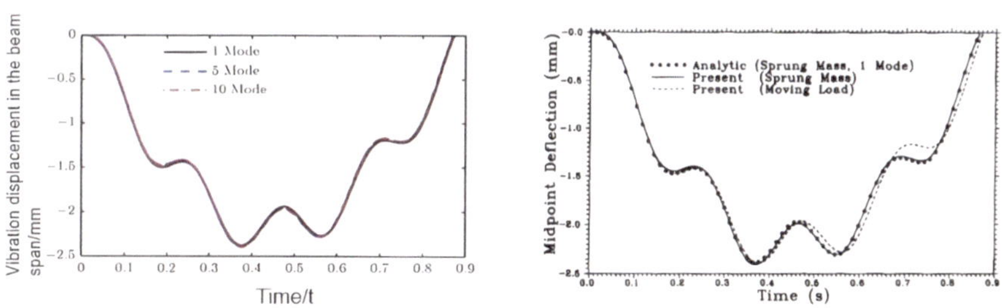


(a) Results calculated by this program

(b) Results in the literature [71]

**Fig. 3.19** Comparison of vibration displacements in beam spans

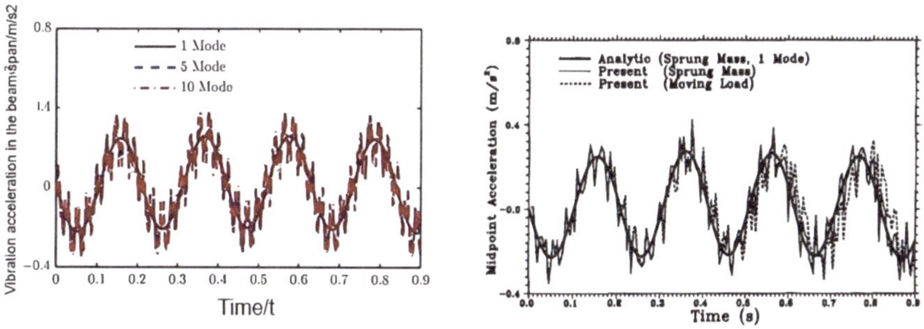


(a) Results calculated by this program

(b) Results in the literature [71]

**Fig. 3.20** Comparison of beam mid-span vibration acceleration

## 3.3 A Spatial Dynamic Analysis Model for Vehicle-Track Systems

### 3.3.1 Track Modeling

1. **High-Speed Railroad Track Structures**

High-speed railroad track type is divided into ballasted track and ballastless track, and the selection of track type is influenced by main technical standards, engineering geological conditions and other aspects.

Ballasted track is bulk structure, has the advantages of short repair time when damaged, high efficiency of automation and mechanized maintenance, simple adjustment of track height and geometric state, etc. When the line is affected by the deformation of traffic settlement, it can be adjusted at the right time according to the actual situation, in order to meet the requirements of railroad operation, and it is the preferred type of track for the bad geological areas such as active fracture zones.

Ballastless track by its own structural limitations, can not be like ballast track, dialing operation, its adjustment capacity is basically the adjustment amount of the fastener system, therefore, in the line level, height and direction of the adjustment amount is very limited to the settlement deformation of the lower foundation of the requirements of the higher. When settlement deformation exceeding its adjustment amount occurs, the track structure will not be able to be adjusted, and other measures must be taken, such as grouting, top beam, slope adjustment, speed limitation, or re-construction of ballastless track and other measures, in order to meet the requirements of safe operation of high-speed railroads.

At present, China's high-speed railroads with speeds of 250 km/h and below are mainly ballasted, while high-speed railroads with speeds above 250 km/h are mainly ballasted.

Ballastless track mainly consists of rails, elastic fasteners, track slabs, concrete base (or support layer) and so on. Compared with ballasted track, ballastless track has good structural stability, easy to maintain track geometry, uniform stiffness, good structural continuity, low structural height, low maintenance workload, low cost and long service cycle. At present, the main ballastless track structure types adopted in China are CRTS I plate type and double block type, CRTS II plate type and CRTS III plate type, and their main technical characteristics are shown in Table 3.4.

Cross section of CRTS III slab track track for roadbed, bridge and tunnel section is shown in Fig. 3.21.

2. **Types of Track Models**

The modeling of the track can be divided into the lumped parameter track model and the distributed parameter track model. The lumped-parameter rail model equates the role of the rail as the last element of the vehicle suspension. It is assumed that the deformation of the rail under the wheels is hindered by a linear spring and a viscous damping parallel to it. This model is characterized by simple model and convenient

**Table 3.4** Comparison table of technical characteristics of different forms of ballastless track technology

| Category | Prefabricated | | | Cast-in-place subgrade type |
|---|---|---|---|---|
| | Unit structure | | Longitudinally connected structures | The subgrade, tunnel sections longitudinally connected; Bridge superstructure units |
| | CRTS I type plate | CRTS III plate type | CRTS II plate type | CRTS I type double block type |
| The main structural components | Prefabricated track slabs<br>CA mortar adjustment layer<br>Concrete base and convex retaining platform | Prefabricated track slabs<br>Self-compacting concrete adjustment layer<br>Concrete base | Prefabricated track slabs<br>Mortar adjustment layer<br>Supporting layer or concrete base | Prefabricated double-block sleepers<br>Concrete subgrade slabs<br>Supporting layer or concrete base |

calculation, but the equivalent parameters are difficult to be determined accurately, and the scope of application has great limitations, which is suitable for qualitative analysis of wheel-rail system dynamics.

The distributed parameter track model can be divided into continuous elastic foundation beam model and continuous elastic discrete point support beam model. The former takes the under-rail foundation as a uniformly distributed whole foundation; while the latter describes the under-rail structure as a series of discrete elastic-damping point support systems separated by sleeper spacing. The former focuses on reflecting the most basic characteristics of the track system, the overall effect; while the latter can further describe the local effect of each sleeper support point, which can more objectively reflect the actual structural characteristics of the track system.

(1) Rigid track model

The track structure is regarded as a rigid foundation, i.e., the influence of the track structure on the dynamic response of the vehicle is ignored. The model can be used to calculate the ideal conditions, line geometry conditions (including line geometry and line smoothness) on the vehicle dynamic response of the analysis, applicable to the dynamic evaluation of the line design program and the study of line parameters.

(2) Lumped-parameter track model

Equate the rail action as the last element of the vehicle suspension. It is assumed that the rail deformation under the wheels is hindered by a linear spring and a viscous damping parallel to it. The model can consider the effect of the overall transverse vertical stiffness of the rail on the dynamic performance of the traveling vehicle.

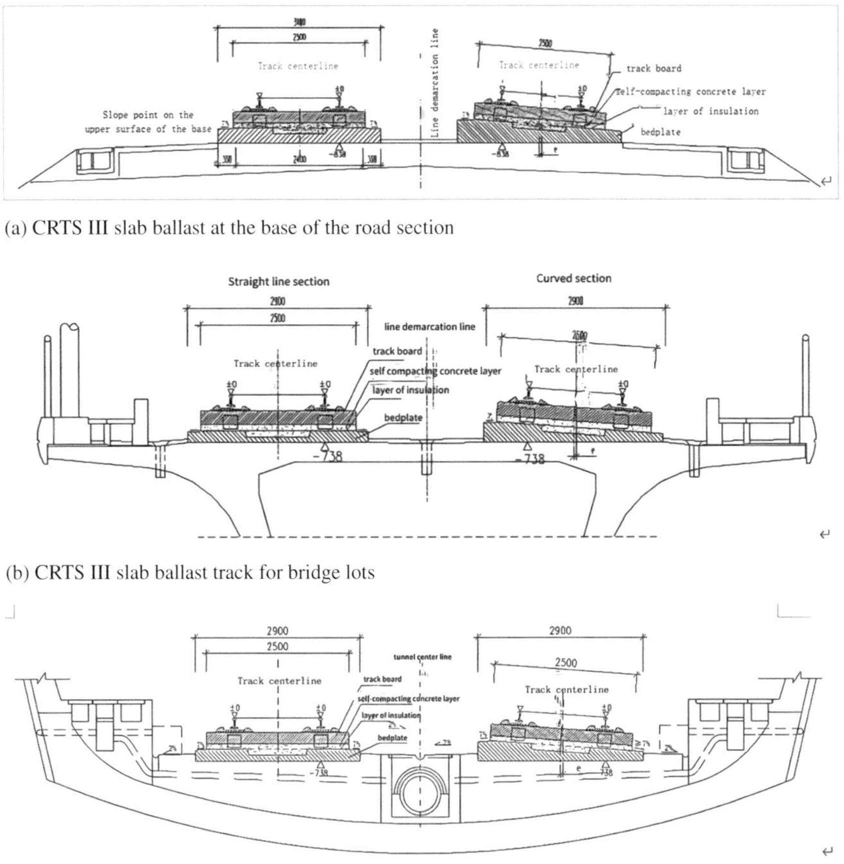


(a) CRTS III slab ballast at the base of the road section

(b) CRTS III slab ballast track for bridge lots

(c) CRTS III slab ballast track in the tunnel section

**Fig. 3.21** CRTS III type slab track section drawing

(3) Single-layer track model

The model treats the rails as Euler beams on continuous elastic discrete point supports in both transverse and vertical directions, and takes into account the effects of rail vibration, under-rail stiffness, and rail transverse restraining stiffness, see Fig. 3.22. The model can be used for the buried track structure of sleeper. The model can be used for the buried rail structure, and other rail structures can also be approximated using the model. The model is applicable to most of the conditions of the line system dynamic analysis problems, including the overall rigidity of the track and the line geometry under the combined influence of the safety and comfort of travel.

(4) Three-layer ballasted track modeling

The model is designed for the use of conventional ballasted track structures, which can take into account the parametric vibrations of rails, sleepers and roadbeds, as well

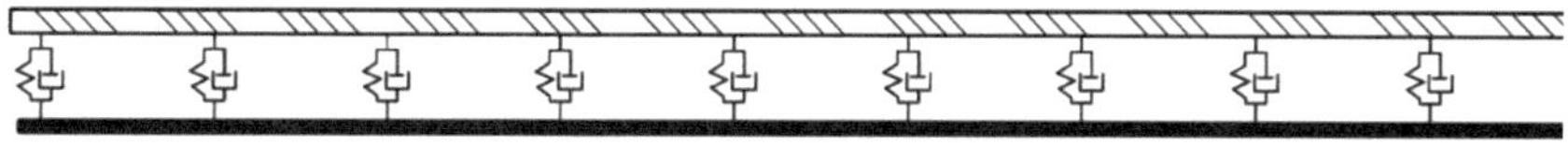

(a) Side view of the single-level track model

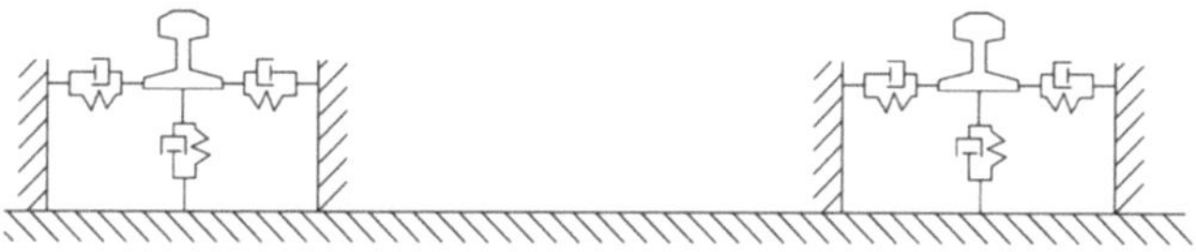

(b) Front view of the single-level track pattern

**Fig. 3.22** Single level track modeling

as the effects of physical properties of rails, quality of sleepers, fastener stiffness, subgrade characteristics and subgrade stiffness, see Fig. 3.23.

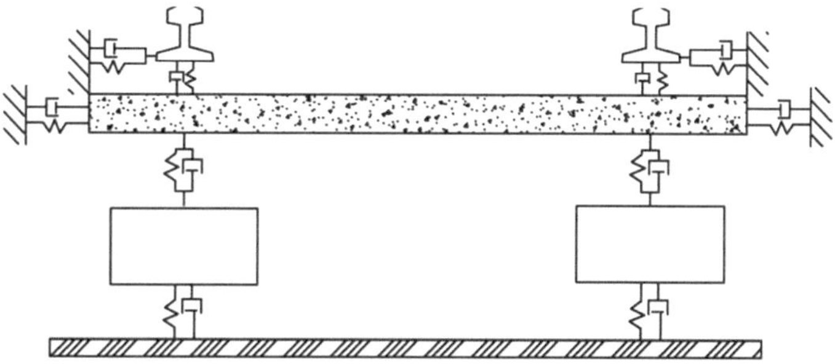

(a) Side view of the three-level ballasted track model

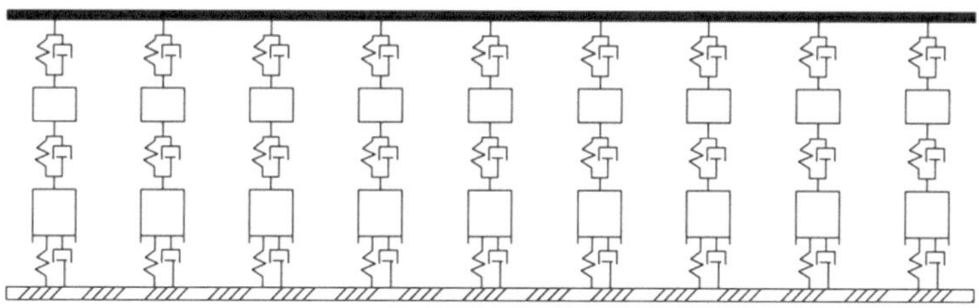

(b) Front view of the three-level ballasted track model

**Fig. 3.23** A model of a three-level ballasted track

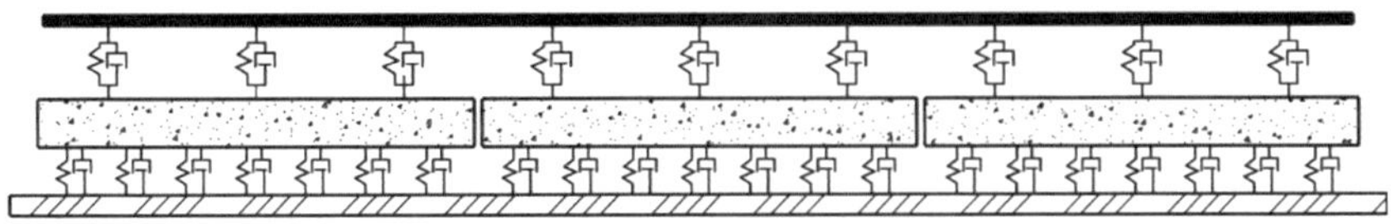

(a) Side view of the slab track model

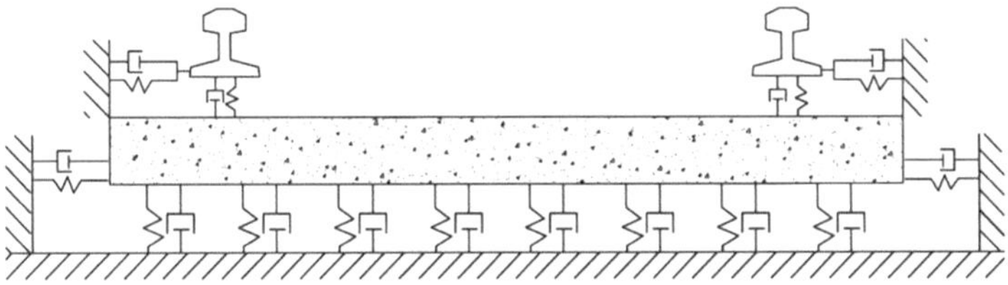

(b) Front view of the slab-on-rail model

**Fig. 3.24** Slab track modeling

(5) Slab track modeling

The vibration of the slab track is mainly reflected in the vibration of the rails and the track slab. The model will be the left and right rails are regarded as discrete elastic points supporting the Euler beam on the foundation. Track slab due to the role of the lower mortar, and the concrete foundation is a face contact state, and the track slab length and width dimensions are much larger than its thickness, and thus in the vertical direction will be track slab according to the elastic foundation on the rectangular elastic sheet of equal thickness; And transverse is considered as a rigid body movement, see Fig. 3.24.

The model has many degrees of freedom, large computational volume, and is suitable for the comparative analysis of the effects of track structure parameters and line parameters on the dynamic response of the system, or the dynamic analysis of a track structure type under specific line conditions.

## 3. **The Dynamic Equations of the Track Structure**

(1) Vibration equations for rails modeled as lumped-parameter tracks

The transverse vertical equations for the left and right rails are

$$\begin{aligned} m_r\ddot{y}_{rL} + c_{ry}\dot{y}_{rL} + k_{ry}y_{rL} &= -F_{Ly} - N_{Ly} \\ m_r\ddot{y}_{rR} + c_{ry}\dot{y}_{rR} + k_{ry}y_{rR} &= -F_{Ry} - N_{Ry} \\ m_r\ddot{z}_{rL} + c_{rz}\dot{z}_{rL} + k_{rz}z_{rL} &= F_{Lz} + N_{Lz} \\ m_r\ddot{z}_{rR} + c_{rz}\dot{z}_{rR} + k_{rz}z_{rR} &= F_{Rz} + N_{Rz} \end{aligned} \tag{3.3.1}$$

When the lumped-parameter model is used, the effects of rail inertia and rail damping are usually ignored, and only the overall rail stiffness is considered.

(2) Vibration equations for rails

Rail vertical and transverse can be regarded as a discrete point support on the Euler beam or Timoshenko beam, its transverse, vertical left and right two strands of the vibration equation of the rail have the same form, the vibration equation is shown in Eqs. (3.2.12) and (3.2.13).

(3) The vibration equation of the track slab

Vehicle load is transferred to the track slab through the left and right fastener nodes. Because the thickness of the track slab is much smaller than the length and width of the track slab, the track slab can be regarded as an elastic thin plate in the vertical direction, and the transverse bending stiffness of the track slab is very large, which can be simplified as a rigid body.

$$\begin{aligned}&\frac{\partial^4 w(x,y,t)}{\partial x^4}+2\frac{\partial^4 w(x,y,t)}{\partial x^2\partial y^2}+\frac{\partial^4 w_{\mathrm{s}}(x,y,t)}{\partial y^4}+\frac{K_{\mathrm{s}}}{D_{\mathrm{s}}}w(x,y,t)\\&+\frac{C_{\mathrm{s}}}{D_{\mathrm{s}}}\frac{\partial w(x,y,t)}{\partial t}+\frac{\rho_{\mathrm{s}}h_{\mathrm{s}}}{D_{\mathrm{s}}}\frac{\partial^2 w(x,y,t)}{\partial t^2}=\frac{1}{D_{\mathrm{s}}}\sum_{i=1}^{N_{\mathrm{P}}}P_{\mathrm{rz}i}(t)\delta(x-x_{\mathrm{P}i})\delta(y-y_{\mathrm{P}i})\end{aligned} \tag{3.3.2}$$

$N_p$ represents the total number of fastener nodes on both rails of the track slab; $P_{rzi}$ is the vertical force at the i-th rail fastener node on the track slab; $w(x, y, t)$ is the vertical deflection of the track slab; $x_{pi}$ and $y_{pi}$ denote the positions of the iii-th rail fastener node on the track slab; $h_s$ is the thickness of the track slab; $\rho_s$ is the mass density of the track slab; $C_s$ is the damping coefficient of the track slab; $K_s$ is the stiffness coefficient of the foundation; $E_s$ is the Young's modulus of the track slab; $D_s$ is the bending stiffness of the track slab; $v_s$ is the Poisson's ratio of the track slab.

$$D_{\mathrm{s}}=\frac{E_{\mathrm{s}}h_{\mathrm{s}}^3}{12\left(1-\upsilon_{\mathrm{s}}^2\right)} \tag{3.3.3}$$

For the four sides of the rectangular plate for the simply supported, fixed, free, elastic boundary of one of the general cases, the use of the two-way beam function combination level approximation method can achieve more satisfactory results, most of the rectangular plate vibration analysis in the project can be used in this method. Therefore, the $(i, j)$ order vibration mode of the track slab can be expressed as:

$$W_{ij}(x,y)=\sum_{m=1}^{N_{\mathrm{x}}}\sum_{n=1}^{N_{\mathrm{y}}}A_{mn}^{ij}X_m(x,y)Y_n(x,y) \tag{3.3.4}$$

In the formula, the $A_{mn}^{ij}$ are coefficients, which are used to adjust the combination of beam functions of different orders to approximate the real solution of the track slab vibration pattern.

$N_x, N_y$ are the cutoff modal orders in the length and width directions of the track slab, respectively; the $X_m(x, y)$, $Y_n(x, y)$ are the beam vibration pattern functions in the length and width directions of the track slab, respectively.

The beam vibration mode function in the length direction of the track slab is given by

$$\begin{cases} X_1 = 1 \\ X_2 = \sqrt{3}\left(1 - \dfrac{2x}{L_s}\right) \\ X_m = \cosh(\alpha_m x) + \cos(\alpha_m x) - \beta_m(\sinh(\alpha_m x) + \sin(\alpha_m x)) \quad (m > 2) \end{cases} \tag{3.3.5}$$

Frequency coefficients $\alpha$ for:

$$\begin{cases} \alpha_1 L_s = 0 \\ \alpha_2 L_s = 0 \\ \alpha_3 L_s = 4.73002 \\ \alpha_m L_s = \dfrac{2m-3}{2}\pi \quad (m \geq 4) \end{cases} \tag{3.3.6}$$

Vibration coefficients $\beta$ for:

$$\begin{cases} \beta_3 = 0.982502 \\ \beta_m = \dfrac{\cosh(\alpha_m L_s) - \cos(\alpha_m L_s)}{\sinh(\alpha_m L_s) - \sin(\alpha_m L_s)} \quad (m \geq 4) \end{cases} \tag{3.3.7}$$

The track slab is a rectangular plate with four free edges, and its boundary conditions are asymmetrical about the diagonal. The vibration mode is predominantly characterized by the $X_m Y_n$ term. Therefore, the expression for the vibration mode of the track slab can be simplified as:

$$W_{mn}(x, y) = X_m(x) Y_n(y) \tag{3.3.8}$$

Equation (3.3.2) constitutes a fourth-order partial differential equation governing the deflection surface function $w(x, y, t)$. By applying the separation of variables method, the solution for track slab deflection is postulated as:

$$w(x, y, t) = \sum_{m=1}^{N_x} \sum_{n=1}^{N_y} X_m(x) Y_n(y) T_{mn}(t) \tag{3.3.9}$$

In the formula, the $T_{mn}(t)$ are the canonical coordinates of the track slab.
Substituting Eq. (3.3.9) into Eq. (3.3.2), we derive the following expression:

$$
\begin{aligned}
&\sum_{m=1}^{N_x}\sum_{m=1}^{N_y} X_m^{(4)}(x)Y_n(y)T_{mn}(t) + 2\sum_{m=1}^{N_x}\sum_{m=1}^{N_y} X_m''(x)Y_n''(y)T_{mn}(t) \\
&+ \sum_{m=1}^{N_x}\sum_{m=1}^{N_y} X_m(x)Y_n^{(4)}(y)T_{mn}(t) + \frac{K_s}{D_s}\sum_{m=1}^{N_x}\sum_{n=1}^{N_y} X_m(x)Y_n(y)T_{mn}(t) \\
&+ \frac{C_s}{D_s}\sum_{m=1}^{N_x}\sum_{n=1}^{N_y} X_m(x)Y_n(y)\dot{T}_{mn}(t) + \frac{\rho_s h_s}{D_s}\sum_{m=1}^{N_x}\sum_{n=1}^{N_y} X_m(x)Y_n(y)\ddot{T}_{mn}(t) \\
&= \frac{1}{D_s}\sum_{i=1}^{N_P} P_{rzi}(t)\delta(x - x_{Pi})\delta(y - y_{Pi})
\end{aligned}
\tag{3.3.10}
$$

In Eq. (3.3.10), both sides are multiplied by the vibration pattern function of the track slab, and the area of the plate is divided. Using the orthogonality of the vibration pattern function of the beam, we can get the second-order ordinary differential equations of the track slab about the canonical coordinates

$$
\begin{aligned}
&\ddot{T}_{mn}(t) + \frac{C_s}{\rho_s h_s}\dot{T}_{mn}(t) + \left(\frac{K_s}{\rho_s h_s} + \frac{D_s}{\rho_s h_s}\frac{B_2B_3 + 2B_4B_5 + B_1B_6}{B_1B_2}\right)T_{mn}(t) \\
&= \frac{1}{\rho_s h_s B_1 B_2}\sum_{i=1}^{N_P} P_{rzi}(t)X_m(x_{Pi})Y_n(y_{Pi})
\end{aligned}
\tag{3.3.11}
$$

In the formula, the $m, n$ is a positive integer, $m = 1, 2, \ldots, N_x, n = 1, 2, \ldots N_y$;

$$
\begin{cases}
B_1 = \int_0^{L_s} X_m^2(x)dx \\
B_2 = \int_0^{W_s} Y_n^2(y)dy \\
B_3 = \int_0^{L_s} X_m^{(4)}(x)X_m(x)dx \\
B_4 = \int_0^{L_s} X_m''(x)X_m(x)dx \\
B_5 = \int_0^{W_s} Y_n''(y)Y_n(y)dy \\
B_6 = \int_0^{W_s} Y_n^{(4)}(y)Y_n(y)dy
\end{cases}
\tag{3.3.12}
$$

In the formula, the $L_s$ is the length of the track slab; The $W_s$ is the width of the track slab.

Lateral movement of the track slab:

$$\rho_{\mathrm{s}} L_{\mathrm{s}} W_{\mathrm{s}} h_{\mathrm{s}} \ddot{y}_{\mathrm{s}} = \sum_{i=1}^{N_{\mathrm{P}}} P_{\mathrm{ry}i} - F_{\mathrm{sy}} \tag{3.3.13}$$

$P_{\mathrm{ry}i}$ denotes the lateral force at the i-th rail fastening node on the track slab. $F_{\mathrm{sy}}$ is the lateral reaction force on the side of the lower bearing surface of the track slab.

4. **Slab Track Finite Element Model**

The slab track comprises rails, rubber mats, fasteners, reaction plates, track boards, CA mortar or self-compacting concrete filler layer, and a concrete foundation. The rails are positioned on the track slab by the fasteners with equal spacing, and the length of the track slab is fixed, which is continuously laid along the line direction. Between the track slab and the concrete foundation is filled with special mortar, which is in face contact, and the concrete foundation is connected with the lower structure rigidly and incorporated into the lower structure.

The vibration pattern of slab track is mainly manifested as the deformation of the rail and track slab in vertical and lateral directions. The rail is regarded as an infinite beam supported by continuous elastic points in both vertical and transverse directions, which is consistent with the model of rail in bolster-embedded rail. Track slab due to the role of the lower mortar, and concrete foundation is a surface contact state. The length and width of the track slab are much larger than its thickness, so the vertical direction will be simplified as an elastic foundation on the elastic thin plate, and its deformation is mainly reflected in the vertical displacement, bending, and warping. In the transverse direction, the deformation of the track slab is mainly the bending of the plate, which is regarded as a continuously supported equal cross-section beam.

In the following, the rail and track slab are modeled and analyzed separately, and the equations are established.

(1) The equations of motion of the rails

Since the cross-section of the rail is symmetric about the vertical axis, the lateral and vertical deformations of the rail can be considered independent of each other under the external load, and they can be considered as beam units. Each node has 4 degrees of freedom and each unit has 8 degrees of freedom. The $\boldsymbol{u}_r^e$ can be expressed as:

$$\boldsymbol{u}_r^e = \left[Z_{ri}, Y_{ri}, \beta_{yi}, \beta_{zi}, Z_{rj}, Y_{rj}, \beta_{yj}, \beta_{zj}\right]^{\mathrm{T}} \tag{3.3.14}$$

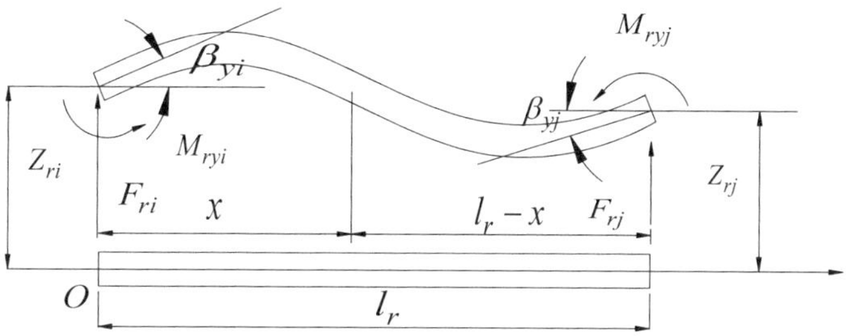


**Fig. 3.25** Beam unit deformation

The unit node equivalent load $F_r^e$ is expressed as:

$$F_r^e = \left[F_{ri},\ F_{ryi},\ M_{ryi},\ M_{rzi},\ F_{rj},\ F_{ryj},\ M_{ryj},\ M_{rzj}\right]^{\mathrm{T}} \tag{3.3.15}$$

Among them, $Z_r$ denotes vertical displacement; $Y_r$ denotes lateral displacement; $\beta_y$ denotes the rotation about the y-axis; $\beta_z$ denotes the rotation about the z-axis; $F_r$ denotes the vertical force acting on the node; $F_{ry}$ denotes the lateral force acting on the node; $M_{ry}$ denotes the bending moment about the y-axis acting on the node; $M_{rz}$ denotes the bending moment about the z-axis acting on the node; $i, j$ denotes the node number; and $l_r$ represents the length of the rail element. Taking the vertical element as an example, its element displacement and element load are illustrated in Fig. 3.25.

① Shaped function

$N_i (i = 1, 2, 3, 4)$ is a shape function which, in the beam unit, represents the shape of a beam bent into when a beam fixed at both ends produces only one unit displacement.

$$\begin{aligned} N_1 &= 1 - \frac{3}{l_r^2}x^2 + \frac{2}{l_r^3}x^3,\ N_2 = -x + \frac{2}{l_r}x^2 - \frac{1}{l_r^2}x^3, \\ N_3 &= \frac{3}{l_r^2}x^2 - \frac{2}{l_r^3}x^3,\ N_4 = \frac{1}{l_r}x^2 - \frac{1}{l_r^2}x^3 \end{aligned} \tag{3.3.16}$$

The displacement of any point on the beam is $(Z_{ri}, Y_{ri})$

$$\begin{Bmatrix} Z_{ri} \\ Y_{ri} \end{Bmatrix} = \begin{bmatrix} N_1 & 0 & N_2 & 0 & N_3 & 0 & N_4 & 0 \\ 0 & N_1 & 0 & N_2 & 0 & N_3 & 0 & N_4 \end{bmatrix} \{u_r\}^e \tag{3.3.17}$$

② Unit quality matrix

$$m_r^e = \int_0^{l_r} m_r N^{\mathrm{T}} N dx \tag{3.3.18}$$

The mass matrix of the rail unit is obtained by integrating.

$$\boldsymbol{m}_r^e = \frac{m_r l_r}{420}\begin{bmatrix} 156 & 0 & 22l_r & 0 & 54 & 0 & -13l_r & 0 \\ 0 & 156 & 0 & 22l_r & 0 & 54 & 0 & -13l_r \\ 22l_r & 0 & 4l_r^2 & 0 & 13l_r & 0 & -3l_r^2 & 0 \\ 0 & 22l_r & 0 & 4l_r^2 & 0 & 13l_r & 0 & -3l_r^2 \\ 54 & 0 & 13l_r & 0 & 156 & 0 & -22l_r & 0 \\ 0 & 54 & 0 & 13l_r & 0 & 156 & 0 & -22l_r \\ -13l_r & 0 & -3l_r^2 & 0 & -22l_r & 0 & 4l_r^2 & 0 \\ 0 & -13l_r & 0 & -3l_r^2 & 0 & -22l_r & 0 & 4l_r^2 \end{bmatrix}_{8\times 8} \tag{3.3.19}$$

Among them $m_r$ is the mass per unit length of rail (kg/m).

③ Unit stiffness matrix

$$\begin{aligned} \boldsymbol{k}_r^e = & \int_0^{l_r} EI_y\begin{bmatrix} \ddot{N}_1 & 0 & \ddot{N}_2 & 0 & \ddot{N}_3 & 0 & \ddot{N}_4 & 0 \end{bmatrix}^T\begin{bmatrix} \ddot{N}_1 & 0 & \ddot{N}_2 & 0 & \ddot{N}_3 & 0 & \ddot{N}_4 & 0 \end{bmatrix}dx \\ & + \int_0^{l_r} EI_z[0 \ \ \ddot{N}_1 \ \ 0 \ \ \ddot{N}_2 \ \ 0 \ \ \ddot{N}_3 \ \ 0 \ \ \ddot{N}_4]^T[0 \ \ \ddot{N}_1 \ \ 0 \ \ \ddot{N}_2 \ \ 0 \ \ \ddot{N}_3 \ \ 0 \ \ \ddot{N}_4]dx \end{aligned} \tag{3.3.20}$$

After integration the rail cell stiffness matrix is obtained as

$$\boldsymbol{k}_r^e = \frac{E}{l_r^3}\begin{bmatrix} 12I_y & 0 & 6l_rI_y & 0 & -12I_y & 0 & 6l_rI_y & 0 \\ 0 & 12I_z & 0 & 6l_rI_z & 0 & -12I_z & 0 & 6l_rI_z \\ 6l_rI_y & 0 & 4l_r^2I_y & 0 & -6l_rI_y & 0 & 2l_r^2I_y & 0 \\ 0 & 6l_rI_z & 0 & 4l_r^2I_z & 0 & -6l_rI_z & 0 & 2l_r^2I_z \\ -12I_y & 0 & -6l_rI_y & 0 & 12I_y & 0 & -6l_rI_y & 0 \\ 0 & -12I_z & 0 & -6l_rI_z & 0 & 12I_z & 0 & -6l_rI_z \\ 6l_rI_y & 0 & 2l_r^2I_y & 0 & -6l_rI_y & 0 & 4l_r^2I_y & 0 \\ 0 & 6l_rI_z & 0 & 2l_r^2I_z & 0 & -6l_rI_z & 0 & 4l_r^2I_z \end{bmatrix}_{8\times 8} \tag{3.3.21}$$

Among them $I_y$, $I_z$ are the moment of inertia of the rail section on the y-axis and z-axis, respectively.

④ Unit damping matrix

In the actual use of the environment, the structural damping of the rail material will have an impact on the vibration pattern of the rail and should be considered for the attenuation of vibration. The determination of structural damping in the actual engineering is very difficult, so in the structural dynamic analysis, usually used

Rayleigh damping form. The structural damping is expressed as a linear combination of mass and stiffness.

Therefore, the damping array of the rail unit is expressed as

$$\boldsymbol{c}_r^e = \alpha \boldsymbol{m}_r^e + \beta \boldsymbol{k}_r^e \tag{3.3.22}$$

The coefficients $\alpha$ and $\beta$ take different values depending on the structural configuration, material properties, and other factors.

⑤ Positioning of the vehicle on the rails

As the vehicle travels along the track at a velocity v, the wheel-induced force on the rail moves forward accordingly. In Fig. 3.26, $ox$ denotes a fixed coordinate system attached to the rail, with its origin at the starting point of the rail, referred to as the fixed coordinate system. $o'x'$ represents a moving coordinate system attached to the vehicle, with its origin located at the position of the fourth wheelset. The transformation between the two coordinate systems is given by $x = x' + x_0 + vt$, where $t$ denotes the running time. Since the positioning principle is the same for all wheelsets, the figure takes the second wheelset as an example to illustrate the positioning method.

At time $t$, the fixed coordinate of the ith wheelset is:

$$x_{wi} = x_0 + vt + l_i \tag{3.3.23}$$

At time $t$, the fixed coordinate of the contact point between the $i$th wheelset and the rail is:

$$x_{ri} = (k_i - 1)l_r + x_i \tag{3.3.24}$$

Among them $l_i$ are the coordinates of each wheelset in the moving coordinate system. $x_0$ are the fixed coordinates of the fourth pair at the starting moment. $k_i$

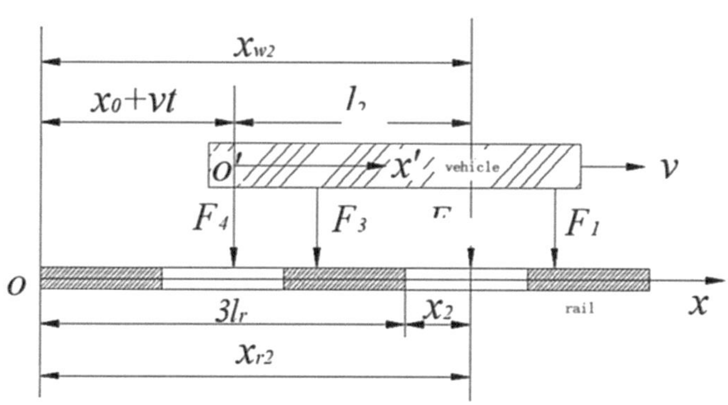


**Fig. 3.26** The coordinate relationship between the vehicle and the track

is the rails on with No $i$, the number of the rail unit in which the contact point of the wheelset is located. $x_i$ is the point of contact in para $k_i$ coordinates in a cell. $l_i$ represents the coordinate of each wheelset in the moving coordinate system; $x_0$ is the fixed coordinate of the fourth wheelset at the initial time; $k_i$ denotes the index of the rail element where the contact point between the ith wheelset and the rail is located; and $x_i$ is the local coordinate of the contact point within the $k_i$th rail element.

Since $x_{wi} = x_{ri}$, $x_i$ can be determined accordingly, and the corresponding shape function for positioning the wheel–rail contact force on the track can be derived based on Eq. (3.3.16).

⑥ Rail unit loads

The load $\boldsymbol{F}_{rk}^{e}$ of the $k$th rail element includes the wheel–rail interaction force $F_k$ and the spring-damping force $R_k$ from the substructure beneath the rail. Track irregularities are taken into account within the wheel–rail interaction force.

$$\begin{aligned}\boldsymbol{F}_{rk}^{e} &= F_k + R_k = \sum_{i=1}^{4} \delta(x_{wi} - x_{ri}) \\ &\left[N_1 F_i(t),\ N_1 F_{yi}(t),\ N_2 F_i(t),\ N_2 F_{yi}(t),\ N_3 F_i(t),\ N_3 F_{yi}(t),\ N_4 F_i(t),\ N_4 F_{yi}(t)\right]^{\mathrm{T}} \\ &+ \left[R_{z(k-1)},\ R_{y(k-1)},\ 0,\ 0,\ R_{zk},\ R_{yk},\ 0,\ 0\right]\end{aligned} \tag{3.3.25}$$

Among them. $F_i(t)$ is the vertical component of the wheel-rail force; $F_{yi}(t)$ is the transverse component of the wheel-rail force; δ is the Dirac function

$$\delta(x_{wi} - x_{ri}) = \begin{cases} 1 & x_{wi} - x_{ri} = 0 \\ 0 & \text{other} \end{cases} \tag{3.3.26}$$

When there is a rigid foundation under the rails

$$\begin{aligned} R_{z(k-1)} &= k_{rz} z_{r(k-1)} + c_{rz} \dot{z}_{r(k-1)}, \quad R_{zk} = k_{rz} z_{rk} + c_{rz} \dot{z}_{rk} \\ R_{y(k-1)} &= k_{ry} y_{r(k-1)} + c_{ry} \dot{y}_{r(k-1)}, \quad R_{yk} = k_{ry} y_{rk} + c_{ry} \dot{y}_{rk} \end{aligned} \tag{3.3.27}$$

When the rail is under the slab track, it is also necessary to consider the influence of the displacement and speed of the track slab at the corresponding position of the fastener.

⑦ the equations of motion of the rails

The mass, stiffness, damping, and force matrices of the rail elements have been derived in the preceding sections. Based on the deformation compatibility conditions and static equilibrium conditions, these element matrices are assembled into the global mass, stiffness, damping, and force matrices of the rail according to the 'degree-of-freedom matching' principle in structural mechanics. These global matrices are then substituted into the equation of motion for the rail:

$$M_r\ddot{u}_r + K_r u_r + C_r\dot{u}_r = F_r \tag{3.3.28}$$

Among them $M_r$, $K_r$, $C_r$, $F_r$ represent the total rail mass, stiffness, damping, and force matrices, respectively; These matrices are obtained by assembling the corresponding rail element matrices, namely the mass matrix (Eq. 3.2.19), stiffness matrix (Eq. 3.2.21), damping matrix (Eq. 3.2.22), and force matrix (Eq. 3.2.25).

(2) The equation of motion of the track slab

Due to the presence of mortar, the bottom surface of the track slab is in surface contact with the concrete foundation. Given that the length-to-width ratio of the track slab is not large and its thickness is much smaller than its length and width, the slab can be simplified in the vertical direction as a thin plate beam on an elastic foundation. The following derivation focuses on the vibration equation of a vertically oriented rectangular thin plate element representing the track slab.

Considering that the slab's thickness is much smaller than its other two dimensions, and that the deflection is much smaller than the thickness, the bending behavior of the elastic thin plate is analyzed based on finite element theory using the following three assumptions:

- The motion in the thickness direction is neglected;
- No in-plane displacement occurs parallel to the mid-surface of the plate;
- The normal to the mid-surface remains normal after deformation.

The track slab is discretized into a finite number of rectangular plate elements, with fasteners serving as the nodal points. An arbitrary plate element is selected for dynamic analysis, and a coordinate system is established as shown in Fig. 3.27. The thin plate element has in-plane dimensions of $2a \times 2b$, and four nodes: *i, j, k* and *l*. According to the bending deformation characteristics of thin plates, the deformation and internal forces are fully determined by the mid-surface deflection $z$. Therefore, each rectangular thin plate element possesses 12 degrees of freedom in total, including 4 nodal displacements and 8 nodal rotations.

The unit displacement is expressed as:

$$u_b^e = \left[z_{bi},\ \theta_{bxi},\ \theta_{byi},\ z_{bj},\ \theta_{bxj},\ \theta_{byj},\ z_{bk},\ \theta_{bxk},\ \theta_{byk},\ z_{bl},\ \theta_{bxl},\ \theta_{byl}\right]^{\mathrm{T}} \tag{3.3.29}$$

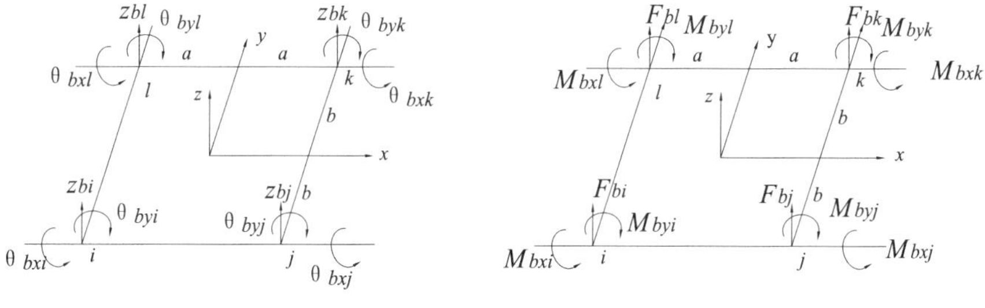


**Fig. 3.27** Nodal displacements and nodal forces in a rectangular thin plate cell

The unit node equivalent load $\boldsymbol{F}_r^e$ is expressed as:

$$\boldsymbol{F}_r^e = \left[F_{bi},\ M_{bxi},\ M_{byi},\ F_{bj},\ M_{bxj},\ M_{byj},\ F_{bk},\ M_{bxk},\ M_{byk},\ F_{bl},\ M_{bxl},\ M_{byl}\right]^{\mathrm{T}} \tag{3.3.30}$$

Here, $z_b$ denotes the vertical displacement; $\theta_x$ denotes the rotation about the x-axis; $\theta_y$ denotes the rotation about the y-axis; $F_b$ denotes the vertical force acting on the node; $M_{bx}$ denotes the bending moment about the x-axis acting on the node; and $M_{by}$ denotes the bending moment about the y-axis acting on the node.

① Shaped function

According to the bending deformation characteristics of thin plates, the deformation and internal force of a thin plate depend entirely on the deflection of the center face $z(x, y)$, expressed as:

$$\begin{aligned} z(x, y) = {} & a_1 + a_2x + a_3y + a_4x^2 + a_5xy + a_6y^2 + a_7x^3 \\ & + a_8x^2y + a_9xy^2 + a_{10}y^3 + a_{11}x^3y + a_{12}xy^3 \end{aligned} \tag{3.3.31}$$

Equation $\alpha_i$ (i = 1,2, … 0.12) are the coefficients to be determined, the displacements of the unit nodes are

$$\boldsymbol{u}_r^e = \left[z_{bi},\ \theta_{bxi},\ \theta_{byi},\ z_{bj},\ \theta_{bxj},\ \theta_{byj},\ z_{bk},\ \theta_{bxk},\ \theta_{byk},\ z_{bl},\ \theta_{bxl},\ \theta_{byl}\right]^{\mathrm{T}} \tag{3.3.32}$$

By substituting this into Eq. (3.3.31), the 12 equations are solved simultaneously, and the resulting expression is

$$\begin{aligned} z(x, y) = {} & N_iz_{bi} + N_{xi}\theta_{bxi} + N_{yi}\theta_{byi} + N_jz_{bj} + N_{xj}\theta_{bxj} + N_{yj}\theta_{byj} \\ & + N_kz_{bk} + N_{xk}\theta_{bxk} + N_{yk}\theta_{byk} + N_lz_{bl} + N_{xl}\theta_{bxl} + N_{yl}\theta_{byl} \end{aligned} \tag{3.3.33}$$

Further abbreviated as $z(x, y) = \boldsymbol{N}^{\mathrm{T}}\boldsymbol{u}_r^e$

$$\boldsymbol{N} = \left[N_i,\ N_{xi},\ N_{yi},\ N_j,\ N_{xj},\ N_{yj},\ N_k,\ N_{xk},\ N_{yk},\ N_l,\ N_{xl},\ N_{yl}\right]^{\mathrm{T}} \tag{3.3.34}$$

Among them

$$\begin{aligned} \left[N_i, N_{xi}, N_{yi}\right] &= \frac{1}{16}X_1Y_1[X_1Y_1 - X_2Y_2 + 2X_1X_2 + 2Y_1Y_2,\quad 2bY_1Y_2,\quad -2aX_1X_2] \\ \left[N_j, N_{xj}, N_{yj}\right] &= \frac{1}{16}X_2Y_1[X_2Y_1 - X_1Y_2 + 2X_1X_2 + 2Y_1Y_2,\quad 2bY_1Y_2,\quad 2aX_1X_2] \\ \left[N_k, N_{xk}, N_{yk}\right] &= \frac{1}{16}X_2Y_2[X_2Y_2 - X_1Y_1 + 2X_1X_2 + 2Y_1Y_2,\quad -2bY_1Y_2,\quad 2aX_1X_2] \\ \left[N_l, N_{xl}, N_{yl}\right] &= \frac{1}{16}X_1Y_2[X_1Y_2 - X_2Y_1 + 2X_1X_2 + 2Y_1Y_2,\quad -2bY_1Y_2,\quad -2aX_1X_2] \end{aligned}$$

$$X_1 = 1 - \frac{x}{a} \quad X_2 = 1 + \frac{x}{a} \quad Y_1 = 1 - \frac{y}{b} \quad Y_2 = 1 + \frac{y}{b} \tag{3.3.35}$$

② Unit quality matrix

$$\boldsymbol{m}_b^e = \iint \boldsymbol{N}^{\mathrm{T}} t \rho \boldsymbol{N} \mathrm{d}x\mathrm{d}y = t\rho \iint \boldsymbol{N}^{\mathrm{T}} \boldsymbol{N} \mathrm{d}x\mathrm{d}y \tag{3.3.36}$$

Among them, $t$ is the thickness of the track slab unit, $\rho$ is the density of the track slab unit, which is organized to obtain the following equation.

$$\boldsymbol{m}_b^e = \frac{abt\rho}{3150}\begin{bmatrix}
1727 & 461b & -461a & 613 & 199b & 274a & 197 & -116b & 116a & 613 & -274b & -199a \\
 & 160b^2 & -126ab & 199b & 80b^2 & 84ab & 116b & -60b^2 & 56ab & 274b & -120b^2 & -84ab \\
 & & 160a^2 & -274a & -84ab & -120a^2 & -116a & 56ab & -60a^2 & -199a & 84ab & 80a^2 \\
 & & & 1727 & 461b & 461a & 613 & -274b & 199a & 197 & -116b & -116a \\
 & & & & 160b^2 & 126ab & 274b & -120b^2 & 84ab & 116b & -60b^2 & -56ab \\
 & & & & & 160a^2 & 199a & -84ab & 80a^2 & 116a & -56ab & -60a^2 \\
 & & & & & & 1727 & -461b & 461a & 613 & -199b & -274a \\
 & & & & & & & 160b^2 & -126ab & -199b & 80b^2 & 84ab \\
 & & \text{Sym} & & & & & & 160a^2 & 274a & -84ab & -120a^2 \\
 & & & & & & & & & 1727 & -461b & -461a \\
 & & & & & & & & & & 160b^2 & 126ab \\
 & & & & & & & & & & & 160a^2
\end{bmatrix} \tag{3.3.37}$$

③ Unit stiffness matrix

The strain energy of a rectangular thin plate unit on an elastic basis consists of two parts, one of which is the deformation energy of the plate unit and the other is the elastic potential energy of the supporting spring. The corresponding unit stiffness matrix is also divided into two terms, namely.

$$\boldsymbol{k}_b^e = \boldsymbol{k}_{b1}^e + \boldsymbol{k}_{b2}^e \tag{3.3.38}$$

Among them, $\boldsymbol{k}_{b1}^e$ represents the deformation energy of the plate unit, $\boldsymbol{k}_{b2}^e$ represents the elastic potential energy of the support spring

$$\boldsymbol{k}_{b1}^e = \iint \boldsymbol{B}^{\mathrm{T}} \boldsymbol{D} \boldsymbol{B} \mathrm{d}x\mathrm{d}y \tag{3.3.39}$$

$$\boldsymbol{D} = \frac{Et^3}{12(1-\nu^2)}\begin{bmatrix} 1 & \nu & 0 \\ \nu & 1 & 0 \\ 0 & 0 & \frac{1-\nu}{2} \end{bmatrix} = \boldsymbol{D}_0 \begin{bmatrix} 1 & \nu & 0 \\ \nu & 1 & 0 \\ 0 & 0 & \frac{1-\nu}{2} \end{bmatrix} \tag{3.3.40}$$

$\boldsymbol{D}_0 = \frac{Et^3}{12(1-v^2)}$ is the bending stiffness of the plate, the $E$ is the modulus of elasticity, the $v$ is the Poisson's ratio of the material.

$$\boldsymbol{B} = \begin{bmatrix} \frac{\partial^2 N_i}{\partial x^2} & \frac{\partial^2 N_{xi}}{\partial x^2} & \frac{\partial^2 N_{yi}}{\partial x^2} & \cdots & \frac{\partial^2 N_{yl}}{\partial x^2} \\ \frac{\partial^2 N_i}{\partial y^2} & \frac{\partial^2 N_{xi}}{\partial y^2} & \frac{\partial^2 N_{yi}}{\partial y^2} & \cdots & \frac{\partial^2 N_{yl}}{\partial y^2} \\ 2\frac{\partial^2 N_i}{\partial x \partial y} & 2\frac{\partial^2 N_{xi}}{\partial x \partial y} & 2\frac{\partial^2 N_{yi}}{\partial x \partial y} & \cdots & 2\frac{\partial^2 N_{yl}}{\partial x \partial y} \end{bmatrix}_{3\times 12} \tag{3.3.41}$$

Bring the formula (3.3.40), (3.3.41) into (3.3.39) to obtain the cell stiffness matrix

$$\boldsymbol{k}_{b1}^e = \frac{Et^3}{360(1-v^2)} \begin{bmatrix} a_1 & b_1 & c_1 & a_2 & b_2 & c_2 & a_3 & b_3 & c_3 & a_4 & b_4 & c_4 \\ & d_1 & -f_1 & b_2 & d_2 & 0 & -b_3 & d_3 & 0 & -b_4 & d_4 & 0 \\ & & e_1 & -c_2 & 0 & e_2 & -c_3 & 0 & e_3 & c_4 & 0 & e_4 \\ & & & a_1 & b_1 & -c_1 & a_4 & b_4 & -c_4 & a_3 & b_3 & -c_3 \\ & & & & d_1 & f_1 & -b_4 & d_4 & 0 & -b_3 & d_3 & 0 \\ & & & & & e_1 & -c_4 & 0 & e_4 & c_3 & 0 & e_3 \\ & & & & & & a_1 & -b_1 & -c_1 & a_2 & -b_2 & -c_2 \\ & & & & & & & d_1 & -f_1 & -b_2 & d_2 & 0 \\ & & \text{Sym} & & & & & & e_1 & c_2 & 0 & e_2 \\ & & & & & & & & & a_1 & -b_1 & c_1 \\ & & & & & & & & & & d_1 & f_1 \\ & & & & & & & & & & & e_1 \end{bmatrix}_{12\times 12} \tag{3.3.42}$$

Among them

$$\begin{aligned} & a_1 = 21 - 6v + \frac{30(a^4+b^4)}{a^2b^2},\ a_2 = -21 + 6v + \frac{15(a^4-2b^4)}{a^2b^2}, \\ & a_3 = 21 - 6v - \frac{15(a^4+b^4)}{a^2b^2},\ a_4 = -21 + 6v + \frac{15(b^4-2a^4)}{a^2b^2}, \\ & b_1 = 3(1+4v)b + \frac{30a^2}{b},\ b_2 = -3(1+4v)b + \frac{15a^2}{b},\ b_3 = 3(1-v)b - \frac{15a^2}{b}, \\ & b_4 = -3(1-v)b - \frac{30a^2}{b} c_1 = -3(1+4v)a - \frac{30b^2}{a}, \\ & c_2 = 3(1-v)a + \frac{30b^2}{a},\ c_3 = -3(1-v)a + \frac{15b^2}{a},\ c_4 = 3(1+4v)a - \frac{15b^2}{a}, \\ & d_1 = 8(1-v)b^2 + 40a^2,\ d_2 = -8(1-v)b^2 + 20a^2,\ d_3 = 2(1-v)b^2 + 10a^2, \\ & d_4 = -2(1-v)b^2 + 20a^2 e_1 = 8(1-v)a^2 + 40b^2,\ e_2 = -2(1-v)a^2 + 20b^2, \\ & e_3 = 2(1-v)a^2 + 10b^2,\ e_4 = -8(1-v)a^2 + 20b^2,\ f_1 = 30vab \end{aligned} \tag{3.3.43}$$

$v$ is Poisson's ratio, taken as 0.3

$$\boldsymbol{k}_{b2}^{e} = k_{rz} \iint \boldsymbol{N}^{\mathrm{T}} \boldsymbol{N} \mathrm{d}x \mathrm{d}y = \frac{k_{rz}}{t\rho} \boldsymbol{m}_{b}^{e} \tag{3.3.44}$$

So far $\boldsymbol{k}_{b1}^{e}$, $\boldsymbol{k}_{b2}^{e}$ have been obtained, so that the unit stiffness matrix of the thin plate can be obtained $\boldsymbol{k}_{b}^{e}$.

④ Unit damping matrix

The damping matrix of the track slab also consists of two parts, one is the material viscous damping of the thin plate cell $\boldsymbol{c}_{b1}^{e}$, taking the Rayleigh damping of the structure, and the other the support damping under the slab $\boldsymbol{c}_{b2}^{e}$.

$$\begin{aligned} \boldsymbol{c}_{b1}^{e} &= \alpha \boldsymbol{m}_{b}^{e} + \beta \boldsymbol{k}_{b}^{e}, \\ \boldsymbol{c}_{b2}^{e} &= \frac{c_{rz}}{t\rho} \boldsymbol{m}_{b}^{e} \end{aligned} \tag{3.3.45}$$

The values of $\alpha$ and $\beta$ are determined based on the structure and material properties of the concrete slab.

⑤ Track slab loads

(1) Equivalent loads for plate units

When a concentrated force $\boldsymbol{Q}$ is acting on the unit, the unit load vector is

$$\boldsymbol{F}^{e} = \boldsymbol{N}^{\mathrm{T}} \boldsymbol{Q} \tag{3.3.46}$$

(2) Force analysis of the track slab

The forces on the track slab unit are divided into two main parts:

Forces acting on rail fasteners.

Supporting force of cement mortar under the track slab.

The force diagram of the track slab is shown in Fig. 3.28. The forces are equivalently applied to the nodes, forming the track slab's force matrix.

⑥ The equation of motion for the vertical direction of the track slab

The unit mass, stiffness, damping and force matrix of the track slab have been solved above. Using the deformation coordination conditions and static equilibrium conditions, in accordance with the "right number" law to assemble the total mass of the track slab, the total stiffness, the total damping, and the total force matrix, brought into the track slab equation of motion.

$$\boldsymbol{M}_{b} \ddot{\boldsymbol{u}}_{b} + \boldsymbol{K}_{b} \boldsymbol{u}_{b} + \boldsymbol{C}_{b} \dot{\boldsymbol{u}}_{b} = \boldsymbol{F}_{b} \tag{3.3.47}$$

Among them, $\boldsymbol{M}_{b}$, $\boldsymbol{K}_{b}$, $\boldsymbol{C}_{b}$, $\boldsymbol{F}_{b}$ represent the global vertical mass, stiffness, damping, and force matrices of the track slab, respectively. These are obtained by

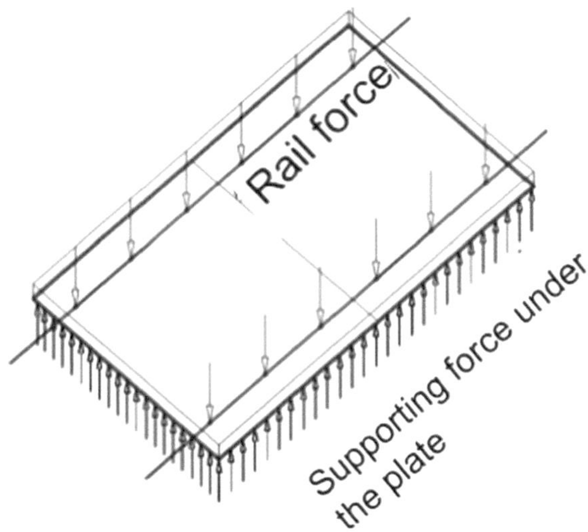


**Fig. 3.28** The force diagram of the track slab

assembling the corresponding vertical element matrices: the mass matrix (Eq. 3.3.47), stiffness matrix (Eq. 3.3.42), damping matrix (Eq. 3.3.45), and force matrix (Eq. 3.3.46).

⑦ The equation of motion of the track slab in the transverse direction

In the lateral direction, the track slab is modeled as a continuously supported beam element with a uniform cross-section. Each node has two degrees of freedom: lateral displacement $Y_b$ and rotation about the z-axis $\theta_{bz}$. The nodal displacement of any beam element is expressed as $\boldsymbol{u}_{by}^{e} = \left[Y_{bi},\ \theta_{bzi},\ Y_{bj},\ \theta_{bzi}\right]$, and the nodal force is represented as $\boldsymbol{F}_{by}^{e} = \left[F_{bi},\ M_{bzi},\ F_{bj},\ M_{bzi}\right]$.

Among them $F_{by}$ denotes the lateral force on the node; $M_{bz}$ represents the bending moment about the z-axis acting on the node.

The element stiffness, mass, and damping matrices of the lateral elements of the track slab share the same form as the corresponding matrices of the vertical rail elements. However, differences arise in the selection of parameters such as flexural rigidity, unit mass, and Rayleigh damping coefficients. Additionally, it is essential to account for the effects of the supporting springs and dampers beneath the slab, the influence of rail fasteners on the lateral elements of the track slab, and the lateral constraints on the sides of the slab.

The differential equation governing the lateral vibration of the track slab is given by:

$$\boldsymbol{M}_{by}\ddot{\boldsymbol{u}}_{by} + \boldsymbol{K}_{by}\boldsymbol{u}_{by} + \boldsymbol{C}_{by}\dot{\boldsymbol{u}}_{by} = \boldsymbol{F}_{by} \tag{3.3.48}$$

Among them, $\boldsymbol{M}_{by}$, $\boldsymbol{K}_{by}$, $\boldsymbol{C}_{by}$, $\boldsymbol{F}_{by}$ represent the mass, stiffness, damping, and force matrix of the track slab transverse direction, respectively.

### 3.3.2 Bridge Modeling

1. **Finite Element Modeling of Bridges**

(1) The matrix of mechanical properties of the space beam unit

A two-node spatial isoparametric straight beam element is used to model the spatial beam. Each node has three rotational and three translational degrees of freedom. Warping and distortion of the beam element are not considered (Fig. 3.29).

1) The form function

In order to perform element analysis, the deflection at any point of the beam element must first be expressed in the form of a specific function. Using shape functions and nodal displacements, the axial displacement $u$, vertical displacement $v$, lateral displacement $w$, and torsional rotation $\theta_x$ of the beam can be expressed as:

$$u = \boldsymbol{N}_u \boldsymbol{u}^e \tag{3.3.49}$$

$$v = \boldsymbol{N}_v \boldsymbol{u}^e \tag{3.3.50}$$

$$w = \boldsymbol{N}_w \boldsymbol{u}^e \tag{3.3.51}$$

$$\theta_x = \boldsymbol{N}_\theta \boldsymbol{u}^e \tag{3.3.52}$$

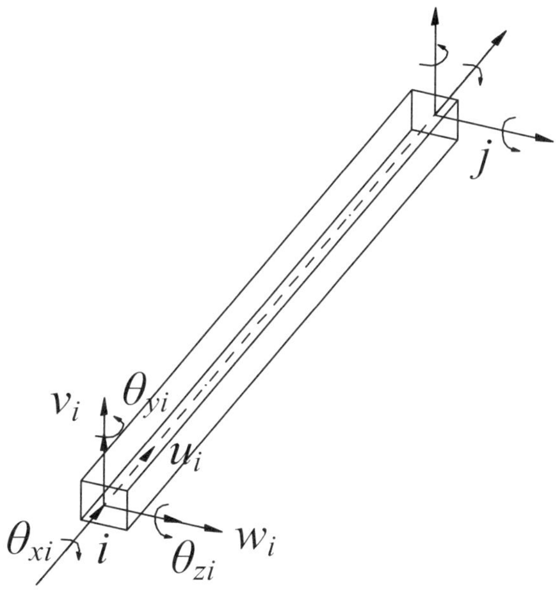


**Fig. 3.29** Equivalent parameter straight beam element

In the equation

$$
\begin{aligned}
\boldsymbol{N}_u &= \left[N_1\ 0\ 0\ 0\ 0\ 0\ N_2\ 0\ 0\ 0\ 0\ 0\right] \\
\boldsymbol{N}_v &= \left[0\ N_3\ 0\ 0\ 0\ N_4\ 0\ N_5\ 0\ 0\ 0\ N_6\right] \\
\boldsymbol{N}_w &= \left[0\ 0\ N_3\ 0\ -N_4\ 0\ 0\ 0\ N_5\ 0\ -N_6\ 0\right] \\
\boldsymbol{N}_\theta &= \left[0\ 0\ 0\ N_1\ 0\ 0\ 0\ 0\ 0\ N_2\ 0\ 0\right]
\end{aligned}
$$

Among them.

$$
N_1 = 1 - \frac{x}{L},\quad N_2 = \frac{x}{L},\quad N_3 = 1 - \frac{3}{l_e^2}x^2 + \frac{2}{l_e^3}x^3,\quad N_4 = -x + \frac{2}{l}x^2 - \frac{1}{l^2}x^3,
$$

$$
N_5 = \frac{3}{l^2}x^2 - \frac{2}{l^3}x^3,\quad N_6 = \frac{1}{l_e}x^2 - \frac{1}{l_e^2}x^3
$$

2) The unit stiffness matrix

The unit stiffness matrix can be calculated by the following equation

$$
\boldsymbol{k}^e = \int\limits_v \boldsymbol{B}^{\mathrm{T}}\boldsymbol{D}\boldsymbol{B}\mathrm{d}v \tag{3.3.53}
$$

The stiffness matrix of a spatial beam unit in local coordinates considering shear deformation can be expressed as:

$$
\boldsymbol{k}^e = \begin{bmatrix}
\frac{EA}{l} & 0 & 0 & 0 & 0 & 0 & -\frac{EA}{l} & 0 & 0 & 0 & 0 & 0 \\
 & \frac{12EI_z}{l^3(1+b_z)} & 0 & 0 & 0 & \frac{6EI_z}{l^2(1+b_z)} & 0 & \frac{-12EI_z}{l^3(1+b_z)} & 0 & 0 & 0 & \frac{6EI_z}{l^2(1+b_z)} \\
 & & \frac{12EI_y}{l^3(1+b_y)} & 0 & \frac{-6EI_y}{l^2(1+b_y)} & 0 & 0 & 0 & \frac{-12EI_y}{l^3(1+b_y)} & 0 & \frac{-6EI_y}{l^2(1+b_y)} & 0 \\
 & & & \frac{GI_p}{l} & 0 & 0 & 0 & 0 & 0 & -\frac{GI_p}{l} & 0 & 0 \\
 & & & & \frac{EI_y(4+b_y)}{l(1+b_y)} & 0 & 0 & 0 & \frac{6EI_y}{l^2(1+b_y)} & 0 & \frac{EI_y(2-b_y)}{l^2(1+b_y)} & 0 \\
 & & & & & \frac{EI_z(4+b_z)}{l(1+b_z)} & 0 & \frac{-6EI_z}{l^2(1+b_z)} & 0 & 0 & 0 & \frac{EI_z(2-b_z)}{l^2(1+b_z)} \\
 & & & & & & \frac{EA}{l} & 0 & 0 & 0 & 0 & 0 \\
 & & & & & & & \frac{12EI_z}{l^3(1+b_z)} & 0 & 0 & 0 & \frac{-6EI_z}{l^2(1+b_z)} \\
 & & & \text{Sym} & & & & & \frac{12EI_y}{l^3(1+b_y)} & 0 & \frac{6EI_y}{l^2(1+b_y)} & 0 \\
 & & & & & & & & & \frac{GI_p}{l} & 0 & 0 \\
 & & & & & & & & & & \frac{EI_y(4+b_y)}{l(1+b_y)} & 0 \\
 & & & & & & & & & & & \frac{EI_z(4+b_z)}{l(1+b_z)}
\end{bmatrix}
$$

In the formula, the $b_y = \frac{12kEI_y}{GA_z l^2} = \frac{24(1+\mu)A}{A_z}\left(\frac{r_y}{l}\right)^2, b_z = \frac{12kEI_z}{GA_y l^2} = \frac{24(1+\mu)A}{A_y}\left(\frac{r_z}{l}\right)^2, I_y, I_z$ denote the second moment of area about the $y$-axis and $z$-axis, the $I_p$ is the torsional moment of inertia, the $r_y, r_z$ are the radius of gyration. When $\frac{r_y}{l}, \frac{r_z}{l}$ are sufficiently small, setting $b_y = b_y = 0$ yields the stiffness matrix of a spatial beam element neglecting shear deformation,$A_y, A_z$ denote the shear areas in the $y$-direction and

$z$-direction, $A$ is the cross sectional area of the beam element, $l$ is the element length, $E$ is the elastic modulus of the material, the $G$ is the shear modulus of the material.

3) The unit quality matrix

The coordination quality matrix can be calculated by the following equation

$$\boldsymbol{m}^e = \int_v \rho \boldsymbol{N}^{\mathrm{T}} \boldsymbol{N} \mathrm{d}v \tag{3.3.54}$$

The spatial beam cell mass matrix can be expressed as follows

$$\boldsymbol{m}^e = \frac{\rho l}{420}\begin{bmatrix} 140 & 0 & 0 & 0 & 0 & 0 & 70 & 0 & 0 & 0 & 0 & 0 \\ & 156 & 0 & 0 & 0 & 22l & 0 & 54 & 0 & 0 & 0 & -13l \\ & & 156 & 0 & -22l & 0 & 0 & 0 & 54 & 0 & 13l & 0 \\ & & & \frac{140I_p}{A} & 0 & 0 & 0 & 0 & 0 & \frac{70I_p}{A} & 0 & 0 \\ & & & & 4l^2 & 0 & 0 & 0 & -13l & 0 & -3l^2 & 0 \\ & & & & & 4l^2 & 0 & 13l & 0 & 0 & 0 & -3l^2 \\ & & & & & & 140 & 0 & 0 & 0 & 0 & 0 \\ & & & & & & & 156 & 0 & 0 & 0 & -22l \\ & & & \text{Sym} & & & & & 156 & 0 & 22l & 0 \\ & & & & & & & & & \frac{140I_p}{A} & 0 & 0 \\ & & & & & & & & & & 4l^2 & 0 \\ & & & & & & & & & & & 4l^2 \end{bmatrix}$$

4) Structural damping matrix

The damping problem is relatively complex, and it is quite difficult to determine the damping matrix precisely. In practical analysis, the damping matrix is usually approximated based on experimental data, using the energy dissipation of the structure during vibration to estimate its value. The structural damping matrix is commonly simplified as a linear combination of the mass and stiffness matrices, known as Rayleigh damping, expressed as:

$$\boldsymbol{C}^e = \alpha \boldsymbol{M}^e + \beta \boldsymbol{K}^e \tag{3.3.55}$$

The coefficients $\alpha$ and $\beta$ can be determined based on the structural vibration frequencies $\omega_1$, $\omega_2$ and the corresponding damping ratios $\xi_1$, $\xi_2$, as follows:

$$\alpha = \frac{2(\xi_1\omega_2 - \xi_2\omega_1)}{\omega_2^2 - \omega_1^2}\omega_1\omega_2$$
$$\beta = \frac{2(\xi_2\omega_2 - \xi_1\omega_1)}{\omega_2^2 - \omega_1^2} \tag{3.3.56}$$

2. **Comprehensive Modal Modeling Method for Bridges**

Since the finite element geometric model used in bridge modeling is a multi-degree-of-freedom system with a large number of computational degrees of freedom, the workload can be considerable. To reduce the number of degrees of freedom, the modal synthesis technique—i.e., the generalized coordinate discretization method—is introduced into the beam-pier model. First, the natural frequencies and mode shapes of the structure are obtained. Then, using the orthogonality of the mode shapes, the coupled equations of motion for hundreds of nodes are decoupled and transformed into a set of independent modal equations.

When the generalized coordinate discretization (modal model) is applied to the bridge model, the motion of any cross-section of the beam in three directions can be represented as a superposition of modal functions, namely:

$$\begin{cases} Y_b(x) = \sum\limits_{n=1}^{N_b} q_n \varphi_h^n(x) \\ \theta_b(x) = \sum\limits_{n=1}^{N_b} q_n \varphi_\theta^n(x) \\ Z_b(x) = \sum\limits_{n=1}^{N_b} q_n \varphi_v^n(x) \end{cases} \tag{3.3.57}$$

In this expression, $\phi_h^n$, $\phi_\theta^n$, and $\phi_v^n$ represent the horizontal, torsional, and vertical components, respectively, of the n-th mode shape of the beam-pier system; $N_b$ is the number of mode shapes considered; and $q_n$ denotes the generalized coordinate, i.e., the modal amplitude.

The modal analysis is performed for the entire bridge system, and the mode shapes between nodes are determined using Lagrange interpolation based on the nodal mode shape values.

If the mode shapes of the beam-pier system are normalized according to $\phi_n^{\mathrm{T}} m \phi_n = 1$, then the n-th order modal equation of the bridge can be written as:

$$\ddot{q}_n + 2\xi_n \omega_n \dot{q}_n + \omega_n^2 q_n = F_n \tag{3.3.58}$$

where: $\xi_n$ and $\omega_n$ are the damping ratio and circular frequency of the nth order vibration mode of the bridge, respectively; $F_n$ is the generalized force.

3. **Bridge Modal Analysis**

The 32 m span bridge designed for a speed of 250 km/h is a two-lane prestressed concrete simply supported box girder, with a two-lane railroad deck width of 12.2 m, a total girder length of 32.7 m, a girder height of 8 m, and a construction height from the rail bottom to the bottom of the girder of 3.17 m. The curved lot layout and girder cross-section are shown in Fig. 3.30.

The 32-m two-lane single-box, single-compartment simply supported girder modeled using finite elements is shown in Fig. 3.31, and the computational parameters are as follows Table 3.5 shown.

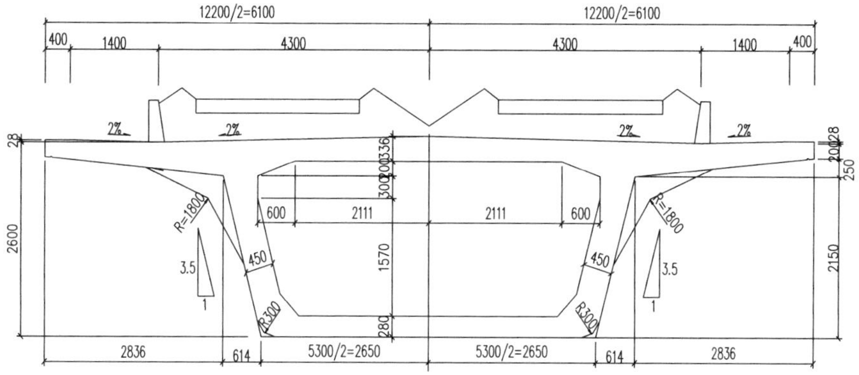


Fig. 3.30 32m span beam section

Fig. 3.31 Finite element modeling of bridges

**Table 3.5** Parameters of 32 m simply supported girder bridge girders

| Serial number | Components | Parameters | Values |
|---|---|---|---|
| 1 | Simply supported beams | Modulus of elasticity | $3.40 \times 10^{10}$ Pa |
| | | Poisson's ratio | 0.2 |
| | | Density | 2500 kg/m$^3$ |
| 2 | Phase II constant load | Mass per linear meter | 15 t |
| 3 | Bridges | Mass per linear meter | 20.85 t |
| 4 | Standoffs | Vertical stiffness | $5 \times 10^{10}$ N/m |
| | | Lateral stiffness | $5 \times 10^{15}$ N/m |
| | | Longitudinal stiffness | $5 \times 10^{10}$ N/m |

**Table 3.6** The first 15 orders of frequency of the bridge modal analysis

| Order | Frequency/HZ | Primary oscillator form |
|---|---|---|
| 1 | 4.97285 | Dropping to bend |
| 2 | 13.0973 | Transverse bending, twisting about the longitudinal axis |
| 3 | 16.0573 | Transverse bending, twisting about the longitudinal axis |
| 4 | 17.1025 | Dropping to bend |
| 5 | 23.5148 | Dropping to bend |
| 7 | 35.2089 | Dropping to bend |
| 8 | 41.9198 | Transverse bending, twisting about the longitudinal axis |
| 9 | 44.2916 | Lateral bending |
| 12 | 70.4461 | Dropping to bend |
| 13 | 72.6213 | Transverse bending, twisting about the longitudinal axis |
| 15 | 75.9881 | Twisting around the longitudinal axis |

The model undergoes a free vibration analysis, and the natural frequencies (partial) are obtained as shown in Table 3.6. The mode shapes for each order are depicted in Fig. 3.32. Among them, the first vertical frequency is 4.97 Hz, and the first lateral frequency is 13.1 Hz.

## 3.4 Modeling of Track Irregularities

Track irregularities are one of the main excitation sources of the vehicle-track system. They are the primary factor that causes vibrations in structures such as vehicles and tracks. The quality of train operation and the force state of the track structure are directly related to the type, amplitude, wavelength, and period of track irregularities.

In general, track irregularities can be classified into two main categories: deterministic disturbances and non-deterministic disturbances. Deterministic disturbances can be further classified into three main types based on their nature: impulse-type, harmonic-type, and dynamic-type disturbances. Impulse-type disturbances include wheel flats, low joints, rail misalignments, etc., which typically input into the wheel-rail system as impact forces. Harmonic-type disturbances include track sine and cosine irregularities, vertical misalignment, and wheel eccentricities, which manifest as displacement disturbances and force disturbances. Dynamic-type disturbances include sudden changes in the subgrade stiffness, hanging slabs, track ballast consolidation, and fastener loosening, which are mainly reflected as changes in the parameters of the model. Non-deterministic disturbances mainly refer to random track geometric irregularities.

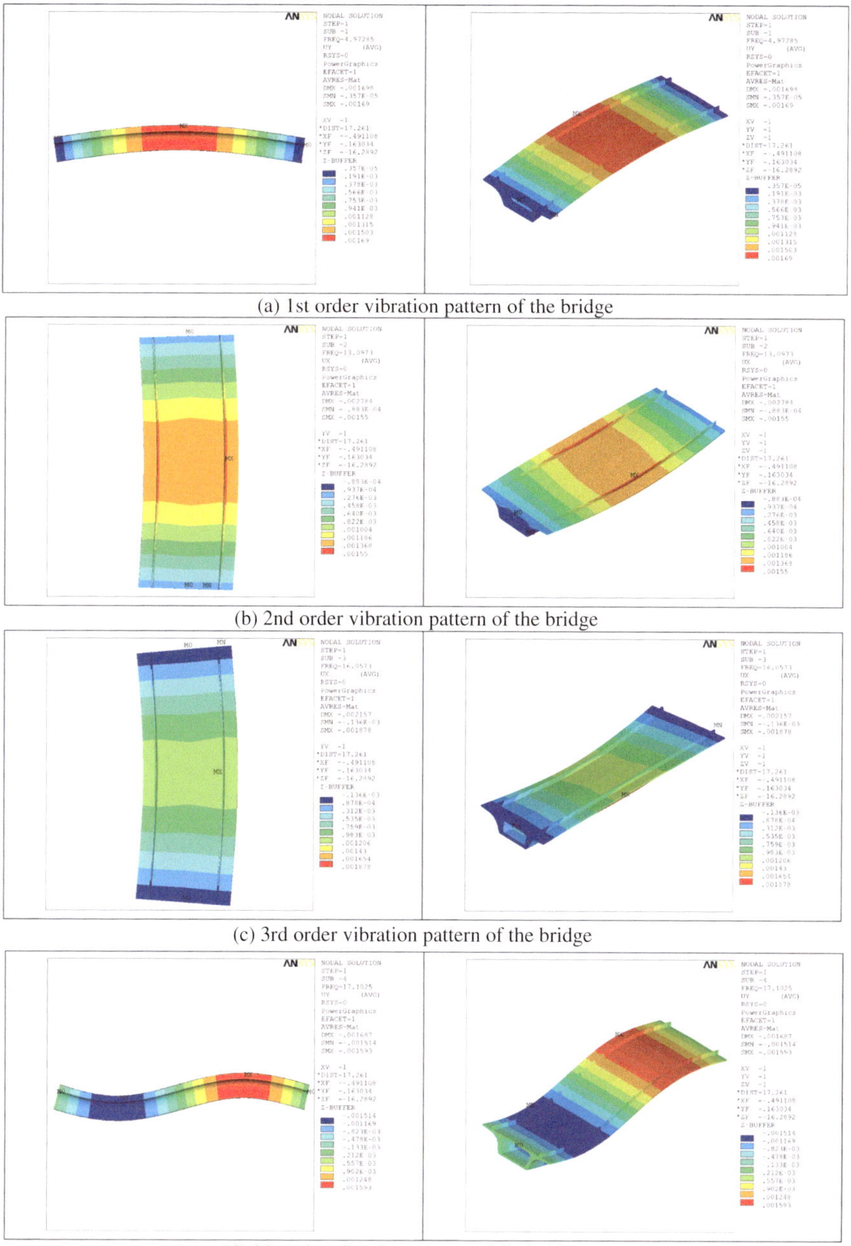

(a) 1st order vibration pattern of the bridge

(b) 2nd order vibration pattern of the bridge

(c) 3rd order vibration pattern of the bridge

(d) 4th order vibration pattern of the bridge

**Fig. 3.32** Bridge vibration patterns

### 3.4.1 Description of the Geometrical Irregularities of the Track

The basic parameters of track geometric irregularities are the vertical deviations of the top surfaces of the left and right rails, representing the elevation irregularities, and the lateral deviations of the rails relative to the centerline, representing the track irregularities. From these, four basic types of track irregularities can be derived: elevation, horizontal, track, and gauge irregularities, as shown in Fig. 3.33. L represents the wavelength of the irregularity, and A represents the amplitude of the track irregularity.

(1) Vertical irregularities

Vertical irregularities refer to the vertical positional deviations of the rail top surface along the track direction, which can be expressed as:

$$z_t = \frac{1}{2}(z_l + z_r) \tag{3.4.1}$$

In the formula, the $z_l$, $z_r$ are the vertical coordinates of the left and right rails, respectively.

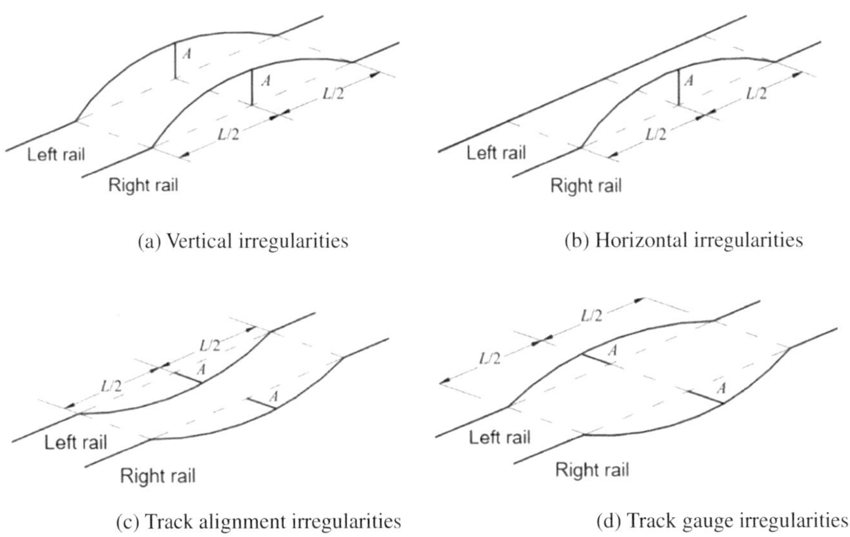


**Fig. 3.33** The basic form of geometrical unevenness of the track

(2) Horizontal irregularities

Horizontal irregularities refer to the relative height difference between the top surfaces of the left and right rails, which can be expressed as:

$$\Delta z_t = z_l - z_r \tag{3.4.2}$$

(3) Track alignment irregularities

Track alignment irregularities refer to the deviation between the actual track centerline and the theoretical track centerline, which can be expressed as:

$$y_t = \frac{1}{2}(y_l + y_r) \tag{3.4.3}$$

In the formula, the $y_l$, $y_r$ are the horizontal coordinates of the left and right rails, respectively.

(4) Track gauge irregularities

Track gauge irregularities refer to the deviation between the actual track gauge and the theoretical track gauge, measured at a position 16 mm below the rail top, which can be expressed as:

$$g_t = y_l - y_r - g_0 \tag{3.4.4}$$

In the formula, the $g_0$ is the nominal gauge.

Vertical and horizontal irregularities are vertical irregularities, reflecting the vertical displacement of the track centerline. Directional and track gauge irregularities are lateral irregularities, reflecting the horizontal displacement difference between the left and right rails. Additionally, there are twisted irregularities (e.g., triangular pits) and composite irregularities where multiple vertical and lateral irregularities occur at the same location. All of these can be described as combinations of the four basic types of irregularities mentioned above.

### 3.4.2 Track Random Irregularity Model

1. **High-Speed Railway Track Irregularity Spectrum**

The geometric state of the actual track is influenced by various factors, such as the condition of the rails, ballast, and subgrade, leading to significant randomness. Power spectral density is commonly used as the most important and widely used statistical function for track irregularities.

When using the power spectral density function for track irregularities, it is important to note that track irregularities are a function of distance. Usually, spatial frequency $F$ (cycle/m) or spatial wavenumber $\Omega$ (rad/m) is used instead of frequency f (1/s) or angular frequency $\omega$(rad/s), with the relationship given by:

$$\begin{cases} \Omega = 2\pi F = \omega/V \\ F = f/V \end{cases} \tag{3.4.5}$$

Britain, Japan, Germany, the United States are based on their own railroad measurements, analysis of the proposed national track spectrum, such as the United States six-level spectrum, Germany high and low interference spectrum. On the basis of a large number of high-speed railroad track inspection data, China has put forward the ballastless track and ballasted track unevenness spectrum of high-speed railroad.

(1) China High-Speed Railway Ballastless Track Irregularity Spectrum

1) Applicable range

This spectrum applies to ballastless tracks of high-speed railways with speeds ranging from 300 km/h to 350 km/h. It is not applicable for analyzing short-wavelength track irregularities with wavelengths less than 2 m.

2) Spatial frequency range

The spatial frequency range for the China high-speed railway ballastless track irregularity spectrum is from 0.005 (1/m) to 0.5 (1/m), corresponding to a track irregularity wavelength range of 2 m to 200 m.

3) Fitting formula

The track irregularity spectrum is segmented and fitted using a power function formula.

$$S(f) = \frac{A}{f^k} \tag{3.4.6}$$

where

$f$—spatial frequency.

$A$, $k$-fitting formula coefficients (Tables 3.7, 3.8 and 3.9).

**Table 3.7** Fitting coefficients of the average spectrum of high-speed railway ballastless track irregularities

| Project | Para | | Para. 2 | | Para. 3 | | Para. 4 | |
|---|---|---|---|---|---|---|---|---|
| | A | k | A | k | A | k | A | k |
| Track gauge irregularities | 5.4978E−02 | 0.8282 | 5.0701E−03 | 1.9307 | 1.8778E−03 | 4.5948 | – | – |
| Horizontal irregularities | 3.6148E−03 | 1.7278 | 4.3685E−02 | 1.0461 | 4.5867E−03 | 2.0939 | – | – |
| Track alignment irregularities | 3.9513E−03 | 1.8670 | 1.1047E−02 | 1.5354 | 7.5633E−03 | 2.8171 | – | – |
| Vertical irregularities | 1.0544E−05 | 3.3891 | 3.5588E−03 | 1.9271 | 1.9784E−03 | 1.3643 | 3.9488E−04 | 3.4516 |

**Table 3.8** Segment points of the spatial frequency (1/m) and corresponding wavelengths (m) for high-speed railway ballastless track irregularities

| Project | Para | | Para. 2 | | Para. 3 | |
|---|---|---|---|---|---|---|
| | Spatial frequency | Spatial wavelengths | Spatial frequency | Spatial wavelengths | Spatial frequency | Spatial wavelengths |
| Track gauge irregularities | 0.1090 | 9.2 | 0.2938 | 3.4 | – | – |
| Horizontal irregularities | 0.0258 | 38.8 | 0.1163 | 8.6 | – | – |
| Track alignment irregularities | 0.0450 | 22.2 | 0.1234 | 8.1 | – | – |
| Elevation irregularities | 0.0187 | 53.5 | 0.0474 | 21.1 | 0.1533 | 6.5 |

**Table 3.9** The conversion coefficient from track irregularity spectrum to percentile spectrum

| Percentile | 10.0 | 20.0 | 25.0 | 30.0 | 50.0 | 60.0 | 63.2 | 70.0 | 75.0 | 80.0 | 90.0 |
|---|---|---|---|---|---|---|---|---|---|---|---|
| The conversion factor C | 0.105 | 0.223 | 0.288 | 0.357 | 0.693 | 0.916 | 1.000 | 1.204 | 1.386 | 1.609 | 2.303 |

(2) The track irregularity spectrum of China's high-speed ballasted track

1) Applicable range

It is applicable to the ballasted track of high-speed railroad with line speed 200–250 km/h, and is not applicable to the analysis of short-wave irregularity of the track with wavelength less than 2 m.

2) Spatial frequency range

The spatial frequency range of the ballast track irregularity spectrum of China's high-speed railroad is 0.01–0.5(1/m), and the corresponding track irregularity wavelength range is 2–100 m.

3) Fitting of formula

The track irregularity spectrum is fitted in segments using a power function formula.

$$S(f) = \frac{A}{f^k} \tag{3.4.7}$$

where

$f$—spatial frequency.

$A$, $k$—fitting formula coefficients (Tables 3.10, 3.11 and 3.12).

**Table 3.10** The fitting coefficients for the average spectrum of track irregularities in high-speed rail ballasted tracks

| Project | Para | | Para. 2 | | Para. 3 | | Para. 4 | |
|---|---|---|---|---|---|---|---|---|
| | A | k | A | k | A | k | A | k |
| Track gauge irregularities | 1.3635E−01 | 0.6903 | 2.1384E−02 | 1.5653 | 8.1641E−04 | 4.1305 | – | – |
| Horizontal irregularities | 9.9714E−03 | 1.6103 | 5.5743E−02 | 1.1336 | 1.6833E−03 | 3.1315 | – | – |
| Track alignment irregularities | 2.5934E−04 | 2.6607 | 1.3245E−01 | 0.8041 | 7.1498E−04 | 3.2979 | – | – |
| Vertical irregularities | 8.1981E−02 | 1.6284 | 4.6880E−05 | 3.4194 | 3.8215E−02 | 1.2306 | 2.4609E−04 | 4.1663 |

**Table 3.11** The segmented spatial frequencies (1/m) and corresponding wavelengths (m) for track irregularities in high-speed rail ballasted tracks

| Project | Para | | Para. 2 | | Para. 3 | |
|---|---|---|---|---|---|---|
| | Spatial frequency | Spatial wavelengths | Spatial frequency | Spatial wavelengths | Spatial frequency | Spatial wavelengths |
| Track gauge irregularities | 0.1204 | 8.3 | 0.2800 | 3.6 | – | – |
| Horizontal irregularities | 0.0270 | 37.0 | 0.1735 | 5.8 | – | – |
| Track alignment irregularities | 0.0348 | 28.7 | 0.1232 | 8.1 | – | – |
| Vertical irregularities | 0.0155 | 64.5 | 0.0468 | 21.4 | 0.1793 | 5.6 |

**Table 3.12** Conversion factor from track inhomogeneity spectrum to percentile spectrum

| Percentile | 10.0 | 20.0 | 25.0 | 30.0 | 50.0 | 60.0 | 63.2 | 70.0 | 75.0 | 80.0 | 90.0 |
|---|---|---|---|---|---|---|---|---|---|---|---|
| The conversion factor C | 0.105 | 0.223 | 0.288 | 0.357 | 0.693 | 0.916 | 1.000 | 1.204 | 1.386 | 1.609 | 2.303 |

(3) German trackl spectrum

The German track spectrum is divided into "low interference" and "high interference" spectra. The "low interference" spectrum is applicable to high-speed rail in Germany with speeds above 250 km/h, while the "high interference" spectrum is used for regular German railways. Additionally, the units for the power spectral density of vertical, direction, and track gauge irregularities are m$^2$/(rad/m), while for horizontal irregularities, due to the use of angular measurements, the power spectral density is expressed in 1/(rad/m).

1) Vertical irregularities

$$S_{\mathrm{v}}(\Omega) = \frac{A_{\mathrm{v}}\Omega_{\mathrm{c}}^2}{(\Omega^2 + \Omega_{\mathrm{r}}^2)(\Omega^2 + \Omega_{\mathrm{c}}^2)} \tag{3.4.8}$$

2) Track alignment irregularities

$$S_{\mathrm{a}}(\Omega) = \frac{A_{\mathrm{a}}\Omega_{\mathrm{c}}^2}{(\Omega^2 + \Omega_{\mathrm{r}}^2)(\Omega^2 + \Omega_{\mathrm{c}}^2)} \tag{3.4.9}$$

3) Horizontal irregularities

$$S_c(\Omega) = \frac{a^{-2} A_v \Omega_c^2 \Omega^2}{(\Omega^2 + \Omega_r^2)(\Omega^2 + \Omega_c^2)(\Omega^2 + \Omega_s^2)} \tag{3.4.10}$$

4) Track gauge irregularities

$$S_g(\Omega) = \frac{A_g \Omega_c^2 \Omega^2}{(\Omega^2 + \Omega_r^2)(\Omega^2 + \Omega_c^2)(\Omega^2 + \Omega_s^2)} \tag{3.4.11}$$

where

$S_v(\Omega)$, $S_a(\Omega)$, $S_g(\Omega)$—Vertical, Track alignment, and track gauge irregularities power spectral density ($m^2$/(rad/m)).
$S_c(\Omega)$—horizontal irregularities power spectral density (1/(rad/m)).
$A_v$, $A_a$, $A_g$—Roughness constant ($m^2$-rad/m).
$\Omega$—Spatial frequency (rad/m).
$\Omega_c$, $\Omega_r$, $\Omega_s$—Truncation frequency (rad/m).
a—half of the distance between the left and right rolling circle (m), generally 0.75 m.

For roughness coefficients and truncation frequencies, see Table 3.13, where A g is a reference value based on a trial calculation of gauge irregularities varying from − 3 to 3 mm.

## 2. Conversion of Track Irregularity Spectrum to Time-Domain Samples

In dynamic simulation analysis, using stochastic track irregularities as system excitation can better reflect the influence of substructure on the dynamic response of the system. The time-domain samples of stochastic irregularities can be obtained in two ways: one is based on measured track data, and the other—commonly used in simulations—is to statistically derive the track irregularity spectrum from extensive measurement data, and then perform time–frequency conversion to generate time-domain samples of random irregularities.

Track random irregularities can be regarded as a stationary Gaussian stochastic process. Various methods are available to generate time-domain samples of irregularities from a given power spectral density (PSD) of track irregularities, such

**Table 3.13** German track spectrum roughness coefficients and cutoff frequencies

| Track level | $\Omega_c$ rad/m | $\Omega_r$ rad/m | $\Omega_s$ rad/m | $A_a$ $m^2$·rad/m | $A_v$ $m^2$·rad/m | $A_g$ $m^2$·rad/m |
|---|---|---|---|---|---|---|
| Low interference | 0.8246 | 0.0206 | 0.4380 | $2.119 \times 10^{-7}$ | $4.032 \times 10^{-7}$ | $5.32 \times 10^{-8}$ |
| High interference | 0.8246 | 0.0206 | 0.4380 | $6.125 \times 10^{-6}$ | $1.08 \times 10^{-6}$ | $1.032 \times 10^{-7}$ |

as the trigonometric series method, the second-order filtering method, the Auto-Regressive (AR) model method, or the Auto-Regressive Moving Average (ARMA) model method.

Using the trigonometric series method, the PSD of track irregularities is transformed into time-domain samples. The time-domain realization of random track irregularities can be expressed as follows:

$$w(x) = \sum_{k=1}^{N} \sqrt{2S(\omega_k)\Delta\omega} \cos(\omega_k x + \phi_k) \tag{3.4.12}$$

In the formula, the $S(\omega_k)$ is the irregularity power spectral density, the $\omega_k (k = 1, N)$ are the upper and lower limits of the cutoff frequency, respectively. $\Delta\omega$ is the frequency interval bandwidth, $\phi_k$ represents the phase corresponding to the $k$-th frequency, which are mutually independent and uniformly distributed in the interval $[0, 2\pi]$.

(2) Numerical Simulation Example of Track Irregularities

Using the track irregularity spectrum of the Chinese high-speed ballastedless track and the power spectral density of the German low-disturbance track (high-speed railway) as examples, numerical simulations of random track irregularity samples were carried out based on the aforementioned simulation method. The results are shown in the following figures.

Figure 3.34 presents a time-domain simulation sample of Chinese high-speed ballastedless track irregularity for wavelengths ranging from 2 to 120 m, while Fig. 3.35 shows a corresponding sample of German low-disturbance track vertical irregularity within the same wavelength range.

By comparing the Chinese and German high-speed track spectra, it can be observed that for wavelengths between 2 and 120 m, the directional irregularity amplitude of the Chinese high-speed ballastedless track varies within the range of − 2 to 2 mm, and the vertical irregularity amplitude ranges from − 2.5 to 2.5 mm. In contrast, the German low-disturbance track shows directional irregularity amplitudes ranging from − 6 to 5 mm and vertical irregularity amplitudes from − 7.5 to 7.5 mm. These results indicate that the Chinese high-speed ballastedless track irregularity spectrum demonstrates superior performance compared to the German low-disturbance track spectrum.

Figure 3.36a presents the time-domain sample of vertical irregularity based on the German low-interference spectrum at a speed of 300 km/h. Figure 3.36b shows the power spectral density estimated from the simulation result in Fig. 3.36a using the periodogram method, which exhibits good agreement with the original analytical spectrum.

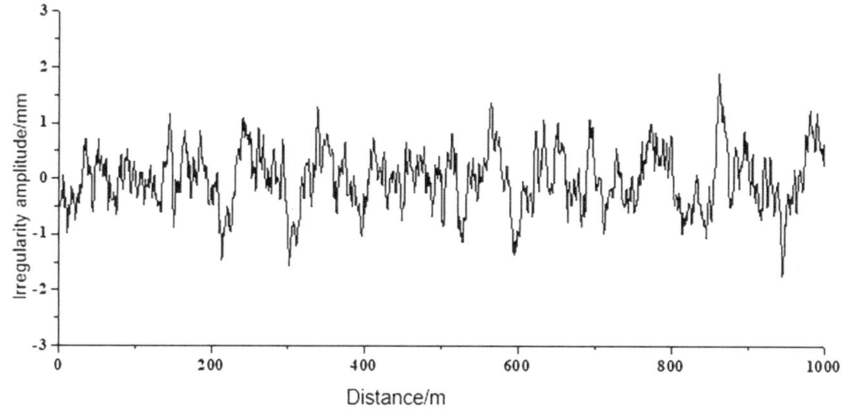


(a) Lateral alignment irregularity of the left rail

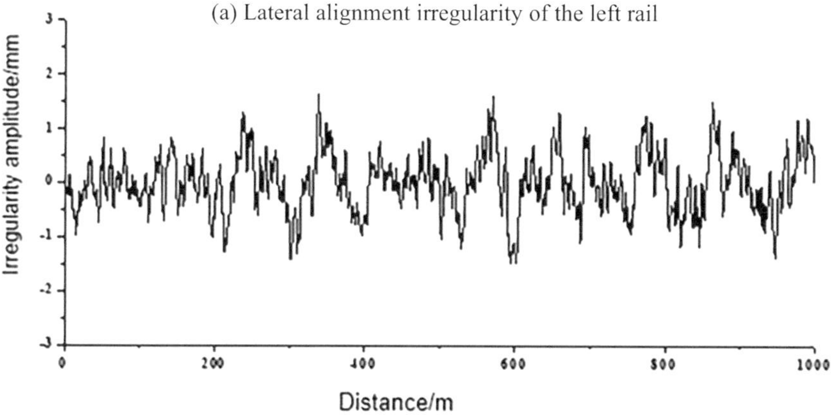


(b) Lateral alignment irregularity of the right rail

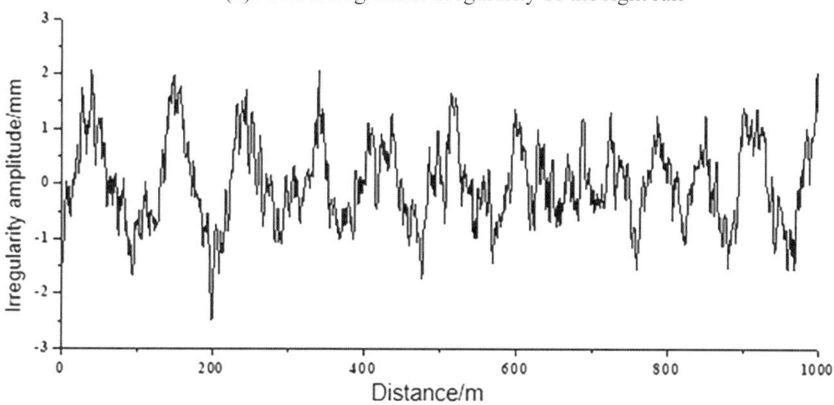


(c) Vertical irregularity of the left rail

**Fig. 3.34** Sample time-domain simulation of ballast track irregularity for high-speed railroads in China

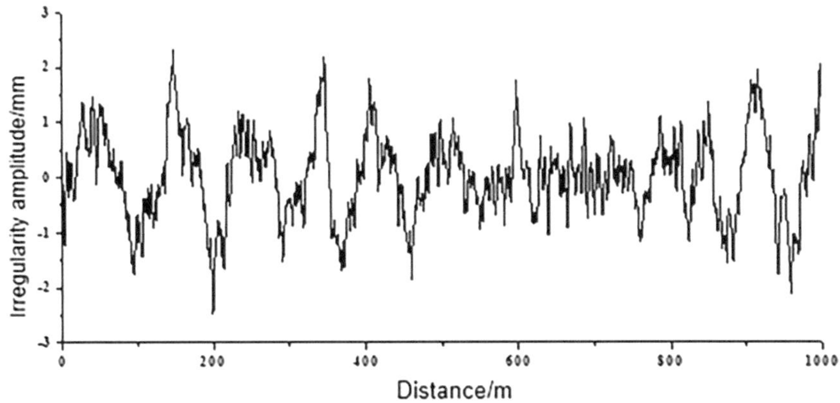


(d) Vertical irregularity of theright rail

**Fig. 3.34** (continued)

## 3.5 Solution of the System Equations

### *3.5.1 Numerical Solution of the Differential Equations*

The motion equations of the system can ultimately be formulated as a system of second-order differential equations. There are two main categories of methods to solve these motion equations: mode superposition and direct integration. For linear systems, the mode superposition method is commonly used to solve the system's dynamic response. The process can be divided into three steps: The system's equations of motion can ultimately be formulated as a system of second-order differential equations. The methods to solve these equations can be divided into two main categories: Mode Superposition and Direct Integration. For linear systems, the Mode Superposition method can be used to solve the system's dynamic response in three steps: first, solve the generalized eigenvalue problem; then, solve the n uncoupled equations of motion, which can be done using direct integration or the Duhamel integral; finally, superimpose the responses of each mode to obtain the overall system response. For nonlinear systems, there is no orthogonality between modes, and thus, the Mode Superposition method cannot be applied.

This section focuses on the Direct Integration Method for solving the equations of motion of linear and nonlinear systems, which involves directly integrating the equations of motion without applying any transformations. Essentially, the Direct Integration Method is based on two fundamental concepts: first, that at any given time within the solution domain $0 \leq t \leq T$, the equation of motion must be approximately satisfied under certain conditions; and second, within a specific $\Delta t$ region, approximate functional forms for displacement $\boldsymbol{u}$, velocity $\dot{\boldsymbol{u}}$, and acceleration $\ddot{\boldsymbol{u}}$ are assumed.

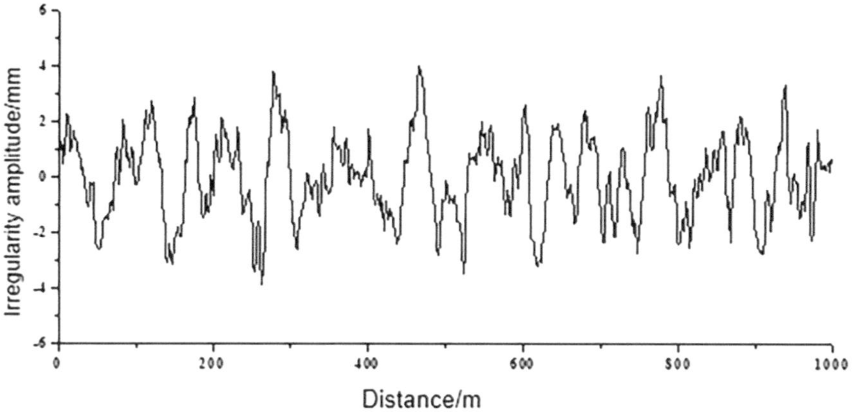


(a) Lateral alignment irregularity of the left rail

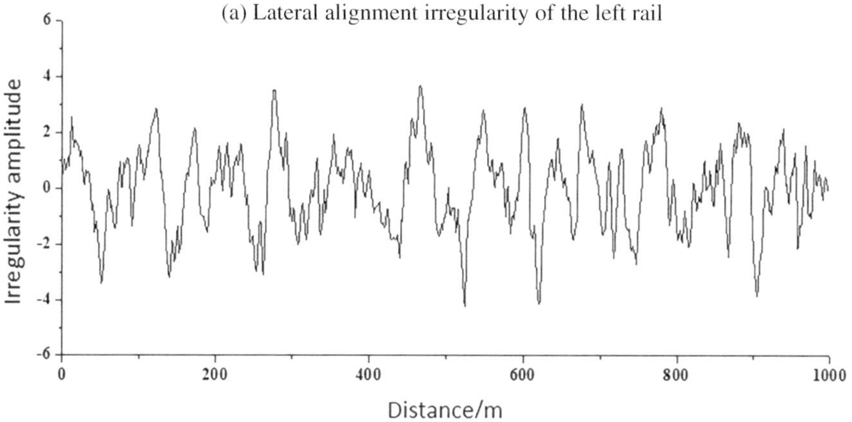


(b) Lateral alignment irregularity of the right rail

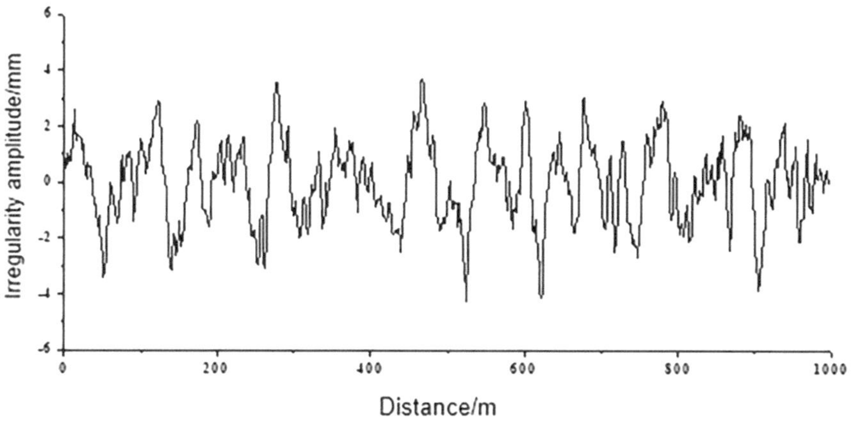


(c) Vertical irregularity of the left rail

**Fig. 3.35** A sample of German time-domain simulations of low-disturbance track irregularities

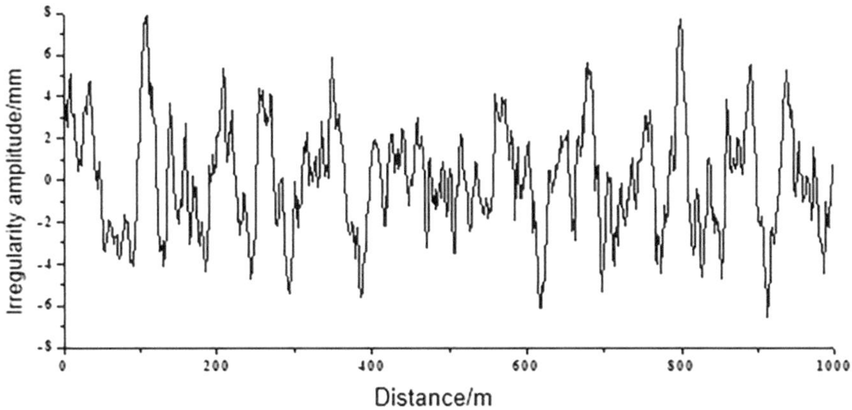


(d) Vertical irregularity of the right rail

**Fig. 3.35** (continued)

Assuming that the displacement $\boldsymbol{u}_0$ and velocity $\dot{\boldsymbol{u}}_0$ are known at time $t = 0$, the solution domain is divided into nnn time intervals. Given that solutions at times $0, \Delta t, 2\Delta t, \ldots, t$ have been obtained, the goal is to calculate the solution at time $t + \Delta t$. Once an algorithm for calculating the solution at time $t + \Delta t$ is established, solutions at all discrete time steps can be determined. Therefore, this method is also referred to as the step-by-step integration method. The step-by-step integration method is an approximate solution method, and each integration step introduces an error. The sources of errors are mainly of two types: One type arises from neglecting higher-order small terms in the approximations for velocity and acceleration, referred to as truncation errors. These errors are inherent to the method itself and increase with the time step $\Delta t$. The other type of error is rounding errors, which occur due to the finite word length of the computer. Any numbers that exceed the machine's capacity are rounded off. Although the rounding errors in each step are small, these errors may accumulate during the solution process and potentially cause the results to deviate significantly. This is the stability issue of the direct integration method. If, regardless of the size of $\Delta t$, the integration results do not grow unbounded for any given initial conditions, the method is termed unconditionally stable. On the other hand, if the time step $\Delta t$ must be smaller than a certain critical value $\Delta t_{\text{cr}}$ to prevent unbounded growth of the results, the method is considered conditionally stable.

1. **Central Difference Method**

The system's equation of motion at time $t$ can be written as the following:

$$\boldsymbol{M}\ddot{\boldsymbol{u}}_t + \boldsymbol{C}\dot{\boldsymbol{u}}_t + \boldsymbol{K}\boldsymbol{u}_t = \boldsymbol{F}_t \tag{3.5.1}$$

Expanding the displacement function by Taylor series, we can get the front and back difference equations

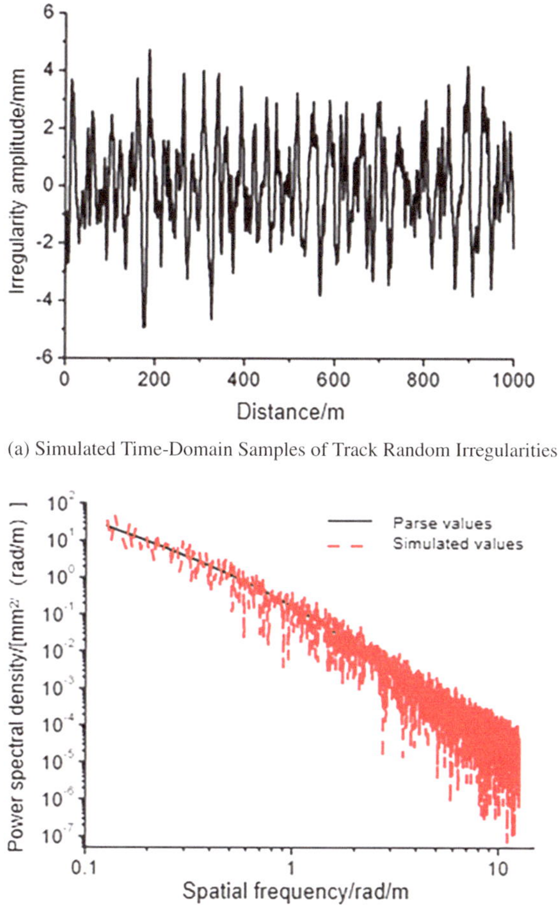


(a) Simulated Time-Domain Samples of Track Random Irregularities

(b) Comparison of Simulated and Analytical Track Spectrum Values

**Fig. 3.36** Comparison of track spectrum conversion to time domain samples

$$\boldsymbol{u}_{t+\Delta t} = \boldsymbol{u}_t + \dot{\boldsymbol{u}}_t \Delta t + \frac{1}{2}\ddot{\boldsymbol{u}}_t \Delta t^2 + \cdots \tag{3.5.2}$$

$$\boldsymbol{u}_{t-\Delta t} = \boldsymbol{u}_t - \dot{\boldsymbol{u}}_t \Delta t + \frac{1}{2}\ddot{\boldsymbol{u}}_t \Delta t^2 + \cdots \tag{3.5.3}$$

By subtracting and adding the above equations and neglecting higher-order small quantities, we can approximate the velocity and acceleration at time $t$ using the displacements at times$t - \Delta t$, t, and $t + \Delta t$.

$$\dot{\boldsymbol{u}}_t = \frac{1}{2\Delta t}(\boldsymbol{u}_{t+\Delta t} - \boldsymbol{u}_{t-\Delta t}) \tag{3.5.4}$$

$$\ddot{\boldsymbol{u}}_t = \frac{1}{\Delta t^2}(\boldsymbol{u}_{t+\Delta t} - 2\boldsymbol{u}_t + \boldsymbol{u}_{t-\Delta t}) \tag{3.5.5}$$

To solve for the displacement at time $t + \Delta t$, we substitute the velocity and acceleration approximations at time $t$ from Eqs. (3.5.4) and (3.5.5) into the motion equation at time $t$ (3.5.1). After simplifying, we obtain the following equation for the displacement at time $t + \Delta t$:

$$\widehat{\boldsymbol{M}}\boldsymbol{u}_{t+\Delta t} = \widehat{\boldsymbol{F}}_t \tag{3.5.6}$$

Among them

$$\widehat{\boldsymbol{M}} = \frac{1}{\Delta t^2}\boldsymbol{M} + \frac{1}{2\Delta t}\boldsymbol{C} \tag{3.5.7}$$

$$\widehat{\boldsymbol{F}}_t = \boldsymbol{F}_t - \left(\boldsymbol{K} - \frac{2}{\Delta t^2}\boldsymbol{M}\right)\boldsymbol{u}_t - \left(\frac{1}{\Delta t^2}\boldsymbol{M} - \frac{1}{2\Delta t}\boldsymbol{C}\right)\boldsymbol{u}_{t-\Delta t} \tag{3.5.8}$$

It can be seen that with the known displacements at time $t - \Delta t$ and $t$, namely $\boldsymbol{u}_{t-\Delta t}$ and $\boldsymbol{u}_t$, the displacement at time $t + \Delta t$ can be determined, and the velocity $\dot{\boldsymbol{u}}_t$ and acceleration $\ddot{\boldsymbol{u}}_t$ can be obtained from Eqs. (3.5.4) and (3.5.5). This method, where the displacement $\boldsymbol{u}_{t+\Delta t}$ at the current step is determined using the known values $\boldsymbol{u}_{t-\Delta t}$ and $\boldsymbol{u}_t$ from the previous two steps, is called the two-step method. As shown by the equation, it is not sufficient to only provide the initial values $\boldsymbol{u}_0$ and $\dot{\boldsymbol{u}}_0$ to determine $\ddot{\boldsymbol{u}}_{\Delta t}$; additional information, namely $\boldsymbol{u}_{-\Delta t}$, must also be specified, indicating the "starting point" issue in this method. By applying the backward difference formula (3.5.3) and setting $t = 0$, the solution can be derived.

$$\boldsymbol{u}_{-\Delta t} = \boldsymbol{u}_0 - \dot{\boldsymbol{u}}_0\Delta t + \frac{1}{2}\ddot{\boldsymbol{u}}_0\Delta t^2 \tag{3.5.9}$$

The algorithm for solving the equations of motion step-by-step by the central difference method can be summarized as follows.

(1) Initial calculations

① Form the stiffness matrix $\boldsymbol{K}$ and quality matrices $\boldsymbol{M}$, damping matrix $\boldsymbol{C}$;
② Given $\boldsymbol{u}_0$ and $\dot{\boldsymbol{u}}_0$, and solving for $\ddot{\boldsymbol{u}}_0$;
③ Selecting the time step $\Delta t$, and calculating the constant of integration

$$a_0 = \frac{1}{\Delta t^2},\ a_1 = \frac{1}{2\Delta t},\ a_2 = 2a_0,\ a_3 = \frac{1}{a_2}$$

④ Calculation $\boldsymbol{u}_{-\Delta t}$ : $\boldsymbol{u}_{-\Delta t} = \boldsymbol{u}_0 - \Delta t\dot{\boldsymbol{u}}_0 + c_3\ddot{\boldsymbol{u}}_0$;
⑤ Formation of an effective quality matrix: $\hat{\boldsymbol{M}} = a_0\boldsymbol{M} + a_1\boldsymbol{C}$;
⑥ Right $\hat{\boldsymbol{M}}$ Triangulation. $\hat{\boldsymbol{M}} = \boldsymbol{L}\boldsymbol{D}\boldsymbol{L}^{\mathrm{T}}$.

(2) Calculating the response at each time step

① Calculating the effective load at time t: $\hat{\boldsymbol{F}}_t$ : $\hat{\boldsymbol{F}}_t = \boldsymbol{F}_t - (\boldsymbol{K} - a_2\boldsymbol{M})\boldsymbol{u}_t - (a_0\boldsymbol{M} - a_1\boldsymbol{C})\boldsymbol{u}_{t-\Delta t}$
② Calculate the displacement at time $t + \Delta t$ : $\Delta \boldsymbol{u}_{t+\Delta t}$ : $\boldsymbol{LDL}^{\mathrm{T}}\boldsymbol{u}_{t+\Delta t} = \hat{\boldsymbol{F}}_t$
③ Calculate the velocity and acceleration at time t

$$\dot{\boldsymbol{u}}_t = a_1(\boldsymbol{u}_{t+\Delta t} - \boldsymbol{u}_{t-\Delta t}) \tag{3.5.10}$$

$$\ddot{\boldsymbol{u}}_t = a_0(\boldsymbol{u}_{t+\Delta t} - 2\boldsymbol{u}_t + \boldsymbol{u}_{t-\Delta t}) \tag{3.5.11}$$

This method for solving $\boldsymbol{u}_{t+\Delta t}$ uses the motion equation at time $t$, where the stiffness matrix $\boldsymbol{K}$ does not appear on the left-hand side of the recursive formula (3.5.6). Therefore, the central difference method does not require the triangular decomposition of matrix $\boldsymbol{K}$. This integration format is referred to as explicit integration.

When the mass matrix $\boldsymbol{M}$ is diagonal, and the damping matrix $\boldsymbol{C}$ can be ignored or is also diagonal, there is no need for matrix inversion during the solution process. Additionally, at each step when calculating the load vector $\hat{\boldsymbol{F}}_t$, the stiffness matrix $\boldsymbol{K}$ of a nonlinear system changes with each step. The modified stiffness matrix $\boldsymbol{K}$ is only multiplied by $\boldsymbol{u}_t$, which results in a relatively low computational cost.

The central difference method is conditionally stable with a time step that $\Delta t$ need to be met

$$\Delta t \leq \Delta t_{\mathrm{cr}} = \frac{T_n}{\pi}\left(\sqrt{1+\xi^2} - \xi\right) \tag{3.5.12}$$

In the formula, the $T_n$ is the minimum intrinsic period of the system. When $T_n$ is very little. $\Delta t$ is also necessarily small, with many computational steps and potentially large computational time.

2. **Newmark-β Method**

The Newmark-$\beta$ method is a method for calculating the response of a system based on an assumed pattern of acceleration changes within a time increment. According to Lagrange's median theorem, put, the velocity at the moment $t + \Delta t$ is expressed as

$$\dot{\boldsymbol{u}}_{t+\Delta t} = \dot{\boldsymbol{u}}_t + \ddot{\boldsymbol{u}}_t^* \Delta t \tag{3.5.13}$$

In the equation, $\boldsymbol{u}_t^*$ represents the value of acceleration $\ddot{\boldsymbol{u}}$ at a certain point within the interval $(0,\Delta t)$. The Newmark-$\beta$ method approximates by assuming:

$$\ddot{\boldsymbol{u}}_t^* = (1-\gamma)\ddot{\boldsymbol{u}}_t + \gamma\ddot{\boldsymbol{u}}_{t+\Delta t} \tag{3.5.14}$$

So, the speed of the moment $t + \Delta t$ is

$$\dot{\boldsymbol{u}}_{t+\Delta t} = \dot{\boldsymbol{u}}_t + (1 - \gamma)\ddot{\boldsymbol{u}}_t \Delta t + \gamma \ddot{\boldsymbol{u}}_{t+\Delta t} \Delta t \tag{3.5.15}$$

Deriving the displacement from the Taylor series

$$\boldsymbol{u}_{t+\Delta t} = \boldsymbol{u}_t + \dot{\boldsymbol{u}}_t \Delta t + \left(\frac{1}{2} - \beta\right)\ddot{\boldsymbol{u}}_t \Delta t^2 + \beta \ddot{\boldsymbol{u}}_{t+\Delta t} \Delta t^2 \tag{3.5.16}$$

In Eqs. (3.5.15) and (3.5.16), $\gamma$ and $\beta$ are parameters related to accuracy and stability. When $\gamma > 1/2$, algorithmic damping occurs; when $\gamma < 1/2$, negative damping arises, leading to a gradual increase in amplitude during the integration process. The critical value $\gamma = 1/2$ is commonly used. Under this condition, different values of $\beta$ correspond to specific methods: $\beta = 0$ represents the constant acceleration method (i.e., the central difference method); $\beta = 1/4$ corresponds to the average acceleration method; and $\beta = 1/6$ to the linear acceleration method.

The Newmark-$\beta$ method is satisfied for each step of the integral, the equations of motion for the moments of $t + \Delta t$

$$\boldsymbol{M}\ddot{\boldsymbol{u}}_{t+\Delta t} + \boldsymbol{C}\dot{\boldsymbol{u}}_{t+\Delta t} + \boldsymbol{K}\boldsymbol{u}_{t+\Delta t} = \boldsymbol{F}_{t+\Delta t} \tag{3.5.17}$$

From Eqs. (3.5.15) and (3.5.16), writing expressions for the velocity and acceleration at the moment of $t + \Delta t$

$$\dot{\boldsymbol{u}}_{t+\Delta t} = \frac{\gamma}{\beta \Delta t}(\dot{\boldsymbol{u}}_{t+\Delta t} - \dot{\boldsymbol{u}}_t) + \left(1 - \frac{\gamma}{\beta}\right)\dot{\boldsymbol{u}}_t + \left(1 - \frac{\gamma}{2\beta}\right)\Delta t \ddot{\boldsymbol{u}}_t \tag{3.5.18}$$

$$\ddot{\boldsymbol{u}}_{t+\Delta t} = \frac{1}{\beta \Delta t^2}(\dot{\boldsymbol{u}}_{t+\Delta t} - \dot{\boldsymbol{u}}_t) - \frac{1}{\beta \Delta t}\dot{\boldsymbol{u}}_t - \left(\frac{1}{2\beta} - 1\right)\ddot{\boldsymbol{u}}_t \tag{3.5.19}$$

Substituting Eqs. (3.5.18) and (3.5.19) into Eq. (3.5.17) yields:

$$\hat{\boldsymbol{K}}\boldsymbol{u}_{t+\Delta t} = \hat{\boldsymbol{F}}_{t+\Delta t} \tag{3.5.20}$$

Among them

$$\hat{\boldsymbol{K}} = \boldsymbol{K} + \frac{1}{\beta \Delta t^2}\boldsymbol{M} + \frac{\gamma}{\beta \Delta t}\boldsymbol{C} \tag{3.5.21}$$

$$\begin{aligned} \hat{\boldsymbol{F}}_{t+\Delta t} = \boldsymbol{F}_{t+\Delta t} &+ \boldsymbol{M}\left[\frac{1}{\beta \Delta t^2}\boldsymbol{u}_t + \frac{1}{\beta \Delta t}\dot{\boldsymbol{u}}_t + \left(\frac{1}{2\beta} - 1\right)\ddot{\boldsymbol{u}}_t\right] \\ &+ \boldsymbol{C}\left[\frac{\gamma}{\beta \Delta t}\boldsymbol{u}_t + \left(\frac{\gamma}{\beta} - 1\right)\dot{\boldsymbol{u}}_t + \left(\frac{\gamma}{2\beta} - 1\right)\Delta t \ddot{\boldsymbol{u}}_t\right] \end{aligned} \tag{3.5.22}$$

The Newmark-$\beta$ method step-by-step algorithm for solving the equations of motion can be summarized as follows.

(1) Initial calculations

① Form the stiffness matrix $\boldsymbol{K}$ and quality matrices $\boldsymbol{M}$, damping matrix $\boldsymbol{C}$;
② Given $\boldsymbol{u}_0$ and $\dot{\boldsymbol{u}}_0$, and solving for $\ddot{\boldsymbol{u}}_0$;
③ Selecting the time step $\Delta t$ and the parameters $\gamma$ and $\beta$, and calculating the constant of integration

$$a_0 = \frac{1}{\beta\Delta t^2}, a_1 = \frac{\gamma}{\beta\Delta t}, a_2 = \frac{1}{\beta\Delta t}, a_3 = \frac{1}{2\beta} - 1,$$
$$a_4 = \frac{\gamma}{\beta} - 1, a_5 = \Delta t\left(\frac{\gamma}{2\beta} - 1\right), a_6 = \Delta t(1-\gamma), a_7 = \gamma\Delta t$$

④ Formation of effective stiffness matrix: $\hat{\boldsymbol{K}} = \boldsymbol{K} + a_0\boldsymbol{M} + a_1\boldsymbol{C}$;
⑤ Triangular decomposition of $\hat{\boldsymbol{K}} : \hat{\boldsymbol{K}} = \boldsymbol{LDL}^{\mathrm{T}}$

(2) Calculating the response at each time step

① Computation of the effective load at time $t + \Delta t$:

$$\hat{\boldsymbol{F}}_{t+\Delta t} = \boldsymbol{F}_{t+\Delta t} + \boldsymbol{M}(a_0\boldsymbol{u}_t + a_2\dot{\boldsymbol{u}}_t + a_3\ddot{\boldsymbol{u}}_t) + \boldsymbol{C}(a_1\boldsymbol{u}_t + a_4\dot{\boldsymbol{u}}_t + a_5\ddot{\boldsymbol{u}}_t)$$

② Solve for the displacement $\boldsymbol{u}_{t+\Delta t}$ at time $t + \Delta t$:

$$\boldsymbol{LDL}^{\mathrm{T}}\boldsymbol{u}_{t+\Delta t} = \hat{\boldsymbol{F}}_{t+\Delta t}$$

③ Compute the acceleration and velocity at time $t + \Delta t$:

$$\ddot{\boldsymbol{u}}_{t+\Delta t} = a_0(\boldsymbol{u}_{t+\Delta t} - \boldsymbol{u}_t) - a_2\dot{\boldsymbol{u}}_t - a_3\ddot{\boldsymbol{u}}_t \tag{3.5.23}$$

$$\dot{\boldsymbol{u}}_{t+\Delta t} = \dot{\boldsymbol{u}}_t + a_6\ddot{\boldsymbol{u}}_t + a_7\ddot{\boldsymbol{u}}_{t+\Delta t} \tag{3.5.24}$$

The Newmark-$\beta$ method solves the motion equation at time $t+\Delta t$, and is therefore an implicit integration scheme. The effective stiffness matrix $\hat{\boldsymbol{K}}$ includes the stiffness matrix $\boldsymbol{K}$, so solving for $\boldsymbol{u}_{t+\Delta t}$ requires inverting $\hat{\boldsymbol{K}}$. When $\gamma \geq \frac{1}{2}$ and $\beta \geq \frac{1}{4}\left(\gamma + \frac{1}{2}\right)^2$, the Newmark-β method is unconditionally stable. For nonlinear systems, the stiffness matrix is time-dependent, and $\hat{\boldsymbol{K}}$ must be inverted at each step.

3. **Wilson-θ Method**

The Wilson-$\theta$ method is based on the linear acceleration method, introducing a parameter $\theta > 1$. Within the time domain $t \sim t + \theta\Delta t$, it assumes that acceleration varies linearly, and requires the system to satisfy the equation of motion at time $t + \theta\Delta t$.

$$\boldsymbol{M}\ddot{\boldsymbol{u}}_{t+\theta\Delta t} + \boldsymbol{C}\dot{\boldsymbol{u}}_{t+\theta\Delta t} + \boldsymbol{K}\boldsymbol{u}_{t+\theta\Delta t} = \boldsymbol{F}_{t+\theta\Delta t} \tag{3.5.25}$$

During the time period $t \sim t + \theta\Delta t$, at any point in time $t \sim \tau(0 \leq \tau \leq \theta\Delta t)$, the acceleration:

$$\ddot{\boldsymbol{u}}_{t+\tau} = \ddot{\boldsymbol{u}}_t + \frac{\tau}{\theta\Delta t}(\ddot{\boldsymbol{u}}_{t+\theta\Delta t} - \ddot{\boldsymbol{u}}_t) \tag{3.5.26}$$

Thus, the velocity and displacement at any time $t + \tau$ are given by:

$$\dot{\boldsymbol{u}}_{t+\tau} = \dot{\boldsymbol{u}}_t + \ddot{\boldsymbol{u}}_t\tau + \frac{\tau^2}{2\theta\Delta t}(\ddot{\boldsymbol{u}}_{t+\theta\Delta t} - \ddot{\boldsymbol{u}}_t) \tag{3.5.27}$$

$$\boldsymbol{u}_{t+\tau} = \boldsymbol{u}_t + \dot{\boldsymbol{u}}_t\tau + \frac{1}{2}\ddot{\boldsymbol{u}}_t\tau^2 + \frac{\tau^3}{6\theta\Delta t}(\ddot{\boldsymbol{u}}_{t+\theta\Delta t} - \ddot{\boldsymbol{u}}_t) \tag{3.5.28}$$

Letting $\tau = \theta\Delta t$, the acceleration and velocity at time $t + \theta\Delta t$ are obtained and substituted into Eq. (3.5.25). Considering that the external force $\boldsymbol{F}_{t+\tau}$ also varies linearly within the interval $t \sim t + \theta\Delta t$, we obtain:

$$\hat{\boldsymbol{K}}\boldsymbol{u}_{t+\theta\Delta t} = \hat{\boldsymbol{F}}_{t+\theta\Delta t} \tag{3.5.29}$$

Among them

$$\hat{\boldsymbol{K}} = \boldsymbol{K} + \frac{6}{\theta^2\Delta t^2}\boldsymbol{M} + \frac{3}{\theta\Delta t}\boldsymbol{C} \tag{3.5.30}$$

$$\begin{aligned}\hat{\boldsymbol{F}}_{t+\theta\Delta t} &= \boldsymbol{F}_t + \theta(\boldsymbol{F}_{t+\Delta t} - \boldsymbol{F}_t) + \boldsymbol{M}\left(\frac{6}{\theta^2\Delta t^2}\boldsymbol{u}_t + \frac{6}{\theta\Delta t}\dot{\boldsymbol{u}}_t + 2\ddot{\boldsymbol{u}}_t\right) \\ &+ \boldsymbol{C}\left(\frac{3}{\theta\Delta t}\boldsymbol{u}_t + 2\dot{\boldsymbol{u}}_t + \frac{\theta\Delta t}{2}\ddot{\boldsymbol{u}}_t\right)\end{aligned} \tag{3.5.31}$$

After obtaining the displacement $\boldsymbol{u}_{t+\theta\Delta t}$ at time $t + \theta\Delta t$ from Eq. (3.5.29), the acceleration, velocity, and displacement at time $t + \Delta t$ can be calculated by substituting $\tau = \Delta t$ into Eqs. (3.5.26) through (3.5.28).

The algorithm of Wilson-$\theta$ method for solving the equation of motion can be summarized as follows:

(1) Initial calculations

① Form the stiffness matrix $\boldsymbol{K}$ and quality matrices $\boldsymbol{M}$, damping matrix $\boldsymbol{C}$;
② Given $\boldsymbol{u}_0$ and $\dot{\boldsymbol{u}}_0$, and solving for $\ddot{\boldsymbol{u}}_0$;
③ Selecting the time step $\Delta t$ and the parameters $\theta$, and calculating the constant of integration:

$$a_0 = \frac{6}{\theta^2 \Delta t^2},\ a_1 = \frac{3}{\theta \Delta t},\ a_2 = 2a_1,\ a_3 = \frac{\theta \Delta t}{2},\ a_4 = \frac{a_0}{\theta}$$
$$a_5 = -\frac{a_2}{\theta},\ a_6 = 1 - \frac{3}{\theta},\ a_7 = \frac{\Delta t}{2},\ a_8 = \frac{\Delta t^2}{6} \tag{3.5.32}$$

④ Formation of effective stiffness matrix: $\hat{\boldsymbol{K}} = \boldsymbol{K} + a_0\boldsymbol{M} + a_1\boldsymbol{C}$;
⑤ Perform triangular decomposition on matrix $\hat{\boldsymbol{K}}$ : $\hat{\boldsymbol{K}} = \boldsymbol{LDL}^{\mathrm{T}}$.

(2) Calculating the response at each time step

① Calculate the effective load at time $t + \theta \Delta t$

$$\widehat{\boldsymbol{F}}_{t+\Delta t} = \boldsymbol{F}_t + \theta(\boldsymbol{F}_{t+\Delta t} - \boldsymbol{F}_t) + \boldsymbol{M}(a_0\boldsymbol{u}_t + a_2\dot{\boldsymbol{u}}_t + 2\ddot{\boldsymbol{u}}_t) + \boldsymbol{C}(a_1\boldsymbol{u}_t + 2\dot{\boldsymbol{u}}_t + a_3\ddot{\boldsymbol{u}}_t)$$

② Solve for the displacement $\boldsymbol{u}_{t+\theta\Delta t}$ at time $t + \theta\Delta t$ : $\boldsymbol{LDL}^{\mathrm{T}}\boldsymbol{u}_{t+\theta\Delta t} = \hat{\boldsymbol{F}}_{t+\theta\Delta t}$
③ Compute the acceleration, velocity, and displacement at time $t + \Delta t$.

$$\ddot{\boldsymbol{u}}_{t+\Delta t} = a_4(\boldsymbol{u}_{t+\theta\Delta t} - \boldsymbol{u}_t) + a_5\dot{\boldsymbol{u}}_t + a_6\ddot{\boldsymbol{u}}_t \tag{3.5.33}$$

$$\dot{\boldsymbol{u}}_{t+\Delta t} = \dot{\boldsymbol{u}}_t + a_7(\ddot{\boldsymbol{u}}_{t+\Delta t} + \ddot{\boldsymbol{u}}_t) \tag{3.5.34}$$

$$\dot{\boldsymbol{u}}_{t+\Delta t} = \boldsymbol{u}_t + t\dot{\boldsymbol{u}}_t + a_8(\ddot{\boldsymbol{u}}_{t+\Delta t} + 2\ddot{\boldsymbol{u}}_t) \tag{3.5.35}$$

Similar to the Newmark-$\beta$ method, the Wilson-$\theta$ method is an implicit integration scheme. For nonlinear systems, if the stiffness matrix $\boldsymbol{K}$ varies at each step, matrix inversion is required at every time step. In addition, both methods are self-starting. When $\theta \geq \left(1 + \sqrt{3}\right)/2$, the Wilson-$\theta$ method is unconditionally stable.

## 4. Generalized α Method (Generalized α Method)

The generalized $\alpha$ method comprehensively extends the design concept of the Newmark-$\beta$ method. The weighted representation of acceleration is more flexible, and the generalized α method eliminates the influence of higher-order mode responses by introducing non-positive parameters and using an incremental step form at each time step, thus improving computational accuracy. Compared to the Newmark-$\beta$ method, the generalized $\alpha$ method allows for larger time step sizes, resulting in higher computational efficiency.

$$\boldsymbol{u}_{t+1} = \boldsymbol{u}_t + h\dot{\boldsymbol{u}}_t + h^2(\varepsilon\ddot{\boldsymbol{u}}_t + \beta\ddot{\boldsymbol{u}}_{t+1}) \tag{3.5.36}$$

$$\dot{\boldsymbol{u}}_{t+1} = \dot{\boldsymbol{u}}_t + (\mu\ddot{\boldsymbol{u}}_t + \gamma\ddot{\boldsymbol{u}}_{t+1})h \tag{3.5.37}$$

The equilibrium equation is not limited to a specific moment; it uses a linear combination of forces at previous and subsequent time steps.

$$(1-\alpha)\boldsymbol{M}\ddot{\boldsymbol{u}}_{t+1}+\alpha\boldsymbol{M}\ddot{\boldsymbol{u}}_t+(1-\delta)\boldsymbol{C}\dot{\boldsymbol{u}}_{t+1}+\delta\boldsymbol{C}\dot{\boldsymbol{u}}_t \\ +(1-\eta)\boldsymbol{K}\boldsymbol{u}_{t+1}+\eta\boldsymbol{K}\boldsymbol{u}_t=(1-\eta)\boldsymbol{F}_{t+1}+\eta\boldsymbol{F}_t \tag{3.5.38}$$

Here, $h$ is the time step, and the subscript $t$ represents the $t$-th time step. In the algorithm equation, there are a total of 7 design parameters, which can be determined based on the performance requirements of the algorithm. Compared to traditional algorithms, it offers greater design flexibility, leading to more possibilities. The final seven design parameters are shown in Table 3.14, where $\rho$ is used to adjust the dissipation performance of the algorithm. When $\rho = 1$, there is no dissipation, and as $\rho$ decreases from 1 to 0, the dissipation capability gradually increases.

Generalized $\alpha$ Method Solution Steps:

1. Initial value calculation

(1) Formation of a systematic matrix $\boldsymbol{K}, \boldsymbol{M}$ and $\boldsymbol{C}$;
(2) Determining the initial conditions that $\boldsymbol{u}_0$, $\dot{\boldsymbol{u}}_0$, $\ddot{\boldsymbol{u}}_0$ where the initial acceleration is calculated as in the center difference method;
(3) Selecting the time step $h$ and the algorithm parameters $\rho$, Calculate the constant of integration according to the design parameter equations in Table 3.14.

$$\begin{aligned}
&m_1=\beta(1-\eta)h^2,\ m_2=\gamma(1-\delta)h,\ m_3=(1-\alpha),\ m_4=\beta\eta h^2,\ m_5=h(1-\alpha),\\
&m_6=(\varepsilon(1-\alpha)-\alpha\beta)h^2,\ m_7=(\gamma(1-\delta)-\beta)h^2,\ m_8=(1-\delta)(\varepsilon\gamma-\beta\mu)h^3,\\
&m_9=-\beta\eta h^2,\ m_{10}=\gamma/(\beta h),\ m_{11}=(\beta-\gamma)/\beta,\ m_{12}=(\beta\mu-\varepsilon\gamma)h/\beta,\\
&m_{13}=1/(\beta h^2),\ m_{14}=-1/(\beta h),\ m_{15}=-\varepsilon/\beta.
\end{aligned} \tag{3.5.39}$$

(4) Formation of equivalent stiffness matrix.

$$\hat{\boldsymbol{K}}=m_1\boldsymbol{K}+m_2\boldsymbol{C}+m_3\boldsymbol{M} \tag{3.5.40}$$

**Table 3.14** Algorithmic parameters of the generalized $\alpha$ method

| | CH-α methodology $\rho \in [0, 1]$ | The HHT-α method $\rho \in [1/2, 1]$ | WBZ-α method $\rho \in [0, 1]$ |
|---|---|---|---|
| $\alpha$ | $\frac{2\rho-1}{1+\rho}$ | 0 | $\frac{\rho-1}{1+\rho}$ |
| $\delta$ | $\frac{\rho}{1+\rho}$ | $\frac{1-\rho}{1+\rho}$ | 0 |
| $\eta$ | $\frac{\rho}{1+\rho}$ | $\frac{1-\rho}{1+\rho}$ | 0 |
| $\varepsilon$ | $\frac{\rho^2+2\rho-1}{(1+\rho)^2}$ | $\frac{\rho^2+2\rho-1}{(1+\rho)^2}$ | $\frac{\rho^2+2\rho-1}{(1+\rho)^2}$ |
| $\beta$ | $\frac{1}{(1+\rho)^2}$ | $\frac{1}{(1+\rho)^2}$ | $\frac{1}{(1+\rho)^2}$ |
| $\mu$ | $\frac{3\rho-1}{2(1+\rho)}$ | $\frac{3\rho-1}{2(1+\rho)}$ | $\frac{3\rho-1}{2(1+\rho)}$ |
| $\gamma$ | $\frac{3-\rho}{2(1+\rho)}$ | $\frac{3-\rho}{2(1+\rho)}$ | $\frac{3-\rho}{2(1+\rho)}$ |

(5) Triangulation of the stiffness matrix.

$$\hat{\boldsymbol{K}} = \boldsymbol{LDL}^{\mathrm{T}} \tag{3.5.41}$$

2. Loop for each time step

(1) Calculate the equivalent load at the next moment

$$\begin{aligned}\widehat{\boldsymbol{F}}_t &= m_1\boldsymbol{F}_t + m_4\boldsymbol{F}_{t-h} + \boldsymbol{M}(m_3\boldsymbol{u}_{t-h} + m_5\dot{\boldsymbol{u}}_{t-h} + m_6\ddot{\boldsymbol{u}}_{t-h}) \\ &\quad + \boldsymbol{C}(m_2\boldsymbol{u}_{t-h} + m_7\dot{\boldsymbol{u}}_{t-h} + m_8\ddot{\boldsymbol{u}}_{t-h}) + m_9\boldsymbol{K}\boldsymbol{u}_{t-h}\end{aligned} \tag{3.5.42}$$

(2) Calculate the displacement at the next moment

$$(\boldsymbol{LDL}^{\mathrm{T}})\boldsymbol{u}_t = \hat{\boldsymbol{F}}_t \tag{3.5.43}$$

(3) Calculate the velocity and acceleration at the next moment

$$\begin{aligned}\dot{\boldsymbol{u}}_t &= m_{10}(\boldsymbol{u}_t - \boldsymbol{u}_{t-h}) + m_{11}\dot{\boldsymbol{u}}_{t-h} + m_{12}\ddot{\boldsymbol{u}}_{t-h} \\ \ddot{\boldsymbol{u}}_t &= m_{13}(\boldsymbol{u}_t - \boldsymbol{u}_{t-h}) + m_{14}\dot{\boldsymbol{u}}_{t-h} + m_{15}\ddot{\boldsymbol{u}}_{t-h}\end{aligned} \tag{3.5.44}$$

(4) Assign the results of the computation to the previous moment for the next recursive computation, respectively

$$\boldsymbol{u}_{t-h} = \boldsymbol{u}_t,\ \dot{\boldsymbol{u}}_{t-h} = \dot{\boldsymbol{u}}_t,\ \ddot{\boldsymbol{u}}_{t-h} = \ddot{\boldsymbol{u}}_t \tag{3.5.45}$$

(5) Return to step (1)

Any numerical integration method is influenced by the chosen time step and the characteristics of the dynamic load. The accuracy of direct integration methods primarily depends on the ratio between the time step and the natural period. For unconditionally stable implicit integration methods, the time step is determined by the desired accuracy of the problem and is controlled by truncation error. In typical structural dynamic problems, low-frequency components dominate the structural response, so it is typically sufficient to accurately capture the contribution of the first $P$ natural modes of the system. For unconditionally stable schemes, the time step $\Delta t$ is typically taken as approximately $T_p/20$, where $T_p$ is the $P$-th natural period of the system. Compared with the central difference method, unconditionally stable schemes allow for significantly larger time steps—up to 1000 times larger. Moreover, using larger time steps can help filter out the influence of spurious high-order modes on the system response. Although the central difference method has a smaller critical time step, it requires significantly less computational effort per time step than implicit schemes.

### 3.5.2 Solution Methods for the Vehicle–Track System Equations

The dynamic interaction analysis model of the vehicle–track system essentially constitutes a time-varying differential equation. In general, depending on the form of the established system equations, two primary solution methods are currently employed: The first method consolidates all degrees of freedom of the vehicle and the track structure into a unified system of equations, which is then solved synchronously using a direct integration method. The second method divides the vehicle and track structure system into two subsystems—the vehicle and the track (rail or bridge)—at the wheel–rail contact interface. The motion equations for the vehicle and the track structure are established separately, and the interaction between them is expressed via the wheel–rail contact relationship and the equilibrium of interaction forces. An iterative procedure is used to solve for the system response.

In the first method, the coefficient matrices of the system equations vary with the vehicle's position along the track. As a result, the mass, stiffness, and damping matrices must be regenerated and decomposed at each time step. Furthermore, as the number of vehicles increases, the number of coupled degrees of freedom grows, substantially increasing the computational cost. In the case of nonlinear equations, iterations are additionally required within each time step, making the solution process even more complex.

In contrast, when the vehicle and track systems are modeled using separate motion equations, the geometric compatibility condition and the equilibrium of interaction forces at the wheel–rail interface are satisfied through an iterative procedure. During the computation, the mass, stiffness, and damping matrices on the left-hand side of the equations remain constant, and only the right-hand terms need to be updated at each time step. This reduces the computational workload and facilitates the treatment of wheel–rail nonlinearities.

In the decoupled iterative method, the entire vehicle–track system is partitioned at the wheel–rail contact into two subsystems: the vehicle and the track. The coupling of vehicle and track vibrations is governed by two conditions at the wheel–rail interface: the geometric compatibility between the displacements of the wheelset and the rail, and the static equilibrium of the interaction forces.

The geometric compatibility condition in vehicle–track dynamic interaction analysis manifests as the relative displacement between the wheelset and rail, derived from the rail displacement and considering the effect of track irregularities. This relative displacement determines the wheel–rail contact geometry, from which the wheel–rail contact forces acting on the vehicle are computed, forming the right-hand side load terms in the vehicle motion equation. The static equilibrium condition ensures that the wheel–rail forces acting on the wheelset are balanced by equal and opposite forces acting on the rail, which in turn form the right-hand side load terms in the track structure motion equation.

Thus, the vibration analysis of the entire vehicle–track system can be realized by linking the vehicle and track subsystems through these two coupling conditions—geometric compatibility and static equilibrium—forming the iterative procedure of the vehicle–track system.

The motion equations for the vehicle and track can ultimately be expressed in the following form:

$$\begin{cases} \boldsymbol{M}_{\mathrm{v}}\ddot{\boldsymbol{u}}_{\mathrm{v}} + \boldsymbol{C}_{\mathrm{v}}\dot{\boldsymbol{u}}_{\mathrm{v}} + \boldsymbol{K}_{\mathrm{v}}\boldsymbol{u}_{\mathrm{v}} = \boldsymbol{F}_{\mathrm{v}} \\ \boldsymbol{M}_{\mathrm{t}}\ddot{\boldsymbol{u}}_{\mathrm{t}} + \boldsymbol{C}_{\mathrm{t}}\dot{\boldsymbol{u}}_{\mathrm{t}} + \boldsymbol{K}_{\mathrm{t}}\boldsymbol{u}_{\mathrm{t}} = \boldsymbol{F}_{\mathrm{t}} \end{cases} \tag{3.5.46}$$

In the equations, $\boldsymbol{u}_v$ and $\boldsymbol{u}_t$ represent the vectors of degrees of freedom for the vehicle and the track structure, respectively. $M_v$ $K_v$ $C_v$ and $M_t$ $K_t$ $C_t$ denote the mass, damping, and stiffness matrices of the vehicle and the track structure, respectively. $F_v$ and $F_t$ are the total external loads acting on the vehicle and track structure, which are related to the motion of the vehicle and track structure as well as the geometric characteristics of the track. The motion equations for the track and the vehicle can be solved separately using direct integration methods. The basic solution procedure is illustrated in Fig. 3.37.

### 3.5.3 A Novel Two-Step Time Integration Method

The wheel–rail dynamic interaction exhibits strong nonlinear characteristics, and the governing equations of motion are typically solved using time-domain numerical integration methods. These methods assume that the structural response within a small time step is linear, and the motion variables are discretized at specific time points. Therefore, the differential equations do not need to be satisfied at all times, but only at these discrete time points. In solving system dynamic responses, attention is usually focused on the contribution of lower modes, as higher modes may introduce computational noise and should be filtered out. Conditionally stable integration methods require the time step size to be inversely proportional to the highest frequency of the discretized system, which in turn forces very small time steps to ensure accuracy for the lower modes. In contrast, unconditionally stable algorithms determine the time step based on stability conditions, significantly improving computational efficiency. An effective numerical integration method must satisfy the criteria of convergence, sufficient accuracy, good stability, and high computational efficiency.

This section presents a novel two-step numerical integration method specifically designed for the iterative decoupled solution of the vehicle–track system's equations of motion. To evaluate the accuracy and efficiency of the proposed method, it is compared against two established methods from the Newmark family: the Newmark-$\beta$ method and the generalized-$\alpha$ method. Both are unconditionally stable and widely

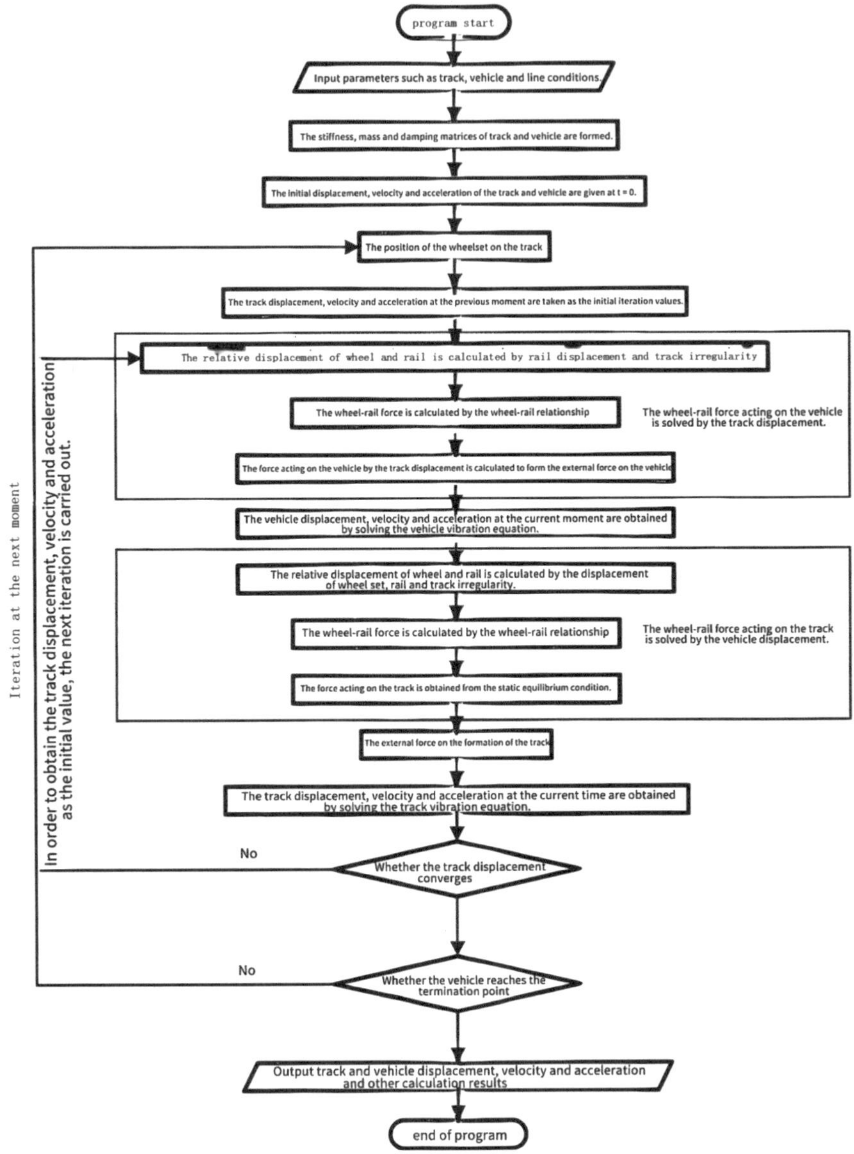


**Fig. 3.37** Flowchart of the vehicle–track system solution procedure

used in solving coupled vehicle–track equations. When appropriate wheel–rail interaction models and reasonably chosen time steps are employed, both the Newmark-$\beta$ and generalized-$\alpha$ methods deliver acceptable accuracy and computational efficiency.

1. **Implementation of the Two-Step Numerical Integration Method**

The novel two-step method is a composite time integration scheme suitable for solving both linear and nonlinear problems in structural dynamics. A key advantage of this method is that it remains stable even in cases where the Newmark method yields unstable solutions. Unlike the one-step Newmark-$\beta$ and generalized-$\alpha$ methods, the two-step approach solves the vehicle–track coupled dynamics in two stages. Instead of directly computing the response at time step $i + 1$, the method first evaluates the intermediate response at $i + 1/2$, and then uses the known responses at time $i$ and $i + 1/2$ to determine the response at $i + 1$.

In this framework, the vehicle and track are treated as two independent subsystems interacting through the wheel–rail contact force. Their mutual influence is governed by constraint equations that define the compatibility and equilibrium of the interaction. The equations of motion consist of two parts: one set governing the dynamic response of the vehicle–track system, and another describing the constraint equations that enforce the wheel–rail interaction forces.

The formulation is as follows:

$$\boldsymbol{M}^v\ddot{\boldsymbol{u}}^v + \boldsymbol{C}^v\dot{\boldsymbol{u}}^v + \boldsymbol{K}^v\boldsymbol{u}^v + \boldsymbol{L}^v\boldsymbol{\lambda} = \boldsymbol{P}_v \tag{3.5.47}$$

In the formula, the $\boldsymbol{M}^v, \boldsymbol{C}^v, \boldsymbol{K}^v$ are the vehicle mass, damping, and stiffness matrices, respectively; the $\boldsymbol{L}^v$ is the effect of wheel-rail forces on the vehicle. $\boldsymbol{P}_v$ is the matrix of external forces on the vehicle; $\ddot{\boldsymbol{u}}^v$, $\dot{\boldsymbol{u}}^v$, $\boldsymbol{u}^v$ are Acceleration, velocity, and displacement of vehicle motion in each degree of freedom, respectively; and $\boldsymbol{\lambda}$ is the wheel track force.

$$\boldsymbol{M}^t\ddot{\boldsymbol{u}}^t + \boldsymbol{C}^t\dot{\boldsymbol{u}}^t + \boldsymbol{K}^t\boldsymbol{u}^t + \boldsymbol{L}^t(t)\boldsymbol{\lambda} = \boldsymbol{P}_t \tag{3.5.48}$$

In the formula, $\boldsymbol{M}^t$, $\boldsymbol{C}^t$, $\boldsymbol{K}^t$ represent the mass, damping, and stiffness matrices of the vehicle, respectively. $\boldsymbol{L}^t(t)$ represents the influence of the wheel–rail force on the vehicle.; $\boldsymbol{P}_t$ is the matrix of track external forces; $\ddot{\boldsymbol{u}}^t$, $\dot{\boldsymbol{u}}^t$, $\boldsymbol{u}^t$ are acceleration, velocity, and displacement of the motion at each node of the track, respectively.

In order to solve the dynamic response of the vehicle-track system more accurately, the elastic deformation between the wheels and rails is taken into account in the wheel-rail constraint equation. According to the Hertz theory, the linear contact relationship between the wheels and rails is considered, then the constraint equation can be expressed as follows:

$$\boldsymbol{L}^v\boldsymbol{u}^v + (\boldsymbol{L}^t(t))^T\boldsymbol{u}^t + \boldsymbol{\rho}(t) = \boldsymbol{\lambda}/\boldsymbol{k}_t \tag{3.5.49}$$

In the equation, ppp denotes the track irregularity at the wheel–rail contact point.

According to the two-step method, the computation at time step $i + 1/2$ adopts the trapezoidal rule assumption, that is,

$$\dot{\boldsymbol{u}}_{n+\frac{1}{2}} = \dot{\boldsymbol{u}}_n + \frac{\Delta t}{4}(\ddot{\boldsymbol{u}}_n + \ddot{\boldsymbol{u}}_{n+\frac{1}{2}}) \tag{3.5.50}$$

$$\boldsymbol{u}_{n+\frac{1}{2}} = \boldsymbol{u}_n + \frac{\Delta t}{4}(\dot{\boldsymbol{u}}_n + \dot{\boldsymbol{u}}_{n+\frac{1}{2}}) \tag{3.5.51}$$

In the formula, the $\Delta t$ is the calculation time step.

Putting the equation, the (3.5.50) and (3.5.51) Substituting into the equation, the (3.5.47)–(3.5.49), which simplifies to give the equation of motion of the car track for n + 1/2 time steps as

$$\overline{\boldsymbol{M}}^{v}_{n+1}\tilde{\boldsymbol{u}}^{v}_{n+1} = \boldsymbol{a}^{v}_{n+1} \tag{3.5.52}$$

$$\overline{\boldsymbol{M}}^{t}_{n+1}\tilde{\boldsymbol{u}}^{t}_{n+1} = \boldsymbol{a}^{t}_{n+1} \tag{3.5.53}$$

$$\left(\frac{1}{\boldsymbol{k}} + \boldsymbol{MM}^{v}_{n+1} + \boldsymbol{MM}^{t}_{n+1}\right)\boldsymbol{\lambda}_{n+1} = \overline{\boldsymbol{\rho}}_{n+1} \tag{3.5.54}$$

Among them.

$$\begin{aligned}
&\overline{\boldsymbol{M}}_{n+1} = \frac{9}{\Delta t^2}\boldsymbol{M} + \frac{3}{\Delta t}\boldsymbol{C} + \boldsymbol{K},\ \tilde{\boldsymbol{u}}^{v}_{n+1} = \boldsymbol{u}^{v}_{n+1} + (\overline{\boldsymbol{M}}^{v}_{n+1})^{-1}(\boldsymbol{L}^{v})^{\mathrm{T}}\boldsymbol{\lambda}_{n+1},\\
&\tilde{\boldsymbol{u}}^{t}_{n+\frac{1}{2}} = \boldsymbol{u}^{t}_{n+\frac{1}{2}} + \left(\overline{\boldsymbol{M}}^{t}_{n+\frac{1}{2}}\right)^{-1}\left(\boldsymbol{L}^{t}\left(t_{n+\frac{1}{2}}\right)\right)^{\mathrm{T}}\boldsymbol{\lambda}_{n+\frac{1}{2}},\\
&\boldsymbol{MM}^{v}_{n+\frac{1}{2}} = \boldsymbol{L}^{v}\left(\overline{\boldsymbol{M}}_{n+\frac{1}{2}}\right)^{-1}(\boldsymbol{L}^{v})^{\mathrm{T}},\\
&\boldsymbol{MM}^{t}_{n+\frac{1}{2}} = \boldsymbol{L}^{t}\left(t_{n+\frac{1}{2}}\right)\left(\overline{\boldsymbol{M}}_{n+\frac{1}{2}}\right)^{-1}\left(\boldsymbol{L}^{t}\left(t_{n+\frac{1}{2}}\right)\right)^{\mathrm{T}},\\
&\overline{\boldsymbol{\rho}}_{n+\frac{1}{2}} = \boldsymbol{L}^{v}\tilde{\boldsymbol{u}}^{v}_{n+\frac{1}{2}} + \boldsymbol{L}^{t}\left(t_{n+\frac{1}{2}}\right)\tilde{\boldsymbol{u}}^{t}_{n+\frac{1}{2}} + \boldsymbol{\rho}\left(t_{n+\frac{1}{2}}\right),\\
&\boldsymbol{a}_{n+\frac{1}{2}} = \boldsymbol{P}_{n+\frac{1}{2}} + \boldsymbol{C}\left(\frac{4}{\Delta t}\boldsymbol{u}_n + \dot{\boldsymbol{u}}_n\right) + \boldsymbol{M}\left(\frac{16}{\Delta t^2}\boldsymbol{u}_n + \frac{8}{\Delta t}\dot{\boldsymbol{u}}_n + \ddot{\boldsymbol{u}}_n\right)
\end{aligned}$$

In the computation at time step $n + 1$, the three-point backward difference scheme is adopted, that is,

$$\dot{\boldsymbol{u}}_{n+1} = \frac{1}{\Delta t}\boldsymbol{u}_n - \frac{4}{\Delta t}\boldsymbol{u}_{n+\frac{1}{2}} + \frac{3}{\Delta t}\boldsymbol{u}_{n+1} \tag{3.5.55}$$

$$\ddot{\boldsymbol{u}}_{n+1} = \frac{1}{\Delta t}\dot{\boldsymbol{u}}_n - \frac{4}{\Delta t}\dot{\boldsymbol{u}}_{n+\frac{1}{2}} + \frac{3}{\Delta t}\dot{\boldsymbol{u}}_{n+1} \tag{3.5.56}$$

Putting the equation, the (3.5.55) and (3.5.56) Substituting into the equation, respectively (3.5.47)–(3.5.49), the simplification yields the equation of motion of the car track at $n + 1$ time steps as

$$\overline{\boldsymbol{M}}^{v}_{n+1}\tilde{\boldsymbol{u}}^{v}_{n+1} = \boldsymbol{a}^{v}_{n+1} \tag{3.5.57}$$

$$\overline{\boldsymbol{M}}^{t}_{n+1}\tilde{\boldsymbol{u}}^{t}_{n+1} = \boldsymbol{a}^{t}_{n+1} \tag{3.5.58}$$

$$\left(\frac{1}{k_t}+\boldsymbol{MM}_{n+1}^v+\boldsymbol{MM}_{n+1}^t\right)\boldsymbol{\lambda}_{n+1}=\overline{\boldsymbol{\rho}}_{n+1} \tag{3.5.59}$$

Among them.

$$\overline{\boldsymbol{M}}_{n+1}=\frac{9}{\Delta t^2}\boldsymbol{M}+\frac{3}{\Delta t}\boldsymbol{C}+\boldsymbol{K},\ \tilde{\boldsymbol{u}}_{n+1}^v=\boldsymbol{u}_{n+1}^v+\left(\overline{\boldsymbol{M}}_{n+1}^v\right)^{-1}\left(\boldsymbol{L}^v\right)^{\mathrm{T}}\boldsymbol{\lambda}_{n+1},$$

$$\tilde{\boldsymbol{u}}_{n+1}^t=\boldsymbol{u}_{n+1}^t+\left(\overline{\boldsymbol{M}}_{n+1}^t\right)^{-1}\left(\boldsymbol{L}^t(t_{n+1})\right)^{\mathrm{T}}\boldsymbol{\lambda}_{n+1},\ \boldsymbol{MM}_{n+1}^v=\boldsymbol{L}^v\left(\overline{\boldsymbol{M}}_{n+1}\right)^{-1}\left(\boldsymbol{L}^v\right)^{\mathrm{T}},$$

$$\boldsymbol{MM}_{n+1}^t=\boldsymbol{L}^t(t_{n+1})\left(\overline{\boldsymbol{M}}_{n+1}\right)^{-1}\left(\boldsymbol{L}^t(t_{n+1})\right)^{\mathrm{T}},\ \overline{\boldsymbol{\rho}}_{n+1}=\boldsymbol{L}^v\tilde{\boldsymbol{u}}_{n+1}^v+\boldsymbol{L}^t(t_{n+1})\tilde{\boldsymbol{u}}_{n+1}^t\boldsymbol{\rho}(t_{n+1}),$$

$$\boldsymbol{a}_{n+1}=\boldsymbol{P}_{n+1}+\boldsymbol{C}\left(\frac{4}{\Delta t}\boldsymbol{u}_{n+\frac{1}{2}}-\frac{1}{\Delta t}\boldsymbol{u}_n\right)$$

$$+\boldsymbol{M}\left(\frac{12}{\Delta t^2}\boldsymbol{u}_{n+\frac{1}{2}}-\frac{3}{\Delta t^2}\boldsymbol{u}_n+\frac{4}{\Delta t}\dot{\boldsymbol{u}}_{n+\frac{1}{2}}-\frac{1}{\Delta t}\dot{\boldsymbol{u}}_n\right)$$

As a result, the solution steps of the new two-step numerical integration method are summarized as follows (Table 3.15):

The novel two-step time integration method adopts a linearized wheel–rail contact model and incorporates the time-varying coupling relationship induced by the vehicle–track interaction through the formulation of constraint Eqs. (3.5.54) and (3.5.59). This approach ensures that the equivalent system matrices remain invariant throughout the solution process, thereby eliminating the need for iterative computation. As a result, the model configuration remains consistent during the solution of the coupled equations, and the algorithm exhibits a high degree of modularity.

2. **Analysis of the Performance of Numerical Integration Methods**

In the wheel-rail model under the linearized contact assumption, the normal force between the wheel and rail can be incorporated into the equivalent stiffness matrix and the equivalent additional load as supplementary matrices, forming the global motion equation. In this case, the vehicle and track subsystems can be directly solved as a unified system, a method referred to as the direct coupling method. When adopting a nonlinear approximation for the wheel-rail contact, the interaction between the two subsystems is realized by imposing additional constraints at the wheel-rail contact points. Due to the presence of higher-order terms in the calculation model for the normal force, iterative calculations are required at each time step until equilibrium conditions (either displacement compatibility or wheel-rail force balance) are satisfied. This method is referred to as the iterative method.

In general, the direct coupling method offers high computational efficiency; However, the coupled matrices vary with time, which is disadvantageous for extracting substructural data. Moreover, under conditions of severe irregularities or prolonged wheel-rail separation, the direct coupling method may not achieve sufficient accuracy in predicting system responses. Conversely, the iterative method provides a more accurate solution for the vehicle and track motion equations under

**Table 3.15** The computational steps of the new two-step numerical integration method

| Step 1 |
|---|
| (1) Assembly $\overline{M}_{n+\frac{1}{2}}$, $\overline{M}_{n+1}$ Matrix |
| (2) Determine the computational time step $\Delta t$, external moments $P_v$, $P_t$, as well as the irregularity of the track $\rho(t)$ |
| (3) Calculate the distance traveled by the vehicle $x\left(t_{n+\frac{1}{2}}\right) = vt_{n+\frac{1}{2}}$ and $x(t_{n+1}) = vt_{n+1}$, of which $v$ For the vehicle speed, determine the matrix of the track shape function $L^v\left(t_{n+\frac{1}{2}}\right)$, $L^t\left(t_{n+\frac{1}{2}}\right)$; |
| (4) Determine the initial displacement and velocity of the vehicle and track $u_0^v$, $u_0^t$, $\dot{u}_0^v$, $\dot{u}_0^t$; |
| (5) According to Eqs. (3.5.52) and (3.5.53), solving $\tilde{u}^v_{n+\frac{1}{2}}$, $\tilde{u}^t_{n+\frac{1}{2}}$; |
| (6) According to Eq. (3.5.54), calculating the wheel-track force for $n + 1/2$ time steps $\lambda_{n+\frac{1}{2}}$; |
| (7) Order $u^v_{n+\frac{1}{2}} = \tilde{u}^v_{n+\frac{1}{2}} - \left(\overline{M}^v_{n+\frac{1}{2}}\right)^{-1}(L^v)^{\mathrm{T}}\lambda_{n+\frac{1}{2}}$, $\dot{u}^v_{n+\frac{1}{2}} = \frac{4}{\Delta t}\left(u^v_{n+\frac{1}{2}} - u^v_n\right) - \dot{u}^v_n$, $\ddot{u}^v_{n+\frac{1}{2}} = \frac{4}{\Delta t}\left(\dot{u}^v_{n+\frac{1}{2}} - \dot{u}^v_n\right) - \ddot{u}^v_n$, The vehicle displacement, velocity, and acceleration responses are calculated for the $n + 1/2$ moments; |
| (8) According to Eq $u^t_{n+\frac{1}{2}} = \tilde{u}^t_{n+\frac{1}{2}} + \left(\overline{M}^t_{n+\frac{1}{2}}\right)^{-1}\left(L^t\left(t_{n+\frac{1}{2}}\right)\right)^{\mathrm{T}}\lambda_{n+\frac{1}{2}}$, $\dot{u}^t_{n+\frac{1}{2}} = \frac{4}{\Delta t}\left(u^t_{n+\frac{1}{2}} - u^t_n\right) - \dot{u}^t_n$, $\ddot{u}^t_{n+\frac{1}{2}} = \frac{4}{\Delta t}\left(\dot{u}^t_{n+\frac{1}{2}} - \dot{u}^t_n\right) - \ddot{u}^t_n$, calculated to obtain the track displacement, velocity, and acceleration response at the n + 1/2 moment; |
| Step 2 |
| (9) According to Eq. (3.5.57) Sum formula (3.5.58), solving $\tilde{u}^v_{n+1}$, $\tilde{u}^t_{n+1}$; |
| (10) According to Eq. (3.5.59), calculating the wheel-track force for $n + 1$ time steps $\lambda_{n+1}$; |
| (11) Again, according to Eq $u^v_{n+1} = \tilde{u}^v_{n+1} - \left(\overline{M}^v_{n+1}\right)^{-1}(L^v)^T\lambda_{n+1}$, $\dot{u}^v_{n+1} = \frac{1}{\Delta t}u^v_n - \frac{4}{\Delta t}u^v_{n+\frac{1}{2}} + \frac{3}{\Delta t}u^v_{n+1}$, $\ddot{u}^v_{n+1} = \frac{1}{\Delta t}\dot{u}^v_n - \frac{4}{\Delta t}\dot{u}^v_{n+\frac{1}{2}} + \frac{3}{\Delta t}\dot{u}^v_{n+1}$, the vehicle displacement, velocity, and acceleration responses are calculated for the n + 1 moment |
| (12) Make $u^t_{n+1} = \tilde{u}^t_{n+1} - \left(\overline{M}^t_{n+1}\right)^{-1}\left(L^t(t_{n+1})\right)^{\mathrm{T}}\lambda_{n+1}$, $\dot{u}^t_{n+1} = \frac{1}{\Delta t}u^t_n - \frac{4}{\Delta t}u^t_{n+\frac{1}{2}} + \frac{3}{\Delta t}u^t_{n+1}$, $\ddot{u}^t_{n+1} = \frac{1}{\Delta t}\dot{u}^t_n - \frac{4}{\Delta t}\dot{u}^t_{n+\frac{1}{2}} + \frac{3}{\Delta t}\dot{u}^t_{n+1}$, the track displacement, velocity, and acceleration responses are calculated for the $n + 1$ moment |

nonlinear wheel-rail contact conditions, albeit at the cost of reduced computational efficiency due to the need for multiple iterations per time step.

In this section, the advantages of the newly proposed two-step numerical integration method for solving the vehicle-track coupled equations will be validated by comparing it with the Newmark-β method and the generalized-α method under different wheel-rail contact conditions.

(1) Irregularity conditions of ballastless track

Kinetic models such as Fig. 3.38 shown. The vehicle contains three degrees of freedom: body vertical displacement, the $u_1$ and the vertical displacement of the bogie $u_2$ and the vertical displacement of the wheelset $u_3$ The rails are simplified to Euler beams. The car body and bogie are made of stiffnesses of $k_2$, the damping is $c_2$ The bogie is connected to the wheelset by a spring damper with a stiffness of $k_1$, the damping is $c_1$ The spring damper connection of the The vehicle is traveling at a speed that $v = 117$ km/h traveling along the track at a constant speed. It is assumed that the vehicle-track coupling model is symmetrical along the line, so only the semi-vehicle model is considered. The parameters of the vehicle-track model are as follows Table 3.16 shown.

Under the condition of smooth track, when the linear contact relationship between wheels and rails is considered, the dynamic response of the rail coupled system is solved by the Newmark $\beta$-method (NML) and the generalized $\alpha$-method (APL), respectively, and the results are compared with those of the new two-step numerical integration method (Two-step).

Such as Fig. 3.39 shown, the dynamic responses of the rail system obtained by NML, APL and the two-step method are basically the same under smooth conditions. Among Fig. 3.39 The maximum error in (c) is calculated as follows. The dynamic model is illustrated in Fig. 3.38. The vehicle possesses three degrees of freedom: vertical displacement of the car body $\boldsymbol{u}_1$, vertical displacement of the bogie $\boldsymbol{u}_2$, and

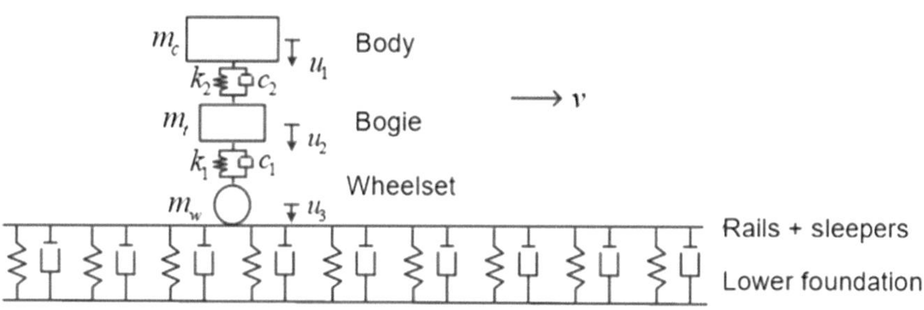


**Fig. 3.38** A two-dimensional vehicle-rail coupling model

**Table 3.16** Two-dimensional vehicle-rail coupling model parameters

| Parameters/units | Value |
|---|---|
| Static wheel weight/N | 82,000 |
| Mass of wheelset/kg | 500 |
| Modulus of elasticity of rail EI/(N-m$^2$) | 4,864,500 |
| Rail mass/(kg/m) | 119 |
| Rail stiffness/(N/m$^2$) | 200 |
| Rail damping/(N-s/m$^2$) | 30 |
| Hertz contact constant G/(N/m$^{3/2}$) | $10^{11}$ |

vertical displacement of the wheelset $\boldsymbol{u}_3$. The rail is simplified as an Euler beam. The car body and the bogie are connected via a spring-damper system with stiffness $\boldsymbol{k}_2$ and damping coefficient $\boldsymbol{c}_2$, while the bogie and the wheelset are connected via another spring-damper system with stiffness $\boldsymbol{k}_1$ and damping coefficient $\boldsymbol{c}_1$. The vehicle moves uniformly along the track at a constant speed of $v = 117$ km/h. Assuming the vehicle–track coupled model is symmetric along the track's centerline, only a half-vehicle model is considered. The parameters for the vehicle–track model are listed in Table 3.16.

Under smooth track conditions and assuming a linear wheel–rail contact relationship, the dynamic response of the vehicle–track coupled system is computed using the Newmark-$\beta$ method (NML) and the generalized-$\alpha$ method (APL), and the results are compared with those obtained using the newly proposed two-step numerical integration method (Two-step).

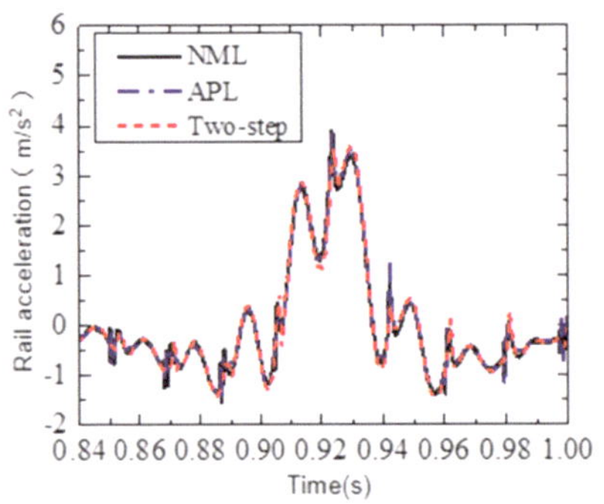


(a) The acceleration at the midpoint of the rail

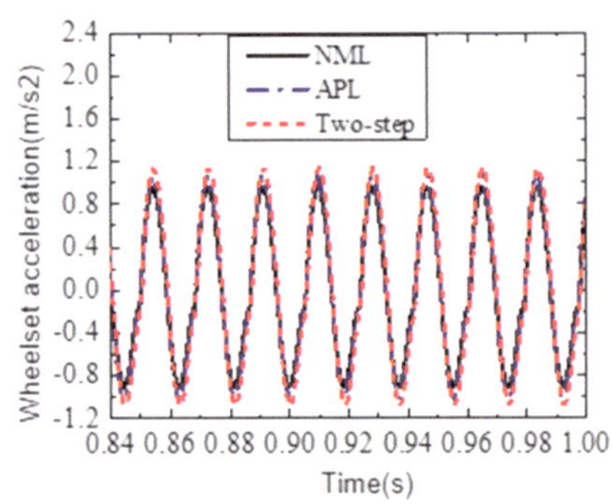


(b) Wheelset acceleration

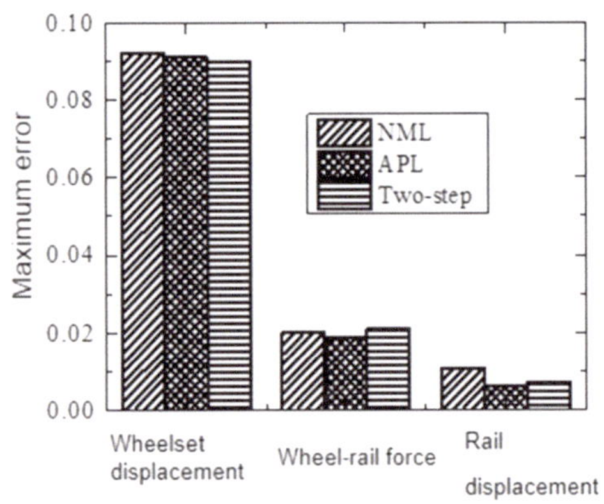


(c) Maximum error in wheelset displacement, rail displacement and wheel-rail force

**Fig. 3.39** Dynamic response of the vehicle–track system under irregular slab track conditions

As shown in Fig. 3.39, under smooth conditions, the dynamic responses of the vehicle–track system computed by the NML, APL, and Two-step methods are generally consistent. In particular, the maximum error illustrated in Fig. 3.39c is calculated using the following formula:

$$T_r = [(u - T_s)/T_s]_{\max} \tag{3.5.60}$$

$u$ is the system response. $T_s$ is the corresponding system static response.

If $u$ is the displacement, and $T_s$ is defined as

$$T_s{=}(m_c + m_t + m_w)g/K^{\mathrm{t}} \tag{3.5.61}$$

In the formula, the $K^t$ is the track stiffness.

If $u$ is the acceleration, and $T_s$ is Defined as

$$T_s = g \tag{3.5.62}$$

If $u$ is the wheel-trackig force, $T_s$ is Defined as.

$$T_s = (m_c + m_t + m_w)g \tag{3.5.63}$$

However, the computational time required by the three methods differs significantly, as shown in Fig. 3.40. The computation time of the Two-step method is markedly shorter than that of the APL, and substantially less than that of the NML. Specifically, the computation time of the Two-step method is only 5.6% and 61.8% of that required by the NML and APL, respectively. It should be noted that the time integration step size for the Newmark-β method is $10^{-5}$ s, for the generalized-α method is $10^{-4}$ s, and for the Two-step method is $10^{-3}$ s. Figure 3.41 presents the computed rail acceleration results using a time step of $10^{-3}$ s. As can be seen, the responses obtained from the NML and APL methods exhibit abnormal high-frequency oscillations. This phenomenon arises because, with a time step of $10^{-3}$ s, the solutions provided by the Newmark-β and generalized-α methods fail to satisfy the dynamic equilibrium of the system equations. Consequently, numerical errors accumulate over time steps, leading to a degradation in overall solution accuracy. Moreover, the smaller the time step required to achieve a stable solution, the longer the computation time and the higher the computational cost. In contrast, the Two-step method supports the use of a larger time integration step, significantly reducing computation time compared to the Newmark-$\beta$ and generalized-$\alpha$ methods.

(2) Wheel flat spot irregularity conditions

The presence of track irregularities directly affects the dynamic interaction between the wheel and rail. When the track irregularities are large or the interaction time is long, instantaneous separation between the wheel and rail may occur, causing impact when the wheel re-engages with the track. This can affect the operational safety and

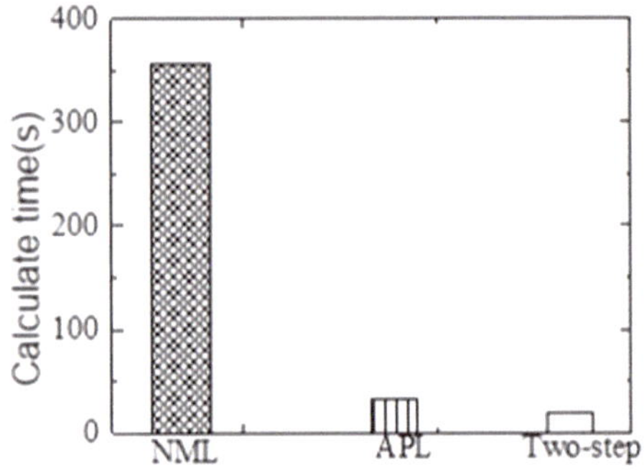


**Fig. 3.40** Computation time of NML, APL, and two-step methods under ballastless track irregularity conditions

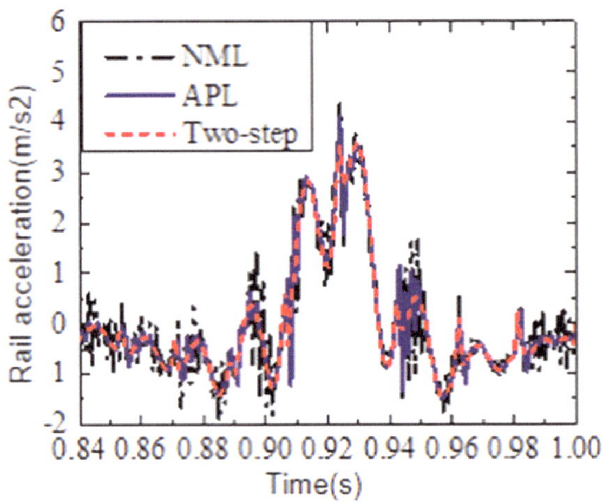


**Fig. 3.41** The acceleration at the midpoint of the rail for an integration step of $10^{-3}$ s under smooth conditions

accelerate track damage. Therefore, accurately calculating the wheel-rail separation time and the magnitude of the wheel-rail force is of significant importance. This section validates the accuracy of numerical integration methods in solving under wheel-rail separation conditions by applying wheel flat spot irregularity.

Since the position of the flat spot on the wheel as it runs on the track is difficult to determine, its impact on the track is challenging to compute. To simulate the situation where a wheel has a flat spot during operation, an equivalent track irregularity is constructed, as shown in Fig. 3.42. The shape of the irregularity is determined through extensive test data, and the depth of the equivalent track irregularity is based on the wheel radius, making it suitable for various vehicle types. The formula for calculating the equivalent irregularity is as follows:

$$p(x_p) = \frac{D}{2}\left(1 - \cos\left(\frac{2\pi}{l}x_p\right)\right),\ D = \frac{l^2}{16r} \tag{3.5.64}$$

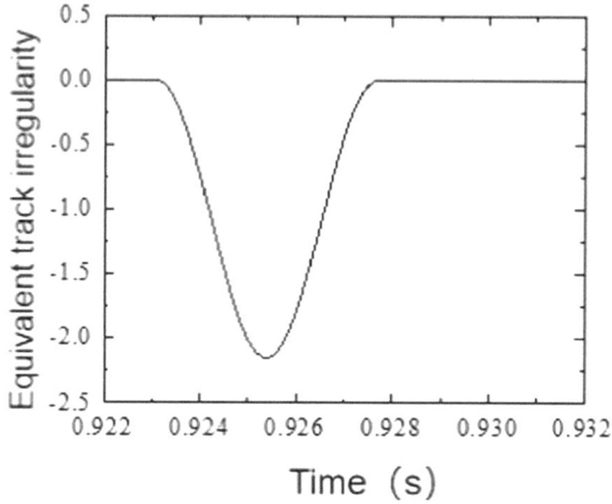


**Fig. 3.42** Equivalent track irregularity

In the formula, $x_p$ represents the longitudinal position of the contact point; $D$ is the maximum depth of the equivalent irregularity, 2.15 mm; $l$ is the length of the equivalent irregularity, 150 mm; and $r$ is the wheel diameter. By adjusting the wheel diameter, this equivalent track irregularity can be adapted to various vehicle types.

To further validate the stability of the results obtained from the Two-Step method, the Newmark-$\beta$ method was applied, considering both linear contact (referred to as NML) and nonlinear contact (referred to as NMN) between the wheel and track. The vehicle-track model from the previous section was used for solving. It is worth noting that the impact of the equivalent track irregularity on the vehicle-track coupling model shown in Fig. 3.38 has been studied and verified by Newton and Clark (referred to as NCM), and their results were calibrated against field test data [74]. As a result, the NCM outcomes serve as the accurate solution for validating the numerical integration methods' accuracy.

Figure 3.43 shows the dynamic response of the vehicle-track system when the equivalent wheel flat spot irregularity is applied. As seen in Fig. 3.43a, when the vehicle moves through the equivalent flat spot irregularity, wheel-rail separation and rebound occur. During rebound, the wheel-rail forces increase instantaneously. The comparison between the wheel-rail force responses obtained from the Two-Step method, NMN, and NML with the NCM results shows that, except for NML, both the Two-Step method and NMN are very consistent with NCM. It should be noted that the results provided by NCM have already filtered out high-frequency modal responses. Therefore, compared to the wheel-rail force results from the Two-Step method and NMN, the NCM results do not exhibit oscillations after 0.929 s.

To further validate the accuracy of the Two-Step method, Fig. 3.43b, c present the acceleration responses of the rail and wheelset. It is evident that when the wheel rebounds, the wheel-rail acceleration increases significantly, indicating a pronounced

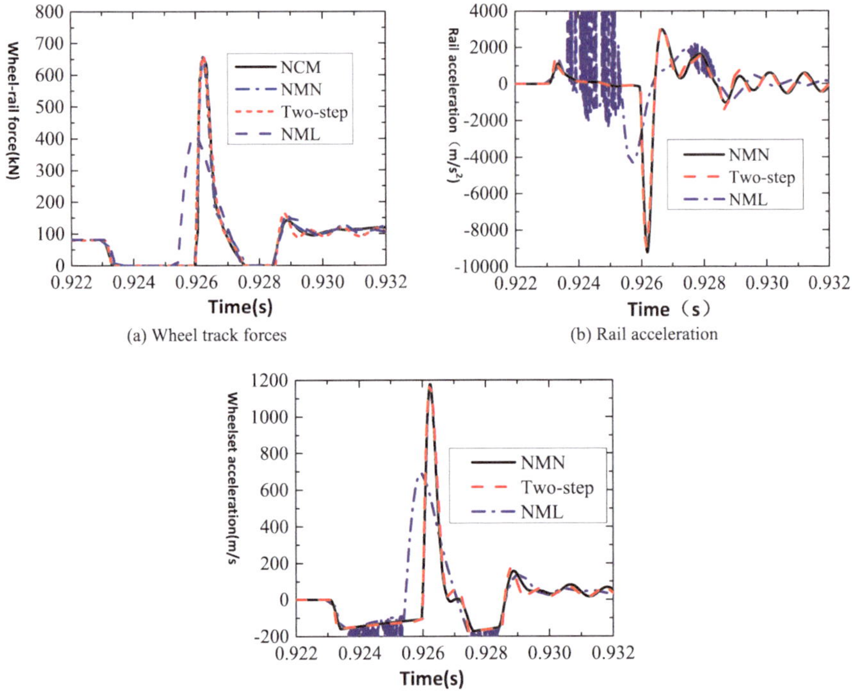


(a) Wheel track forces

(b) Rail acceleration

(c) Wheelset acceleration

**Fig. 3.43** The dynamic response of the vehicle-track system when the equivalent wheel flat spot irregularity is applied

impact on the vehicle-track system. At the same time, the results from the Two-Step method and NMN are nearly identical, whereas the NML-calculated wheel-rail acceleration response shows abnormal oscillations.

This indicates that when large deformations and prolonged wheel-rail contact separation occur, the results from the linear contact Newmark-β method (NML) become unstable. However, the Two-Step method remains stable as long as an appropriate time step is chosen while meeting the calculation precision. The time step for NML and NMN is $10^{-5}$ s, while the time step for the Two-Step method is $10^{-4}$ s. As shown in Fig. 3.44, compared to NML and NMN, the Two-Step method, due to its larger time step and the linear contact model, does not require iteration and significantly reduces the computation time, thereby improving the computational efficiency.

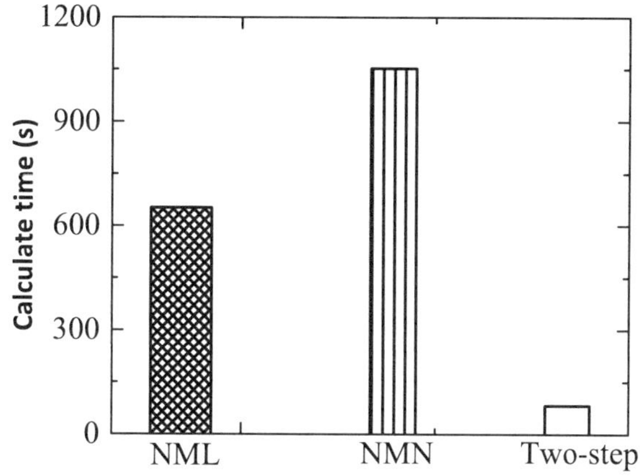


**Fig. 3.44** Computation time of NML, NMN, and the two-step method under wheel flat spot conditions

### *3.5.4 Simulation Method for Long-Distance Railway Lines*

For the vehicle–track system dynamics, the excitation input is typically modeled by moving the vehicle while keeping the track fixed. However, simulating a moving vehicle often requires a sufficiently long segment of track, which significantly increases the degrees of freedom and computational load as the track length grows. To address these issues, researchers have proposed the sliding window method, which captures the motion of the vehicle and the affected track structure by using a computational window that moves along the track. This approach greatly reduces the computational domain of the vehicle–track model while still accounting for relevant track structural characteristics.

The basic idea of the sliding window method is to reflect the track structure through a fixed-length window, the window sliding instead of the vehicle movement, the window sliding distance and the vehicle moving distance is the same, so as to ensure that the amount of calculations within the window is unchanged. The sliding window method has the following two characteristics: (1) only the vibration inside the window is considered in the solution, and the vibration of the track structure outside the window is assumed to be 0; (2) the state of the track structure inside the window after the window slides is the same as that before the window slides. The core idea of the sliding window method is to represent the track structure within a fixed-length window, where the motion of the window replaces the actual movement of the vehicle. The sliding distance of the window corresponds to the distance traveled by the vehicle, thereby ensuring a constant computational load within the window. The sliding window method has the following two key features: (1) During the computation, only the vibration within the window is considered, and the vibration of the track structure outside the window is assumed to be zero. (2) After sliding, the internal state of the track structure within the window remains the same as before the slide.

The length of the selected sliding window should be an integer multiple of the slab length, and the boundaries of the window should lie at the joints between slabs, as shown in Fig. 3.45. Within each sub-window, the vehicle is modeled as a simplified moving vehicle. Starting from the m-th window, whenever the vehicle has moved the length of one track slab, the sliding window shifts forward by one slab length in the direction of vehicle motion, and the computational domain transitions from the m-th to the (m + 1)-th window. During the transition, the overlapping part of the m-th and (m + 1)-th windows (indicated in red in Fig. 3.45) maintains the same track vibration state, while the non-overlapping part (indicated in blue) assumes zero vibration.

The following takes the vertical displacement of the rail as an example to illustrate the vibration state of the track structure during the transition from the m-th sub-window to the (m + 1)-th sub-window. Since the rail model is constructed using the modal superposition method, the vertical displacement within the m-th sub-window satisfies:

$$Z_r(x,t) = \sum_{k=1}^{N_Z} Z_k(x) q_{Vk}(t), x \in [x_a, x_b] \tag{3.5.65}$$

That is:

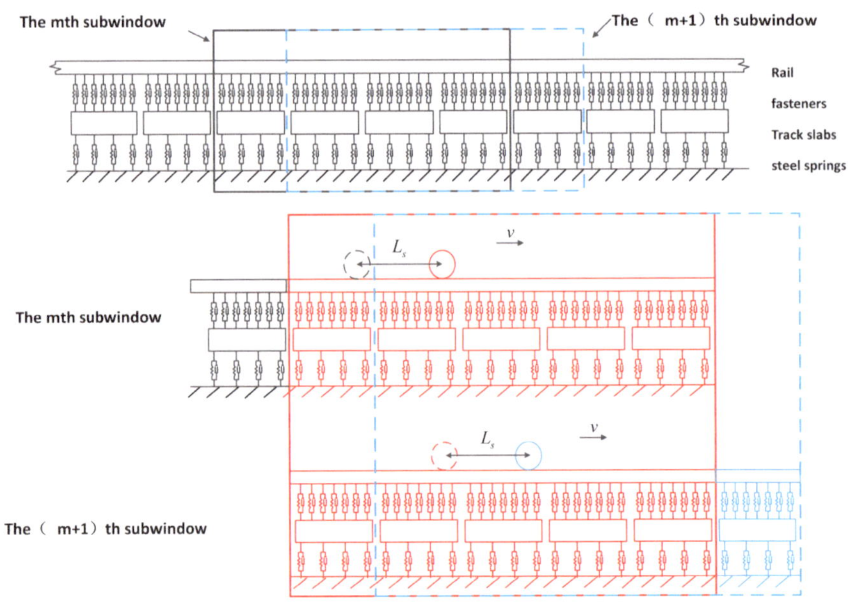


**Fig. 3.45** Schematic diagram of the sliding window method

$$\begin{bmatrix} Z_r(x_a,t) \\ Z_r(x_{a+1},t) \\ \vdots \\ Z_r(x_b,t) \end{bmatrix} = \begin{bmatrix} Z_1(x_a) & Z_2(x_a) & \cdots & Z_{N_Z}(x_a) \\ Z_1(x_{a+1}) & Z_2(x_{a+1}) & \cdots & Z_{N_Z}(x_{a+1}) \\ \vdots & \vdots & & \vdots \\ Z_1(x_b) & Z_2(x_b) & \cdots & Z_{N_Z}(x_b) \end{bmatrix} \begin{bmatrix} q_{V1}(t) \\ q_{V2}(t) \\ \vdots \\ q_{VN_Z}(t) \end{bmatrix} \tag{3.5.66}$$

When the window shifts, the above equation is transformed in the (m + 1)th sub-window as follows:

$$\begin{bmatrix} Z_r(x_{a+m},t) \\ Z_r(x_{a+m+1},t) \\ \vdots \\ Z_r(x_{b+m},t) \end{bmatrix} = \begin{bmatrix} Z_1(x_{a+m}) & Z_2(x_{a+m}) & \cdots & Z_{N_Z}(x_{a+m}) \\ Z_1(x_{a+m+1}) & Z_2(x_{a+m+1}) & \cdots & Z_{N_Z}(x_{a+m+1}) \\ \vdots & \vdots & & \vdots \\ Z_1(x_{b+m}) & Z_2(x_{b+m}) & \cdots & Z_{N_Z}(x_{b+m}) \end{bmatrix} \begin{bmatrix} p_{V1}(t) \\ p_{V2}(t) \\ \vdots \\ p_{VN_Z}(t) \end{bmatrix} \tag{3.5.67}$$

According to the characteristics of the modal functions of beams, if the boundary support conditions remain unchanged, the modal functions remain invariant, i.e.,

$$\begin{bmatrix} Z_1(x_a) & Z_2(x_a) & \cdots & Z_{N_Z}(x_a) \\ Z_1(x_{a+1}) & Z_2(x_{a+1}) & \cdots & Z_{N_Z}(x_{a+1}) \\ \vdots & \vdots & & \vdots \\ Z_1(x_b) & Z_2(x_b) & \cdots & Z_{N_Z}(x_b) \end{bmatrix} = \begin{bmatrix} Z_1(x_{a+m}) & Z_2(x_{a+m}) & \cdots & Z_{N_Z}(x_{a+m}) \\ Z_1(x_{a+m+1}) & Z_2(x_{a+m+1}) & \cdots & Z_{N_Z}(x_{a+m+1}) \\ \vdots & \vdots & & \vdots \\ Z_1(x_{b+m}) & Z_2(x_{b+m}) & \cdots & Z_{N_Z}(x_{b+m}) \end{bmatrix} \tag{3.5.68}$$

According to the sliding window method characteristic, without considering the vibration of the track structure in the non-overlapping part between the neighboring windows, then

$$Z_r(x_{b+1},t) = Z_r(x_{b+2},t) = \cdots = Z_r(x_{b+m},t) = 0 \tag{3.5.69}$$

After determining the rail displacement after window sliding, the inverse matrix of the modal function is used to solve for the new generalized coordinates of the rail, i.e.

$$\begin{bmatrix} p_{V1}(t) \\ p_{V2}(t) \\ \vdots \\ p_{VN_Z}(t) \end{bmatrix} = \begin{bmatrix} Z_1(x_{a+m}) & Z_2(x_{a+m}) & \cdots & Z_{N_Z}(x_{a+m}) \\ Z_1(x_{a+m+1}) & Z_2(x_{a+m+1}) & \cdots & Z_{N_Z}(x_{a+m+1}) \\ \vdots & \vdots & & \vdots \\ Z_1(x_{b+m}) & Z_2(x_{b+m}) & \cdots & Z_{N_Z}(x_{b+m}) \end{bmatrix}^{-1} \begin{bmatrix} Z_r(x_{a+m},t) \\ Z_r(x_{a+m+1},t) \\ \vdots \\ Z_r(x_{b+m},t) \end{bmatrix} \tag{3.5.70}$$

For the track slab structure, during the window sliding process, a full track slab length is moved. This is equivalent to no longer considering the vibration of the leftmost track slab in the window before sliding. After sliding, the degrees of freedom of the rightmost track slab in the window are "activated," but its initial vibration is zero.

## 3.6 System Dynamics Analysis Software Development and Verification

### *3.6.1 Software Development*

#### 3.6.1.1 General Content and Requirements of Software Development

1. **Software Development Process and Content**

Figure 3.46 shows the main stages of simulation software development. It is, in fact, a cyclical process, where the issues discovered during the technical support phase serve as input for conceptual development. Software development is an ongoing process of improvement.

The general workflow of the railway vehicle-track system dynamics analysis software includes several main steps as shown in Fig. 3.47.

(1) Question posed

This stage mainly involves clarifying the problems to be solved, boundary conditions, and functional requirements.

(2) Modeling

Modeling is the core of dynamic analysis software. As a specialized program for railway vehicle-track system dynamics analysis, the model should have broad applicability and flexibility. For example, it should include modeling modules for vehicles, track alignment, track irregularities, and bridges, which can be flexibly combined.

(3) Computational Solution

Through modeling, the system dynamics equations are formed. One of the main issues in railway track system dynamics is obtaining the time history of each state

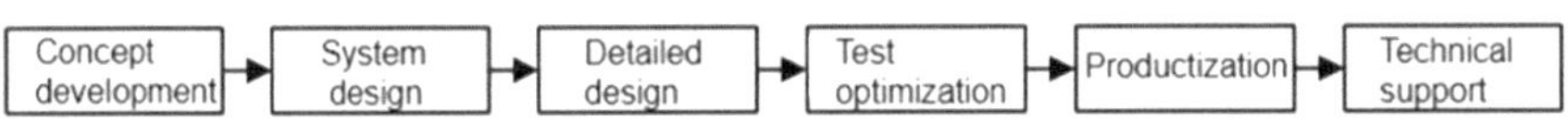


**Fig. 3.46** Simulating the software development process

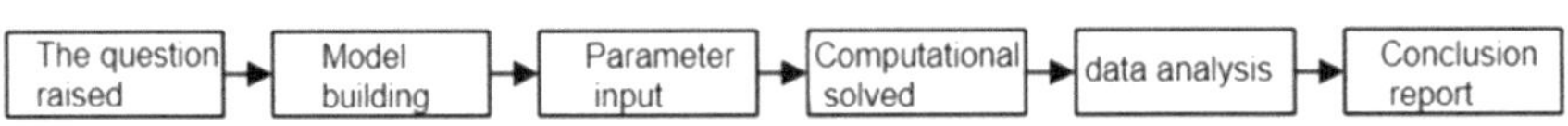


**Fig. 3.47** The main steps of power simulation analysis

variable from the motion equations given specific initial conditions. Different numerical integration methods have different applicable conditions for solving multi-body system dynamics equations, so the choice of an appropriate numerical integration method is crucial for dynamic simulation. Different numerical algorithms should be provided for different analysis problems to enhance the software's applicability.

(4) Data Analysis

After completing the simulation calculations, data analysis and processing can provide intuitive results. This typically includes statistical analysis, vehicle safety and ride comfort analysis, and the calculation and analysis of dynamic characteristics of track and bridge structures. Results should be presented in tabular and graphical formats.

(5) Conclusion Report

After completing the computational analysis, various forms of result data output and conclusion reports should be provided. Good simulation software design can improve software efficiency, providing user-friendly interfaces for pre-processing and post-processing. The pre-processing program can reduce the workload of data input, including vehicle, track, and bridge structure parameters, track geometry, track irregularities, etc. The geometric relationship of various wheel-rail profiles can also be pre-calculated and expressed as functions of lateral displacement and yaw angle.

The post-processing program mainly includes functions like data reporting, plotting, and generating reports, providing analysis reports on the dynamic characteristics and performance of vehicles and track structures through tables, 2D/3D graphs.

Verification and optimization of simulation software are of great significance. Common verification methods include theoretical analysis, real-world data validation, and comparison with similar software.

2. **Software General Requirements**

A software to become an effective and reliable tool, generally need to meet the reliability, efficiency, ease of operation, portability, integrity, compatibility, help documentation, technical support and other requirements.

- Reliability: The software needs to be fully validated and have stable computational capabilities so that correct results can be obtained from physically meaningful input data.
- Completeness: the software should include the main content of the power analysis of the railroad vehicle—line system, with a wider scope of application and adaptability, reducing the dependence on other computer software.
- Efficiency: mainly refers to the computation time and data access speed, modeling and algorithm design need to consider the computational efficiency.
- Ease of operation: the software should be user-friendly interface, minimize the requirements for input data, as far as possible to help input dialog box, and can check the input of logical errors.

- Portability: Reduce the platform operating system and hardware requirements, in order to facilitate portability, the program should be written in commonly used high-level language, reducing the direct operation involving the operating system.
- Compatibility: It is useful to have compatibility with other major applications (e.g., finite element or existing simulation programs), not only with respect to program data management, but also with respect to data conversions, such as relative and absolute coordinate transformations, and conversions of different graphs and topologies. The following are some examples of compatibility issues
- Help Documentation: The software should be accompanied by complete documentation, background papers, and user manuals. The documentation contains information on modeling, approximations and assumptions, and scope of application.
- Technical Support: The software needs to provide continuous upgrades and corrections and other related technical support services.

#### 3.6.1.2 Development and Main Functions of the Dynamic Analysis Software

With the advancement of multibody system dynamics theory and computer software technology, multibody dynamic simulation software has developed rapidly. In the field of railway vehicle system dynamics analysis, several widely used multibody simulation tools include MSC.ADAMS, SIMPACK, NUCARS, and VAMPIRE. Each of these packages has its own characteristics, scope of application, and functionality. However, they mainly focus on analyzing vehicle dynamic performance, while considering the dynamic characteristics of track structures and the interaction between vehicles and infrastructure to a lesser extent, often with high modeling complexity.

Based on the theoretical framework of railway infrastructure system dynamics introduced earlier, the author has developed a specialized software package titled Train-Structure Dynamic Interaction Analysis System (TSDIANS). This software is designed for evaluating and optimizing railway alignment parameters, dynamic assessment of alignment schemes, matching infrastructure and alignment in complex terrain, and analyzing vibrations and parameter design of track and bridge structures.

The software allows users to select different track and bridge models depending on the specific object and problem under investigation. It adopts a decoupled iterative solution strategy for the vehicle-infrastructure system, coupling them via the wheel-rail interaction, thereby significantly improving computational efficiency. By gradually releasing and resetting the mass, stiffness, and damping matrices of the substructure, and reusing them, the software effectively addresses the challenge of simulating long-distance alignment dynamics involving substructures.

The software consists of a main vehicle–track system computation module and two post-processing modules for data analysis and visualization.

(1) Data Input and Pre-processing Module

This module defines the data input formats for vehicle, track, bridge structures, and alignment parameters, offering both graphical user interface and batch input modes, and performs data organization and preprocessing.

(2) Simulation Module

The simulation module includes five fundamental submodules: vehicle, alignment geometry, track irregularity, track structure, and bridge structure. These submodules can be flexibly combined based on the analysis task at hand, enabling four typical types of dynamic simulations: vehicle–alignment interaction, vehicle–alignment–bridge interaction, vehicle–track interaction, and vehicle–track–bridge interaction.

(3) Post-processing Module

The post-processing module includes functions for data analysis and output. The data analysis functions cover statistical analysis, ride comfort evaluation, operational safety assessment, track irregularity evaluation, structural vibration analysis, and digital signal processing.

The output functions support various forms of results, including data tables, graphical plots, and comprehensive analysis reports.

The overall architecture and function modules of the software are illustrated in Fig. 3.48.

In Fig. 3.49, the software mapping interface is given.

## *3.6.2 Software Validation*

### 3.6.2.1 Comparison with Beijing-Tianjin Intercity Test Data

The comparison with the Beijing-Tianjin Intercity test data is based on the April 28, 2010 test data of CRH3 trains operating on the downstream line of the Beijing-Tianjin Intercity.

1. **Time-Domain Contrast**

Section JJK59 to JJK62 of Beijing-Tianjin Intercity Railway is selected for comparative analysis. This section includes one round curve and one vertical curve, and the traveling speed is 243 km/h. The specific line conditions are shown in the following Fig. 3.50.

Figures 3.51 and 3.52 show the comparisons of the lateral and vertical accelerations of the car body, respectively. It can be observed from both figures that there is a high degree of consistency in both the magnitude and the variation trend of the accelerations.

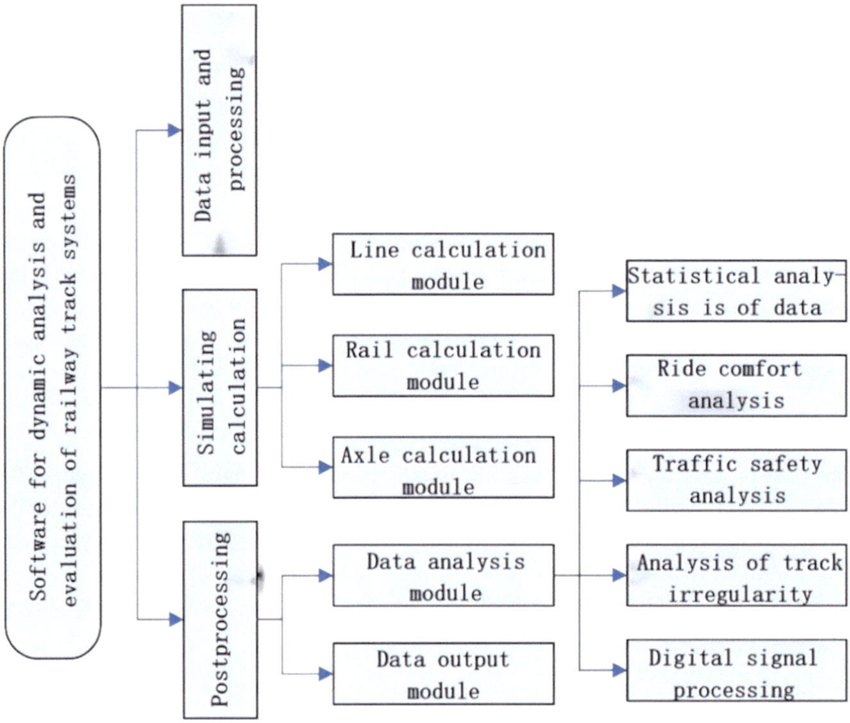


Fig. 3.48 Software function modules

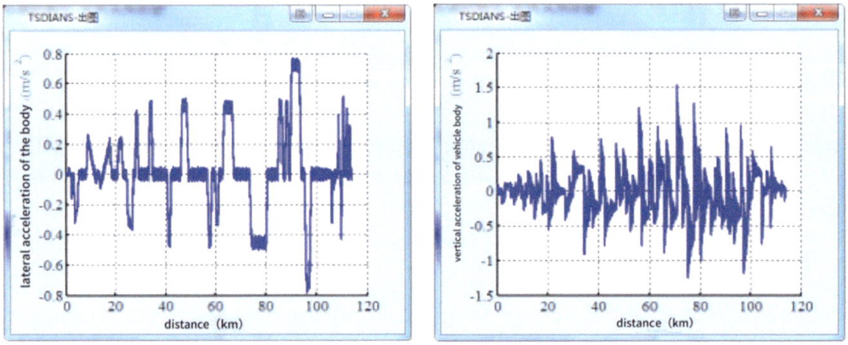


Fig. 3.49 The main interface of the software

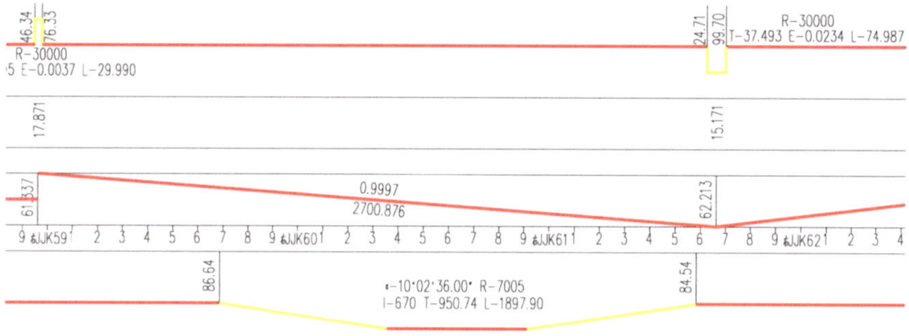

**Fig. 3.50** Schematic diagram of the local horizontal and vertical alignment of the Beijing–Tianjin intercity railway

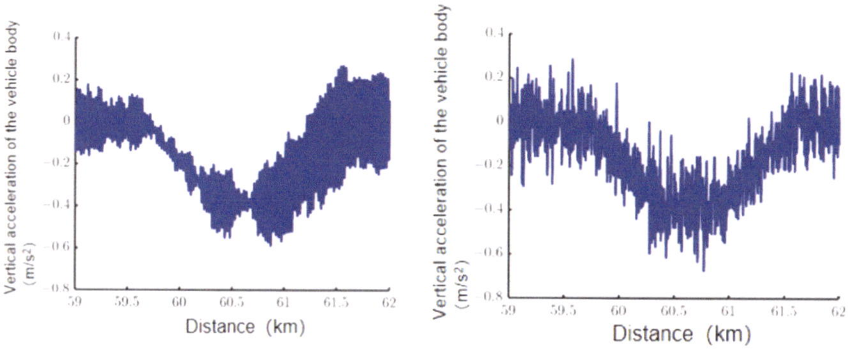


(a) Calculated results of this program (b) Measured data

**Fig. 3.51** Comparison of lateral acceleration of the car body

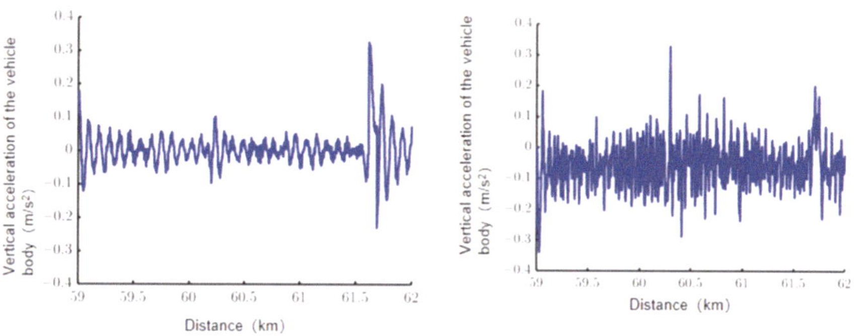


(a) Calculated results of this program (b) Measured data

**Fig. 3.52** Comparison of car body vertical acceleration

## 2. **Frequency Domain Comparison**

Figure 3.53a, b can be seen that the measured main frequency of transverse vibration of the car body is about 0.4–0.7 Hz. In the simulation results, the main frequency of transverse body vibration is 0.52 Hz.

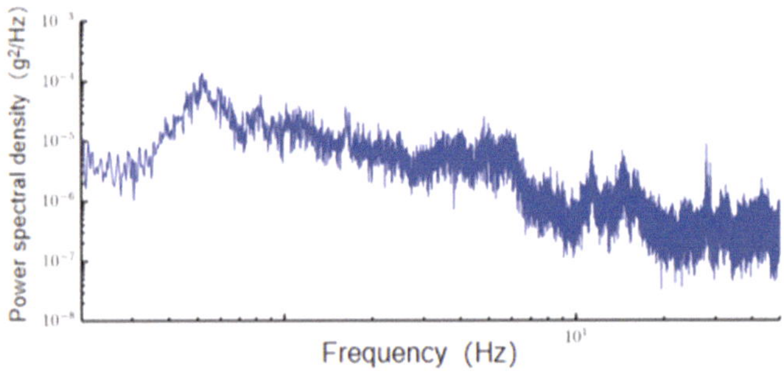


(a) Measured data (April 28, 2010)

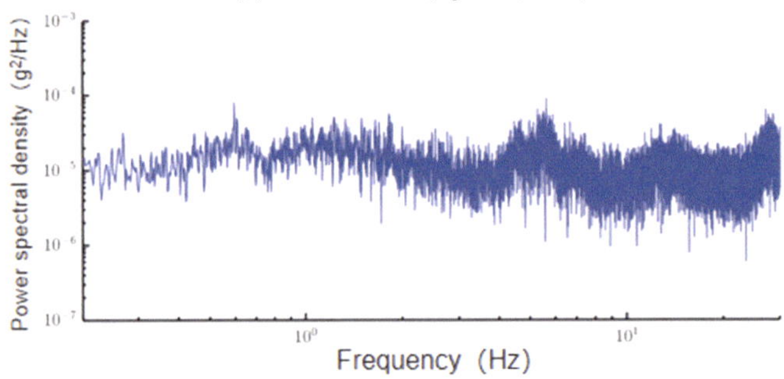


(b) Measured data (May 31, 2010)

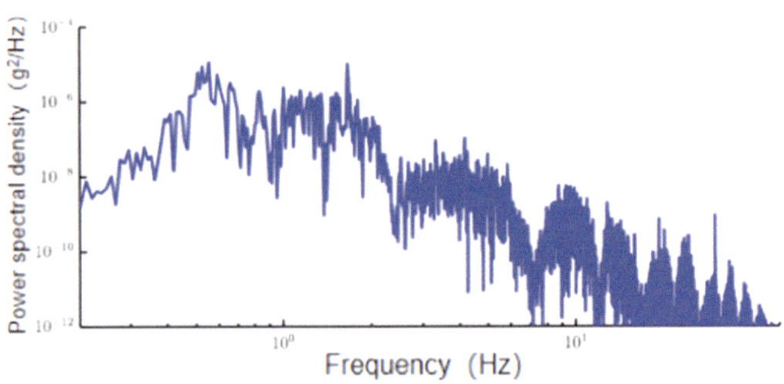


(c) The results of the calculations of this procedure

**Fig. 3.53** Comparison of transverse acceleration power spectra of the car body

From Fig. 3.54a, b, it can be observed that the dominant frequency of the measured vertical vibration of the car body is approximately 0.6–0.8 Hz. In the simulation results, the dominant frequency of the vertical vibration is 0.72 Hz.

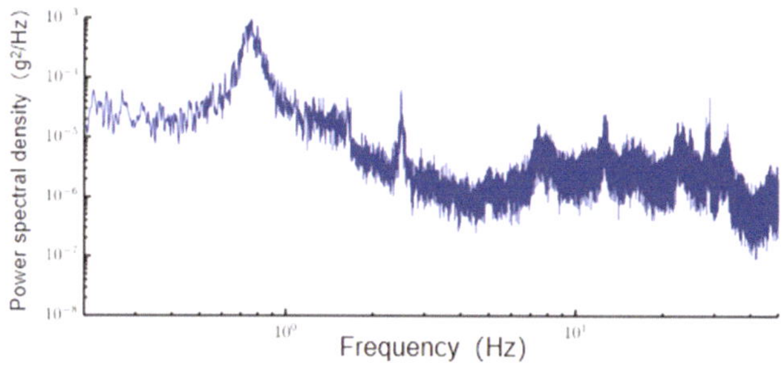


(a) Measured data (April 28, 2010)

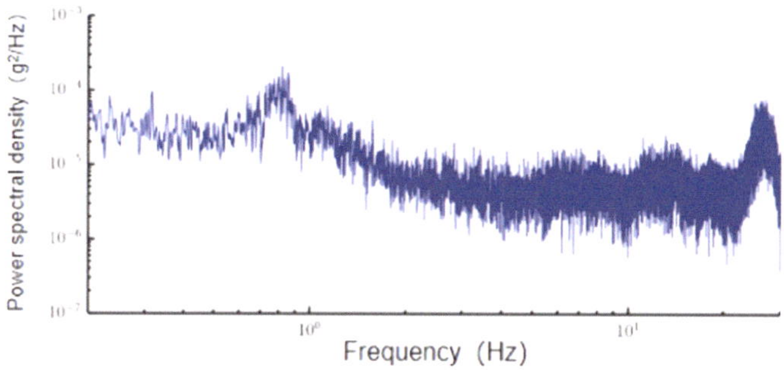


(b) Measured data (May 31, 2010)

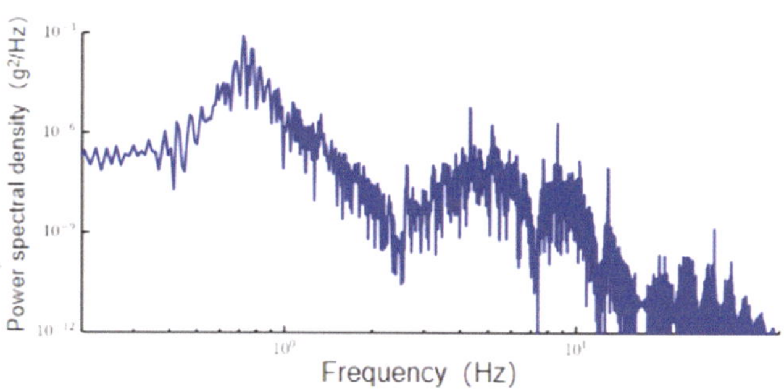


(c) The results of the calculations of this procedure

**Fig. 3.54** Comparison of power spectra of car body vertical acceleration

In actual operation, due to factors such as vehicle passenger capacity, the dominant frequencies of the car body's lateral and vertical vibrations may vary to some extent. From the comparison between the simulation results and the measured data, it can be seen that the dominant frequency of the car body vibration and the frequency distribution of the acceleration power spectrum obtained from the simulation calculations are in good agreement with the measured results. This indicates that the simulation results can accurately reflect the vibration characteristics of the actual vehicle.

#### 3.6.2.2 Comparison with General Dynamics Software

The multi-body system simulation software package SIMPACK (Simulation of Multi-body systems PACKAGE) is a mainstream general-purpose multi-body system dynamics analysis software, applicable in fields such as automotive, railway, engine, and various mechanical engineering sectors. Its railway module, SIMPACK/Rail, has been widely used in railway vehicle design and dynamic performance prediction. This section further verifies the developed software by comparing it with the results from the SIMPACK software.

The vehicle parameters used in the calculations in this section are shown in Table 3.17. The first test condition in the model validation is used to examine the reliability of the vehicle curve passing calculation. The specific conditions are as follows: curve radius of 7000 m, running speed of 250 km/h, and cant of 65 mm. The second test condition in the model validation is used to examine the reliability of the vehicle's lateral dynamic response, where the conditions are: after a 50 m straight section, 5 consecutive lateral sinusoidal track irregularities are set with peak-to-peak amplitude of 6.35 mm and wavelength of 11.88 mm, and the train running speed is 22.5 m/s. Figure 3.55 shows the specific settings.

Figure 3.56 shows the time history curves of wheelset roll angle, derailment coefficient, and wheel load reduction rate for the first test condition, calculated using the program developed in this paper and the SIMPACK software. From the figure, it can be observed that the amplitudes and waveform patterns of the roll angle of each wheelset, derailment coefficient on both sides of the wheelset, and the wheel load reduction rate are consistent. It can be concluded that under the curve passing condition, the results obtained from the software developed with the theoretical model in this paper are in good agreement with those from the general-purpose dynamics software.

Figure 3.57 shows the lateral displacement time-history curves of the car body and wheelset calculated by the program in this paper and the SIMPACK software under the second working condition. As can be seen from the figure, the results of the two methods are in good agreement.

**Table 3.17** The main dynamic parameters of the vehicle

| Name | Values | Unit |
|---|---|---|
| Body mass ($m_c$) | $4.17 \times 10^4$ | kg |
| Bogie mass ($m_t$) | 3049 | kg |
| Mass of wheelset ($m_w$) | 1780 | kg |
| Body nodding moment of inertia ($I_{cy}$) | $2.1 \times 10^6$ | kg $m^2$ |
| Body roll moment of inertia ($I_{cx}$) | $2.32 \times 10^4$ | kg $m^2$ |
| Body rocker moment of inertia ($I_{cz}$) | $2.08 \times 10^6$ | kg $m^2$ |
| Bogie head nodding moment of inertia ($I_{ty}$) | 2340 | kg $m^2$ |
| Bogie roll moment of inertia ($I_{tx}$) | 1580 | kg $m^2$ |
| Bogie head moment of inertia ($I_{tz}$) | 3930 | kg $m^2$ |
| Nodal moment of inertia of wheelset ($I_{wy}$) | 112 | kg $m^2$ |
| Side-roll moment of inertia of wheelset ($I_{wx}$) | 1140 | kg $m^2$ |
| Moment of inertia of rocking head of wheelset ($I_{wz}$) | 1120 | kg $m^2$ |
| Half of vehicle fixation distance ($l_c$) | 8.75 | m |
| Half of wheelbase ($l_t$) | 1.25 | m |
| Duplex vertical stiffness ($k_{pz}$) | $2.65 \times 10^5$ | N/m |
| Series II vertical damping ($c_{pz}$) | $4.51 \times 10^4$ | N s/m |
| System II lateral stiffness ($k_{py}$) | $1.76 \times 10^5$ | N/m |
| Series II lateral damping ($c_{py}$) | $3.92 \times 10^4$ | N s/m |
| Vertical stiffness of a system ($k_{tz}$) | $5.8 \times 10^5$ | N/m |
| One-series vertical damping ($c_{tz}$) | $1.96 \times 10^4$ | N s/m |
| Transverse stiffness of a system ($k_{ty}$) | $2.35 \times 10^6$ | N/m |

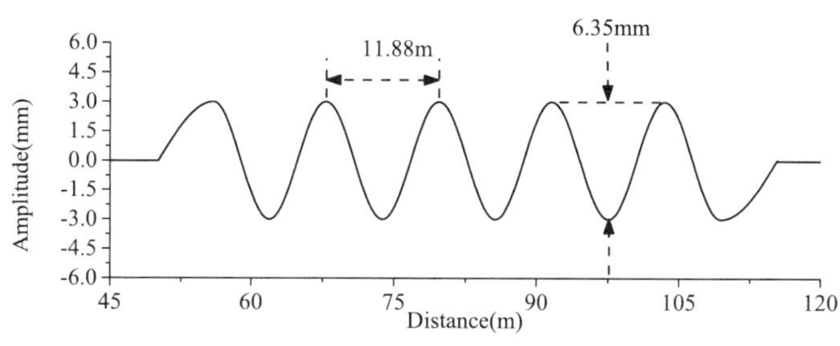


**Fig. 3.55** Schematic diagram of working condition II

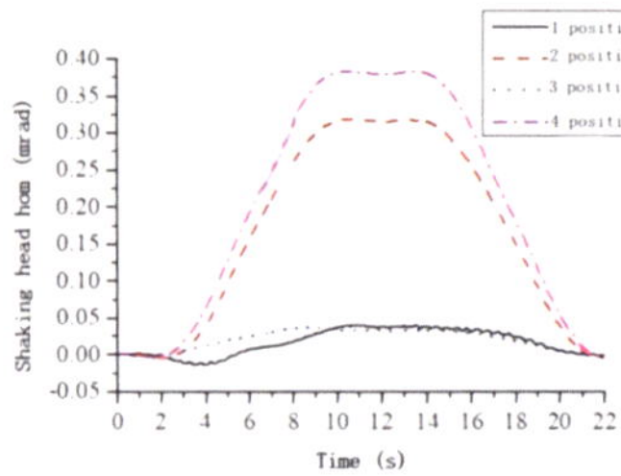


(a) The program in this paper calculates the roll angle

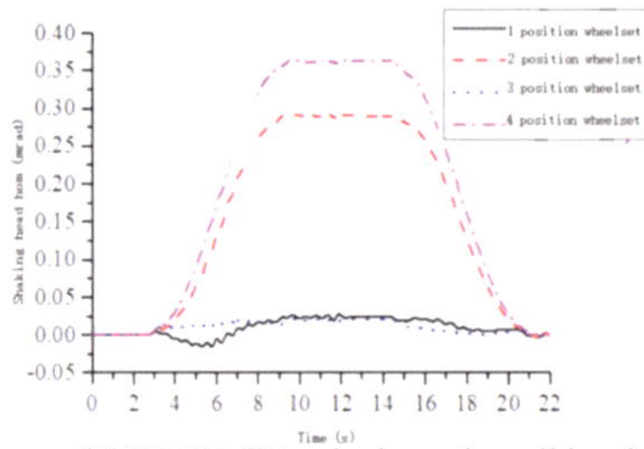


(b) SIMPACK calculates the roll head angle

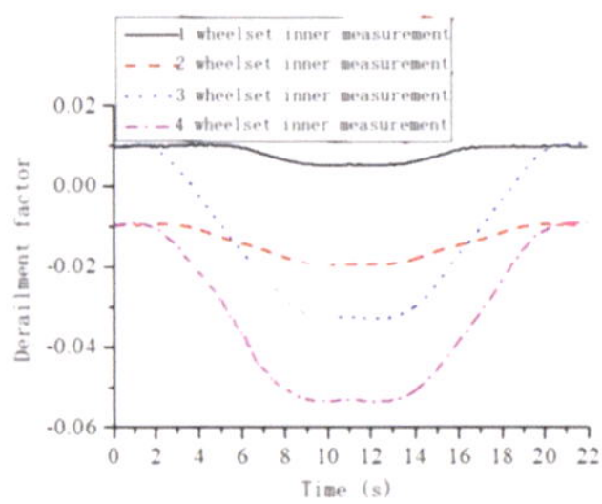


(c) The program in this paper calculates the derailment coefficient

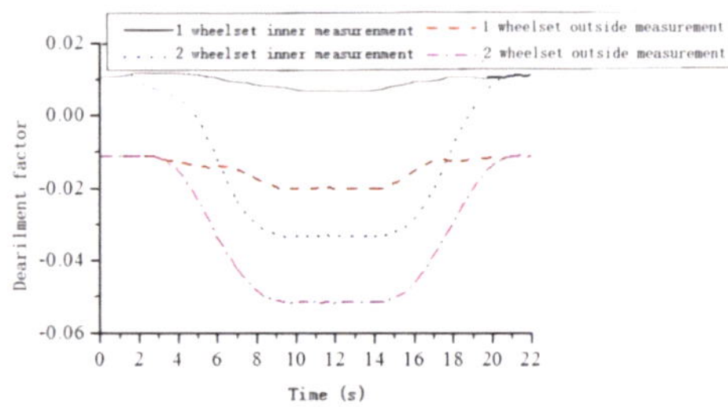


(d) SIMPACK calculates the derailment coefficient

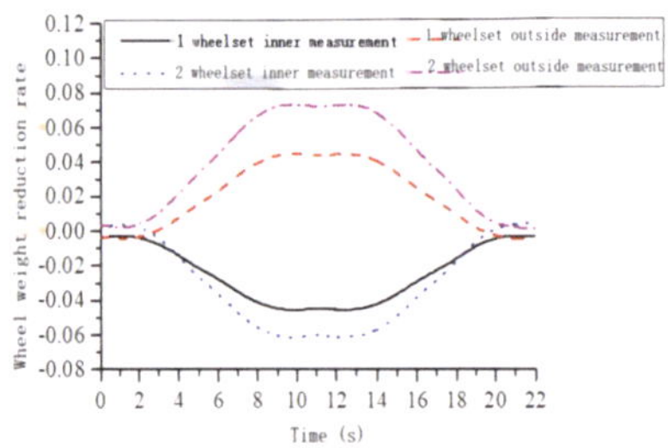


(e) The program in this paper calculates the wheel load reduction rate

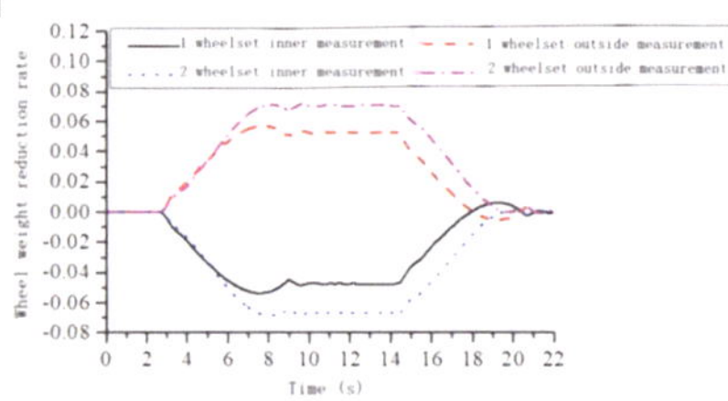


(f) SIMPACK calculates the wheel load reduction rate

**Fig. 3.56** Comparison of dynamic indices in curve sections

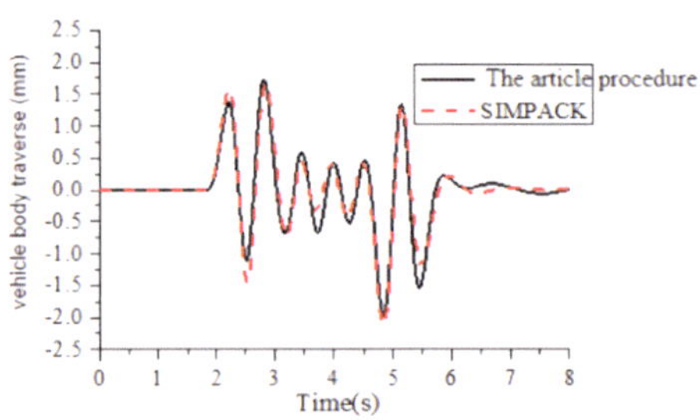


(a) vehicle body traverse

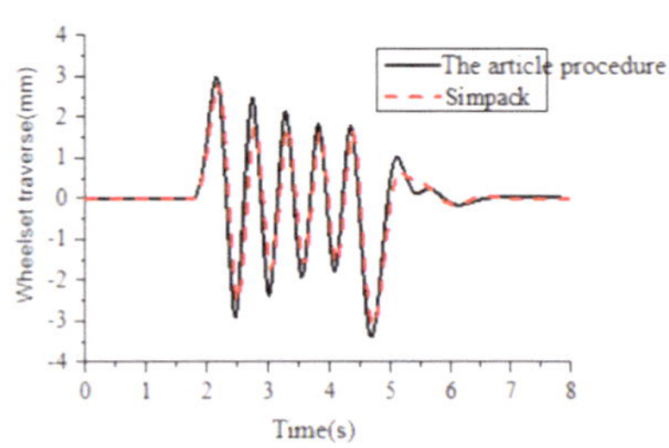


(b) wheelset traverse

**Fig. 3.57** Power response comparison

# Chapter 4
# Evaluation Criteria and Methods for Dynamic Performance

The evaluation criteria and methods for the dynamic performance of vehicles and track structures form the foundation of dynamic simulation analysis. This chapter first introduces the primary evaluation indicators and standards for vehicle dynamics, as well as the dynamic performance evaluation criteria for track, bridges, and other trackside structures in different countries. Furthermore, in response to the research needs of high-speed railway track dynamics, this chapter analyzes the applicability of ride comfort indicators and proposes corresponding evaluation methods based on different research subjects and objectives in track dynamics.

## 4.1 Evaluation Indicators and Methods for Vehicle Dynamics

The dynamic performance of railway vehicles is a critical technical parameter for assessing their operational performance. In the train-track dynamic interaction system, vehicle dynamic performance also serves as an important criterion for determining whether the underlying infrastructure meets operational requirements.

This section introduces railway vehicle dynamic evaluation indicators and standards based on Chinese national standards, including:

*Specification for Dynamic Performance Assessment and Testing Verification of Rolling Stock* (*GB/T 5599-2019*)

*High-Speed Railway Design Code* (*TB 10621-2014*)

*Technical Code for Dynamic Acceptance of High-Speed Railway Projects* (*TB 10761-2013*)

*Testing of High-speed Electric Multiple Unit on Completion of Construction* (Tieyun [2008] No. 28).

© China Railway Publishing House Co., Ltd. 2026

X. Long et al., *Vehicle–Track–Bridge System Subjected to Track Alignment*,
https://doi.org/10.1007/978-981-95-1649-0_4

Additionally, international standards from ISO (International Organization for Standardization), UIC (International Union of Railways), and EN (European Standards) are referenced to present a comprehensive overview of vehicle dynamic evaluation criteria.

### 4.1.1 Evaluation Indicators for Running Safety

The operational safety of railway vehicles is the most fundamental requirement in railway transportation, primarily concerning the risk of derailment and overturning. Running safety is typically evaluated using indicators such as:

1. **Derailment Coefficient**

The derailment coefficient is defined as the ratio of the instantaneous lateral wheel-rail force to the vertical wheel-rail force, used to assess the likelihood of wheel derailment. Currently, the widely adopted derailment coefficient is based on the Nadal Criterion, proposed by the French scholar Nadal in 1908. Nadal derived the derailment coefficient expression based on the equilibrium condition at the wheel-rail contact point when the wheel is in a critical derailment state. Given that $\mu$ is the friction coefficient between the wheel flange and the rail, and $\delta$ is the wheel flange angle, the critical derailment coefficient is expressed as:

$$\frac{Q}{P} = \frac{tan\delta - \mu}{1 + \mu tan\delta} \tag{4.1.1}$$

From Eq. (4.1.1), it can be observed that the critical value of the derailment coefficient depends on the wheel flange angle and the friction coefficient between the wheel and the rail. Specifically: a larger wheel flange angle results in a higher critical derailment coefficient; a higher friction coefficient leads to a lower critical derailment coefficient (Fig. 4.1).

According to the *Specification for Dynamic Performance Assessment and Testing Verification of Rolling Stock* (*GB/T 5599-2019*), the limit values for derailment coefficient assessment are specified in Table 4.1.

In recent years, both China's and international studies have shown that the classical Nadal equation is not entirely comprehensive and is only applicable to low-speed operations. For evaluating the running safety of high-speed trains, derailment assessment criteria must be redefined based on operational experience and research findings from high-speed rail systems worldwide.

International regulations regarding the derailment coefficient include the following:

(1) The International Union of Railways (UIC) specifies $Q/P \le 1.2$.
(2) The European Railway Alliance's EUROCODE adopts $Q/P \le 0.8$.
(3) The German ICE high-speed train follows $Q/P \le 0.8$ in high-speed testing.
(4) Japan's *Railway Structure Design Standards* also apply $Q/P \le 0.8$.

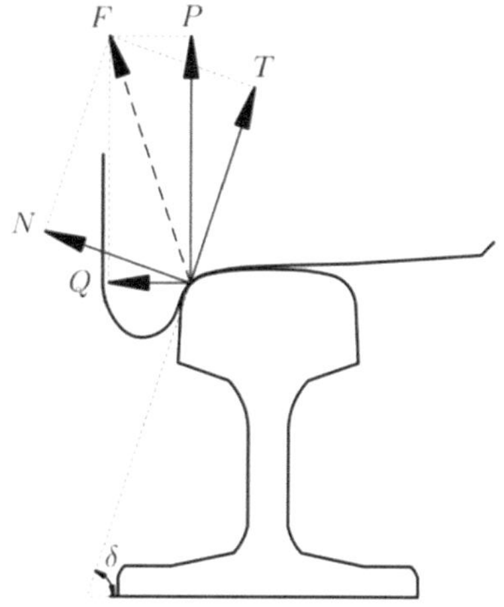


**Fig. 4.1** Force relationships at the wheel-rail contact point

**Table 4.1** Evaluation limits for derailment coefficient

| Type of vehicle | Derailment coefficient $Q/P$ | |
|---|---|---|
| | Curve radius $250\ \text{m} \le R \le 400\ \text{m}$; Lateral passing through 9# and 12# turnouts | Other tracks (curve radius $R > 400\ \text{m}$) |
| Passenger cars and EMUs | $\le 1.0$ | $\le 0.8$ |
| Locomotives | $\le 0.9$ | $\le 0.8$ |
| Freight cars | $\le 1.2$ | $\le 1.0$ |

In China, the derailment coefficient limits specified in national standards include:

(1) *Technical Code for Dynamic Acceptance of High-Speed Railway Projects* (*TB 10761-2013*)
(2) *High-Speed Railway Design Code* (*TB 10621-2014*)
(3) *Code for Design of Railway Bridges and Culverts* (*TB 10002-2017*).

According to these standards, the derailment coefficient limit for high-speed railways, intercity railways, and mixed passenger-freight railways is 0.8. For heavy-haul railways, the locomotive derailment coefficient is 0.8, while freight cars adopt a limit of 1.0.

2. **Wheel Load Reduction Ratio**

The wheel load reduction ratio is a safety indicator used to assess the risk of derailment due to severe unloading of one wheel under conditions where the lateral force on the wheelset is zero or nearly zero. It is used in conjunction with the derailment coefficient to evaluate railway vehicle derailment safety. The wheel load reduction ratio is defined as the ratio of the wheel load reduction amount $\Delta P$ on the unloading side to the average static wheel load $\overline{P}$ of the wheelset, denoted as $\Delta P/\overline{P}$.

According to the *Specification for Dynamic Performance Assessment and Testing Verification of Rolling Stock* (*GB/T 5599-2019*), the evaluation of the wheel load reduction ratio is classified based on speed as follows:

(1) When the test speed is $v \leq 160$ km/h, $\frac{\Delta P}{P} \leq 0.65$;
(2) When the test speed is $v > 160$ km/h, $\frac{\Delta P}{P} \leq 0.80$.

It should be noted that the derailment coefficient and the wheel load reduction ratio are indicators mainly used to evaluate wheel derailment under two different conditions: when the lateral force on the wheelset H > 0 and when H = 0. Both are derived based on the force equilibrium conditions of the wheel. In general, the derailment coefficient should be used as the evaluation indicator for running safety, while the above-mentioned limit values of the wheel load reduction ratio can be regarded as evaluation criteria under static or quasi-static conditions and are not suitable for assessing dynamic wheel unloading.

Dynamic wheel unloading refers to the phenomenon in which, when locomotives or rolling stock operate at high speeds, impact vibrations caused by passing over rail joints, local rail surface irregularities, or turnouts may result in significant transient reductions in wheel load. In high-speed train tests, the dynamic wheel load reduction ratio limit is 0.9 in Germany, 0.9 in the United States, and 0.8 in Japan during Shinkansen speed increases. In China, during speed-increase and high-speed operation tests, a dynamic wheel load reduction ratio limit of 0.8 is generally adopted.

China's *Technical Code for Dynamic Acceptance of High-Speed Railway Projects* (*TB 10761-2013*) stipulates that the limit is 0.8 when using discrete force-measuring wheelsets and 0.65 when using continuous force-measuring wheelsets. The *Code for Design of High Speed Railway* (*TB 10621-2014*) specifies that in vehicle-bridge dynamic analysis, the limit is 0.6, and for speeds above 350 km/h, the limit can be set to 0.8. The *Testing of High-speed Electric Multiple Unit on Completion of Construction* (Tieyun [2008] No. 28) stipulates that the quasi-static limit for the wheel load reduction ratio is 0.65, while the dynamic limit is 0.8.

When adopting the wheel load reduction ratio evaluation standard, it is necessary to consider whether the experimental testing or computational analysis includes high-frequency dynamic components.

3. **Lateral Force on Wheelsets**

Studies on the stability of continuously welded rail have shown that excessive lateral axle force $H$ (the sum of the lateral forces $Q_1$ and $Q_2$ on both sides of a wheelset) is the main cause of track panel lateral displacement, dynamic instability of continuously welded rail, and rail expansion-induced track buckling. The lateral axle force is used to evaluate whether excessive lateral force during vehicle operation may cause track gauge widening or severe track deformation.

The International Union of Railways (UIC) defines the limit value for lateral axle force using the following formula:

$$H \leq \beta(10 + P_0/3) \tag{4.1.2}$$

In this formula, $P_0$ represents the static axle load (kN); the coefficient $\beta$ is 1 for passenger cars, 0.75 for empty freight cars, and 0.8 for loaded freight cars.

The *High-Speed Railway Engineering Dynamic Acceptance Technical Specifications* (*TB10761-2013*) and the Code for Design of High-Speed Railway (*TB10621-2014*) specify the lateral force limit for wheelsets as follows:

$$H \leq 10 + P_0/3 \tag{4.1.3}$$

The *Specification for Dynamic Performance Assessment and Testing Verification of Rolling Stock* (*GB/T 5599-2019*) stipulates that the evaluation of wheelset lateral forces is calculated using the formula:

$$H \leq 15 + P_0/3 \tag{4.1.4}$$

4. **Wheel-Rail Interaction Forces**

(1) Wheel-Rail Lateral Force

Excessive wheel-rail lateral force can cause track gauge widening and severe track deformation. Currently, the direct assessment of wheel-rail lateral force is rarely used in China's railways. In Europe and the U.S., based on experiments, a general limit of 0.4 times the axle load is applied as the allowable limit for lateral force:

$$Q = 0.4P_0 \tag{4.1.5}$$

In the formula, $P_0$ represents the static axle load (kN).

UIC 518 specifies a limit of 60 kN for lateral wheel-rail forces on curves.

(2) Wheel-Rail Vertical Force

Japan has specific regulations for the maximum wheel load in speed-up tests on existing lines, requiring it to be less than the design load of track components (such as concrete ties and track slab). The design load takes into account the dynamic impact caused by wheel flats. Investigations indicate that the dynamic-static load ratio caused by wheel flats is approximately 1.7, and with a safety factor, this value is taken as 2.0. Therefore, the maximum wheel load with a flat spot can be taken as three times the static wheel load, which is used as the design load.

The Shinkansen uses a standard axle load of 170 kN, with a design load of 270 kN; while for existing lines, an axle load of 160 kN is used with a design load of 255 kN.

The German Federal Railway (DB) specifies that, for line load, the non-impact mid-low frequency wheel-rail vertical force should not exceed a limit of 170 kN. The British Railway (BR) sets the limit for the wheel-rail vertical force during transient impacts, such as when vehicles pass over rail joints, at 250 kN.

The UIC 518 standard provides a formula for calculating the vertical force limit:

$$P = 90 + P_0 \tag{4.1.6}$$

In the formula, $P_0$ represents the static wheelset load (kN). UIC 518 also specifies limits for different design speeds. For design speeds exceeding 300 km/h, the vertical wheel-rail force limit is 160 kN, while for curved tracks, the limit is 145 kN.

The Chinese code *Testing of High-speed Electric Multiple Unit on Completion of Construction* (Tieyun [2008] No. 28) states that the wheel-rail vertical force limit for China's high-speed trains is 170 kN.

5. **Bogie Frame Acceleration**

Bogie frame acceleration is commonly used to evaluate lateral stability. The *Specification for Dynamic Performance Assessment and Testing Verification of Rolling Stock* (*GB/T 5599-2019*) and the *High-Speed Railway Engineering Dynamic Acceptance Technical Specifications* (*TB10761-2013*) stipulate that if the peak vibration acceleration of the bogie frame reaches or exceeds 8 m/s$^2$ more than six consecutive times (filtered within the 0.5–10 Hz band), the bogie is considered to have experienced lateral instability.

### 4.1.2 Indicators for Evaluating Ride Comfort

1. **Definition of Comfort**

Passenger comfort, as a comprehensive physiological indicator, reflects the level of fatigue experienced by passengers during their journey and serves as a statistical standard. Broadly speaking, the comfort of a railway train refers to the comprehensive psychological reaction of passengers to the quality of their ride while traveling on the train. Factors such as vibrations, noise, humidity, temperature, seat quality, space design, passenger posture, and mental state all affect passenger comfort. There are individual differences in the perception of comfort, as different people have varying health expectations and sensitivities to the travel environment, which also influences their reactions to comfort. In a broad sense, comfort encompasses various states, including the fulfillment of comfort requirements, physical and mental well-being, and the absence of pain or distress.

Passenger comfort is influenced by various factors, with vibrations in railway vehicles being considered the greatest influencing factor. During the vehicle's operation, vibrations are constant and have a lasting impact on both the train's performance and the passengers' perception of comfort. The narrow sense of comfort discussed in this article refers to the discomfort caused by vibrations generated during train operation, also known as vibration comfort. The impact of vibrations on the human body is determined by multiple factors, including frequency, vibration intensity, direction,

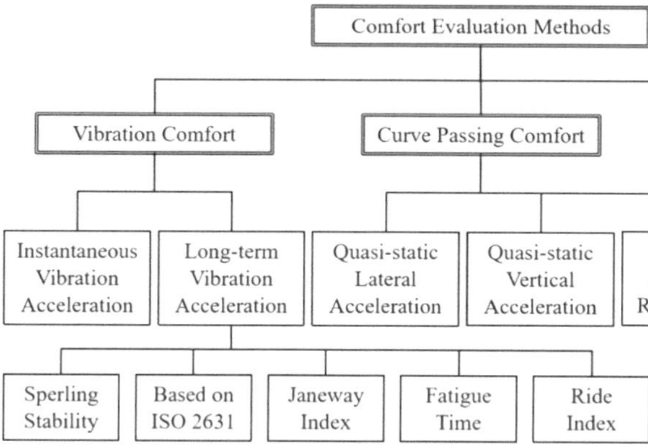


**Fig. 4.2** Classification of comfort evaluation methods

and duration. The evaluation of vibration comfort is one of the main components of overall comfort assessment.

2. **Classification of Comfort Evaluation Indicators**

Passenger comfort is generally evaluated through vibration acceleration and rate of change of acceleration. From the nature of the evaluation indicators, factors affecting comfort include vibration acceleration, quasi-static acceleration, and rate of change of acceleration. From the perspective of the factors influencing vibration, the evaluation can generally be divided into three categories: vibration comfort, curve pass comfort, and acceleration/deceleration comfort, as shown in Fig. 4.2.

3. **Vibration Comfort**

Vibration comfort refers to the impact of vehicle stability on ride comfort. The most direct indicator for evaluating vibration comfort is the vehicle's vibration acceleration. For short-term comfort evaluation, the vibration acceleration magnitude is generally used for assessment. For long-term comfort evaluation, not only the vibration amplitude but also the vibration frequency and duration need to be considered. The ISO 2631 standard focuses on vibrations in the range of 0.5–80 Hz, while the Sperling index focuses on vibrations in the range of 0.5–30 Hz, and the UIC 513 standard focuses on vibrations in the range of 0.4–80 Hz.

(1) Instantaneous Vibration Acceleration Evaluation

Vehicle body vibration acceleration is an important technical indicator for assessing ride comfort. Currently, international standards for vehicle body vibration acceleration vary. The EUROCODE has specified vertical vibration acceleration limits for passenger car bodies, as shown in Table 4.2.

In Japan's Shinkansen slow-speed management objectives, the vertical acceleration limit for vehicle bodies is set at 0.45 g.

**Table 4.2** Vertical vibration acceleration standards for passenger vehicles specified by EUROCODE

| Comfort level | Excellent | Good | Acceptable |
|---|---|---|---|
| Vertical acceleration (g) | 0.1 | 0.13 | 0.2 |

Lateral acceleration is primarily used to evaluate ride comfort when a train passes through curves. In 1963, Japan's Train Speed Survey Committee, based on test results related to ride comfort, established a lateral acceleration standard with a maximum value of 0.08 g, which remains the evaluation standard for ride comfort on curves today. However, on the Shinkansen, the maximum allowable lateral acceleration is 0.09 g, with a maximum operating speed of 270 km/h. When assessing ride comfort on curves, in addition to the lateral acceleration limit, the rate of change of acceleration must also be considered. Even if the lateral acceleration does not exceed the limit, passengers may still experience discomfort if the lateral acceleration changes too rapidly. Therefore, the rate of change of lateral acceleration must be restricted. In general, it should be kept below 0.3 $m/s^3$, and in special cases, it should not exceed 0.4 $m/s^3$.

The *Code for Design of High Speed Railway* (*TB 10621-2014)* specifies that the vertical vibration acceleration of the car body should be 1.3 $m/s^2$ (half-peak value), while the lateral vibration acceleration should be 1.0 $m/s^2$ (half-peak value).

The *Specification for Dynamic Performance Assessment and Testing Verification of Rolling Stock* (*GB/T 5599-2019*), *Testing of High-speed Electric Multiple Unit on Completion of Construction* (Tieyun [2008] No. 28), and UIC 518 specify that the lateral and vertical acceleration limits for the car body are 2.5 $m/s^2$.

The *High-Speed Railway Slab Track Maintenance Regulations* (Tieyun [2012] No. 83) and the *High-Speed Railway Ballasted Track Maintenance Regulations* (Tieyun [2013] No. 29) categorize the permissible deviation management values for track dynamic quality into four levels: routine maintenance, comfort, temporary repair, and speed restriction (160 km/h). The corresponding car body vertical and lateral acceleration standards are shown in Table 4.3.

(2) Long-Term Vibration Acceleration Evaluation

The evaluation standards for railway passenger train vibration comfort vary significantly worldwide. Research on evaluation methods mainly focuses on comfort index evaluation scales and frequency weighting. Differences in vibration frequency and intensity are the main factors affecting human perception of vibration. To reflect the decisive role of these factors, various evaluation methods incorporate frequency weighting of vibration signals, converting physical vibration quantities into perceived vibration levels for different frequency bands. Currently, the primary differences in national railway evaluation results lie in the comfort index evaluation scale, while frequency weighting differences are relatively minor. Most countries develop their vibration comfort evaluation frequency weighting curves based on the ISO 2631

**Table 4.3** Track management corresponding to car body vibration acceleration evaluation standards

| Items | Regular maintenance | Comfort | Temporary repairs | Speed limit (160 km/h) |
|---|---|---|---|---|
| Deviation levels | Level I | Level II | Level III | Level IV |
| Car body vertical acceleration ($m/s^2$) | 1.0 | 1.5 | 2.0 | 2.5 |
| Car body lateral acceleration ($m/s^2$) | 0.6 | 0.9 | 1.5 | 2.0 |

*Note* Car body vertical acceleration is evaluated using a 20 Hz low-pass filter. The Level I and Level II standards for car body lateral acceleration are assessed using values processed with a 0.5–10 Hz band-pass filter, while the Level III and Level IV standards are evaluated using values processed with a 10 Hz low-pass filter

international standard, with slight differences in the frequency range used for evaluation. However, there are differences in classification criteria for evaluation indices, which result in variations in final assessment outcomes.

The fundamental methods for evaluating railway passenger ride comfort generally include the root mean square (RMS) value-based evaluation method and statistical evaluation methods. Different evaluation methods have variations in the assessment process. At present, there is no globally unified standard for evaluating railway passenger train vibration comfort. The major international standards include:Sperling ride quality evaluation index, ISO 2631 Whole-Body Vibration Evaluation Standard, UIC 513 Comfort Evaluation Standard, BS 6841 Whole-Body Vibration Evaluation Standard (UK), EN 12299 Railway Vehicle Vibration Evaluation Standard (EU). China currently adopts the Sperling index and the comfort indicators specified in UIC 513.

① Sperling Ride Quality Index

In the 1940s, German researcher Sperling conducted vibration platform tests to establish comfort standards for railway vehicles. The tests involved 25 subjects, who were exposed to 2–10 min of vibration. Based on the test results, Sperling derived a formula for the ride quality index, which was later refined through further studies to improve accuracy by incorporating a frequency correction factor F(f) (see Table 4.4). Additionally, vertical and lateral vibrations were evaluated separately.

**Table 4.4** Frequency correction factors

| Vertical vibration | | Lateral vibration | |
|---|---|---|---|
| $f$ / Hz | $F(f)$ | $f$ / Hz | $F(f)$ |
| $0.5 \leq f < 5.9$ | $0.325f^2$ | $0.5 \leq f < 5.4$ | $0.8f^2$ |
| $5.9 \leq f < 20.0$ | $400/f^2$ | $5.4 \leq f < 26.0$ | $650/f^2$ |
| $f > 20.0$ | 1 | $f > 26.0$ | 1 |

This formula, known as the Sperling equation, was widely used internationally for a long time and is still in use in many countries, including China.

According to the "Locomotive and Rolling Stock Dynamic Performance Evaluation and Testing Specifications" (*GB 5599-2019*), the Sperling comfort index is calculated using the following formula:

$$W = 3.57\sqrt[10]{\frac{A^3}{f}F(f)} \tag{4.1.7}$$

where $A$ represents the vibration acceleration of the vehicle body (m/s$^2$); $f$ represents the vibration frequency (Hz); and $F(f)$ represents the frequency correction coefficient, as shown in Table 4.4.

The introduction of frequency correction factors $F(f)$ primarily accounts for the human body's varying sensitivity to vibrations at different frequencies. These correction factors differ for vertical and lateral vibrations. Research from both China's and international studies indicates that the human body is most sensitive to lateral vibrations below 2 Hz and to vertical vibrations in the 4–8 Hz range.

Based on Table 4.4, the weighted function curves for vertical and lateral directions $F(f)/f$ are plotted as shown in Fig. 4.3. The chart shows that the frequencies with the highest weight affecting lateral and vertical ride comfort indices are 5.4 Hz and 5.9 Hz, respectively. In vertical vibration acceleration, frequency components above 26 Hz have little impact, while in lateral vibration acceleration, components above 20 Hz have minimal effect.

Since the vibration of the vehicle is random vibration, its acceleration and frequency are constantly changing. In actual evaluation, the acceleration waveform to be analyzed is divided into frequency groups. Based on the acceleration and frequency of each group, the ride quality index $W_i$ for that group is calculated. The index for the entire waveform is calculated using the following formula:

$$W = \sqrt[10]{\left(W_1^{10} + W_2^{10} + \cdots W_N^{10}\right)} \tag{4.1.8}$$

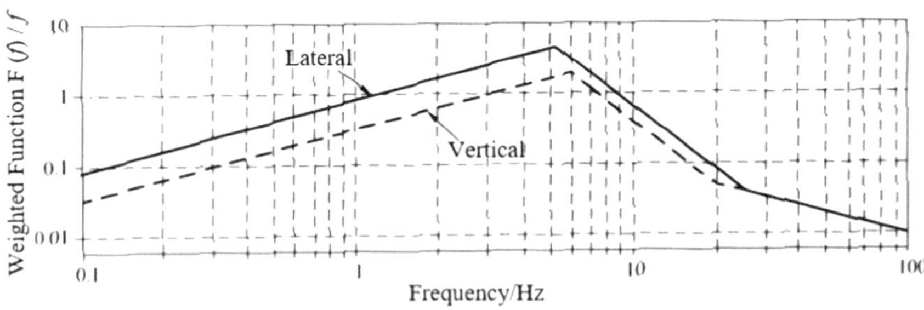


**Fig. 4.3** Sperling index weighted curve

**Table 4.5** Stability level evaluation table for passenger cars and EMUs

| Stability level | Stability index $W$ | Evaluation |
|---|---|---|
| Level 1 | $W \leq 2.50$ | Excellent |
| Level 2 | $2.50 < W \leq 2.75$ | Good |
| Level 3 | $2.75 < W \leq 3.00$ | Acceptable |

The *Specification for Dynamic Performance Assessment and Testing Verification of Rolling Stock* (*GB/T 5599-2019*) and the *High-Speed Railway Engineering Dynamic Acceptance Technical Specifications* (*TB10761-2013*) specify the evaluation grades, as shown in Table 4.5. The Sperling ride quality index should meet the "Good" standard or above.

The evaluation method has the following main issues:

(1) This method evaluates vertical and lateral vibrations of the vehicle separately, without considering their combined effect on the human body. When there is a significant difference between the ride quality levels in the vertical and lateral directions, assessing comfort based solely on one direction can lead to a biased and incomplete evaluation.
(2) The sensitivity of this method is relatively low. If the ride quality index is excessively high within a certain frequency range—exceeding the human tolerance threshold—but remains low in other ranges, the total index calculated for that direction may still appear small. Relying on such a result to judge comfort as acceptable is clearly misleading.
(3) The Sperling method does not reflect the influence of exposure duration in a vibrating environment. If the vibration lasts only for a short period, even uncomfortable vibrations may not cause significant harm to the human body. Therefore, this method cannot quantitatively assess how the duration of exposure affects human comfort or health.
(4) The Sperling method also fails to account for the train's operating speed, which has a direct impact on ride quality. Since ride quality can vary significantly across different speed ranges, evaluating a vehicle's performance based on a fixed speed range may not provide an accurate reflection of its overall ride comfort.

② ISO-2631 Vibration Evaluation Standard.

ISO2631 ISO 2631 is the international standard for evaluating human exposure to vibration. It is an authoritative standard established by ISO/TC108, the International Organization for Standardization/Technical Committee on Mechanical Vibration and Shock, and its SC4—Human Exposure to Mechanical Vibration and Shock Technical Committee, which is globally recognized. ISO 2631-1 (1997) specifies the measurement methods for whole-body vibration, including periodic, random, and transient vibrations, and outlines the main factors that determine the acceptability of vibration exposure. This standard divides the effects of vibration on the human body into four parts: health effects, perceptual effects, comfort effects, and motion sickness. For evaluating low-frequency vibrations of railway vehicles, ISO 2631-4 (2001) provides guidelines for evaluating low-frequency vibrations that may cause motion sickness.

ISO 2631 provides seven frequency weighting curves, such as $W_c$, $W_e$, $W_j$, $W_k$, $W_d$, $W_f$, and $W_b$ for rail vehicles, to evaluate the health, comfort, perceptual, and motion sickness effects in different directions and postures. The selection of these weighting curves is shown in Tables 4.6 and 4.7, and the corresponding frequency weighting curves are shown in Figs. 4.4 and 4.5. Among them, in the vertical vibration comfort assessment of railway vehicles, the specified weighting curve $W_b$ can be used as an acceptable approximation of the general curve $W_k$. The $W_b$ weighting curve is shown in Fig. 4.6.

From the various frequency weighting curves, it can be seen that for evaluating motion sickness, the $W_f$ weighting curve for low-frequency vertical vibration acceleration mainly evaluates vibration accelerations below 4 Hz, with the frequency range most sensitive to motion sickness being between 0.09 Hz and 0.4 Hz. The comfort-sensitive frequency range for the $W_d$, $W_c$, and $W_e$ weighting curves, which are used to evaluate lateral and vertical vibration accelerations, is around 0.4–7 Hz. The $W_k$

**Table 4.6** Guidelines for the application of basic weighting frequency weighting curves

| Frequency weighting | Health | Comfort | Perception | Motion sickness |
|---|---|---|---|---|
| $W_k$ | $z$-axis, seat surface | $z$-axis, seat surface; $z$-axis, standing posture; $x$, $y$, $z$-axes, seated foot; vertical, lying posture(excluding head) | $z$-axis, seat surface; vertical, lying posture; $z$-axis, standing posture | – |
| $W_d$ | $x$-axis, seat surface; $y$-axis, seat surface | $x$-axis, seat surface; $y$-axis, seat surface; $x$, $y$-axes, standing posture; horizontal, lying posture; $y$, $z$-axes, seat-back support | $x$-axis, seat surface; $x$, $y$-axes, standing posture; horizontal, lying posture; $y$-axis, seat surface | – |
| $W_f$ | – | – | – | Vertical direction |

**Table 4.7** Application guidelines for additional frequency weighting curves

| Frequency weighting | Health | Comfort | Perception | Motion sickness |
|---|---|---|---|---|
| $W_c$ | $x$-axis, seat-back support | $x$-axis, seat-back support | $x$-axis, seat-back support | – |
| $W_e$ | – | $r_x$, $r_y$, $y_z$-axes, seat surface | $r_x$, $r_y$, $r_z$-axes, seat surface | – |
| $W_j$ | – | Vertical, lying posture(head) | Vertical, lying posture(head) | – |

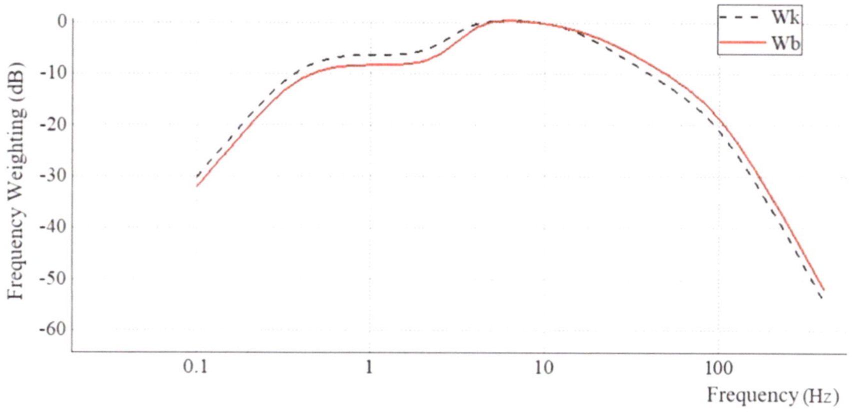


**Fig. 4.4** Frequency weighting curve for basic weighting values

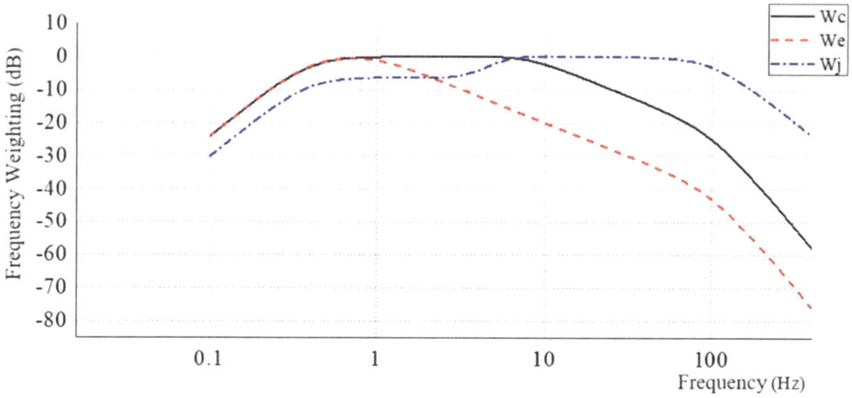


**Fig. 4.5** Frequency weighting curve for additional weighting values

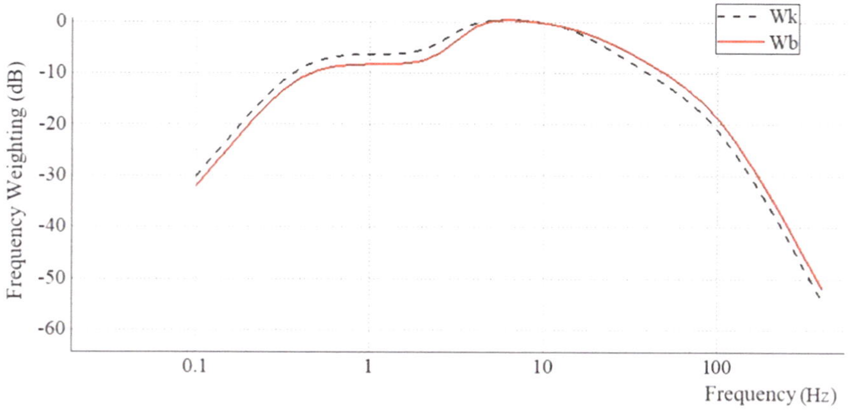


**Fig. 4.6** $W_b$ Frequency weighting curve for railway vehicles

and $W_j$ frequency weighting curves for vertical vibration acceleration comfort evaluation have a wide sensitivity range, with the most significant impact occurring in the 1–40 Hz frequency range.

According to ISO2631-1, vibration evaluation should include the measurement of weighted RMS acceleration as defined in the standard. The recorded acceleration time history $a(t)$ is filtered through a frequency weighting function $w(f)$ to obtain the weighted acceleration time history $a_w(t)$. Weighted RMS acceleration is expressed in $m/s^2$ for translational vibration and in $rad/s^2$ for rotational vibration. The definition of the weighted RMS acceleration is as follows:

$$a_{rms} = \sqrt{\frac{1}{T}\int a_w^2(t)dt} \tag{4.1.9}$$

Here, $a_w(t)$ represents the time history of the weighted acceleration (translational or rotational), and $T$ is the measurement duration (in seconds).

According to the relevant provisions of ISO2631-4, the assessment of comfort in railway vehicles generally uses the weighting factor corresponding to the modified frequency weighting curve $W_b$ for vertical vibration acceleration signals. ISO2631-1 also provides numerical values to approximate potential human responses to various combined vibration magnitudes in public transportation, as shown in Table 4.8.

The Ride Comfort Level (RCL) used in South Korea, and Japan is defined as follows:

$$\text{RCL} = 20\log_{10}(A_{rms}/A_{ref}) \quad (\text{dB}) \tag{4.1.10}$$

For South Korea, $A_{ref}$ is set at $10^{-6}$, while in Japan it is set at $10^{-5}$. The evaluation criteria for comfort levels are detailed in Table 4.9.

③ UIC513 Comfort Evaluation Standard

In order to study the application of the international standard ISO 2631 (1978) to railway vehicles, the European Rail Research Institute (ERRI) established the B153 topic in 1979, which resulted in specific indices for applying the whole-body vibration evaluation method to the vibration assessment of railway vehicles.

**Table 4.8** ISO2631 comfort values in different directions

| Comfort level | Weighted r.m.s. acceleration ($m/s^2$) | Evaluation |
|---|---|---|
| 1 | $\leq 0.315$ | Imperceptible discomfort |
| 2 | 0.315–0.63 | Slightly uncomfortable |
| 3 | 0.5–0.1 | Fairly uncomfortable |
| 4 | 0.8–0.6 | Uncomfortable |
| 5 | 1.25–2.5 | Very uncomfortable |
| 6 | $\geq 2$ | Extremely uncomfortable |

**Table 4.9** RCL evaluation standards for comfort levels (dB)

| Taiwan, South Korea | Japan | Evaluation |
|---|---|---|
| $\leq 103$ | $\leq 83$ | Very comfortable |
| 103–108 | 83–88 | Comfortable |
| 108–113 | 88–93 | Relatively comfortable |
| 113–118 | 93–98 | Uncomfortable |
| $\geq 118$ | $\geq 98$ | Very uncomfortable |

After ten years of research, in 1988, ERRI B153 proposed a draft UIC standard titled "Guidelines for the Vibration Comfort Evaluation of Railway Vehicles." After review, in 1994, the International Union of Railways (UIC) formally issued and implemented the "Guidelines for the Vibration Comfort Evaluation of Railway Vehicles" under the UIC513 standard number. This evaluation index follows the basic evaluation parameters of ISO 2631 when assessing the vibration of railway vehicles, namely using the weighted root mean square (r.m.s.) value of vibration acceleration signals as the basic quantity for evaluation.

In order to fully consider the variability of vibrations on railways and establish a more accurate and practical evaluation method, UIC513 uses statistical analysis methods when determining comfort levels, with the final results obtained using a 95% confidence probability method. For different postures and measurement points, the UIC513 comfort formula provides three methods for measurement, as follows: Simplified Formula for measuring floor surface comfort (Simplified Method):

$$N_{\mathrm{MV}} = 6\sqrt{\left(a_{X\mathrm{P}95}^{W_{\mathrm{d}}}\right)^2 + \left(a_{Y\mathrm{P}95}^{W_{\mathrm{d}}}\right)^2 + \left(a_{Z\mathrm{P}95}^{W_{\mathrm{b}}}\right)^2} \tag{4.1.11}$$

Full formula for measuring floor surface comfort in standing position (Full Method in Standing Position):

$$\mathrm{N}_{\mathrm{VA}} = 4\left(a_{Z\mathrm{P}95}^{W_{\mathrm{b}}}\right) + 2\sqrt{\left(a_{Y\mathrm{A}95}^{W_{\mathrm{d}}}\right)^2 + \left(a_{Z\mathrm{A}95}^{W_{\mathrm{b}}}\right)^2} + 4\left(a_{X\mathrm{D}95}^{W_{\mathrm{d}}}\right) \tag{4.1.12}$$

Full formula for measuring floor surface and seat comfort in seated position (Full Method in Seated Position):

$$N_{\mathrm{VD}} = 3\sqrt{16\left(a_{X\mathrm{P}95}^{W_{\mathrm{d}}}\right)^2 + \left(a_{Y\mathrm{P}95}^{W_{\mathrm{d}}}\right)^2 + \left(a_{Z\mathrm{P}95}^{W_{\mathrm{b}}}\right)^2} + 5\left(a_{Y\mathrm{P}95}^{W_{\mathrm{d}}}\right) \tag{4.1.13}$$

In the formulas, $N_{\mathrm{MV}}$, $N_{\mathrm{VA}}$, and $N_{\mathrm{VD}}$ represent comfort indices, with comfort levels shown in Table 4.10; $a$ represent RMS values of acceleration. Superscripts and $W_{\mathrm{d}}$ correspond to the frequency-weighted values of the weighted curves d and b, respectively. Subscripts $X$, $Y$, $Z$ and denote longitudinal, lateral, and vertical acceleration directions. P refers to the floor surface; A represents the contact point between the seat cushion and the human body; D represents the contact point at the backrest. The numbers 95 and 50 indicate the 95% and 50% probability percentiles, respectively.

The above formulas represent statistical results from 5-min measurement data. During this 5-min period, data is collected in 60 segments of 5 s each. The calculation method is shown in Eq. (4.1.14). Each data group is weighted and filtered using the physiological filter curves ($W_b$, $W_c$, $W_d$) specified in ISO263-1 (as shown in Fig. 4.7), followed by probability processing at 95% or 50% confidence levels. Effective RMS peak values are obtained for statistical participation, and vibration comfort is ultimately determined using the RMS method.

$$a_{X_j}^{W_i}(t) = \sqrt{\frac{1}{T}\int_{t-T}^{t} \ddot{X}_{W_i}^{2}(\tau)\mathrm{d}\tau} \tag{4.1.14}$$

When evaluating ride comfort in a single direction, the simplified comfort formula for each direction can be derived from Eq. (4.1.15):

$$\begin{cases} N_{\mathrm{MV}_x} = 6a_{X\,\mathrm{P}95}^{W_\mathrm{d}} \\ N_{\mathrm{MV}_y} = 6a_{Y\mathrm{P}95}^{W_\mathrm{d}} \\ N_{\mathrm{MV}_z} = 6a_{X\,\mathrm{P}95}^{W_\mathrm{d}} \end{cases} \tag{4.1.15}$$

**Table 4.10** Comfort level table (UIC 513)

| Comfort level | Comfort index | | Evaluation |
|---|---|---|---|
| Level 1 | $N < 1$ | $N_{x,y,z} < 0.6$ | Very comfortable |
| Level 2 | $1 \leq N < 2$ | $0.6 \leq N_{x,y,z} < 1.2$ | Comfortable |
| Level 3 | $2 \leq N < 4$ | $1.2 \leq N_{x,y,z} < 2.3$ | Average |
| Level 4 | $4 \leq N < 5$ | $2.3 \leq N_{x,y,z} < 2.9$ | Uncomfortable |
| Level 5 | $N \geq 5$ | $N_{x,y,z} \geq 2.9$ | Very uncomfortable |

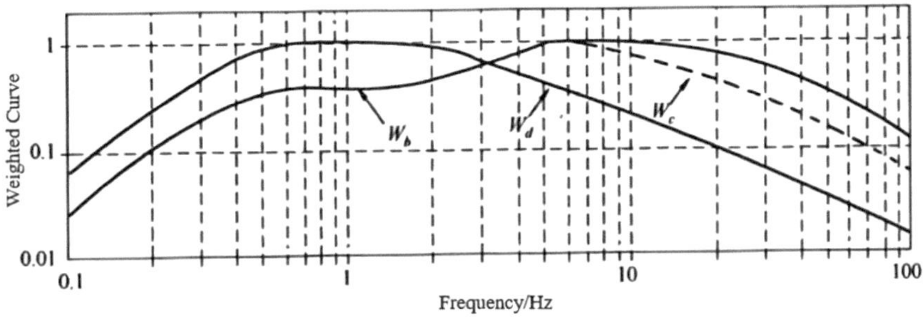


**Fig. 4.7** Weighted curves for UIC513 indicators

**Table 4.11** Comfort level table for passenger cars and EMUs (*GB/T 5599-2019*)

| Comfort level | Japan | Evaluation |
|---|---|---|
| Level 1 | $N_{\mathrm{MV}} < 1.5$ | Very comfortable |
| Level 2 | $1.5 \le N_{\mathrm{MV}} < 2.5$ | Comfortable |
| Level 3 | $2.5 \le N_{\mathrm{MV}} < 3.5$ | Fairly comfortable |
| Level 4 | $3.5 \le N_{\mathrm{MV}} < 4.5$ | Uncomfortable |
| Level 5 | $N_{\mathrm{MV}} \ge 4.5$ | Very uncomfortable |

*Note* Excludes power cars of concentrated-powered EMUs

The relationship between single-direction comfort evaluation criteria and three-direction integrated comfort criteria can be expressed by the following formula:

$$N_{\mathrm{MV}_{x,y,z_{\lim}}} = \sqrt{\frac{N^2_{\mathrm{MV}_{\lim}}}{3}} \tag{4.1.16}$$

Table 4.10 shows the comfort levels defined by the UIC513 evaluation method.

The comfort index levels in the *Specification for Dynamic Performance Assessment and Testing Verification of Rolling Stock* (*GB/T 5599-2019*) are shown in Table 4.11.

China's *Testing of High-speed Electic Multiple Unit on Completion of Construction* (Tieyun [2008] No. 28) specifies comfort indices $N_{\mathrm{MV}} < 2.0$ for high-speed EMUs.

The European standard EN 12299 is based on the RMS value of acceleration. It adopts the average comfort indices $N$ defined by UIC and additionally defines continuous comfort indices $C_{\mathrm{C}}$:

$$\begin{cases} C_{\mathrm{C}x}(t) = a^{W_{\mathrm{d}}}_{X\mathrm{P}}(t) \\ C_{\mathrm{C}y}(t) = a^{W_{\mathrm{d}}}_{Y\mathrm{P}}(t) \\ C_{\mathrm{C}z}(t) = a^{W_{\mathrm{d}}}_{Z\mathrm{P}}(t) \end{cases} \tag{4.1.17}$$

The evaluation criteria for continuous comfort indices are shown in Table 4.12.

**Table 4.12** Continuous comfort index $C_C$ evaluation standards (EN 12299)

| Comfort level | Comfort index(m/s$^2$) | Evaluation |
|---|---|---|
| Level 1 | $C_{\mathrm{C}y}, C_{\mathrm{C}z} < 0.20$ | Very comfortable |
| Level 2 | $0.20 \le C_{\mathrm{C}y}, C_{\mathrm{C}z} < 0.30$ | Comfortable |
| Level 3 | $0.30 \le C_{\mathrm{C}y}, C_{\mathrm{C}z} < 0.40$ | Fairly comfortable |
| Level 4 | $C_{\mathrm{C}y}, C_{\mathrm{C}z} \ge 0.40$ | Uncomfortable |

④ Comfort Indices $P_{CT}$ and $P_{DE}$

The comfort index based on the easing curve, proposed by the European standard EN12299, evaluates passenger comfort. Italy and Switzerland have validated the reasonableness and applicability of this method through extensive testing. The easing curve comfort method primarily focuses on the impact of the easing curve on passenger comfort, with calculations mainly considering the lateral acceleration, lateral impacts, and roll angular velocity of the vehicle in a steady state. The expression is as follows:

$$P_{CT} = 100 \cdot \left\{ \max\left[ \left( A\ddot{y} + B\dddot{y} - C \right), 0 \right] + \left( D\dot{\vartheta} \right)^{E} \right\} \tag{4.1.18}$$

In the formula, $\ddot{y}$ represents the maximum average lateral acceleration per second of the vehicle body, specifically the maximum value of the vehicle's center of mass from the start to the end of the easing curve, and for 1.6 s after the end of the curve. $\dddot{y}$ represents the maximum average lateral rate of change of acceleration per second, specifically the maximum value of the vehicle's center of mass in the range from 1 s before the start of the easing curve to the end of the curve. $\dot{\vartheta}$ represents the maximum average side roll angular velocity of the vehicle body at the vehicle's center of mass within the easing curve range. A – E are coefficients that vary based on the standing or sitting posture. The values for standing posture are as follows: 0.2854 $s^2$/m, 0.2069 $s^3$/m, 0.111, 3.64 s/rad, 2.283; for sitting posture, the values are 0.0897 $s^2$/m, 0.0968 $s^3$/m, 0.059, 0.916 s/rad, 1.626.

The $P_{CT}$ index is measured in percentage (%), with higher values indicating lower comfort. This evaluation index is applicable to the entry into easing curves and reverse curves but not to the exit from easing curves (Fig. 4.8).

The British Standard EN12299 (2009) also proposed a comfort index $P_{DE}$, which can be used to evaluate the comfort of lateral acceleration on independent sections, without considering the cumulative effects from preceding or subsequent sections. Based on the maximum peak-to-peak and average values of lateral acceleration at curvature changes, the comfort index $P_{DE}$ can be calculated as follows

$$P_{DE} = 100 \cdot \left\{ \max\left[ \left( a\ddot{y}_{pp} + b\ddot{y}_{2s} - c \right), 0 \right] \right\} \tag{4.1.19}$$

In the equation, $\ddot{y}_{pp}$ represents the peak-to-peak value of the lateral acceleration of the vehicle's body within 2 s; $\ddot{y}_{2s}$ is the average value of the lateral acceleration of the vehicle's body within 2 s; and a – c are coefficients that vary depending on whether the passenger is standing or sitting. The values for standing posture are 0.1662 $s^2$/m, 0.2701 $s^2$/m, and 0.37, while the values for sitting posture are 0.0846 $s^2$/m, 0.1305 $s^2$/m, and 0.217 (Fig. 4.9).

Similar to the $P_{CT}$ index, $P_{DE}$ expresses the percentage of passengers experiencing discomfort. Higher values indicate lower comfort. This is a relative index used to compare comfort levels under different conditions.

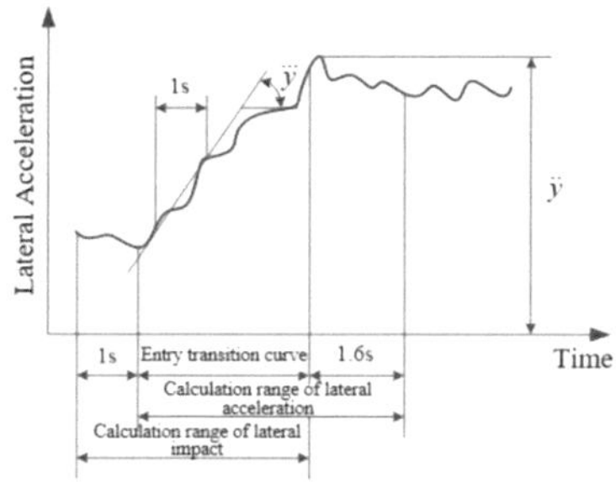


(a) Definitions of carbody lateral acceleration $\ddot{y}$ and the rate of change of carbody lateral acceleration $\dddot{y}$ in $P_{CT}$.

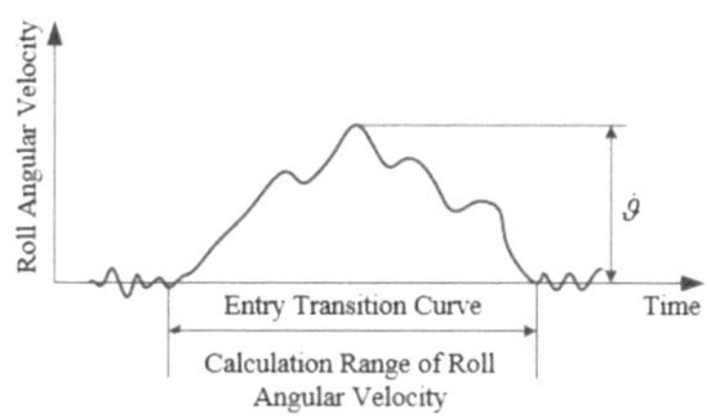


(b) Definition of maximum roll angular velocity $\dot{\vartheta}$ in $P_{CT}$.

**Fig. 4.8** Definition of vibration quantities in $P_{CT}$

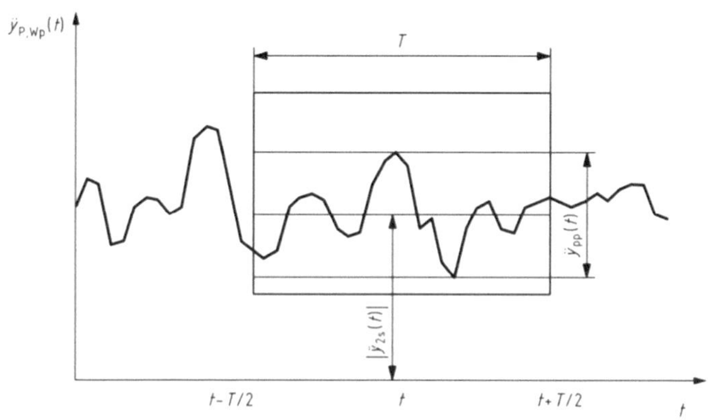


**Fig. 4.9** Definition of vibration quantities in $P_{DE}$

⑤ Selection of Ride Comfort Evaluation Indices

In China, the Sperling ride quality index has long been used as the main evaluation standard for vehicle operation stability and ride comfort, while the UIC 513R comfort standard serves as an auxiliary evaluation index for ride comfort. Currently, international evaluation standards generally tend to align with ISO 2631. European countries such as Germany and France use the EN 12299 ride comfort evaluation standard, which, like UIC 513R, is based on a statistical assessment method of ISO 2631-1. Both calculation methods are similar, with differences in the weighted curves and evaluation index standards. Japan, on the other hand, uses a ride comfort level based on ISO 2631-1 as the primary evaluation index for passenger comfort.

The Sperling ride quality index and ride comfort level evaluate the lateral and vertical ride comfort separately, whereas the UIC 513R ride comfort first calculates the comfort index for each direction separately and then combines the indices to determine the final ride comfort. Therefore, UIC 513R ride comfort considers the comprehensive impact of the longitudinal, lateral, and vertical directions. In general, when studying the impact of track alignment parameters on ride comfort, it is advisable to evaluate ride comfort in each direction separately to understand the effect of track parameters. When evaluating and optimizing track plans for ride comfort, the combined effects of all three directions need to be considered. In vehicle-track dynamic simulations, typically only the lateral and vertical vibrations of the vehicle body are considered. In this case, longitudinal ride comfort can be calculated from the measured longitudinal vibration acceleration, and the ride comfort for the lateral and vertical directions from simulation can be combined to obtain the UIC 513R ride comfort index value.

The above-mentioned ride comfort evaluation indices only consider the impact of lateral, vertical, or longitudinal vibration accelerations on passenger comfort. In reality, when a train passes through a curve, the vehicle's roll, nodding, and other rotational velocities and accelerations also affect passenger ride comfort. Therefore, in the evaluation and analysis of track alignment parameters or track plans for ride comfort, in addition to the above-mentioned ride comfort indices, curve-pass comfort indices should also be incorporated into the analysis.

(3) Motion Sickness Evaluation Index

Track alignment can induce low-frequency vibrations in the vehicle body, and vibrations with frequencies below 0.5 Hz can cause motion sickness, i.e., carsickness. ISO 2631 provides the motion sickness evaluation index and calculation method. While studies have shown that roll and nodding movements may also contribute to motion sickness, the primary motion responsible for inducing motion sickness is vertical motion. The motion sickness evaluation index is based on the vertical vibration in the 0.1–0.5 Hz range at the vehicle floor level. The likelihood of experiencing motion sickness is closely related to the duration of low-frequency vibrations. The calculation formula for the motion sickness degree value ($\mathrm{MSDV}_z$) is as follows:

$$\mathrm{MSDV}_z = \left[ \int_0^T a_{\mathrm{w}}^2(t)\mathrm{d}t \right]^{\frac{1}{2}} \tag{4.1.20}$$

In the formula, $a_{\mathrm{w}}(t)$ represents the frequency-weighted vertical acceleration, and $T$ is the total duration of vibration. The $\mathrm{MSDV}_z$ value is expressed in units of $\mathrm{m/s}^{1.5}$, which can be seen as the product of the weighted RMS acceleration and $T^{1/2}$.

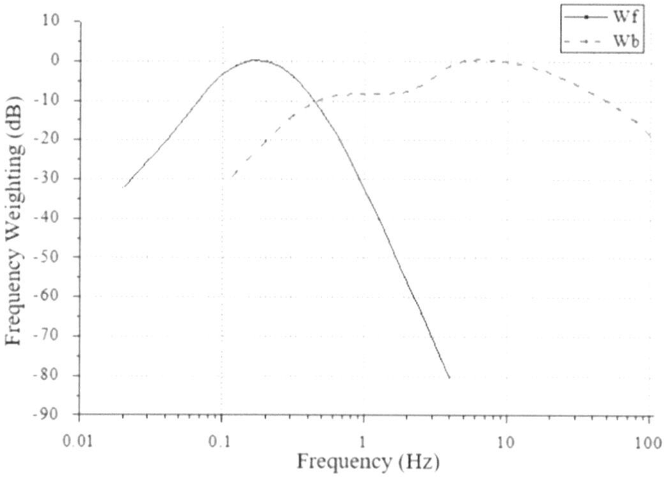


**Fig. 4.10** Frequency weighting curve for vertical acceleration

According to the relevant provisions of ISO 2631-4, when evaluating ride comfort, the vertical vibration acceleration signal is generally weighted using the modified frequency weighting curve $W_b$ and its corresponding weighting factor; in motion sickness evaluations, the vertical vibration acceleration signal is generally weighted using the frequency weighting curve $W_f$ and its corresponding weighting factor, as shown in Fig. 4.10. The impact of low-frequency vibrations varies greatly between individuals. A higher motion sickness value indicates a greater likelihood of experiencing motion sickness symptoms.

4. **Curve Passing Comfort**

When a vehicle passes through a curve, the combined effects of the curve's radius of curvature, cant, and speed cause the vehicle to experience unbalanced accelerations. Key indicators that significantly impact comfort include: unbalanced centrifugal acceleration (cant deficiency), unbalanced acceleration time rate (cant deficiency time rate), cant superelevation slope, cant time rate (roll angle velocity), and quasi-static vertical acceleration on the vertical curve.

(1) Unbalanced Centrifugal Acceleration (Cant Deficiency)

The essence of unbalanced centrifugal acceleration is cant deficiency. The conversion relationship between unbalanced centrifugal acceleration $a_y$ (m/s$^2$) and cant deficiency $h_q$ (mm/s) is as follows:

$$a_y = \frac{h_q}{S} g \tag{4.1.21}$$

In the formula, $S$ represents the distance between the centerlines of the two rails, which is taken as 1500 mm.

In the 1950s, the American Association of Railroads (AAR) conducted tests on representative vehicles from various railways to study comfort during curve passage. The tests led to the following conclusions:

A. The allowable limit for lateral steady-state acceleration is 1.0 m/s$^2$ (cant deficiency of 150 mm).
B. When considering vehicle tilt, the allowable cant deficiency is 76 mm (unbalanced centrifugal acceleration of 0.5 m/s$^2$), but for vehicles with small tilt, it can be allowed up to 102 mm (unbalanced centrifugal acceleration of 0.7 m/s$^2$).
C. The allowable limit for cant deficiency change rate is 0.3 m/s$^3$.

In the early 1960s, the Japan National Railways Technical Research Institute conducted two comfort tests for trains passing through curves and switches. The tests used passenger ratings to establish the relationship between lateral acceleration and curve comfort. Through statistical analysis, the critical value of lateral acceleration that 5% of seated passengers could not tolerate during curve passage was found to be 0.8 m/s$^2$ (cant deficiency of 122 mm). Later, the "Nozomi" high-speed trains on the Tōkaidō Shinkansen, which were entirely seated, allowed an unbalanced centrifugal acceleration limit of 0.9 m/s$^2$ (cant deficiency of 138 mm).

For the French high-speed passenger rail system, TGV, the allowable cant deficiency limit for general sections is 76–90 mm; for difficult sections, it is 90–130 mm (on the TGV southeast line at 270 km/h, 130 mm was used). This corresponds to an unbalanced centrifugal acceleration of 0.5–0.6 m/s$^2$ for general sections and 0.6–0.85 m/s$^2$ for difficult sections.

In Germany, for mixed-use high-speed rail lines, the allowable cant deficiency is 130 mm, while for passenger-only lines with a curve radius greater than 1000 m, the allowable limit is 150 mm.

According to the interoperability technical specifications for the European High-Speed Rail System, for lines with a speed of 250–300 km/h, the general cant deficiency limit is 100 mm, and the maximum limit is 130 mm (ballasted track) or 150 mm (ballastless track). For lines with speeds over 350 km/h, the cant deficiency limit is 80 mm.

For China's high-speed rail lines designed for a speed of 350 km/h, the allowable cant deficiency limits are divided into three categories: good (40 mm), normal (60 mm), and poor (90 mm).

Based on the above standards, the static lateral acceleration limit is generally between 0.4–1.0 m/s$^2$. Considering the influence of vehicle suspension on static lateral acceleration, the actual unbalanced centrifugal acceleration experienced by passengers inside the vehicle is estimated to range between 0.5–1.25 m/s$^2$.

(2) Rate of Change of Unbalanced Centrifugal Acceleration (Rate of Change of Cant Deficiency)

The unbalanced centrifugal rate of change of acceleration (rate of change of cant deficiency) $\beta$ (m/s$^3$) and the rate of change of cant deficiency $b$ (mm/s) are essentially the same measurement of passenger ride comfort, but expressed in different forms. The conversion relationship between the two is given by the following formula:

$$\beta = \frac{b}{S} g \tag{4.1.22}$$

According to measured data from the UK, a rate of change of unbalanced centrifugal acceleration of 0.4 m/s$^3$ approaches the threshold of noticeable discomfort for passengers. The recommended value by British Railways is 0.25 m/s$^3$.

In Japan, the rate of change of unbalanced lateral acceleration is generally required to be less than 0.3 m/s$^3$, with a maximum limit of 0.4 m/s$^3$.

The American Association of Railroads (AAR) determined through experiments that the allowable limit for the rate of change of unbalanced centrifugal acceleration is 0.3 m/s$^3$.

For French high-speed rail lines operating at speeds of 250–300 km/h, the rate of change of cant deficiency is generally 30 mm/s (0.2 m/s$^3$), with a maximum value of 75 mm/s (0.5 m/s$^3$).

According to European Standard EN 13803-1 (2010), for high-speed rail lines designed for speeds of 200–300 km/h, the general limit for the rate of change of cant deficiency is 55 mm/s (0.36 m/s$^3$), with an exceptional limit of 75 mm/s (0.49 m/s$^3$).

The International Union of Railways (UIC) recommends that the rate of change of unbalanced centrifugal acceleration range from 0.17 to 0.60 m/s$^3$, depending on the type of transition curve.

For high-speed railways in China, the limit for the rate of change of cant deficiency is 23 mm/s (0.15 m/s$^3$) under favorable conditions and 38 mm/s (0.25 m/s$^3$) under difficult conditions.

Considering the standards adopted by various countries, the limit for the rate of change of unbalanced centrifugal acceleration (rate of change of cant deficiency) generally ranges from 0.2 to 0.5 m/s$^3$.

(3) Rate of Change of Cant (Roll Angle Velocity)

The rate of change of cant $f$ (mm/s) is the rate of change of cant with respect to time, while the cant gradient $i$ (%) is the rate of change of cant with respect to track length. Essentially, these two are the same. The conversion relationship between the two is as follows:

$$i = \frac{3.6f}{V} \tag{4.1.23}$$

The rate of change of cant reflects both the rate at which the outer rail elevates and the roll angular velocity of the vehicle. At a constant speed, it also indicates the cant gradient, making it an essential parameter for determining the length of transition curves.

For the Tokaido Shinkansen in Japan, the typical rate of change of cant is 38 mm/s, increasing to 53 mm/s under difficult conditions. For the Sanyo Shinkansen and subsequent lines, the typical rate is 28 mm/s, with a maximum of 45 mm/s under challenging conditions.

In the European Standard EN 13803-1 (2010) for high-speed railways designed for speeds of 200–300 km/h, the general limit for cant gradient is 2.25‰, with a maximum limit of 2.5‰. The general limit for the rate of change of cant is 50 mm/s, with an exceptional limit of 60 mm/s.

In summary, the rate of change of cant for high-speed railways abroad generally ranges from 20 to 60 mm/s. Ignoring the effects of the vehicle suspension system, this corresponds to a roll angular velocity of 0.8 to 2.3°/s. The maximum allowable roll angular velocity and angular acceleration for comfort are 5°/s and 15°/s$^2$, respectively.

Chinese high-speed railways adopt a linear cant gradient. To ensure passenger comfort, high requirements are set for the cant gradient and the rate of change of cant. The cant gradient is required to be no less than 0.25‰, while the rate of change of cant is categorized into three levels: 25, 28, and 31 mm/s.

(4) Quasi-Static Vertical Acceleration on Vertical Curves

According to the European standard EN 13803-1 (2010), the general limit value for quasi-static vertical acceleration is 0.22 m/s$^2$, with an exceptional limit of 0.44 m/s$^2$. The allowable value for convex vertical curves can be increased by 10%, while for concave vertical curves, it can be increased by 30%. For railway lines with a high proportion of standing passengers, it is recommended that the value does not exceed 0.1 m/s$^2$. The Japanese Shinkansen sets a quasi-static vertical acceleration limit of 0.33 m/s$^2$. In China's high-speed railways, the allowable quasi-static vertical acceleration is generally 0.4 m/s$^2$, with a maximum permissible value of 0.5 m/s$^2$ in difficult conditions.

Overall, the permissible vertical acceleration values for high-speed railways vary significantly. For convex curves, the range is generally between 0.2 and 0.5 m/s$^2$, while for concave curves, the range is approximately 0.2 to 0.6 m/s$^2$.

5. **Acceleration and Deceleration Comfort**

Acceleration and deceleration comfort focuses on the longitudinal acceleration and its rate of change during train acceleration and braking. These factors are mainly considered in the design of traction and braking systems, as they also impact passenger ride comfort.

(1) General Motors (GM) Test in the United States

Wilson, from General Motors, conducted braking comfort tests at a GM test track and obtained the following results:

0.27 g: A stop that feels comfortable;

0.34 g: A stop that feels uncomfortable;

0.43 g: A stop that feels extremely uncomfortable, equivalent to objects being thrown off the ground, leading to a sense of danger among passengers.

(2) Japanese National Railways (JNR) Test

In the early 1960s, the Railway Technical Research Institute of the former Japanese National Railways conducted two experiments to assess braking comfort. The tests required participants to give a binary judgment (acceptable or unacceptable). Using the threshold at which 5% of passengers found the deceleration unacceptable, the results were:

0.10 g for standing passengers.

0.08 g for seated passengers.

A deceleration rate of change of 0.07 g/s.

## 4.2 Structural Dynamic Performance Evaluation Standards

This section is based on standards such as the Code for Design of Railway Bridges and Culverts (*TB 10002-2017*), the Technical Code for Dynamic Acceptance of High-Speed Railway Engineering (*TB 10761-2013*), the Code for Design of High-Speed Railways (*TB 10621-2014*), and the Code for Design of Intercity Railways (*TB 10623-2014*), as well as research findings from China and abroad. It introduces the dynamic evaluation indicators and standards for track, bridges, and subgrade structures.

### *4.2.1 Track Dynamic Performance Evaluation Indicators*

The track structure consists of key components such as rails, sleepers, fastening components, ballast beds, anti-creep devices, and turnouts. It directly bears the vehicle load and transmits it to the underlying subgrade or bridge/tunnel structures while also guiding the movement of high-speed trains. Track structures are classified into ballasted and ballastless tracks, with ballastless tracks being widely used in China's high-speed rail system. According to the Technical Code for Dynamic Acceptance of High-Speed Railway Engineering (*TB 10761-2013*), dynamic testing of track structures includes indicators related to operational safety, such as: Rail head lateral displacement, Rail vertical displacement, Sleeper vertical displacement, Sleeper lateral displacement, Track slab lateral displacement, Vertical relative displacement between track slab and base, Rail vibration acceleration, Sleeper vibration acceleration, Track slab vibration acceleration.

The evaluation standards for operational safety indicators have already been described in Sect. 4.1.1 and will not be repeated here. The track structure stability indicators are evaluated by measuring the lateral displacement of the rail head and calculating the dynamic track gauge variation, which assesses the fastening system's ability to maintain the gauge and the overall track stability. The track structure stiffness indicators are evaluated by measuring the vertical displacement of the rail to determine whether the elasticity of the fastening system meets design requirements and whether the track stiffness is reasonable and uniform. The vibration testing of track components and foundation involves measuring the vertical vibration accelerations of rails, sleepers, and track slabs to analyze the vibration transmission characteristics of these components under dynamic train loads. The specific evaluation standards for various track dynamic performance indicators are shown in Table 4.13.

A turnout is a track connection device that enables locomotives and rolling stock to switch from one track to another. Its function is to withstand and transmit various loads induced by train operations and to guide the wheels along the predetermined direction on the track. Turnouts are extensively installed in stations. As a crucial track component, turnouts exhibit greater geometric irregularities and non-uniform elastic irregularities of the substructure compared to the mainline track section, making their dynamic interactions with rolling stock more complex. Under high-speed conditions, these dynamic interactions become even more intense, leading to shorter component service life and higher maintenance workload. Therefore, turnouts represent both a weak link in the track system and a critical component for ensuring safety and controlling train speeds when passing through stations.

**Table 4.13** Criteria for evaluating the dynamic performance of track structures

| Test items | Ballastless track | | Ballasted track | |
|---|---|---|---|---|
| | Maximum allowable value | Baseline value | Maximum allowable value | Baseline value |
| Wheel-rail vertical force (kN) | 170 | 120 | 170 | 120 |
| Rail lateral displacement (mm) | 2.0 | 1.5 | 2.0 | 1.5 |
| Sleeper lateral displacement (mm) | – | – | 2.0 | 1.0 |
| Track slab lateral displacement (mm) | 1.0 | 0.5 | – | — |
| Rail vertical displacement (mm) | 2.0 | 1.5 | 2.5 | 2.0 |
| Track slab vertical displacement (Mid-Panel/Panel edge) (mm) | 0.3/0.5 | 0.2/0.4 | – | — |
| Sleeper vertical displacement (mm) | – | – | 2.0 | 1.5 |
| Rail vibration acceleration ($m/s^2$) | 5000 | | 3000 | |
| Sleeper vibration acceleration | – | | 500 | |
| Track slab vibration acceleration | 300 | | – | |

According to the *High-Speed Railway Engineering Dynamic Acceptance Technical Specifications* (*TB10761-2013*), dynamic detection of track structures includes indicators for operational safety, lateral stability of turnouts, vertical wheel-rail force transition and magnitude, and the performance of switching devices.

The lateral stability of turnouts is evaluated by measuring parameters such as the lateral displacement of the rail head, the opening width of the switch rail, and the lateral displacement of the wing rail. The vertical wheel-rail force transition and magnitude refer to the vertical force variations acting on measurement points continuously arranged at switch rail head widths of 10, 20, 30, 40, and 50 mm, thereby determining the transition range and magnitude of the vertical wheel-rail force. The performance of switching devices is assessed by measuring the dynamic force of switch rail traction rods, the dynamic displacement between the switch rail (or crossing rail) and the stock rail (or wing rail) at the tight contact detector position, and witnessing the dynamic forces experienced by traction rods under multi-machine, multi-point control. Additionally, parameters such as switch rail resistance, crossing rail resistance, and foreign object intrusion tests are evaluated to verify whether the actual switching resistance meets design requirements. The dynamic performance standards for turnouts are shown in Table 4.14.

**Table 4.14** Dynamic performance standards for turnouts

| Item | Standard |
|---|---|
| Lateral elastic displacement of rail components (straight/curved passing through turnout) (mm) | $\leq 1.5/3.0$ |
| Vertical displacement of rail (mm) | $\leq 3.0$ |
| Switch rail and closure rail opening width (mm) | $\leq 3.0$ |
| Turnout switching resistance | Less than 1.3–1.5 times the turnout machine output force at each traction point, not exceeding 6 kN |
| Tight contact inspection | A 4 mm gap at traction points within the tight contact section should not lock; a 5 mm gap at the inspection point should not indicate closure |
| Stress on key components in the turnout zone | Below the allowable stress of the material |
| Vertical displacement of switch rail relative to stock rail and closure rail relative to wing rail (mm) | $\leq 1.5$ |
| Wheel-rail force transition in turnout zone | In compliance with design requirements |
| Horizontal displacement of switch rail relative to stock rail at traction points and inspection points (mm) | $\leq 1.5$ |

### 4.2.2 Bridge Dynamic Performance Evaluation Indicators

The main components of bridge engineering include the superstructure, piers, abutments, and foundation. The superstructure, also called the girder, primarily supports the track and connects the two abutments. The abutments bear the backfill pressure and various forces transmitted from the girder, restrict girder displacement, connect the girder with the embankment, and support the track. Piers primarily bear the forces transmitted from the girder and restrict displacement. The foundation distributes the forces transmitted by the piers and abutments to the ground, ensuring the stability of the bridge structure. Excessive bridge vibration within the train-track dynamic interaction system can lead to difficulty in maintaining track conditions, reduced structural fatigue strength, changes in ballast interlocking, and compromised operational safety, making the evaluation of bridge dynamic performance critically important.

Standards such as the *Code for Design on Railway Bridge and Culvert* (*TB10002-2017*), the *Code for Design of High Speed Railway* (*TB10621-2014*), and the *Code for Design of Intercity Railway* (*TB10623-2014*), as well as China's and international research, provide relevant requirements for the evaluation of bridge dynamic performance.

1. **Bridge Deformation**

The vertical deflection of the bridge under static and live train loads (excluding vertical displacement at supports) must comply with the requirements specified in Table 4.15 and design standards.

**Table 4.15** Limits for bridge vertical deflection

| Railway design standards | Design speed (km/h) | $L \leq 40$ m | $40\text{ m} < L \leq 80$ m | $L > 80$ m |
|---|---|---|---|---|
| High-speed railway | 250 | $\leq L/1400$ | $\leq L/1400$ | $\leq L/1000$ |
| | 300 | $\leq L/1500$ | $\leq L/1600$ | $\leq L/1100$ |
| | 350 | $\leq L/1600$ | $\leq L/1900$ | $\leq L/1500$ |
| Intercity railway | 200 | $\leq L/1750$ | $\leq L/1600$ | $\leq L/1200$ |
| | 160 | $\leq L/1600$ | $\leq L/1350$ | $\leq L/1100$ |
| | 120 | $\leq L/1350$ | $\leq L/1100$ | $\leq L/1100$ |
| Mixed passenger and freight railway | 200 | $\leq L/1200$ | $\leq L/1000$ | $\leq L/900$ |
| | 160 | $\leq L/1000$ | $\leq L/900$ | $\leq L/800$ |
| Heavy haul railway | 120 km/h and below | $\leq L/900$ | $\leq L/800$ | $\leq L/700$ |

*Note* 1. The limits in the table apply to double-track simply supported beams with three or more spans. For continuous beams with three or more spans, the vertical deflection limit is 1.1 times the values in the table. For continuous beams with two spans or less, and double-track simply supported beams with two spans or less, the vertical deflection limit is 1.4 times the values in the table

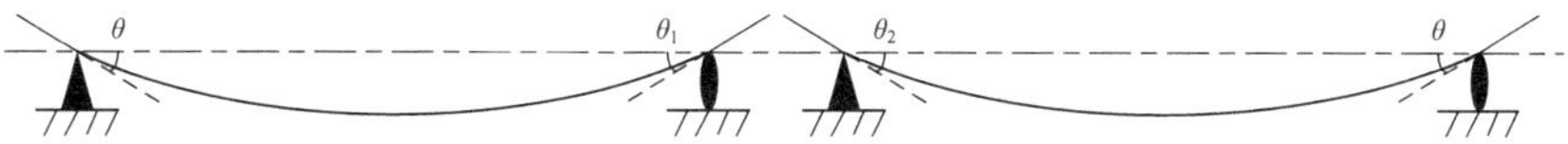


**Fig. 4.11** Illustration of beam end rotation angles

(1) For single-track simply supported or continuous beams, the vertical deflection limit is 0.6 times the corresponding values for double-track bridges.
(2) In the table, L represents the span of the simply supported or continuous beam under consideration.

Under the influence of train lateral sway forces, centrifugal forces, wind, and temperature, the horizontal deflection of the bridge should not exceed 1/4000 of its calculated span. For ballastless track bridges on high-speed and intercity railways, the relative lateral displacement of rail supports on adjacent ends should not exceed 1 mm.

Under vertical static and live train loads, the vertical rotation angle at the bridge ends (as shown in Fig. 4.11) must meet the limits specified in the table. For ballastless track bridges, if the bridge end rotation angle does not meet the limits in Table 4.16, a recalculation of the forces on the track structure and fastening system at the bridge ends is required.

2. **Bridge Natural Frequency**

For design speeds of 200 km/h and above, the vertical natural frequency limits for simply-supported beams are given by:

$$f_v \geq \begin{cases} 80/L & (4\text{m} \leq L \leq 20\text{m}) \\ 23.58\text{L}^{-0.592} & (20\text{m} \leq L \leq 96\text{m}) \end{cases} (\text{Hz}) \tag{4.2.1}$$

where $f_v$ is the vertical natural frequency limit for simply-supported beams (Hz); $L$ is the beam span (m).

For high-speed railways operating 24–26 m trainsets and concrete double-track simply-supported box beams with spans no greater than 32 m, when the natural frequency of the beam meets or exceeds the vertical natural frequency limit, further vehicle-bridge coupled dynamic response analysis may not be required. For 16 m spans, the frequency should not be less than 6.25 Hz (Table 4.17).

3. **Bridge Vibration Acceleration**

The purpose of limiting vertical vibration acceleration of bridges is to prevent excessive wheel-rail forces, ensuring track bed stability and safe train operations. Based on Chinese railway standards and international references, the maximum allowable vertical vibration acceleration at the bridge center is set at 0.35 g.

According to the European standard EUROCODE, when calculating bridge vibration acceleration, at least the third vibration mode of the structure must be considered. Frequencies are calculated up to $f_{\text{lim}} = max[30, \lambda f_0]$(Hz), where $f_0$ is the first-order

**Table 4.16** Beam end rotation angle limits for railways

| Railway design standards | Track type | Location | Limit (rad) | Remarks |
|---|---|---|---|---|
| Railways with a mixed freight and passenger speed of 200 km/h | Ballasted track | Between abutment and bridge | $\theta \leq 3.0^0/_{00}$ | – |
| | | Between adjacent beam spans | $\theta_1 + \theta_2 \leq 3.0^0/_{00}$ | – |
| High-speed railway | Ballasted track | Between abutment and bridge | $\theta \leq 2.0^0/_{00}$ | – |
| | | Between adjacent beam spans | $\theta_1 + \theta_2 \leq 4.0^0/_{00}$ | – |
| | Ballastless track | Between abutment and bridge | $\theta \leq 1.5^0/_{00}$ | Beam end overhang length $\leq 0.55$m |
| | | | $\theta \leq 1.0^0/_{00}$ | 0.55m < Beam end overhang length $\leq 0.75$ m |
| | | Between adjacent beam spans | $\theta_1 + \theta_2 \leq 3.0^0/_{00}$ | Beam end overhang length $\leq 0.55$ m |
| | | | $\theta_1 + \theta_2 \leq 2.0^0/_{00}$ | 0.55 m < Beam end overhang length $\leq 0.75$ m |
| Intercity railway | Ballasted track | Between abutment and bridge | $\theta \leq 3.0^0/_{00}$ | – |
| | | Between adjacent beam spans | $\theta_1 + \theta_2 \leq 6.0^0/_{00}$ | – |
| | Ballastless track | Between abutment and bridge | $\theta \leq 2.1^0/_{00}$ | Beam end overhang length $\leq 0.30$ m |
| | | | $\theta \leq 1.5^0/_{00}$ | 0.30 m < Beam end overhang length $\leq 0.55$ m |
| | | | $\theta \leq 1.0^0/_{00}$ | 0.55m < Beam end overhang length $\leq 0.75$ m |
| | | Between adjacent beam spans | $\theta_1 + \theta_2 \leq 4.2^0/_{00}$ | Beam end overhang length $\leq 0.30$ m |

(continued)

**Table 4.16** (continued)

| Railway design standards | Track type | Location | Limit (rad) | Remarks |
|---|---|---|---|---|
| | | | $\theta_1 + \theta_2 \leq 3.0‰$ | 0.30m < Beam end overhang length $\leq 0.55$ m |
| | | | $\theta_1 + \theta_2 \leq 2.0‰$ | 0.55 m < Beam end overhang length $\leq 0.75$ m |

*Note* The sum of the rotation angles of two adjacent beam spans ($\theta_1 + \theta_2$) must meet the specified limits of this section. Additionally, the rotation angle of each beam span must comply with the limit requirements set for the rotation angles between the abutment and the bridge

**Table 4.17** Vertical natural frequency limits (Hz) for common double-track simply-supported box beams exempt from dynamic calculations

| | Design speed | | |
|---|---|---|---|
| Span (m) | 250 km/h | 300 km/h | 350 km/h |
| 20 | $100/L$ | $100/L$ | $120/L$ |
| 24 | $100/L$ | $120/L$ | $140/L$ |
| 32 | $120/L$ | $130/L$ | $150/L$ |

natural frequency of the structure, $\lambda = 1 \sim 2$. Vertical vibration acceleration limits are specified for:

$$[a]_{\max} \geq \begin{cases} 350\,\text{cm/s}^2 \text{ (Ballasted Track Bridges)} \\ 500\,\text{cm/s}^2 \text{ (Open Deck Bridges)} \end{cases} \tag{4.2.2}$$

According to China's *Code for rating existing railway bridges*, the lateral vibration acceleration of a bridge should not exceed 1.4 m/s$^2$ during train passage. Experiments by the Railway Technical Research Institute of Japan indicate that lateral vibration accelerations of 0.1–0.2 g can lead to train derailment.

### *4.2.3 Evaluation Metrics for Subgrade Dynamic Performance*

Railway subgrade serves as the foundation for tracks, formed through excavation or filling. Its primary roles include meeting the requirements for track installation, bearing loads from tracks and trains, and providing necessary conditions for train operations. The two common forms are embankments and cuttings. Subgrade structures use high-quality layered and compacted materials to ensure sufficient strength and rigidity, maintaining smoothness under adverse climatic conditions such as freezing and rainfall, thus ensuring long-term stability and safety of the

**Table 4.18** Subgrade dynamic performance standards

| Track type | Ballasted track | Ballastless track |
|---|---|---|
| Ballast thickness (m) | $\geq 0.35$ | – |
| Subgrade surface layer thickness (m) | $\geq 0.7$ | $\geq 0.4$ |
| Dynamic load (kPa) | $\leq 0.4P_j$ | $\leq 32.4$ |
| Dynamic deformation (mm) | $\leq 0.1$ | $\leq 0.22$ |
| Vibration acceleration (m/s$^2$) | $\leq 10.0$ | |

*Note* $P_j$ is the static axle load (kN)

railway foundation. Railway lines are composed of bridges, tunnels, and subgrades, with differing connection structures and settlement patterns. To avoid phenomena such as vehicle "bouncing" on highways, transition sections are installed to keep settlement differences within allowable limits. According to the *Technical regulations for dynamic acceptance for high-speed railways construction* (*TB10761-2013*), subgrade dynamic testing items include ballast thickness (ballasted track), subgrade surface layer thickness, moisture content conditions, dynamic load, dynamic deformation, and vibration acceleration. Standards for subgrade dynamic performance are listed in Table 4.18.

## 4.3 Evaluation Method of the Influence of Track Geometry on System Dynamic Performance

With the significant increase in high-speed railway operating speeds, the dynamic interaction between trains and tracks has intensified considerably. Operational practices indicate that vibrations of high-speed trains greatly affect passenger comfort, with low-frequency vibrations induced by track geometry changes under high-smoothness conditions being one of the primary factors influencing riding comfort. Determining reasonable parameters for high-speed railway track geometry and evaluating and optimizing geometry schemes have become pressing issues.

For a long time, the determination of railway track design parameters in China has been based on quasi-static methods. These methods simplify the train into a mass point or rigid body moving along a rigid track, perform force analysis and calculations, and evaluate comfort indices such as unbalanced acceleration and its rate of change on curves to establish standards for track vertical and horizontal alignment parameters. However, these methods neglect the dynamic interactions between vehicles and track structures, fail to accurately analyze the effects of track parameters on operational safety and passenger comfort at high speeds, and cannot assess the interaction between longitudinal track alignments or evaluate the overall track geometry scheme. Thus, these methods have significant limitations.

Today, the dynamics of the train-track/bridge system has been a major research focus in the field of railway engineering, achieving significant progress. This research has been widely applied in the design and inspection of high-speed railway tracks, bridges, subgrades, and other line structures, yielding substantial economic and social benefits. In recent years, Chinese researchers have gradually begun to apply vehicle–track system dynamics theories and methods to the study of railway alignment design, and some progress has been made in this area.

Existing research on track geometry based on vehicle-track system dynamics mainly focuses on the establishment and optimization of dynamic models, with limited attention to methods and indices for evaluating ride comfort specifically tailored to geometry dynamics analysis. Generally, the same evaluation indices and methods used for structural dynamics analysis of tracks and bridges are applied. However, the dynamic responses caused by track geometry differ significantly from those caused by substructure rigidity. First, the vibrations induced by track geometry are primarily in the low-frequency range, resembling long-wavelength irregularities under high-speed conditions. Second, vibration analysis based solely on amplitude metrics makes it difficult to isolate the effects of geometry. Thus, developing ride comfort evaluation indices and methods suited for geometry dynamics analysis has significant theoretical and practical value.

### *4.3.1 Key Factors Influencing Ride Comfort Due to Track Conditions*

The evaluation of operational safety typically involves indices such as derailment coefficient, wheel load reduction ratio, and wheel-rail lateral force, similar to the study of the effects of track structures on system dynamic performance. Therefore, when analyzing track geometry, the key focus is on its impact on ride comfort. To assess the influence of track geometry on dynamic performance, it is also necessary to distinguish between system responses caused by track irregularities and those caused by substructure stiffness.

Figure 4.12 illustrates the intrinsic relationship between track conditions and ride comfort. Track geometry consists of alignment and smoothness. Generally, alignment primarily affects quasi-static acceleration and the rate of change of acceleration. Moreover, at points of abrupt alignment changes, such as transition-start points and curve-transition points, impacts are generated on the vehicle, causing vibrations at the vehicle's natural frequencies.

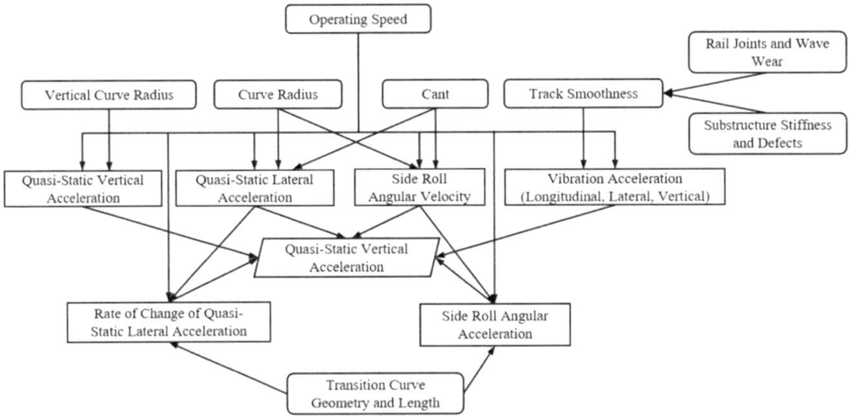


**Fig. 4.12** Key factors influencing ride comfort due to track conditions

## 4.3.2 *Key Issues in Using Ride Comfort Indices for Evaluating Track Geometry*

The influence of track geometry on ride comfort must focus on the following aspects.

1. **Combined Effects of Vertical and Lateral Vibrations**

The Sperling ride quality index and the ISO 2631-1 ride comfort evaluation method separately assess lateral and vertical comfort, without considering the combined physiological and psychological impacts of vertical and lateral vibrations on passengers. Consequently, comfort levels determined from lateral and vertical vibrations under the same track conditions may differ. The ride comfort evaluation method in UIC 513R accounts for the combined effects of vibration accelerations in all directions. Thus, multiple comfort evaluation methods should be combined for a comprehensive assessment. Particularly when the root mean square (RMS) components of acceleration are significant in a specific narrow frequency band, multiple evaluation methods should be employed simultaneously.

2. **Combined Effects of Vibrational and Quasi-Static Accelerations**

In evaluating ride comfort on curves, current Chinese standards do not integrate vehicle vibrational acceleration with lateral quasi-static acceleration, ignoring their interaction. However, British railways have recognized this interaction and have adopted a comfort evaluation standard that combines vehicle vibrational acceleration with lateral quasi-static acceleration.

Ride comfort tests conducted in Japan demonstrate that combining both factors yields a better evaluation of ride comfort than using either alone. When lateral quasi-static acceleration is high, the permissible value for vibrational acceleration decreases, and vice versa. This relationship is shown in Fig. 4.13, where the bold line

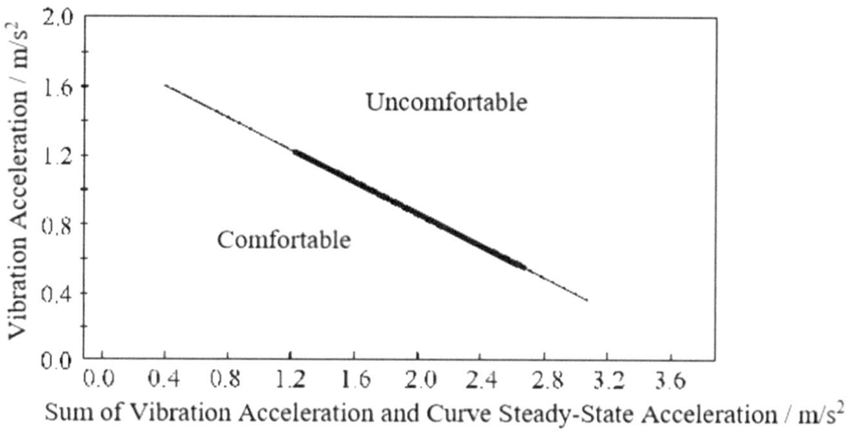


**Fig. 4.13** Critical lines of ride comfort obtained from test results

represents the actual data range obtained during test runs, and the fine line indicates extrapolated values.

Therefore, ride comfort on curves should consider not only lateral quasi-static acceleration but also vibrational components. Lateral quasi-static acceleration is primarily determined by factors such as curve radius, cant of the outer rail, operating speed, and vehicle suspension system characteristics, while vibrational components at high speeds are mainly influenced by track irregularities and variations in structural stiffness.

3. **Effects of Track Smoothness and Geometry**

When using vehicle acceleration to evaluate track conditions, whether measured by track inspection vehicles or directly on trains, the obtained vehicle acceleration includes unbalanced lateral and vertical accelerations. Unbalanced accelerations are primarily caused by track geometry, and their magnitudes can even exceed those caused by track irregularities. Therefore, applying the same acceleration magnitude standard to evaluate track conditions on both straight and curved sections has significant limitations.

The peak-to-peak evaluation method adopted in Japan essentially ignores the effects of low-frequency unbalanced accelerations. As train speeds continue to increase, under-cant deficiency will also grow, causing unbalanced accelerations to exceed permissible acceleration limits. Consequently, the current approach of solely using acceleration to evaluate track quality lacks rationality. When assessing track smoothness, low-frequency components caused by geometry should be filtered out, and when evaluating geometry, the influence of track irregularities should be minimized.

### 4.3.3 Selection of Ride Comfort Evaluation Indices for Track Geometry

Based on the characteristics of system responses induced by track geometry, multiple indices should be utilized for comprehensive analysis. Additionally, appropriate indices should be selected depending on the research objectives and subjects.

1. **Study of Track Design Parameters**

When analyzing reasonable values for the design parameters of the track vertical and horizontal profiles, it is necessary to consider their impacts on operational safety and ride comfort. Comfort evaluation may utilize quasi-static indices for curves, PCT indices for transition curves, PDE indices for curves, as well as the rules of vehicle vibration attenuation and superposition.

2. **Dynamic Performance Evaluation of Track Schemes**

Dynamic performance evaluation of track schemes should include assessments of overall safety and comfort. Comfort evaluation methods include the Sperling ride quality index, UIC 513R, and ISO 2631-1. However, the weighted curves of these ride comfort indices show that their primary impact frequencies range from 0.4 to 80 Hz. Since the quasi-static acceleration induced by geometry approaches 0 Hz, it is necessary to further introduce the concept of motion sickness value (MSV) to focus on vibrations below 0.5 Hz. When analyzing the peak values of vibrational acceleration, effective values, frequency distributions, and cumulative characteristics of acceleration can be used for comprehensive evaluation.

3. **Analysis of Compatibility Between Track Parameters and Substructure**

When analyzing the compatibility of track parameters with substructure, such as in road-bridge transitions or sections with long bridges combined with track geometry, it is necessary to consider not only the structural dynamic performance indices but also the evaluation indices for the aforementioned track parameters and schemes.

Appropriate comfort evaluation schemes should be adopted for different research subjects, as shown in Fig. 4.14.

### 4.3.4 Ride Comfort Evaluation Method for Track Geometry

To address the issues in existing research on railway track geometry, a geometry dynamics design and evaluation method based on vehicle-track dynamic interaction is proposed. The specific steps are as follows:

1. Develop a vehicle-track system dynamic analysis model based on the characteristics of high-speed trains and tracks.

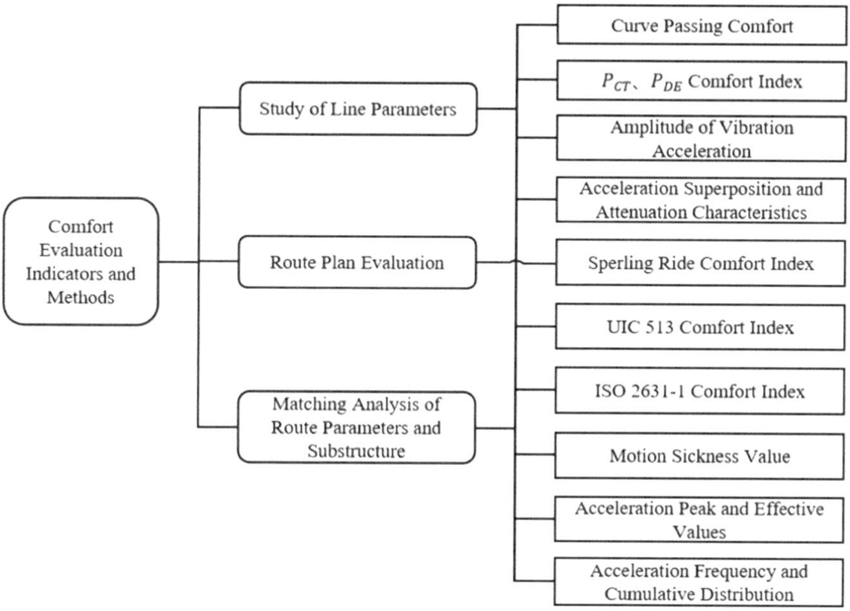


**Fig. 4.14** Comfort evaluation indices for different research subjects

The model consists of two subsystems: the vehicle and the track structure (with a suitable track structure model selected based on the research problem). These are interconnected through the dynamic relationship between the wheel and rail. The model mainly considers: (1) vibration characteristics of the car body, bogie, wheelsets, and suspension system; (2) the influence of track, bridges, and other substructures; (3) dynamic interactions between the vehicle and track; (4) track geometry and track smoothness conditions. The system's equations of motion are solved using numerical integration methods.

2. Based on the results of system dynamic analysis, combine corresponding dynamic evaluation indices and methods to determine track geometry parameter standards and evaluate and optimize track geometry schemes. The specific steps are as follows:

    (1) Obtain system dynamic responses through vehicle-track system dynamic analysis.
    (2) Select appropriate evaluation methods and indices according to different analysis subjects of track geometry schemes and parameters.

(1) Comfort Evaluation of Track Geometry Schemes

The ride comfort evaluation of track geometry schemes should proceed sequentially from an overall comprehensive perspective, through directional evaluations, to specific challenging segments, identifying and optimizing the factors most sensitive to comfort. Vibrational acceleration data should be evaluated using band-weighted RMS calculations according to international standard ISO 2631-1, considering motion sickness and comfort indices. Specific evaluation methods and indices include:

① Overall Evaluation: Frequency distribution chart of RMS acceleration values, cumulative frequency distribution chart, and average comfort index $N_{MV}$.
② Directional Evaluation: Continuous comfort index $C_C$.
③ Specific Challenging Sections: Comfort index $P_{CT}$ for transition curves and instantaneous comfort index $P_{DE}$.

(2) Comfort Evaluation for Track Geometry Parameters.

According to international standard ISO 2631-1 and European standard EN 12299, the comfort evaluation introduces indices such as average comfort index $N_{MV}$, continuous comfort index $C_C$, transition curve comfort index $P_{CT}$, and instantaneous comfort index $P_{DE}$ for comprehensive evaluation. The analysis incorporates vehicle vibration acceleration, lateral impact, side roll angular velocity attenuation, and superposition characteristics, along with RMS acceleration frequency distribution charts and cumulative frequency distribution charts, to assess track geometry parameters.

The specific evaluation methods and indices are as follows:

① Evaluation of dynamic simulation analysis data for single track geometry parameter conditions.

Comfort evaluation methods: Instantaneous vibration acceleration, its attenuation and superposition characteristics, comfort index $P_{CT}$ for transition curves, and instantaneous comfort index $P_{DE}$.

② Extend the representative parameters to form a track scheme for comprehensive evaluation, analyze the sensitivity of overall comfort to these parameters, and determine appropriate parameter values.

Comfort evaluation methods: Frequency distribution chart of RMS acceleration values, cumulative frequency distribution chart, average comfort index $N_{MV}$, and continuous comfort index $C_C$.

③ Based on system dynamic analysis results, determine reasonable track parameters and optimize track geometry schemes to improve track design from the perspective of dynamic performance.

Compared to conventional quasi-static methods for track geometry analysis, this method has the following advantages:

(1) This method addresses the limitations of traditional track geometry evaluation and analysis methods, enabling predictions of dynamic performance under various geometry combinations. It provides critical analytical tools for the dynamic design of high-speed railways, offering significant scientific value and strong guidance for practical engineering design.
(2) The comfort evaluation methods employed in this approach effectively assess low-frequency components caused by geometry, distinguishing comfort issues induced by geometry from those related to track smoothness and substructure stiffness variations. In the comparison of track design parameters and dynamic performance of track schemes, it enhances discriminative power, providing scientific evidence for parameter determination and scheme comparison.

# Chapter 5
# On-Site Testing and Analysis of Track System Dynamics

System dynamics testing and analysis form the basis for the verification of dynamic models and simulation analysis. This chapter first briefly introduces the methods for system dynamics testing. It then discusses the testing and analysis of ride comfort and train-track system dynamics conducted on the Beijing-Tianjin Intercity Railway and the Jinan-Qingdao High-Speed Railway.

## 5.1 System Dynamics Testing Methods

### *5.1.1 Setup of the Dynamics Testing System*

The railway track system dynamics testing system comprises several components, including testing sensors, data acquisition systems, and data analysis systems, as shown in Fig. 5.1.

1. **Vibration Sensors**

Vibration sensors are at the forefront of the dynamics testing system, converting vibrations to be measured into identifiable testing signals. Based on the measurement object (e.g., displacement, velocity, acceleration), vibration sensors are categorized into displacement sensors, velocity sensors, and accelerometer. According to the nature of the sensitive elements, vibration sensors can also be classified as inductive, eddy current, piezoelectric, piezoresistive, etc. Piezoelectric accelerometer are commonly used in vibration measurements.

2. **Data Acquisition System**

In vibration measurement, high-pass or low-pass filtering is often required to eliminate zero drift or high-frequency noise interference and suppress low-frequency or

© China Railway Publishing House Co., Ltd. 2026

X. Long et al., *Vehicle–Track–Bridge System Subjected to Track Alignment*,
https://doi.org/10.1007/978-981-95-1649-0_5

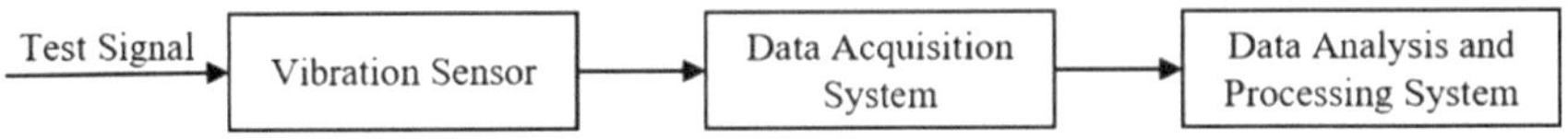


**Fig. 5.1** Dynamics testing system

high-frequency signals. The resulting analog signals are then converted into digital signals for easy analysis and processing, and stored in appropriate media. With advances in electronic technology, these signal processing steps can be achieved using suitable data acquisition systems paired with acquisition software.

3. **Data Analysis and Processing System**

The essence of the data analysis and processing system lies in the use of computer software (e.g., Matlab, Origin, SPSS) to perform statistical analysis on the collected data, including time-domain analysis, frequency-domain analysis, and amplitude analysis testing results are displayed and stored in real time in forms such as graphs and videos.

### *5.1.2 Track Structure Dynamics Testing*

Testing data on track condition detection and track dynamic performance can guide adjustments and optimization of track service conditions while verifying the safety and stability of train operations. Field tests can measure wheel-rail vertical forces and lateral forces when the testing train passes track structure dynamic measurement points at different speeds. From these, indicators such as derailment coefficients, wheel load reduction rates, and wheelset lateral forces can be calculated. The wheel-rail vertical force is calibrated on-site using a vertical force calibration frame. It can also be quasi-statically calibrated and verified when comprehensive inspection trains or testing freight trains pass the measurement points at 5 km/h. The wheel-rail lateral force can be calibrated on-site using a lateral force calibration frame. Specific details can be referred to in the *Track side testing methods of vertical and lateral wheel-rail forces* (*TB/T 2489-2016*). Track vibration acceleration and displacement can be measured by placing accelerometers and displacement transducers at appropriate locations to testing the dynamic response of rails, sleepers, track slabs, and base slabs.

By testing the natural vibration characteristics of the bridge and the dynamic response of the testing train passing the bridge at various speeds, the working condition of the bridge structure during service can be assessed. The tests verify whether the bridge possesses reasonable vertical and lateral stiffness and evaluate the stability and dynamic performance of the bridge under traffic. The vertical and lateral vibration of the bridge structure when the testing train passes is used to analyze the characteristics of vertical and lateral excitation and whether the bridge structure exhibits vertical or lateral resonance. Additionally, methods such as ambient micro-vibration or residual vibration can be used to measure the natural frequencies and damping ratios of the bridge's vertical and lateral responses. The vertical vibration acceleration of the bridge deck during train passage is used to assess whether the ballast on the bridge remains in good condition.

### 5.1.3 Railway Vehicle Dynamics Testing

To meet the objectives of on-site testing, the *Specification for dynamic performance assessment and testing verification of rolling stock* (*GB/T 5599-2019*) can be referenced for testing the acceleration of components such as the vehicle body, bogie frame, and wheelsets during operation. For instance, a lateral accelerometer can be installed on the frame above the axle box to measure lateral vibrations, which are used to assess the lateral stability of the bogie. Additionally, lateral and vertical accelerometers can be installed on the floor of the carriage 1 m laterally offset from the bogie center to measure vehicle vibrations and calculate ride stability indices.

In general, the coordinate system used in vehicle dynamics testing is a right-hand coordinate system, with the train's forward direction as the positive x-axis and upward as the positive z-axis, as shown in Fig. 5.2.

During on-site testing, the maximum testing speed is the greater of 1.1 times the maximum design speed or the maximum design speed plus 10 km/h. When testing the vibration characteristics of the carbody, the vertical and lateral acceleration measurement points can be arranged diagonally on the train's interior floor,

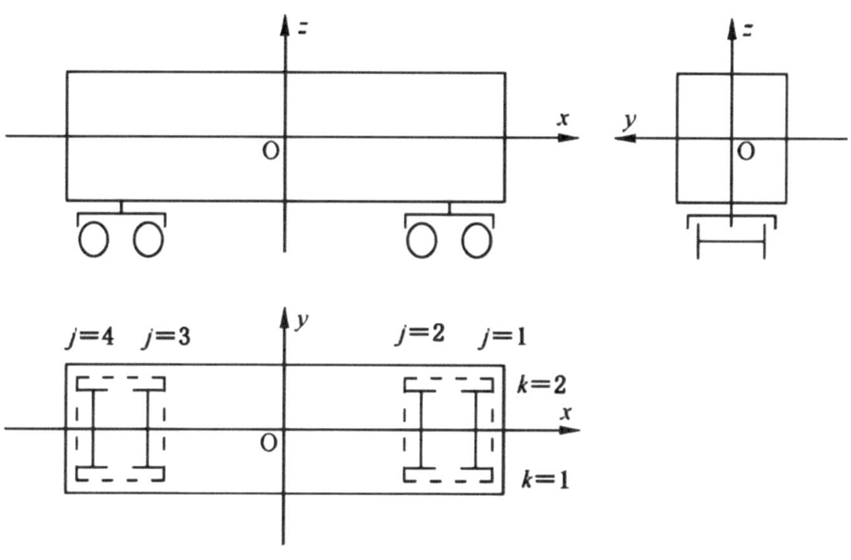


**Fig. 5.2** Vehicle dynamics testing coordinate system

positioned 1000 mm from the center of the first and second bogies toward the car body, as shown in Fig. 5.3. For the driver's cab of the electric multiple unit (EMU), the vertical and lateral acceleration measurement points are typically placed on the floor beneath the driver's seat.

The accelerometers used for vehicle vibration acceleration testing must meet the technical requirements outlined in Table 5.1.

In general, the running quality of vehicles is evaluated using vertical and lateral acceleration data of the vehicle body, while the running stability of vehicles is assessed using the stability index W or the ride comfort index NMV. Detailed methods for processing testing data can be found in the *Specification for dynamic performance assessment and testing verification of rolling stock* (*GB/T 5599-2019*).

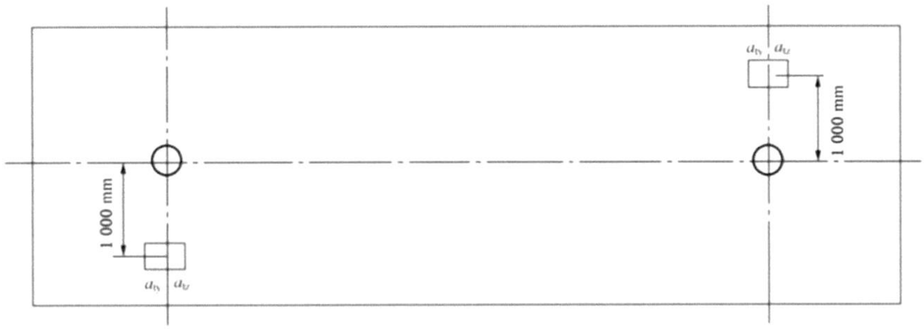


**Fig. 5.3** Layout of vibration acceleration measurement points on EMU car body

**Table 5.1** Technical requirements for vibration accelerometers

| Sensor | Range ($m/s^2$) | Frequency range (Hz) | Amplitude nonlinearity (%) | Lateral sensitivity (%) |
|---|---|---|---|---|
| Vehicle body vibration acceleration[a] | 20–50 | 0–100 | $\leq 1$ | $\leq 3$ |
| Vehicle body vibration acceleration[b] | 10–30 | 0–100 | $\leq 0.2$ | $\leq 1$ |
| Bogie/frame vibration acceleration | 100–500 | 0–100 | $\leq 1$ | $\leq 3$ |

[a] is used to evaluate the stability index $W$ and acceleration $a$
[b] is used to evaluate the comfort index $N_{MV}$

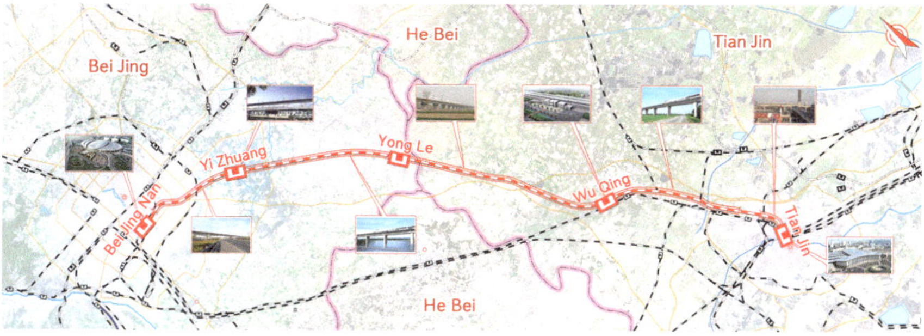


**Fig. 5.4** Schematic layout of the Beijing-Tianjin intercity railway

## 5.2 Ride Comfort Testing and Analysis on the Beijing-Tianjin Intercity Railway

### *5.2.1 Testing Conditions*

The Beijing-Tianjin Intercity Railway runs from Beijing South Station to Tianjin, with a mainline length of 118.3 km and a distance of 116.9 km from the center of Beijing South Station to the center of Tianjin Station. The curved track length is approximately 57.2 km, accounting for about 49% of the total, while the straight track length is about 59.7 km, making up 51%. The track layout is shown in Fig. 5.4. The Beijing-Tianjin Intercity Railway officially broke ground on July 4, 2005, and began operations on August 1, 2008. It is China's first high-speed railway designed for a speed of 350 km/h.

The horizontal and vertical profile design of the Beijing-Tianjin Intercity Railway incorporates large curve radii, extended gradient lengths, and large vertical curve radii. The mainline primarily uses CRTS II slab ballastless tracks. The main design standards for the line include a minimum curve radius of generally 7000 m (5500 m under challenging conditions), a minimum tangent or circular curve length of no less than 280 m (210 m under challenging conditions), a maximum gradient of 20‰, and a minimum gradient length of no less than 900 m (600 m under challenging conditions), with no continuous use of such gradients.

### *5.2.2 Ride Comfort Analysis and Evaluation*

To assess the ride comfort level of the Beijing-Tianjin Intercity Railway and analyze the relationship between track alignment and ride comfort, real vehicle tests were conducted on operating trains. In April 2010, accelerometers and other sensors were installed on a CRH3 EMU operating on the Beijing-Tianjin Intercity Railway's up-line to carry out testing. Figures 5.5 and 5.6 show the measured train speed curve

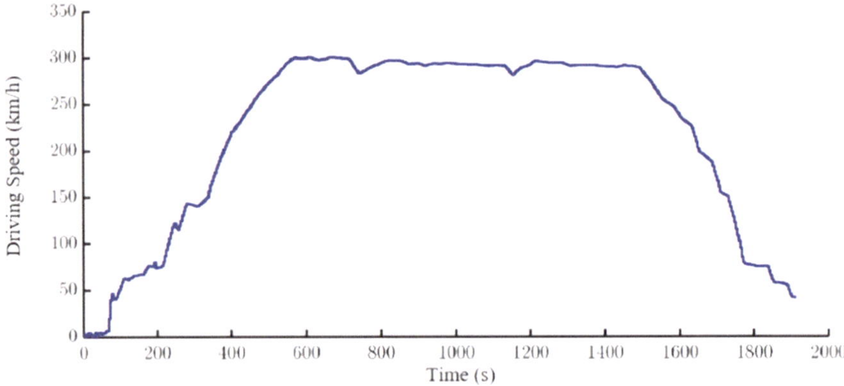


**Fig. 5.5** Measured train speed curve of the CRH3 EMU on the Beijing-Tianjin intercity railway

and the vehicle body's acceleration curves in the longitudinal, lateral, and vertical directions, respectively.

1. **Analysis of Acceleration Amplitude**

Analysis of the measured vehicle body acceleration data shows that the lateral acceleration of the vehicle body is less than 1.0 m/s$^2$ and does not exceed the Grade II ride comfort management threshold. Moreover, in most cases, the acceleration is less than the Grade I ride comfort management threshold of 0.6 m/s$^2$. Table 5.2 lists the locations and track conditions where the lateral acceleration of the vehicle body exceeded 0.6 m/s$^2$. It can be observed that the points where the lateral acceleration of the vehicle body exceeded the limit mainly occurred in station turnout areas and locations with poor alignment of the horizontal alignment and vertical profiles, especially at overlaps or closely spaced horizontal and vertical curves.

Figure 5.7 illustrates a section of track with typical poor alignment of the horizontal alignment and vertical profiles. In this section, the transition curve length is short, the vertical curve is very close to the tangent point of the transition curve, and the vertical curve is also very close to the tangent point of the circular curve. Measured data indicate that near both ends of this curve, the peak lateral acceleration of the vehicle body reached 0.84 m/s$^2$, exceeding the vibration levels caused by station turnout areas. Therefore, the alignment of horizontal alignment and vertical profiles has a significant impact on the lateral vibration of the vehicle body. When the horizontal alignment is poor, efforts should be made to avoid overlap between vertical curves and circular curves, or to maintain a certain distance between vertical curves and the start or end points of transition curves.

Based on the measured data, the vertical acceleration of the vehicle body was less than 2.0 m/s$^2$, staying within the Grade III management threshold, with a maximum value of 1.95 m/s$^2$. Furthermore, in most cases, it was below the Grade II comfort threshold of 1.5 m/s$^2$. Table 5.3 lists the locations and track conditions where the vertical acceleration of the vehicle body exceeded 1.5 m/s$^2$. It is evident that the

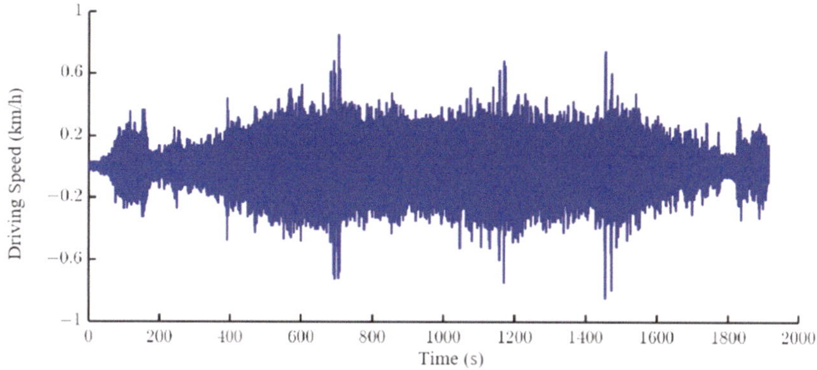


(a) Longitudinal acceleration of the vehicle body.

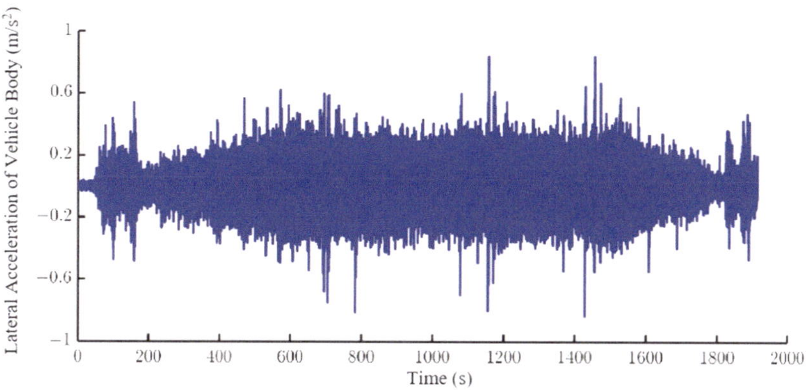


(b) Lateral acceleration of the vehicle body.

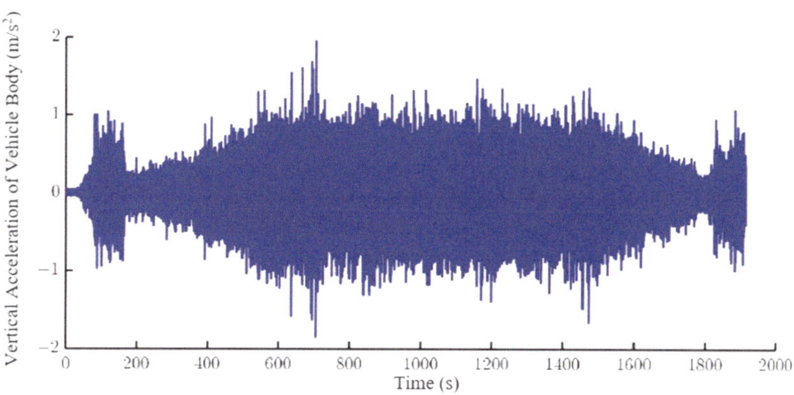


(c) Vertical acceleration of the vehicle body.

**Fig. 5.6** Measured vehicle body acceleration curves of the CRH3 EMU on the Beijing-Tianjin intercity railway

**Table 5.2** Lateral acceleration exceedance points of the vehicle body on the Beijing-Tianjin intercity railway

| Sr. No. | Mileage | Peak value (m/s$^2$) | Description |
|---|---|---|---|
| 1 | K95+394 | 0.62 | On a concave vertical curve, $\Delta i = 4.5‰$ |
| 2 | K85+080 | 0.68 | Wuqing station turnout area |
| 3 | K84+180 | 0.75 | Wuqing station turnout area |
| 4 | K77+876 | 0.81 | On a circular curve near a vertical curve, $R$ : 7005 m, $\Delta i = 1‰$ |
| 5 | K53+804 | 0.70 | On a straight section, cause unknown |
| 6 | K47+500 | 0.84 | Yongle station turnout area |
| 7 | K46+516 | 0.68 | Yongle station turnout area |
| 8 | K25+610 | 0.84 | On a transition curve near a vertical curve, $R$ : 7995 m, $Lh$ : 570 m, $\Delta i = 3.5‰$ |
| 9 | K23+293 | 0.84 | At a straight-transition point near a vertical curve, R : 7995 m, Lh : 570 m, $\Delta i = 4‰$ |
| 10 | K21+993 | 0.67 | Yizhuang station turnout area |

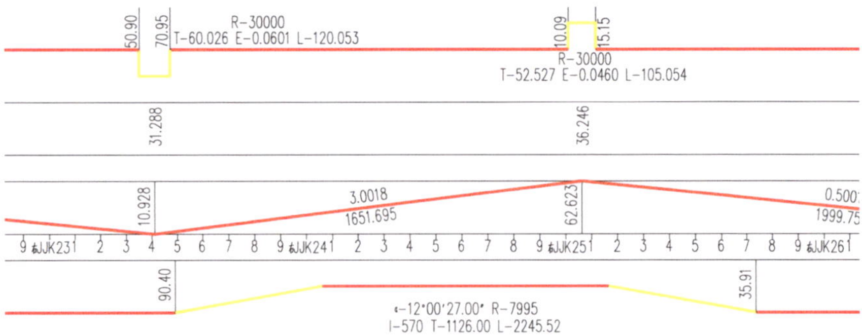


**Fig. 5.7** Local view of the Beijing-Tianjin intercity railway horizontal alignment and vertical profiles -1

exceedance points for vertical acceleration mainly occurred in station turnout areas and on vertical curves with large algebraic differences in gradients. Overall, the magnitude of vertical vibrations was greater than that of lateral vibrations.

Figure 5.8 shows a section of the Beijing-Tianjin Intercity Railway with a large algebraic gradient difference. The algebraic difference in the gradient of the vertical profile reaches $\Delta i = 7.5‰$, and the vertical curve is very close to the transition curves at both ends, resulting in poor overall alignment conditions. The peak vertical acceleration of the vehicle body occurred at the grade change point, reaching 1.58 m/s$^2$. When the algebraic gradient difference is large, the vertical curve length is also longer, resulting in passengers experiencing the combined effects of centripetal (or centrifugal) acceleration and vehicle vibrations caused by track irregularities for a

**Table 5.3** Vertical acceleration exceedance points of the vehicle body on the Beijing-Tianjin intercity railway

| No | Mileage | Peak value (m/s$^2$) | Description |
|---|---|---|---|
| 1 | K89+900 | 1.58 | On a convex vertical curve, $\Delta i = 7.5‰$ |
| 2 | K87+407 | 1.61 | On a concave vertical curve, $\Delta i = 3.5‰$ |
| 3 | K85+350 | 1.68 | Wuqing station turnout area |
| 4 | K85+224 | 1.81 | Wuqing station turnout area |
| 5 | K84+961 | 1.60 | Wuqing station turnout area |
| 6 | K84+218 | 1.95 | Wuqing station turnout area |
| 7 | K21+933 | 1.66 | Yizhuang station turnout area |

longer duration. Therefore, to ensure a good level of ride comfort, sections with large algebraic gradient differences should have improved horizontal alignment standards, preferably straight lines with long tangent lengths. As shown in Fig. 5.8, the tangent sections between the curves are too short, and the vertical curve is very close to the transition curves on both ends. When vertical vibrations on the vertical curve fail to dissipate effectively and are compounded by lateral vibrations caused by the transition curves, the overall ride comfort is severely degraded.

2. **Analysis of Ride Stability and Passenger Comfort**

Filtering and Fourier transform analysis of the data in Fig. 5.6 yield Sperling ride quality index of 1.57 and 1.54 for lateral and vertical directions, respectively, both below the Grade I "Excellent" standard for stability ($W < 2.5$).

According to the weighted curves in ISO 2631, the effective values of lateral and vertical vibration acceleration were calculated, yielding comfort levels (RCL) of 78.7 and 82.6, respectively. Both values fall below the Grade I "Very Comfortable" standard (RCL $\leq 83$).

According to the UIC 513R standard, the ride comfort index (NMV) is 1.66, corresponding to Grade II "Comfortable" ($1 \leq N_{MV} < 2$).

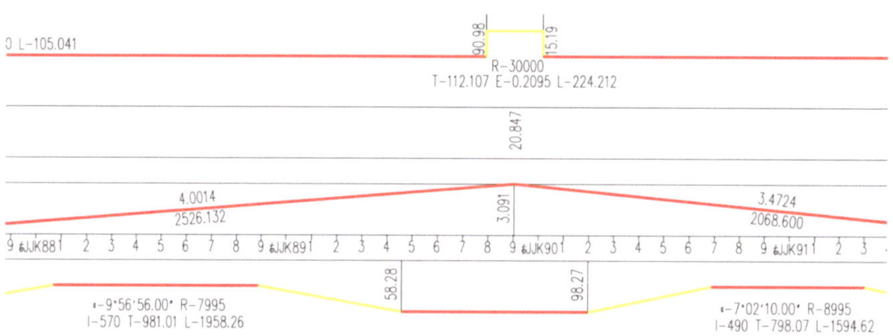


**Fig. 5.8** Local view of the Beijing-Tianjin intercity railway horizontal alignment and vertical profiles -2

During the testing, due to strong wind conditions, the operating CRH3 EMU reached a maximum speed of 300 km/h. Under these conditions, the Beijing-Tianjin Intercity Railway demonstrated good ride comfort.

### 5.2.3 Comparison of the Effects of Track Alignment and Smoothness on Ride Comfort

Based on the track irregularities and vehicle vibration acceleration data collected by the CIT-0 comprehensive inspection train, a comparative analysis was conducted on the effects of track alignment and track irregularities on ride comfort.

Figure 5.9 shows the speed curve of the comprehensive inspection train, with a maximum operating speed of 250 km/h. Figure 5.10 presents the vehicle body acceleration variation curves for the comprehensive inspection train. The frequency of the measured acceleration is close to 0 Hz, which adequately reflects the effects of unbalanced acceleration on horizontal and vertical profile curves. As seen in the figures, the lateral acceleration amplitude of the vehicle body is less than 1.2 m/s$^2$, with most values below 1.0 m/s$^2$. On certain circular curves, the lateral quasi-static acceleration combined with vibration acceleration exceeds 1.0 m/s$^2$. The vertical acceleration amplitude of the vehicle body is less than 0.8 m/s$^2$.

Through the analysis of lateral and vertical acceleration data, the Sperling ride quality index for lateral and vertical directions were found to be 1.89 and 1.58, respectively, both below the "Excellent" comfort level threshold of 2.5. According to UIC513R standards, the ride comfort index (NMV) was calculated as 1.31, corresponding to Grade II "Comfortable".

The section between K59+000 and K62+000 on the Beijing-Tianjin Intercity Railway was selected for a comparative analysis of the effects of track alignment

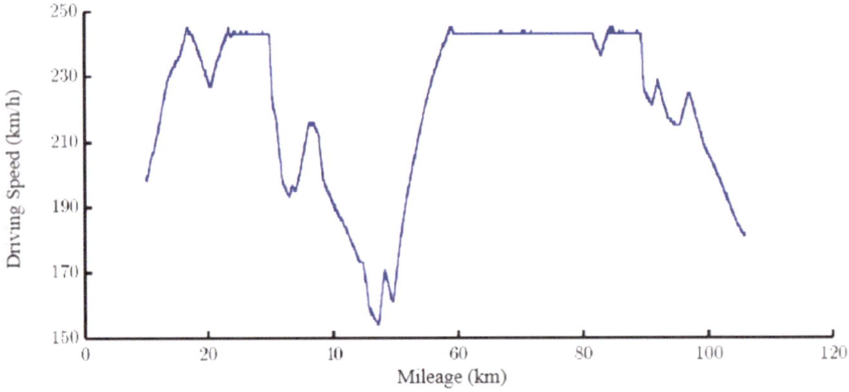


**Fig. 5.9** Speed curve of the CIT-0 comprehensive inspection train on the Beijing-Tianjin intercity railway

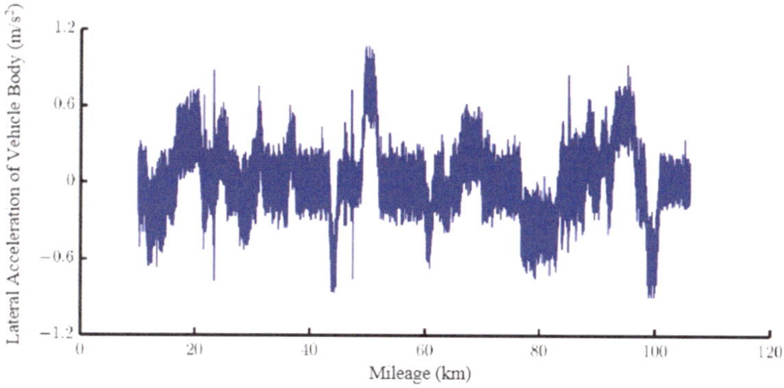


(a) Lateral acceleration of the vehicle body.

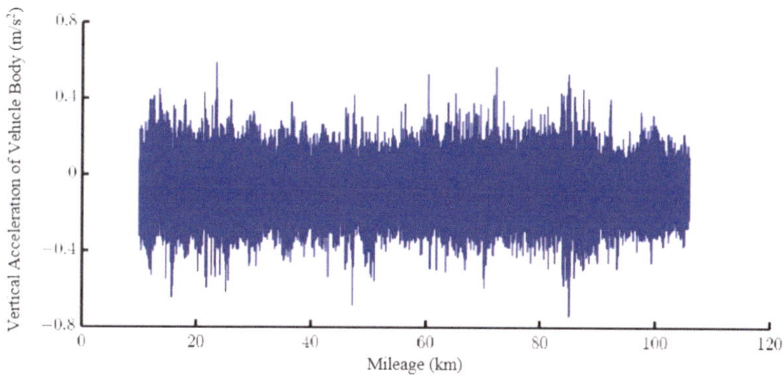


(b) Vertical acceleration of the vehicle body.

**Fig. 5.10** Vehicle body acceleration curves of the CIT-0 comprehensive inspection train on the Beijing-Tianjin intercity railway

and track irregularities on ride comfort. The parameters of this section are shown in Fig. 5.11. This section includes a complete circular curve and a vertical curve, with a train passing speed of 243 km/h.

Figure 5.12 compares the measured lateral acceleration data of the vehicle body in this section with the filtered data at different frequencies. Under ideal alignment conditions, the frequency of unbalanced acceleration is generally close to 0 Hz. When the data is passed through a low-pass filter with a cutoff frequency of 0.3 Hz, the main component is the unbalanced acceleration generated by the vehicle body on the curve. As shown in the figure, the lateral unbalanced acceleration of the vehicle body is approximately 0.38 m/s$^2$. Since the natural frequency of lateral vibration of the vehicle body is generally between 0.5 and 1.5 Hz, components below 0.3 Hz are mainly quasi-static accelerations caused by alignment, while vibrations at the vehicle's natural frequency induced by alignment changes are filtered out.

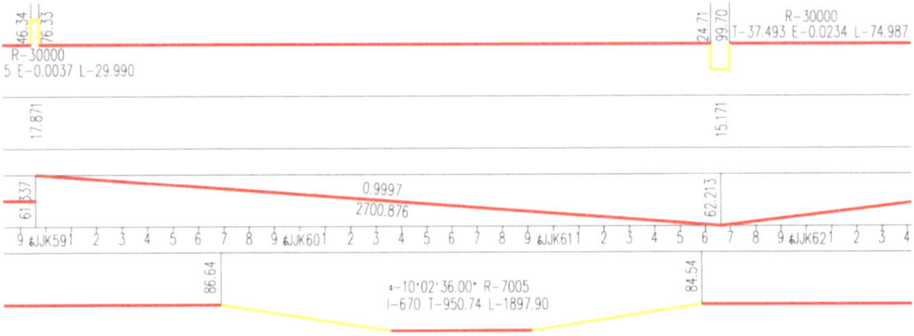

**Fig. 5.11** Local view of the horizontal alignment and vertical profiles on the Beijing-Tianjin intercity railway

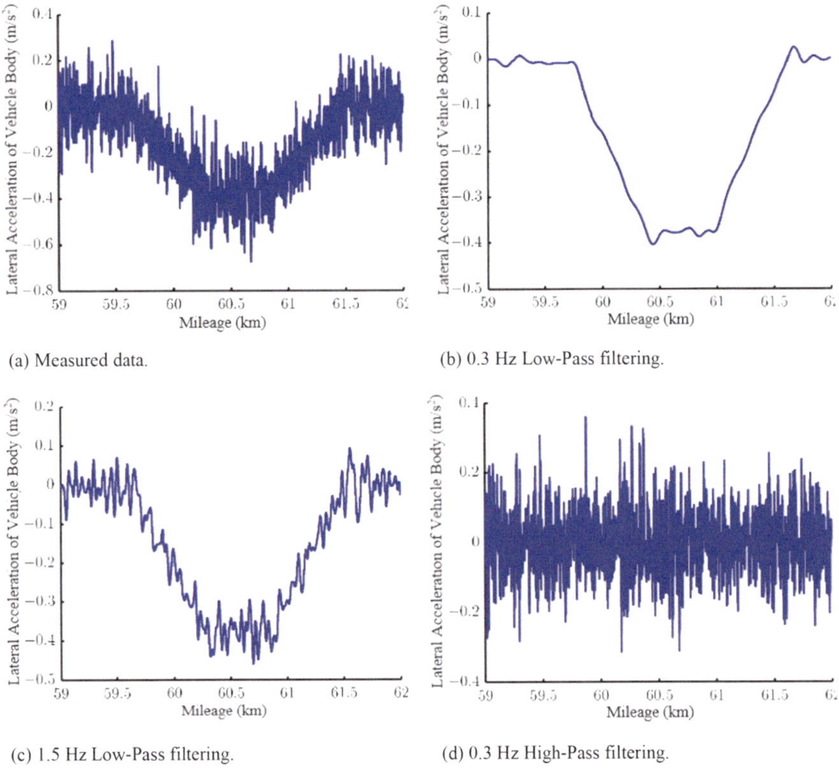

(a) Measured data.

(b) 0.3 Hz Low-Pass filtering.

(c) 1.5 Hz Low-Pass filtering.

(d) 0.3 Hz High-Pass filtering.

**Fig. 5.12** Filtered comparative analysis of measured lateral acceleration data of the vehicle body

Figure 5.12c shows the acceleration curve after low-pass filtering with a cutoff frequency of 1.5 Hz. This data includes the vibration components at the natural frequency of the vehicle body. The natural frequency vibration components of the vehicle body are caused by a combination of factors, including alignment, track irregularities, or structural stiffness. Therefore, it is difficult to completely separate the vibrations caused by alignment from those caused by track irregularities. As shown in the figure, the vehicle acceleration increases significantly at straight-to-transition, transition-to-circular, circular-to-transition, and transition-to-straight points. Especially at the transition-to-straight points, the vehicle acceleration increases significantly and shows periodic attenuation. This indicates that abrupt changes in track alignment cause impacts on the vehicle, significantly increasing its vibration acceleration. Compared with other locations, the vehicle acceleration caused by abrupt alignment changes can reach approximately twice the acceleration caused by track irregularities.

Figure 5.12d shows the acceleration curve after high-pass filtering with a cutoff frequency of 0.3 Hz. This data mainly reflects the vehicle acceleration caused by track irregularities but also includes natural frequency vibrations of the vehicle body caused by abrupt alignment changes. As shown in the figure, the acceleration peak exceeds 0.4 $m/s^2$, with the peaks mainly occurring at transition-to-circular and circular-to-transition points, which are abrupt alignment change locations. It can be observed that the peak lateral acceleration of the vehicle body results from the combined effects of planar alignment changes and track irregularities.

Figure 5.13 presents a comparison of the effects of track alignment and track irregularities on the lateral acceleration of the vehicle body. The allowable under-elevation value for high-speed railway tracks is generally 90 mm, corresponding to an unbalanced centrifugal acceleration of approximately 0.6 $m/s^2$. The measured data indicates that the acceleration peaks caused by track irregularities and the quasi-static acceleration caused by alignment are of the same order of magnitude, with little difference. Lateral acceleration exceedance points for the vehicle body generally occur on planar curves, resulting from the superposition of unbalanced acceleration and vibration acceleration. Therefore, special attention should be paid to the maintenance of track irregularities in curved sections, particularly near alignment transition points such as transition-to-circular and circular-to-transition points.

Figure 5.14 presents the measured vertical acceleration data of the vehicle on this section of the line, along with filtered data at different frequencies. As shown in Fig. 5.14b, the impact of track alignment on vertical acceleration is primarily observed at abrupt changes in superelevation gradients and vertical curves.

Figure 5.14c includes the vehicle's natural frequency vibrations induced by track alignment. The figure shows significant impacts on the vehicle at superelevation gradient transitions and the start and end points of vertical curves. The absence of transition curves on vertical curves results in substantial impacts at their endpoints, whereas horizontal circular curves with transition curves exhibit relatively smaller lateral impacts at transition points. Figure 5.14d illustrates the vertical acceleration components of the vehicle after removing the unbalanced acceleration induced by alignment, mainly including impacts from alignment changes and vibrations caused

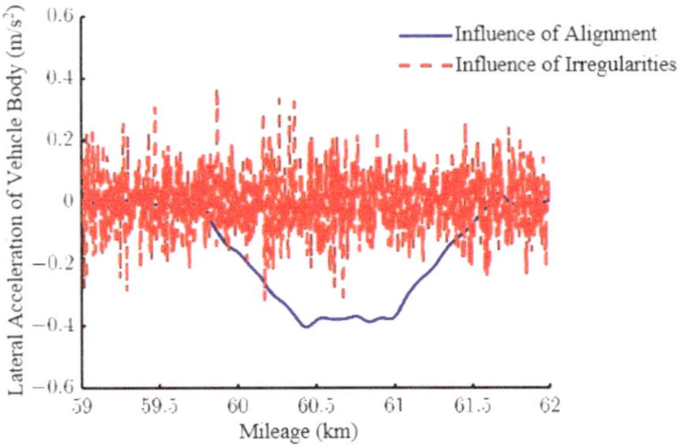


**Fig. 5.13** Comparison of the effects of track alignment and irregularities on lateral acceleration of the vehicle

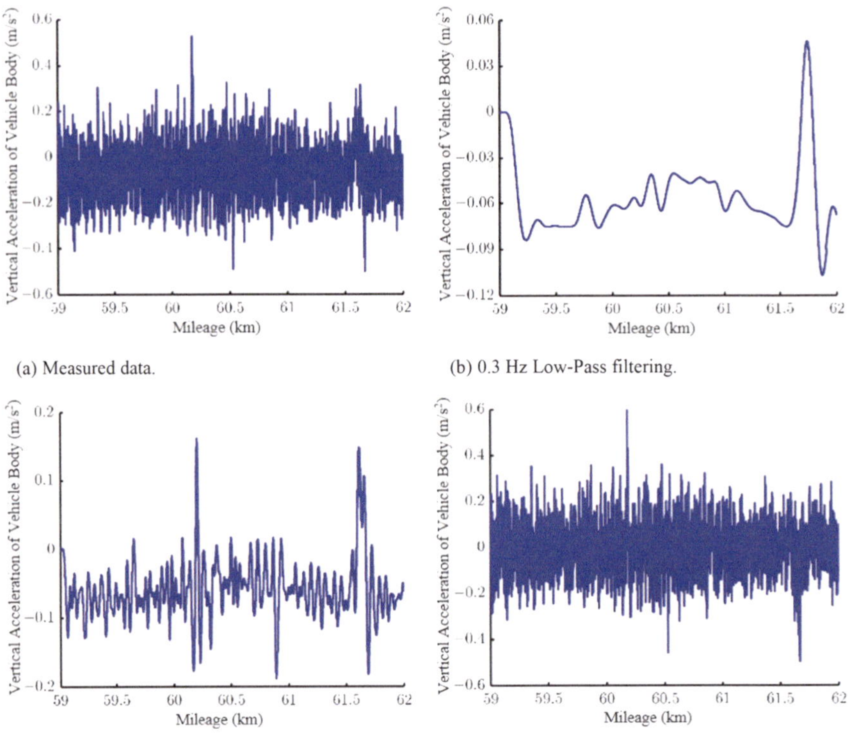


(a) Measured data.

(b) 0.3 Hz Low-Pass filtering.

(c) 1.5 Hz Low-Pass filtering.

(d) 0.3 Hz High-Pass filtering.

**Fig. 5.14** Comparative analysis of filtered measured vertical acceleration data by track irregularities. The figure indicates that vertical acceleration peaks still occur near superelevation gradient transitions and vertical curves, demonstrating that these peaks are primarily the result of combined effects of longitudinal alignment variations and track irregularities. In practice, alignment changes are often accompanied by poor track irregularity conditions, which intensify the wheel-rail dynamic interactions and result in significant vehicle vibrations.

Figure 5.15 compares the effects of track alignment and irregularities on vertical acceleration of the vehicle. The vertical curve radius of the longitudinal profile in this section is 30,000 m, with a measured operating speed of 243 km/h. The theoretical unbalanced vertical acceleration is 0.15 $m/s^2$, while the measured unbalanced vertical acceleration is approximately 0.13 $m/s^2$. When considering the impacts at the start and end points of vertical curves, the absolute value can reach 0.25 $m/s^2$, approximately twice the unbalanced acceleration (as shown in Fig. 5.14b, c). At a speed of 350 km/h, the theoretical unbalanced vertical acceleration will increase to 0.32 $m/s^2$. Considering the impacts at the start and end points of vertical curves as well as track irregularities, the vertical acceleration of the vehicle on vertical curves can easily exceed the management threshold.

Analysis of measured acceleration amplitudes of the CRH3 EMU in actual operation indicates that vertical vibration levels of the vehicle are significantly greater than lateral vibration levels. Most lateral accelerations are below 0.6 $m/s^2$, while most vertical accelerations are below 1.5 $m/s^2$ (both including unbalanced accelerations). Instances where vertical acceleration exceeds the limit are mainly found near vertical curves with gradient algebraic differences exceeding 4‰, apart from turnout areas near stations. The larger the gradient algebraic difference and the longer the vertical curve, the higher the likelihood of unbalanced acceleration on the vertical curve, impacts at curve endpoints, and vibrations caused by significant track irregularities combining to affect the vehicle. Therefore, optimizing the longitudinal alignment design of high-speed railways is an important topic, including considerations of the relationship between gradient algebraic differences and vertical curve radius, transition segment length, and the spacing of vertical transition curves.

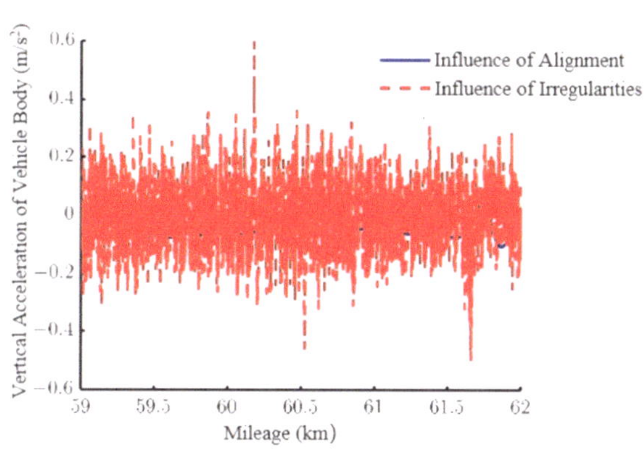


**Fig. 5.15** Comparison of the effects of track alignment and irregularities on vertical acceleration of the vehicle

In summary, the difference in the impact of track alignment and irregularities on vehicle acceleration lies primarily in the vibration frequency. Acceleration caused by alignment is mainly unbalanced acceleration near 0 Hz and natural frequency vibrations due to abrupt alignment changes. In contrast, the vibration acceleration caused by track irregularities typically occurs at frequencies above 0.3 Hz.

The impact of vibrations on ride comfort depends on their amplitude, frequency, and duration. The nature of vehicle vibrations caused by track alignment and irregularities differs significantly in terms of frequency and duration. Therefore, the decision to ignore one factor or consider their combined effects should depend on the specific research question. When studying specific alignment parameter standards for comparative analysis, the influence of track irregularities should be ignored. Conversely, when evaluating track irregularities for maintenance purposes, the effect of unbalanced acceleration on the vehicle should be excluded. In current track irregularity management standards, vehicle acceleration thresholds do not distinguish between straight and curved track sections, applying the same criteria to both. In practice, unbalanced acceleration and alignment-induced acceleration on curves may exceed the specified acceleration thresholds. Therefore, using only vehicle acceleration amplitude is insufficient for accurately evaluating track irregularity conditions.

When assessing the ride comfort of a track alignment plan, the combined effects of alignment and track irregularities must be considered. At this point, the choice of ride comfort evaluation methods and standards becomes particularly important. Common ride comfort indicators, such as the Sperling ride quality index, ISO2631, and UIC513R comfort evaluation indices, primarily account for vehicle vibration accelerations in the 0.4–30 Hz frequency range, disregarding the effects of unbalanced acceleration. Furthermore, when evaluating acceleration vibration amplitudes, relying solely on the maximum or minimum acceleration values of the vehicle fails to account for the impact of vibration duration. Thus, while the Sperling ride quality index, UIC513R ride comfort, and vehicle acceleration amplitude provide insights into ride comfort for specific track alignment plans, they do not comprehensively reflect the overall impact of the alignment plan on ride comfort. Particularly when comparing the impact of different alignment plans on ride comfort, these evaluation indices alone are insufficient for effectively differentiating between the levels of impact. Therefore, in dynamic simulation analyses of track alignment plans, additional indicators such as the root mean square of acceleration, acceleration amplitude frequency, and cumulative distribution characteristics should be used, alongside the aforementioned standards, to further compare the impacts of different alignment plans on ride comfort.

# 5.3 Dynamic Testing and Analysis of the Vehicle-Track System on the Jinan-Qingdao High-Speed Railway

## *5.3.1 Testing Conditions*

The Jinan-Qingdao High-Speed Railway is a high-speed rail line in Shandong Province, China, connecting Jinan and Qingdao. The railway runs east–west and forms the eastern section of the Qingdao-Yinchuan corridor within China's "Eight Vertical and Eight Horizontal" high-speed railway network. The Jinan-Qingdao High-Speed Railway extends from Jinan East Station in the west to Hongdao Station in the east, with a total length of 307.9 km and 11 stations. The design speed is 350 km/h, while the initial operating speed is 300 km/h, as shown in Fig. 5.16.

This testing primarily consists of two main components: bridge dynamics testing under train operation conditions and vehicle vibration testing. The bridge dynamics testing was conducted on a 32-m standard-span box girder located on a straight section and a curved section (with an 8000-m radius circular curve). The vehicle vibration testing was performed on two carriages of a trial-operated EMU between Qingdao North and Weifang North. Field testing provided the dynamic responses of the bridge structure in both straight and curved sections and the vehicle vibration responses at different speeds. This data was further analyzed to explore the relationship between the horizontal alignment and vertical profiles of high-speed railway lines and the dynamic responses of bridges, as well as the influence of alignment parameters on the dynamic performance of high-speed railway operations.

1. **Bridge Testing Conditions**

The tests were conducted on two 32-m standard-span box girders, one located in a straight section (DK099+636.060) and the other in a curved section with an 8000-m radius (DK094+872.960). Specific testing site details are shown in Table 5.4.

Photos of field tests on the straight and curved sections are shown in Fig. 5.17.

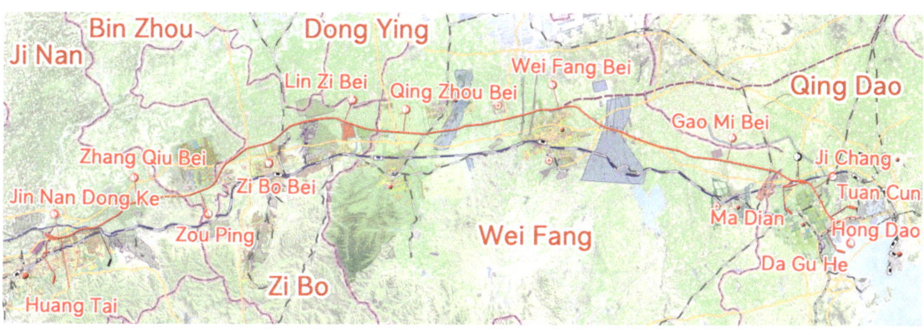


**Fig. 5.16** Jinan-Qingdao high-speed railway

**Table 5.4** Information on bridge testing sites

| Testing site | Small mileage pile number | Large mileage pile number | Small mileage pier number | Large mileage pier number | Pier span | Remarks |
|---|---|---|---|---|---|---|
| Straight Section | DK99+636.060 | DK99+668.760 | 486# | 487# | 32.74m | |
| Curved Section | DK94+872.960 | DK95+900.700 | 339# | 340# | 32.74m | R8000m |

(a) Straight section (b) Curved section

**Fig. 5.17** Field testing locations

For acceleration measurement points on the straight and curved section bridges, eight accelerometers were placed on the bottom of the box girder, as follows:

V1: Vertical acceleration at the large mileage end along the longitudinal centerline of the girder bottom;
V2: Vertical acceleration at the 1/4 point along the longitudinal center line of the girder bottom;
V3: Vertical acceleration at the midspan along the longitudinal centerline of the girder bottom;
V4: Vertical acceleration at the 3/4 point along the longitudinal centerline of the girder bottom;
V5: Vertical acceleration at the small mileage end along the longitudinal centerline of the girder bottom;
V6: Vertical acceleration at the ascending side edge of the girder midspan;
L7: Lateral acceleration at the large mileage end along the longitudinal centerline of the girder bottom;
L8: Lateral acceleration at the midspan along the longitudinal centerline of the girder bottom.

The layout of accelerometers at the measurement points is shown in Fig. 5.18. Partial accelerometer placement is shown in Figs. 5.19 and 5.20.

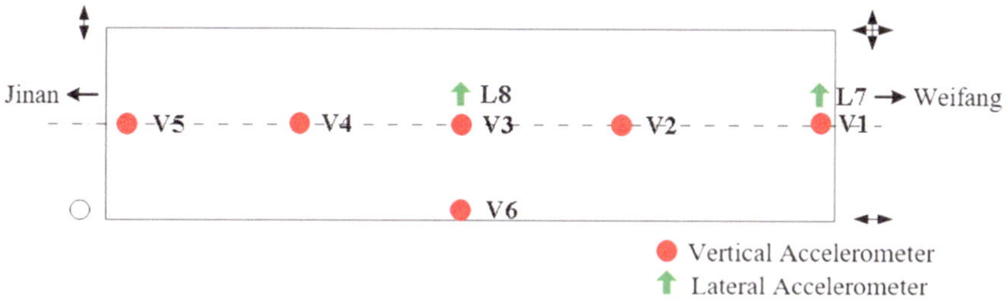


**Fig. 5.18** Layout of accelerometers at measurement points

**Fig. 5.19** Vertical and lateral acceleration at the ends

**Fig. 5.20** Vertical acceleration at the 1/4 point

For displacement measurement points on the straight and curved section bridges, three displacement transducers were placed on the bottom of the box girder, as follows:

DV1: Vertical displacement at the midspan along the longitudinal centerline of the girder bottom;
DV2: Vertical displacement at the small mileage end along the longitudinal centerline of the girder bottom;
DL3: Lateral displacement at the small mileage end along the longitudinal centerline of the girder bottom.

The layout of displacement transducers at the measurement points is shown in Fig. 5.21.

Partial placement of displacement transducers is shown in Fig. 5.22.

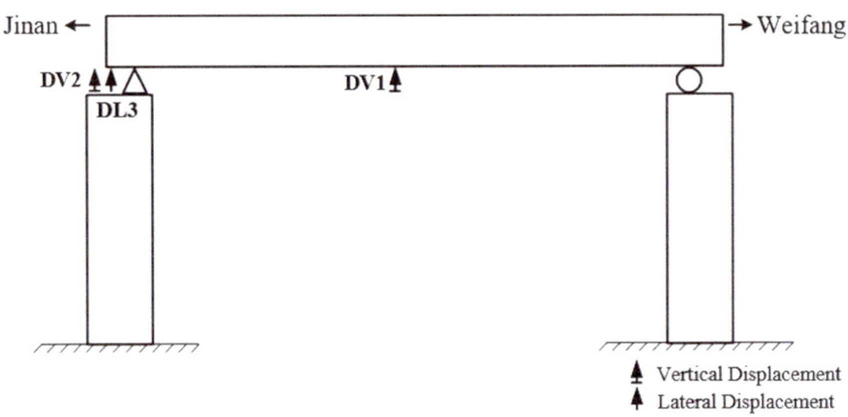


**Fig. 5.21** Layout of displacement transducers at measurement points

**Fig. 5.22** Installation of displacement transducers

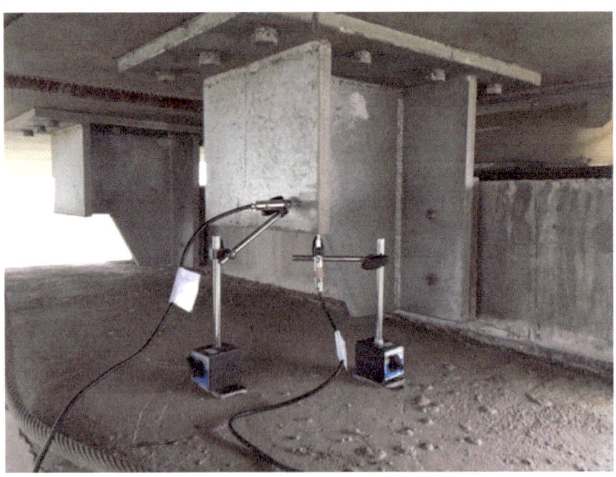

2. **Vehicle Testing Conditions**

Based on the operational plan of the Jinan-Qingdao High-Speed Railway, the field testing train model was the CRH380B EMU (16-car configuration in double formation), as shown in Fig. 5.23.

Vehicle acceleration testing was conducted using two systems: an inertial navigation system (INS) and accelerometers. The INS was installed in the 9th carriage, while accelerometers were placed in the 11th carriage to capture vehicle vibration during train operation.

(1) Inertial Navigation Testing System

The inertial navigation testing equipment was installed in the 9th carriage, with the layout of the INS system's main unit and antenna shown in Fig. 5.24.

**Fig. 5.23** CRH380B double-formation EMU

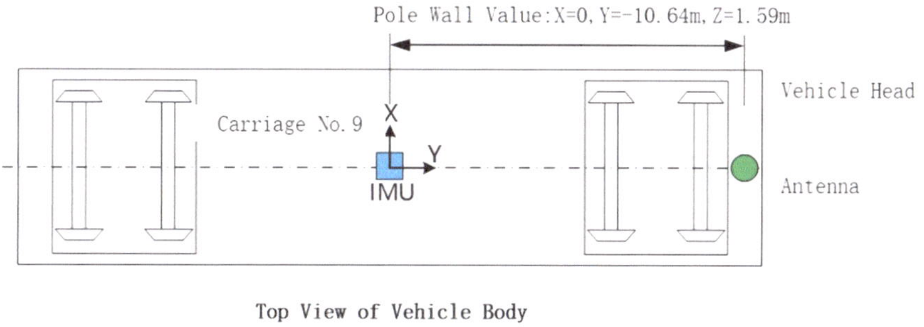


**Fig. 5.24** Layout of the inertial navigation testing system

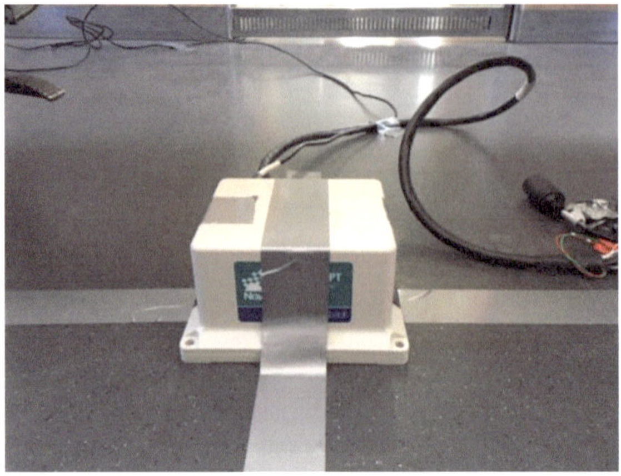

**Fig. 5.25** SPAN-CPT navigation system

**Fig. 5.26** High-precision antenna

The INS testing utilized a high-precision antenna and the SPAN-CPT integrated closed-loop fiber optic navigation system provided by BDStar Navigation, with the specific field layout shown in Figs. 5.25 and 5.26.

(2) Accelerometer Testing System

Eight acceleration measurement points were installed in the 11th carriage to monitor vehicle vibrations, as follows:

V1: Vertical acceleration at the rear end of the longitudinal centerline of the carriage floor;
V2: Vertical acceleration at the 1/4 point of the longitudinal centerline of the carriage floor;
V3: Vertical acceleration at the midspan of the longitudinal centerline of the carriage floor;
V4: Vertical acceleration at the 3/4 point of the longitudinal centerline of the carriage floor;

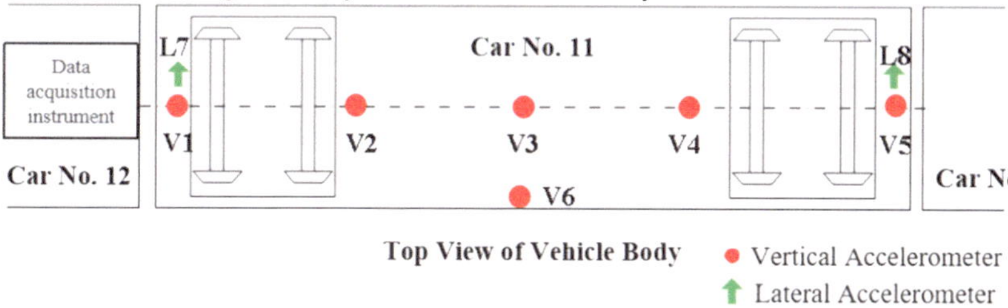


Fig. 5.27 Accelerometer layout on the carriage floor

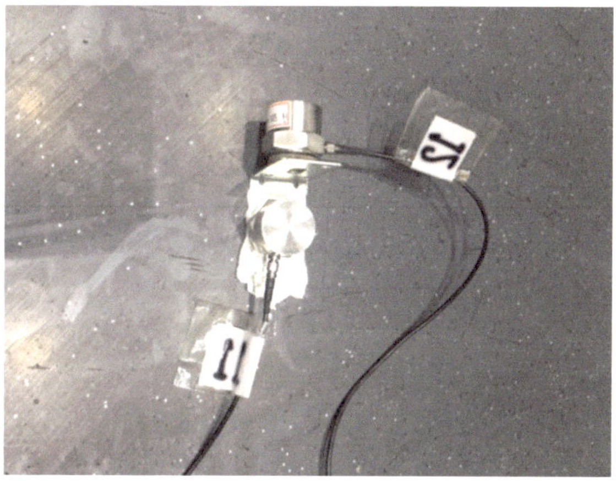

Fig. 5.28 Accelerometer layout

V5: Vertical acceleration at the front end of the longitudinal centerline of the carriage floor;
V6: Vertical acceleration at the edge of the carriage floor midspan;
L7: Lateral acceleration at the rear end of the longitudinal centerline of the carriage floor;
L8: Lateral acceleration at the front end of the longitudinal centerline of the carriage floor.

The accelerometer layout in the 11th carriage is shown in Fig. 5.27.
The vibration accelerometers on the carriage floor are shown in Fig. 5.28.

### *5.3.2 Analysis of Bridge Dynamics Testing Results*

1. **Bridge Vibration Acceleration**

Based on the train operation schedule, a CR400AF EMU was selected to pass through the straight and curved section bridge measurement points at a speed of 300 km/h for bridge acceleration vibration testing.

(1) Straight Section

When the testing train, a CR400AF EMU, passed through the 32-m standard-span bridge in the straight section at a speed of 300 km/h, the acceleration responses at various positions of the box girder bottom are shown in Figs. 5.29, 5.30, 5.31, 5.32, 5.33, 5.34, 5.35 and 5.36.

From Figs. 5.29, 5.30, 5.31, 5.32, 5.33, 5.34, 5.35 and 5.36, the maximum acceleration values at each measurement point when the CR400AF EMU passed through the 32-m standard-span bridge at 300 km/h are as follows:

Maximum vertical acceleration at the large mileage end (V1): 0.7434 m/s$^2$;
Maximum vertical acceleration at the 1/4 point (V2): 2.8092 m/s$^2$;
Maximum vertical acceleration at midspan (V3): 3.1784 m/s$^2$;
Maximum vertical acceleration at the 3/4 point (V4): 2.4049 m/s$^2$;
Maximum vertical acceleration at the small mileage end (V5): 1.0927 m/s$^2$;
Maximum vertical acceleration at the edge of midspan (V6): 1.2212 m/s$^2$;
Maximum lateral acceleration at the large mileage end (L7): 0.6107 m/s$^2$;
Maximum lateral acceleration at midspan (L8): 1.2618 m/s$^2$;

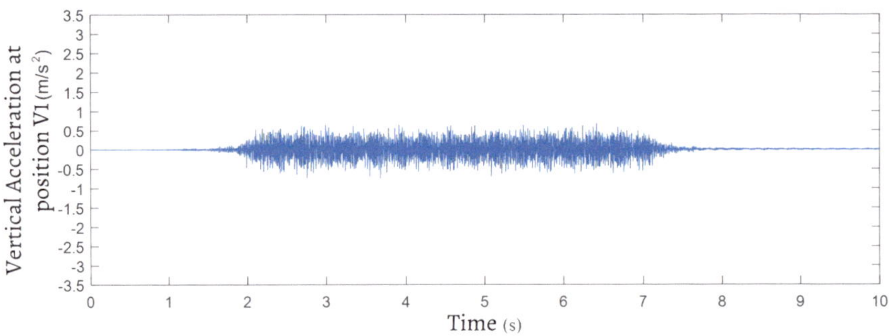


**Fig. 5.29** Time history of vertical acceleration at large mileage end (V1) of the straight section bridge

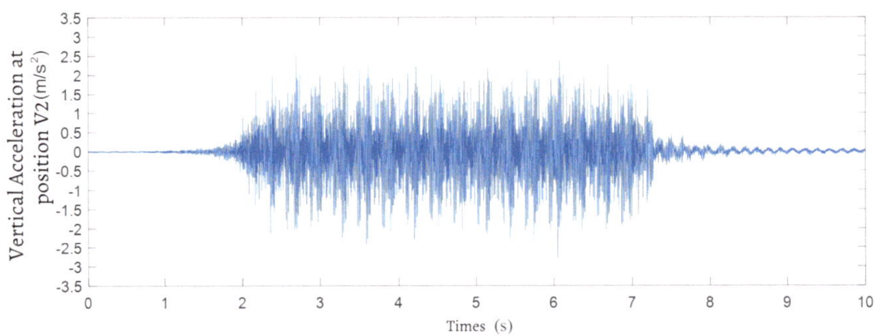


**Fig. 5.30** Time history of vertical acceleration at 1/4 point (V2) of the straight section bridge

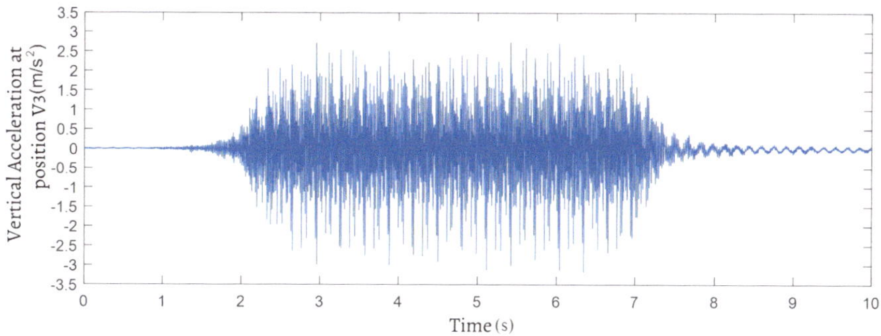


**Fig. 5.31** Time history of vertical acceleration at midspan (V3) of the straight section bridge

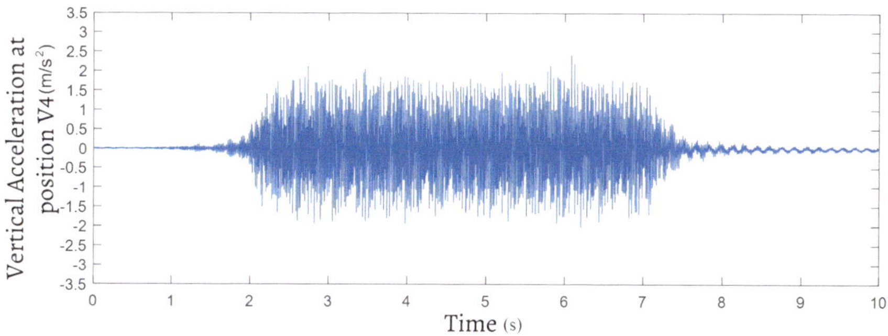


**Fig. 5.32** Time history of vertical acceleration at 3/4 point (V4) of the straight section bridge

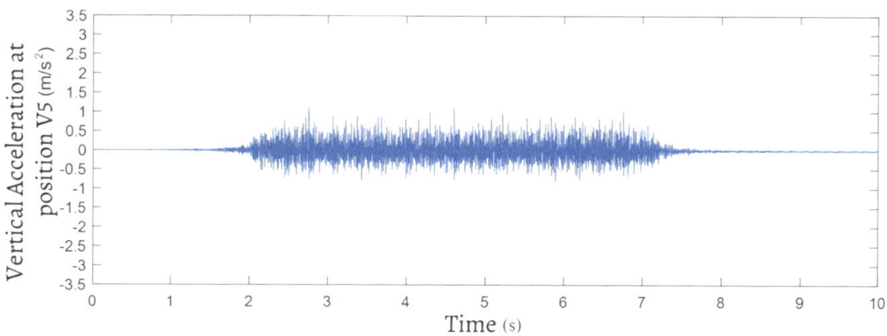


**Fig. 5.33** Time history of vertical acceleration at small mileage end (V5) of the straight section bridge

### (2) Curved Section

When the testing train CR400AF ran at a speed of 300 km/h through a 32-m standard-span bridge in the curved section (radius of 8000 m), the acceleration responses at

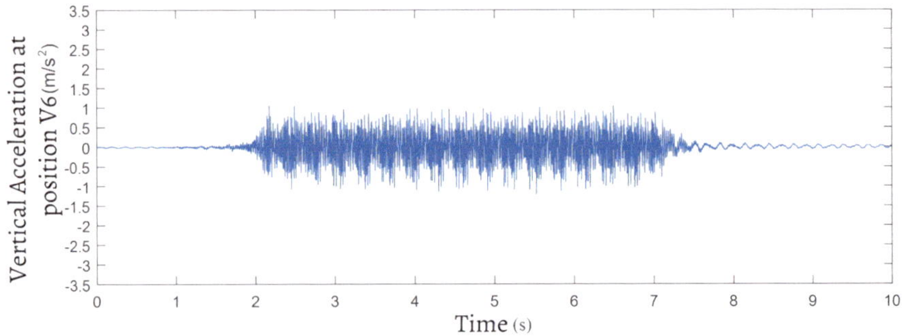


**Fig. 5.34** Time history of vertical acceleration at edge of midspan (V6) of the straight section bridge

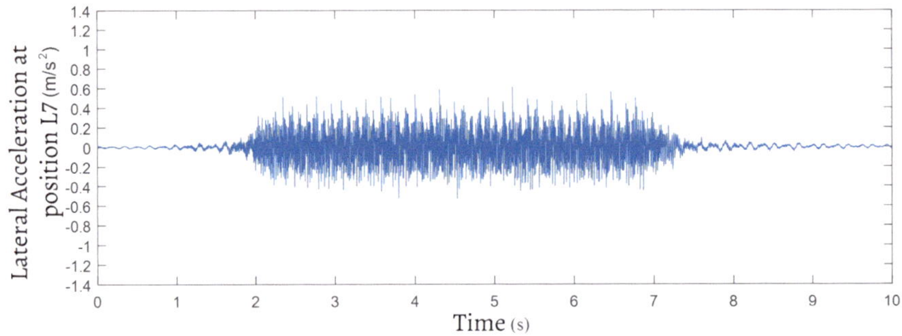


**Fig. 5.35** Time history of lateral acceleration at large mileage end (L7) of the straight section bridge

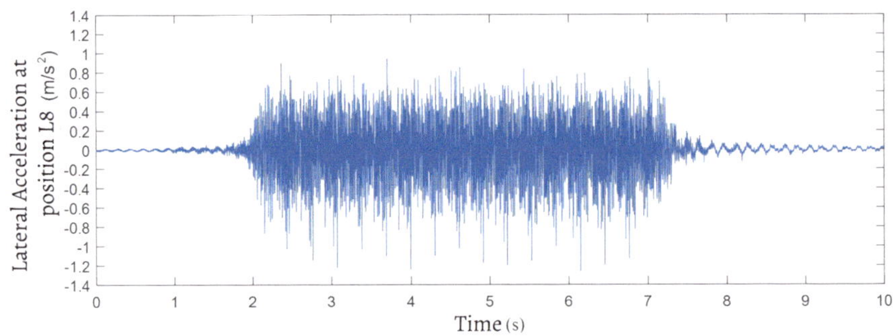


**Fig. 5.36** Time history of lateral acceleration at midspan (L8) of the straight section bridge

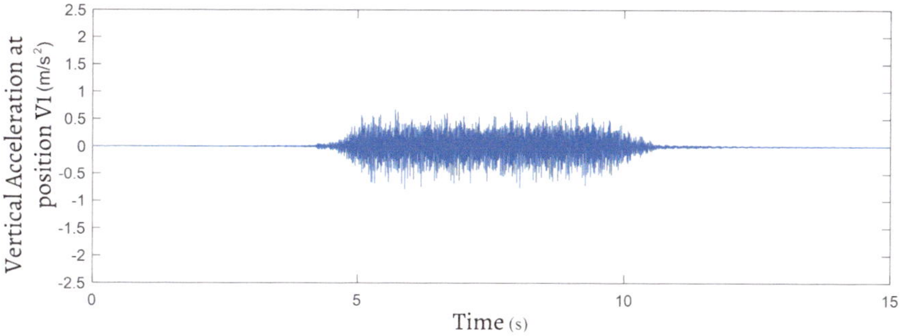


Fig. 5.37 Time history of vertical acceleration at the large mileage end (V1) of the curved section bridge

various positions of the box girder bottom are shown in Figs. 5.37, 5.38, 5.39, 5.40, 5.41, 5.42, 5.43, and 5.44.

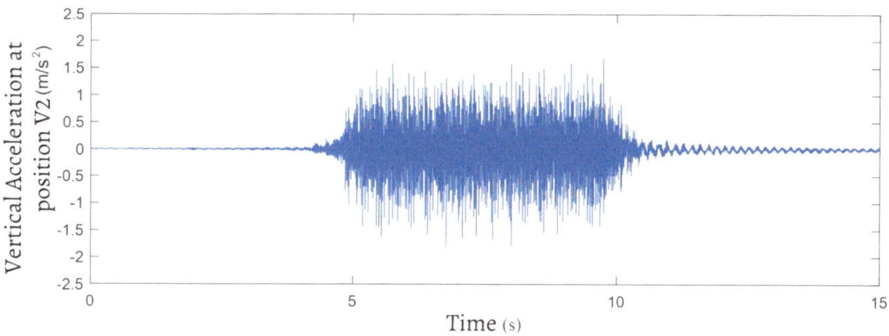


Fig. 5.38 Time history of vertical acceleration at the 1/4 point (V2) of the curved section bridge

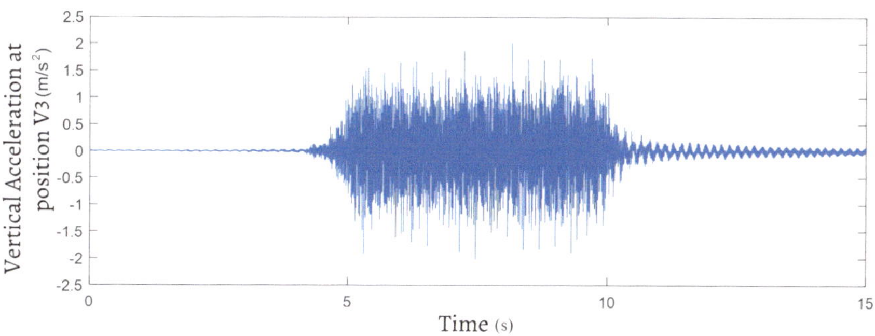


Fig. 5.39 Time history of vertical acceleration at midspan (V3) of the curved section bridge

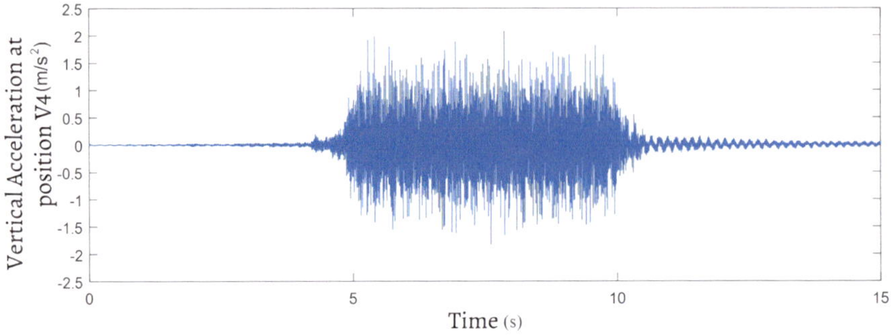


**Fig. 5.40** Time history of vertical acceleration at the 3/4 point (V4) of the curved section bridge

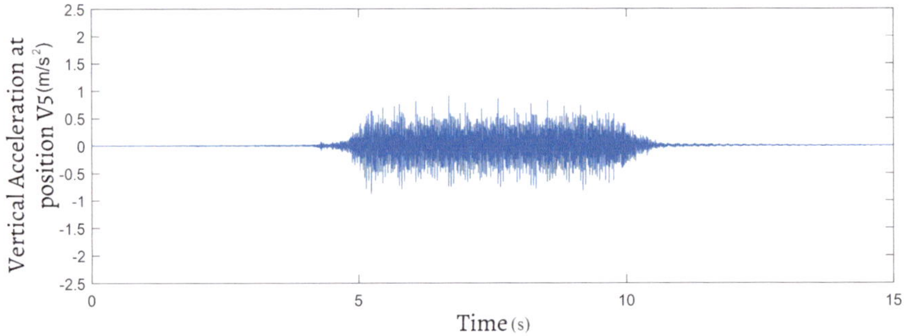


**Fig. 5.41** Time history of vertical acceleration at the small mileage end (V5) of the curved section bridge

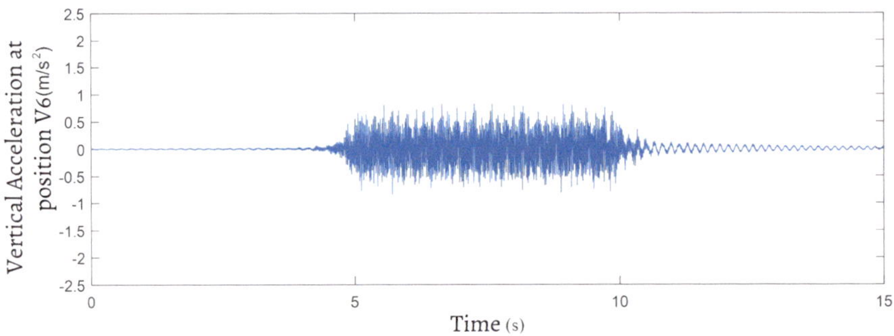


**Fig. 5.42** Time history of vertical acceleration at the midspan edge (V6) of the curved section bridge

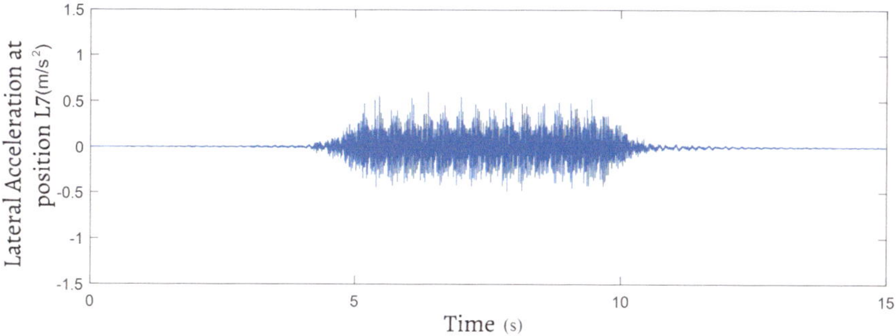


**Fig. 5.43** Time history of lateral acceleration at the large mileage end (L7) of the curved section bridge

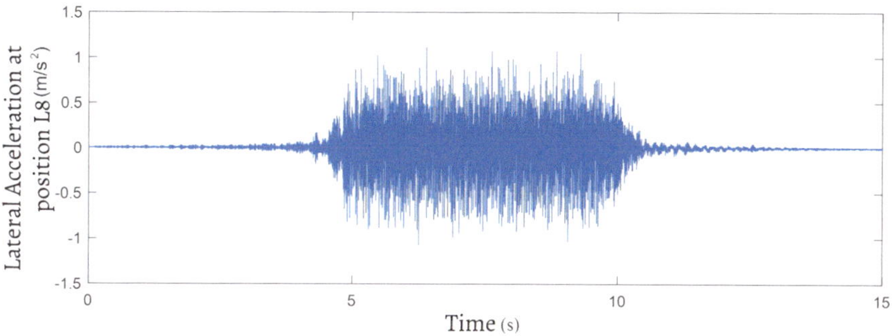


**Fig. 5.44** Time history of lateral acceleration at midspan (L8) of the curved section bridge

From Figs. 5.37, 5.38, 5.39, 5.40, 5.41, 5.42, 5.43, and 5.44, the maximum acceleration values at each measurement point when the CR400AF EMU passed through the 32-m standard-span bridge at 300 km/h are as follows:

Maximum vertical acceleration at the large mileage end (V1): 0.7908 m/s$^2$;
Maximum vertical acceleration at the 1/4 point (V2): 1.8304 m/s$^2$;
Maximum vertical acceleration at midspan (V3): 2.0764 m/s$^2$;
Maximum vertical acceleration at the 3/4 point (V4): 2.0144 m/s$^2$;
Maximum vertical acceleration at the small mileage end (V5): 0.8455 m/s$^2$;
Maximum vertical acceleration at the midspan edge (V6): 1.2212 m/s$^2$;
Maximum lateral acceleration at the large mileage end (L7): 0.5994 m/s$^2$;
Maximum lateral acceleration at midspan (L8): 1.1138 m/s$^2$.

The maximum acceleration values at each measurement point when the CR400AF EMU passed through the 32-m standard-span bridges in the straight and curved sections are summarized in Table 5.5.

From the data in Table 5.5, the following conclusions can be drawn:

**Table 5.5** Maximum acceleration values of the straight and curved section bridges (m/s$^2$)

| Measurement location | Maximum value | |
|---|---|---|
| | Straight section | Curved section |
| Large mileage end vertical (V1) | 0.7434 | 0.7908 |
| 1/4 point vertical (V2) | 2.8092 | 1.8304 |
| Midspan vertical (V3) | 3.1784 | 2.0764 |
| 3/4 point vertical (V4) | 2.4049 | 2.0144 |
| Small mileage end vertical (V5) | 1.0927 | 0.9148 |
| Midspan edge vertical (V6) | 1.2212 | 0.8455 |
| Large mileage end lateral (L7) | 0.6107 | 0.5994 |
| Midspan lateral (L8) | 1.2618 | 1.1138 |

1) The maximum vertical vibration acceleration for the CR400AF EMU on both the straight and curved (radius 8000 m) 32-m standard-span bridges occurred near the midspan, with maximum values of 3.1784 m/s$^2$ and 2.0764 m/s$^2$, respectively. Both values comply with the *Code for Design of High-Speed Railway* (*TB 10621-2014*), which limits vertical vibration acceleration to no more than 5.0 m/s$^2$.
2) The lateral vibration acceleration at the midspan of the 32-m standard-span bridge exceeded that at the ends for both straight and curved sections (radius 8000 m), with maximum values of 1.2618 m/s$^2$ and 1.1138 m/s$^2$, respectively. Both values meet the requirements of the *Code for rating existing railway bridges* (Railway Operations Letter [2004] No.120), which limits lateral vibration acceleration to no more than 1.4 m/s$^2$.
3) A comparison of the vibration accelerations for the CR400AF EMU passing through the 32-m standard-span bridge in the straight and curved sections (radius 8000 m) revealed that vibration accelerations at each measurement point were slightly higher in the straight section than in the curved section, but the differences were not significant. This indicates that with a curve radius of 8000 m, the bridge vibration acceleration is not strongly correlated with curve design parameters, and the curve radius is no longer a key factor affecting bridge vibration acceleration.

2. **Bridge Vibration Displacement**

(1) Straight Section

When the testing train CR400AF ran at 300 km/h through the 32-m standard-span bridge in the straight section, the displacement responses at the bridge bottom measurement points are shown in Figs. 5.45, 5.46 and 5.47.

From Figs. 5.45, 5.46 and 5.47, the maximum displacement values at each measurement point when the CR400AF EMU passed through the 32-m standard-span bridge in the straight section at 300 km/h are as follows:

Maximum vertical displacement at midspan (DV1): 0.6189 mm;

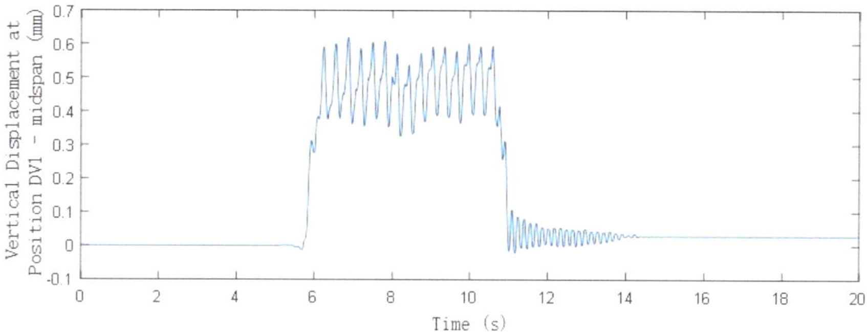


**Fig. 5.45** Time history of vertical displacement at midspan (DV1) of the straight section bridge

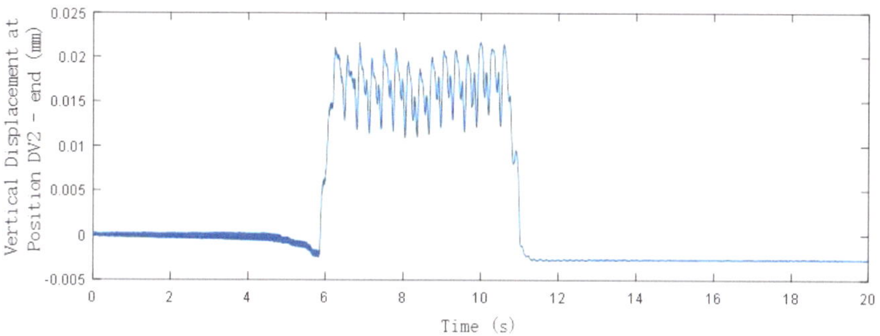


**Fig. 5.46** Time history of vertical displacement at small mileage end (DV2) of the straight section bridge

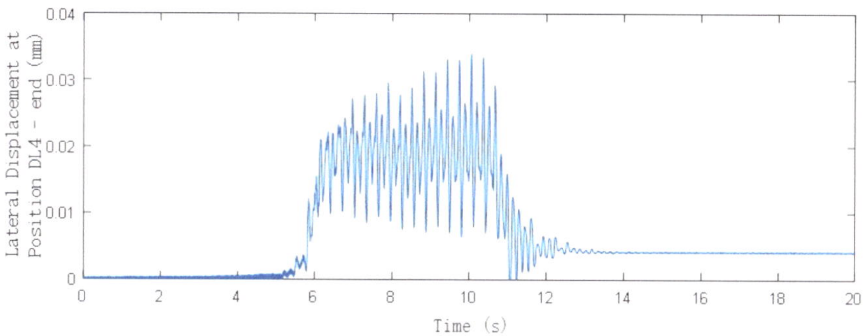


**Fig. 5.47** Time history of lateral displacement at small mileage end (DL3) of the straight section bridge

Maximum vertical displacement at the end (DV2): 0.02145 mm;
Maximum lateral displacement at the end (DL3): 0.03319 mm.

(2) Curved Section

When the CR400AF EMU ran at 300 km/h through the 32-m standard-span bridge in the curved section (radius 8000 m), the displacement responses at the bridge bottom measurement points are shown in Figs. 5.48, 5.49 and 5.50.

From Figs. 5.48, 5.49 and 5.50, the maximum displacement values at each measurement point when the CR400AF EMU passed through the 32-m standard-span bridge in the curved section at 300 km/h are as follows:

Maximum vertical displacement at midspan: 0.5773 mm;
Maximum vertical displacement at the end: 0.0232 mm;
Maximum lateral displacement at the end: 0.0104 mm.

The maximum displacement values at each measurement point when the CR400AF EMU passed through the 32-m standard-span bridges in the straight and curved sections (radius 8000 m) are summarized in Table 5.6.

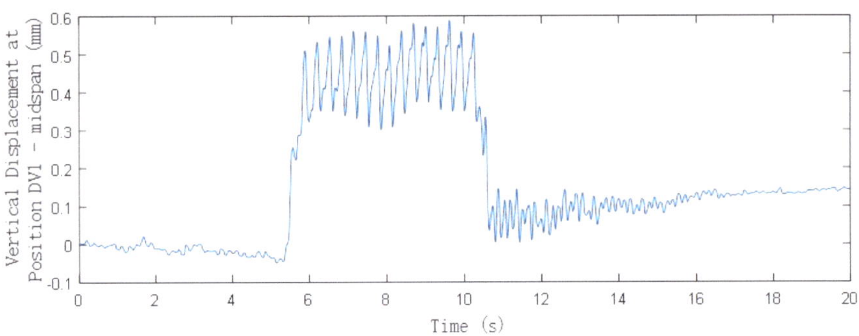


**Fig. 5.48** Time history of vertical displacement at midspan (DV1) of the curved section bridge

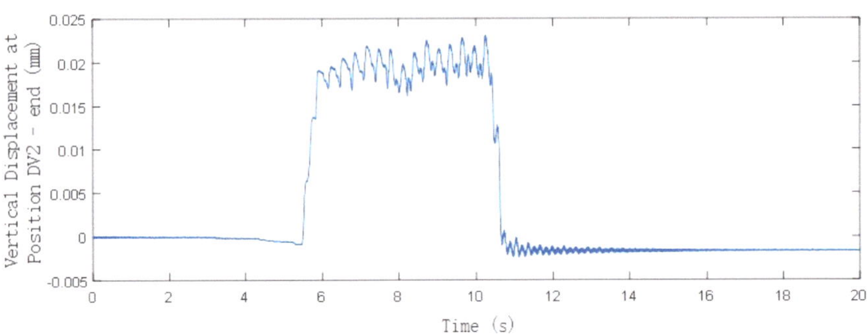


**Fig. 5.49** Time history of vertical displacement at small mileage end (DV2) of the curved section bridge

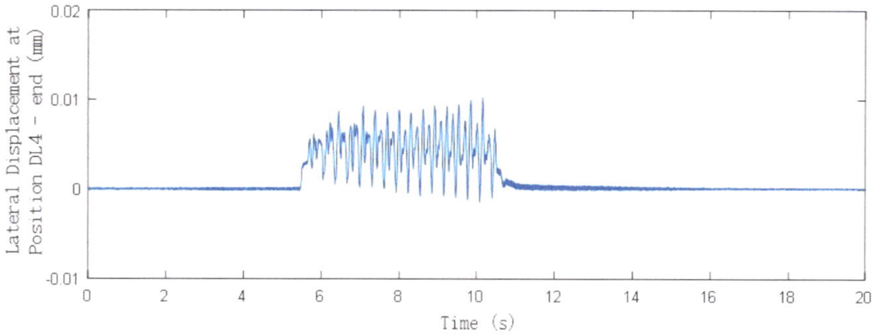


**Fig. 5.50** Time history of lateral displacement at small mileage end (DL3) of the curved section bridge

**Table 5.6** Maximum displacement values of the straight and curved section bridges (mm)

| Measurement Point | Straight section | Curved section |
|---|---|---|
| Vertical displacement at midspan (DV1) | 0.6189 | 0.5773 |
| Vertical displacement at the end (DV2) | 0.0215 | 0.0232 |
| Lateral displacement at the end (DL3) | 0.0332 | 0.0104 |

From the data in Table 5.6, the following conclusions can be drawn:

1) When the CR400AF EMU passed through the 32-m standard-span bridges in the straight and curved sections (radius 8000 m), the maximum vertical displacements were 0.6189 mm and 0.5773 mm, respectively, both well below the limits specified in the *Code for Design of High-Speed Railway* (*TB 10621-2014*).
2) When the CR400AF EMU passed through the 32-m standard-span bridges in the straight and curved sections (radius 8000 m), the maximum lateral displacements were 0.0332 mm and 0.0104 mm, respectively, both significantly lower than the safety limits specified in the *Code for rating existing railway bridges* (Railway Operations Letter [2004] No.120).
3) A comparison of the vibration displacements for the CR400AF EMU passing through the straight and curved sections (radius 8000 m) of the 32-m standard-span bridge shows that vertical displacements at measurement points were slightly higher in the straight section than in the curved section, but the differences were negligible. This suggests that with a curve radius of 8000 m, the bridge vibration displacement is not strongly correlated with the curve design parameters, and the curve radius is no longer a key factor.

3. **Vertical Natural Frequency of the Bridge**

The natural frequency is an essential dynamic characteristic parameter of a bridge and reflects the structure's stiffness. Studies have shown that if the natural frequency of the girder is too low, the train is more likely to induce a significant dynamic response or even resonance when passing over the bridge. The natural vibration characteristics of the bridge can be tested using the free vibration decay method (induced by train excitation). The acceleration testing results of the 32-m standard-span bridges in the straight and curved sections (radius 8000 m) were processed to obtain the bridge acceleration spectrum, as shown in Figs. 5.51 and 5.52.

From Figs. 5.51 and 5.52, the fundamental frequencies of the bridges in the straight and curved sections (radius 8000 m) were determined to be 6.399 Hz and 6.499 Hz, respectively. Both values are higher than the specified limits and the threshold for dynamic calculations in the *Code for Design of High-Speed Railway* (*TB 10621-2014*), with sufficient safety margins.

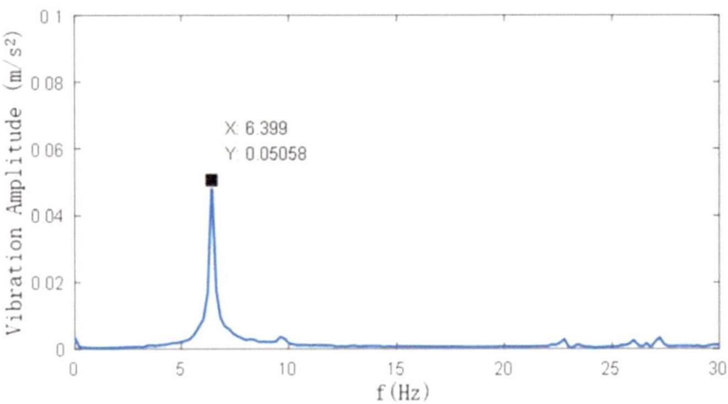


**Fig. 5.51** Acceleration spectrum of the straight section bridge

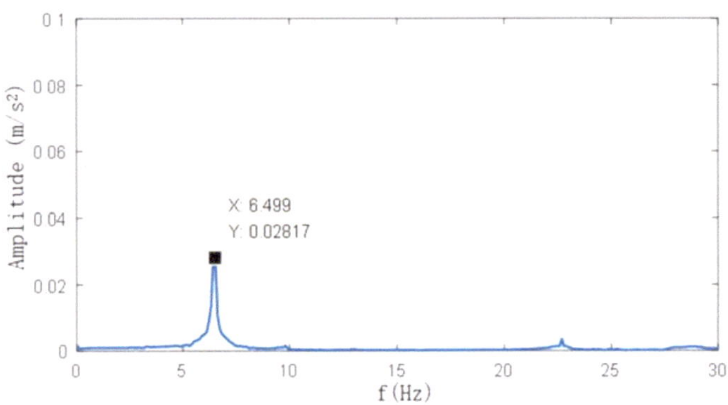


**Fig. 5.52** Acceleration spectrum of the curved section bridge

4. **Summary**

Based on the analysis of bridge vibration acceleration and displacement responses, the following conclusions are drawn:

(1) The 32-m standard-span box girder used in the Jinan-Qingdao Railway exhibited maximum vertical acceleration and displacement values of 3.1784 m/s$^2$ and 0.6189 mm, respectively, under the load of the testing train CR400AF EMU, meeting the requirements of the *Code for Design of High-Speed Railway* (*TB 10621-2014*). The maximum lateral acceleration and displacement values of the bridge were 1.2618 m/s$^2$ and 0.0332 mm, respectively, conforming to the *Code for rating existing railway bridges* (Railway Operations Letter [2004] No.120). This demonstrates that the tested 32 m standard-span bridge has sufficient vertical and lateral stiffness to meet the needs of high-speed EMU operation.
(2) A comparison of the vibration responses of the 32-m standard-span bridge in the straight and curved sections (radius 8000 m) under the load of the CR400AF EMU revealed that the vibration acceleration and displacement of the girder were similar, with no significant differences. This suggests that when the curve radius is 8000 m, the bridge vibration response is not strongly correlated with the curve design parameters, and the curve radius is no longer a key factor.
(3) Regardless of whether the bridge is in a straight or curved section, the measured vertical natural frequency of the 32-m standard-span box girder on the Jinan-Qingdao Railway was approximately 6.4 Hz. This is higher than the specified limits and the threshold for dynamic calculations in the *Code for Design of High-Speed Railway* (*TB 10621-2014*), with a sufficient margin.

### 5.3.3 Analysis of Vehicle Vibration Testing Results

1. **Vehicle Vibration Acceleration**

A CRH380B double-formation EMU was tested in a typical section between Jiaozhou North and Weifang North. According to the *High-Speed Railway Ballastless Track Maintenance Rules* (Railway Operations [2012] No. 83), vertical vibration acceleration was low-pass filtered at 20 Hz, and lateral vibration acceleration was band-pass filtered in the range of 0.5–10 Hz.

(1) Jiaozhou North to Weifang North Direction (Upward)

During the upward operation of the CRH380B double-formation EMU from Jiaozhou North to Weifang North, the vibration acceleration at various measurement points on the carriage floor, along with the track condition and operating speed, are shown in Fig. 5.53.

From Fig. 5.53, the maximum acceleration values at various measurement points in the 11th carriage during the upward operation of the CRH380B double-formation EMU from Jiaozhou North to Weifang North are as follows:

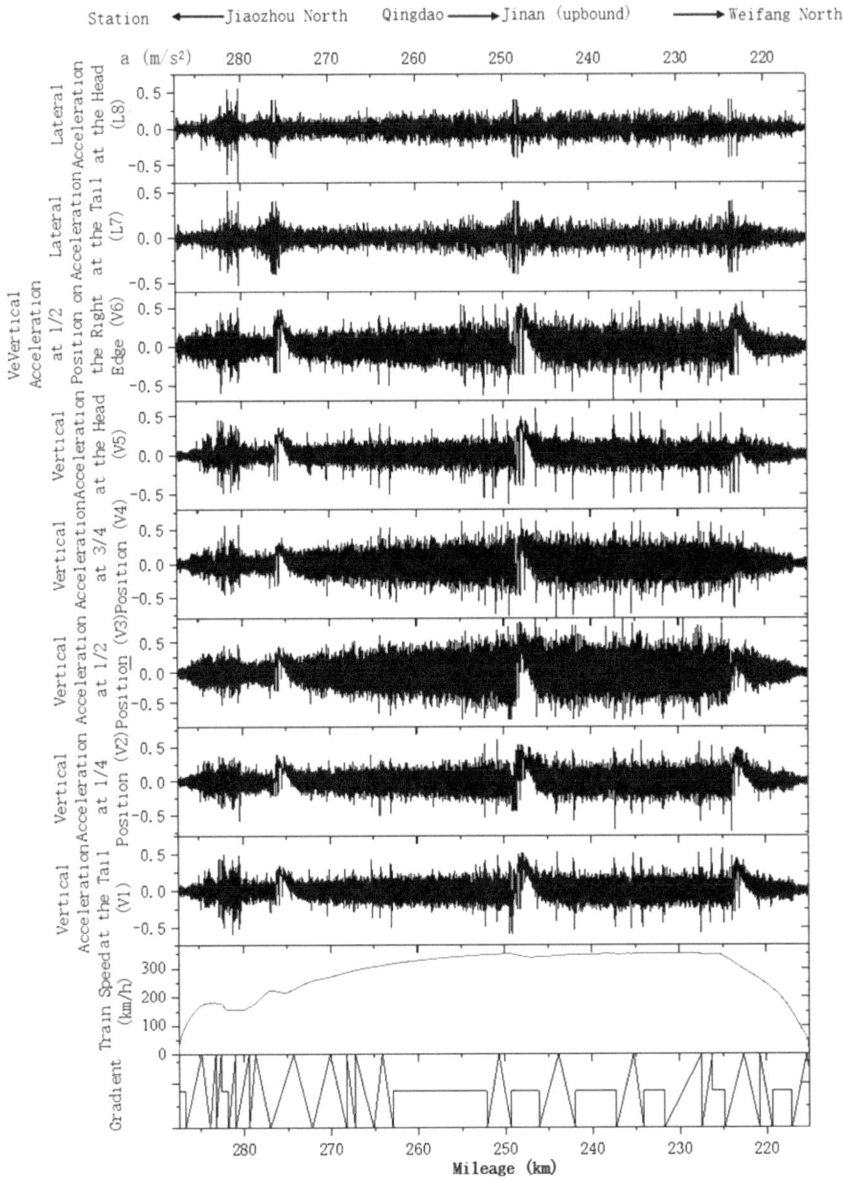


**Fig. 5.53** Vehicle vibration acceleration and section information (upward)

- Vertical acceleration at the rear end of the longitudinal centerline (V1): 0.5994 $m/s^2$;
- Vertical acceleration at the 1/4 point of the longitudinal centerline (V2): 0.7374 $m/s^2$;
- Vertical acceleration at the midspan of the longitudinal centerline (V3): 0.9750 $m/s^2$;
- Vertical acceleration at the 3/4 point of the longitudinal centerline (V4): 0.7995 $m/s^2$;
- Vertical acceleration at the front end of the longitudinal centerline (V5): 0.6364 $m/s^2$;
- Vertical acceleration at the edge of the midspan (V6): 0.7208 $m/s^2$;
- Lateral acceleration at the rear end of the longitudinal centerline (L7): 0.5325 $m/s^2$;
- Lateral acceleration at the front end of the longitudinal centerline (L8): 0.7939 $m/s^2$.

(2) Weifang North to Jiaozhou North Direction (Downward)

During the downward operation of the CRH380B double-formation EMU from Weifang North to Jiaozhou North, the vibration acceleration at various measurement points on the carriage floor, along with the track condition and operating speed, are shown in Fig. 5.54.

From Fig. 5.54, the maximum acceleration values at various measuring points inside the 11th carriage of the CRH380B coupled EMU running from Weifang North to Jiaozhou North (downward direction) are as follows:

- The vertical acceleration at the rear end of the carriage floor along the longitudinal centerline (V1) was 0.7999 $m/s^2$;
- The vertical acceleration at the 1/4 point of the carriage floor along the longitudinal centerline (V2) was 0.6582 $m/s^2$;
- The vertical acceleration at the center of the carriage floor along the longitudinal centerline (V3) was 0.7982 $m/s^2$;
- The vertical acceleration at the 3/4 point of the carriage floor along the longitudinal centerline (V4) was 0.7813 $m/s^2$;
- The vertical acceleration at the front end of the carriage floor along the longitudinal centerline (V5) was 0.6963 $m/s^2$;
- The vertical acceleration at the center edge of the carriage floor (V6) was 0.7012 $m/s^2$;
- The lateral acceleration at the rear end of the carriage floor along the longitudinal centerline (L7) was 0.4500 $m/s^2$;
- The lateral acceleration at the front end of the carriage floor along the longitudinal centerline (L8) was 0.4991 $m/s^2$.

When the CRH380B coupled EMU runs from Jiaozhou North to Weifang North, the maximum acceleration values at various measuring points in the 11th carriage are summarized in Table 5.7.

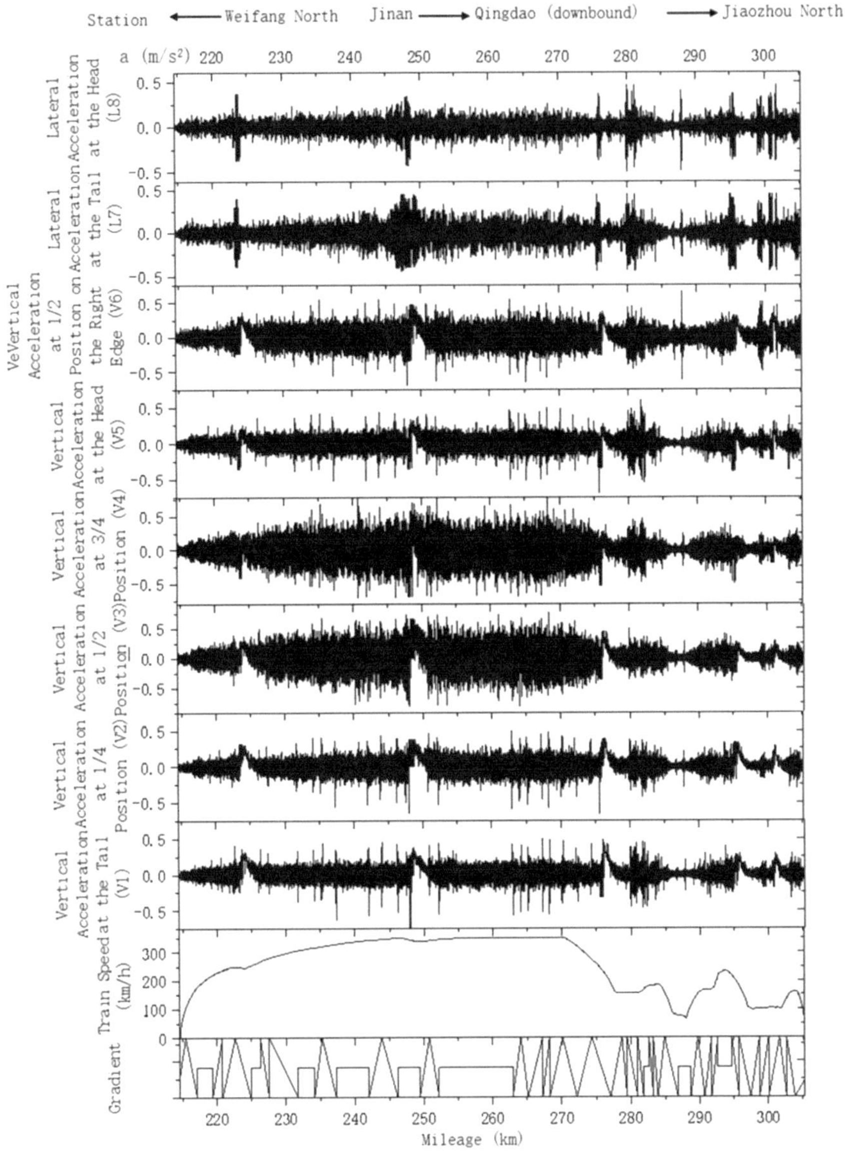


**Fig. 5.54** Car body vibration acceleration and section information (downward direction)

Based on the statistical data in Table 5.7:

1) When the CRH380B coupled EMU runs from Jiaozhou North to Weifang North, the maximum vertical acceleration of the carriage is 0.9750 m/s$^2$, meeting the Grade I standard for vertical acceleration of car bodies in the *Maintenance Rules*

**Table 5.7** Testing results of carriage floor acceleration (m/s$^2$)

| Measuring point location | Jiaozhou North to Weifang North (upward direction) | Weifang North to Jiaozhou North (downward direction) |
|---|---|---|
| Rear vertical (V1) | 0.5994 | 0.7999 |
| 1/4 vertical (V2) | 0.7374 | 0.6582 |
| Center vertical (V3) | 0.9750 | 0.7982 |
| 3/4 vertical (V4) | 0.7995 | 0.7813 |
| Front vertical (V5) | 0.6364 | 0.6963 |
| Center edge vertical (V6) | 0.7208 | 0.7012 |
| Rear lateral (L7) | 0.5325 | 0.4500 |
| Front lateral (L8) | 0.7939 | 0.4991 |

*for Ballastless Track Lines of High-Speed Railways* (Tieyun [2012] No. 83), which represents the optimal level.

2) When the CRH380B coupled EMU runs from Jiaozhou North to Weifang North, the maximum lateral acceleration of the carriage is 0.7939 m/s$^2$, meeting the Grade II standard for lateral acceleration of car bodies in the *Maintenance Rules for Ballastless Track Lines of High-Speed Railways* (Tieyun [2012] No. 83). However, in most sections, it is less than 0.6 m/s$^2$, meeting the Grade I standard.

2. **Influence of Vertical Transition Points on Train Vibration**

When passengers travel on high-speed EMUs, they may sometimes experience sensations of "weightlessness" or "overweight," as well as a "sudden up-and-down" feeling. Investigations have shown that these phenomena mainly occur at vertical transition points located at the top or bottom of slopes. To investigate the effect of vertical transition points on car body vibration during the operation of the CRH380B coupled EMU, the track section between K233+500 and K238+000 was selected for analysis. This section includes three vertical transition points located at K234+487, K235+286, and K237+286, with corresponding algebraic grade differences of 3.5‰, 5.5‰, and 2‰, respectively. During the operation of the CRH380B coupled EMU, the vibration acceleration at various measuring points on the carriage floor, the track conditions, and the operating speed are shown in Figs. 5.55 and 5.56.

From Figs. 5.55 and 5.56, the following conclusions can be drawn:

(1) When the train passes through a vertical transition point, the vertical acceleration at various measuring points on the carriage floor exhibits significant fluctuation, with the central part of the carriage being more affected than other points. This fluctuation dissipates within approximately 500 m after the train passes the transition point.
(2) When the train passes through a vertical transition point, the impact on lateral vibration is less significant compared to vertical vibration. This indicates that vertical transition points primarily affect vertical vibrations of the vehicle, thus influencing passenger comfort to a certain extent.

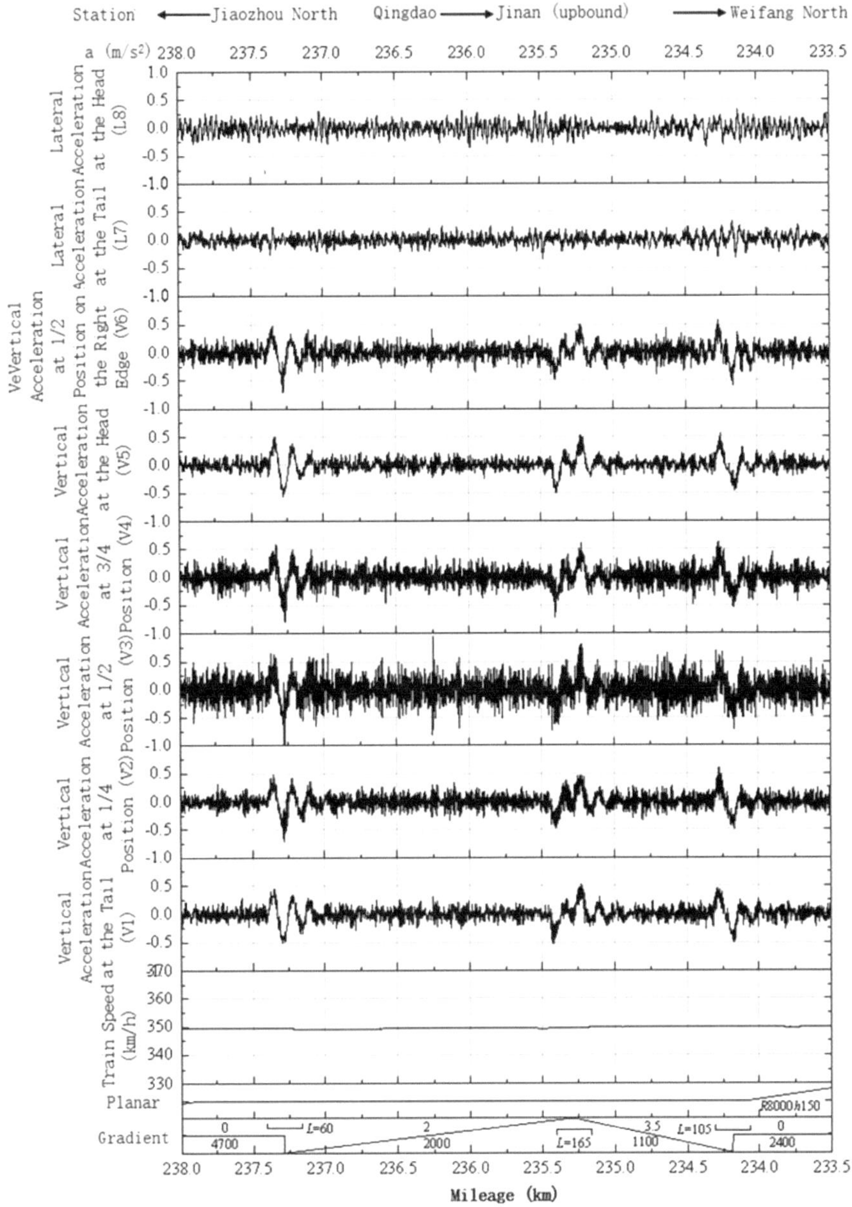


**Fig. 5.55** Influence of vertical transition points on car body vibration acceleration (upward direction)

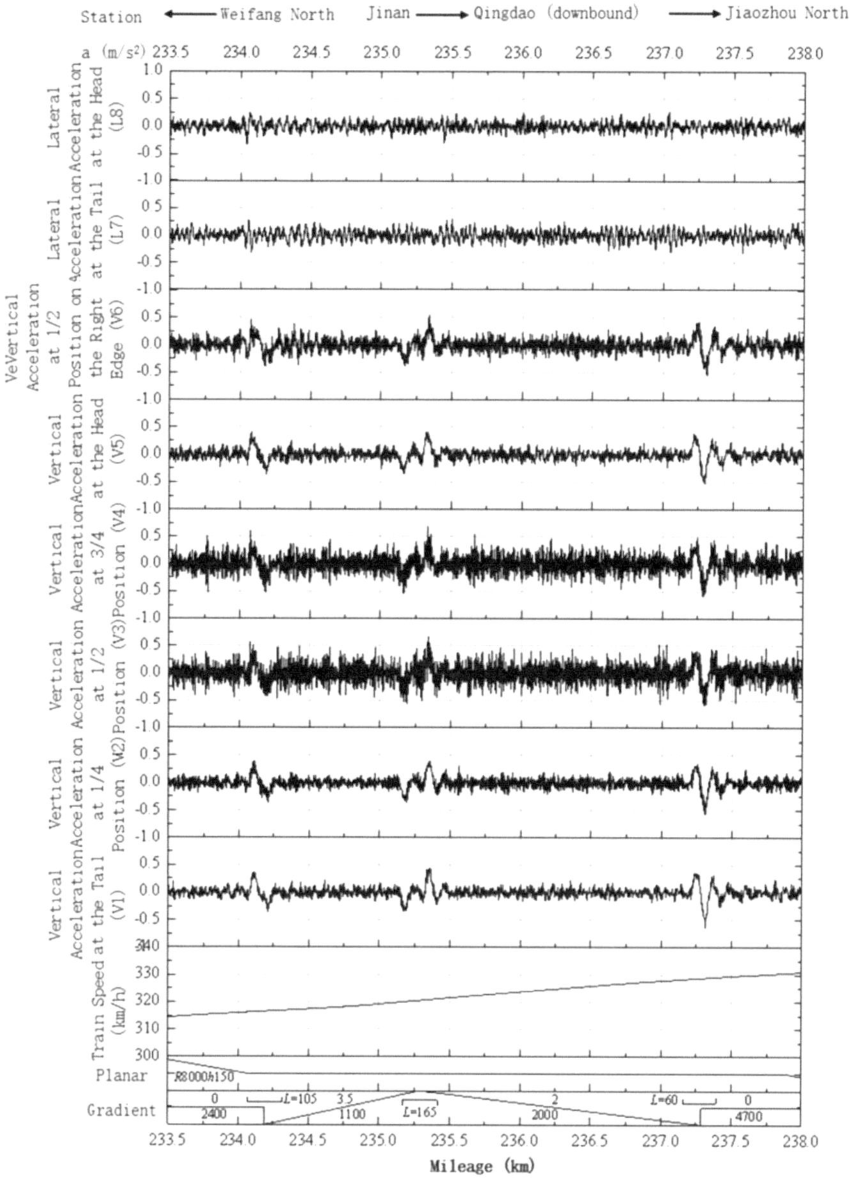


**Fig. 5.56** Influence of vertical transition points on car body vibration acceleration (downward direction)

3. **Acceleration Variation Patterns on Plane Curves**

The dynamic performance of a vehicle is not only related to its intrinsic parameters but also strongly influenced by track conditions. To study the vibration behavior of the car body as the train passes through curves at different speeds, a plane curve with a radius of 8000 m located on a bridge was selected for analysis. The curve was designed with a superelevation of 135 mm and a transition curve length of 590 m. At the center of the curve, the maximum operating speed of the train was 342 km/h, with a cant deficiency of 38 mm, while the minimum speed was 235 km/h, with a cant excess of 56 mm. The acceleration response data of the car body at different speeds while passing through the curve were processed using filtering methods: vertical acceleration was processed with a 20 Hz low-pass filter, and lateral acceleration was processed with a 0.5–10 Hz band-pass filter. The relationships between the peak vertical and lateral accelerations of the car body at different speeds are shown in Figs. 5.57 and 5.58.

From Figs. 5.57 and 5.58, the following conclusions can be drawn:

(1) When the curve radius is relatively large (e.g., R = 8000 m in this case), the cant deficiency caused by speed variation has little effect on the peak vertical acceleration of the car body. The peak vertical acceleration of the car body increases with speed. The influence of the curve radius on the vertical acceleration is not significant when the train passes the curve at different speeds.

(2) The lateral acceleration of the car body shows slight variation when the train passes the curve at different speeds. The minimum peak lateral acceleration occurs at approximately 302 km/h, where the cant deficiency is smallest. When the operating speed deviates from the equilibrium speed corresponding to the

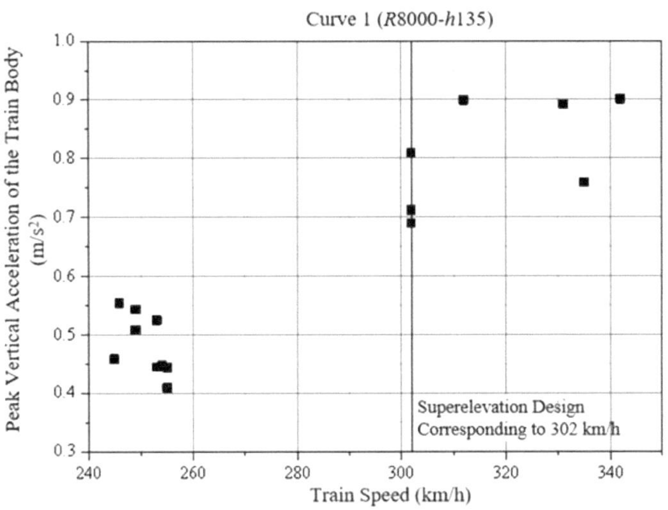


**Fig. 5.57** Comparison of vertical accelerations at different speeds for an 8000 m radius curve

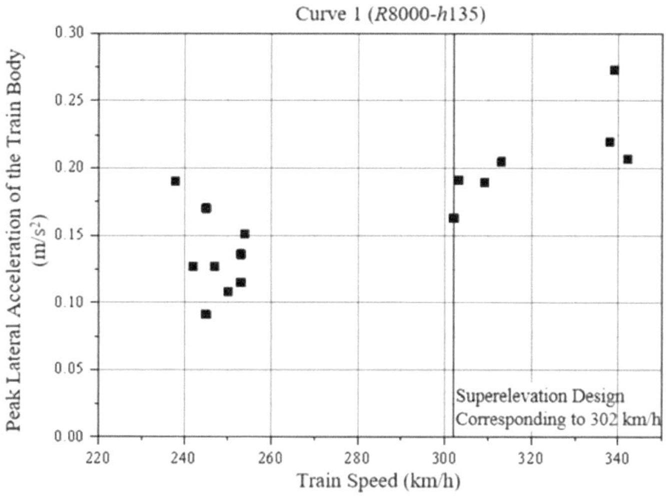


**Fig. 5.58** Comparison of lateral accelerations at different speeds for an 8000 m radius curve

designed superelevation, the lateral acceleration increases to some extent but remains well below the Grade I standard for lateral acceleration of car bodies as specified in the *Maintenance Rules for Ballastless Track Lines of High-Speed Railways* (Tieyun [2012] No. 83), with a significant margin.

4. **Evaluation of Train Ride Stability and Comfort**

Representative track sections were selected, and the ride comfort of the CRH380B coupled EMU was evaluated using the Sperling ride quality evaluation method and the ISO-2631 comfort evaluation method for both upward and downward directions. The processed testing data yielded the Sperling ride quality index at various measuring points in the carriages of the CRH380B coupled EMU, as shown in Table 5.8.

**Table 5.8** Train sperling ride quality index

| Measuring point | Jiaozhou North to Weifang North (upward direction) | Weifang North to Jiaozhou North (downward direction) |
|---|---|---|
| Rear vertical (V1) | 1.3638 | 1.3186 |
| 1/4 vertical (V2) | 1.3726 | 1.3259 |
| Center vertical (V3) | 1.3421 | 1.2849 |
| 3/4 vertical (V4) | 1.4467 | 1.2151 |
| Front vertical (V5) | 1.5183 | 1.3579 |
| Center edge vertical (V6) | 1.7368 | 1.6881 |
| Rear lateral (L7) | 1.8396 | 1.806 |
| Front lateral (L8) | 1.7056 | 1.6747 |

As shown in Table 5.8, the maximum Sperling ride quality index for the vertical and lateral directions of the CRH380B coupled EMU while operating from Jiaozhou North to Weifang North were 1.7368 and 1.8396, respectively. Both values are below the Grade I stability threshold of 2.5, corresponding to an evaluation rating of "Excellent."

By processing the train testing data using the ISO-2631 evaluation method, the ISO-2631 comfort indices at various measuring points in the carriages of the CRH380B coupled EMU were obtained, as shown in Table 5.9.

As shown in Table 5.9, the maximum vertical and lateral ISO-2631 comfort indices for the CRH380B coupled EMU while operating from Jiaozhou North to Weifang North were 0.1456 and 0.1835, respectively. Both values are below the Grade I comfort threshold of 0.315, corresponding to an evaluation rating of "Comfortable."

5. **Summary**

Based on the vibration testing and analysis conducted for the 9th and 11th carriages of the CRH380B coupled EMU, the following conclusions were drawn:

(1) During the operation of the CRH380B coupled EMU on the section between Jiaozhou North and Weifang North, the maximum vertical acceleration of the carriage was 0.9750 m/s$^2$, meeting the Grade I standard for vertical acceleration of car bodies specified in the *Maintenance Rules for Ballastless Track Lines of High-Speed Railways* (Tieyun [2012] No. 83), which corresponds to the optimal level. The maximum lateral acceleration of the carriage was 0.7939 m/s$^2$, meeting the Grade II standard, though in most sections it was below 0.6 m/s$^2$, satisfying the Grade I standard.

(2) When the train passes through an 8000 m radius curve at different speeds, the vertical acceleration of the car body is not significantly affected by the curve radius. However, the lateral acceleration of the car body reaches a minimum at around 302 km/h, where the cant deficiency is smallest. If the train's speed deviates from the equilibrium speed corresponding to the designed superelevation, the lateral vibration acceleration increases slightly but remains well below

**Table 5.9** ISO-2631 comfort indices

| Measuring point | Jiaozhou North to Weifang North (upward direction) | Weifang North to Jiaozhou North (downward direction) |
|---|---|---|
| Rear vertical (V1) | 0.0979 | 0.0823 |
| 1/4 vertical (V2) | 0.1157 | 0.0907 |
| Center vertical (V3) | 0.1382 | 0.1113 |
| 3/4 vertical (V4) | 0.1247 | 0.1048 |
| Front vertical (V5) | 0.1199 | 0.1049 |
| Center edge vertical (V6) | 0.1456 | 0.1394 |
| Rear lateral (L7) | 0.1775 | 0.1835 |
| Front lateral (L8) | 0.1073 | 0.1067 |

the Grade I standard, with a significant margin, as specified in the *Maintenance Rules for Ballastless Track Lines of High-Speed Railways* (Tieyun [2012] No. 83).

(3) When the train passes through a vertical transition point, the vertical acceleration at various measuring points on the carriage floor exhibits significant fluctuations. Compared to vertical vibrations, the lateral vibrations of the car body are less affected by the transition points. This indicates that vertical transition points in the longitudinal profile primarily influence vertical vibrations, thereby affecting passenger comfort to some extent.

(4) During the operation of the CRH380B coupled EMU on the section between Jiaozhou North and Weifang North, both the vertical and lateral Sperling ride quality index were below the Grade I threshold of 2.5, corresponding to an evaluation rating of "Excellent." Similarly, the vertical and lateral ISO-2631 comfort indices were below the Grade I threshold of 0.315, corresponding to an evaluation rating of "Comfortable."

# Chapter 6
# Dynamic Analysis of Track Horizontal and Vertical Alignment Parameters

Traditional research on railway track parameters is based on vehicle dynamics theory, neglecting the dynamic interaction between vehicles and track structures. As the operation speed of high-speed railways increases, the system response induced by track geometry becomes significantly stronger, and the relationship between track geometry and passenger comfort becomes more important. The limitations of traditional track research methods are becoming more evident, mainly in the following aspects: (1) inability to analyze wheel-rail dynamic safety indices and ride comfort indices when trains pass through curves; (2) inability to analyze and evaluate the matching of horizontal and vertical alignments and the mutual influence between successive alignments; (3) inability to separately consider the influence of different subgrade structures. Therefore, applying railway track system dynamics theory to study the reasonable selection of track alignment parameters, as well as evaluating and optimizing track design schemes, has become a more scientific approach. This chapter introduces the application of track system dynamics theory in the study of key horizontal and vertical alignment parameters of high-speed railway tracks, the matching of horizontal and vertical profile parameters, and the applicability of parameters for high-speed railway speed increases.

## 6.1 Horizontal Alignment Parameter Analysis

### *6.1.1 Analysis of Minimum Radius of Circular Curves*

1. **Minimum Curve Radius Analysis**

The minimum radius of a horizontal curve is one of the most important design standards for high-speed railways. It mainly depends on the transportation mode, target speed, passenger comfort, and other factors.

© China Railway Publishing House Co., Ltd. 2026

X. Long et al., *Vehicle–Track–Bridge System Subjected to Track Alignment*,
https://doi.org/10.1007/978-981-95-1649-0_6

The minimum curve radius for a single high-speed train line is mainly determined by the train's maximum operating speed, the sum of the maximum cant and cant deficiency allowance, and other factors.

$$R_{\min} = \frac{11.8\, V_{\max}^2}{[h + h_q]} \tag{6.1.1}$$

The primary characteristic of mixed medium- and high-speed transportation is the operation of trains at different speeds on the same track. The minimum curve radius is primarily determined by factors such as the highest and lowest speeds of the passing trains, and the sum of the permissible cant deficiency and cant overrun.

$$R_{\min} = 11.8 \frac{V_{\max}^2 - V_{\min}^2}{[h_q + h_g]} \tag{6.1.2}$$

From the above formula, it can be seen that the minimum curve radius is closely related to the maximum cant, cant deficiency, and cant excess permissible values. Table 6.1 presents the corresponding standard indicators for high-speed railways in different countries.

Table 6.2 presents the calculated values of the minimum curve radius and corresponding limit values under different speed conditions based on the maximum cant, cant deficiency, and cant excess standards of China's high-speed rail.

2. **Analysis of the Impact of Minimum Curve Radius on Driving Dynamics Performance**

The curve radius is the core parameter of the track plan and has a significant impact on ride comfort. This study performs simulation calculations for a section of approximately 30 km of track. The main technical parameters include a running speed of 350 km/h, with the track plan composed of 14 segments of curves with consistent parameters, a tangent length of 500 m, and a transition curve length of 590 m; the vertical profile is a flat slope, using the German high-speed railway low-disturbance spectrum as the track disturbance source, including track high-low, track alignment, and horizontal irregularities. The curve radii and cant are set as follows: 7000 m and below, 175 mm; 8000 m, 150 mm; 9000 m, 135 mm; 10,000 m, 120 mm; 11,000 m, 105 mm; 12,000 m, 95 mm.

Table 6.3 presents a comparison of the ride comfort indicators for each track scheme using different curve radii. Figure 6.1 shows the corresponding curves for each indicator as the curve radius changes. Overall, when the curve radius is less than 7000 m, the lateral ride comfort indicator decreases as the radius increases. When the curve radius exceeds 7000 m, the lateral ride comfort changes little, and the effect of curve radius on vertical ride comfort is minimal. Furthermore, when the curve radius is 4500 m, the maximum lateral acceleration of the vehicle reaches 1.1 m/s$^2$, and when the curve radius is 4000 m, the lateral comfort level indicator RCLy exceeds 83, surpassing the first-level standard of "very comfortable."

**Table 6.1** Maximum cant, cant deficiency, and cant excess permissible values

| Category | Maximum cant allowable value | Defficient cant allowable value | Excessive cant allowable value |
|---|---|---|---|
| European standard EN 13803-1 (2010) | Self-determined limit value:<br>160 mm<br>Approved allowable value:<br>180 mm<br>200 mm (High-speed railway without freight transport) | Self-determined limit value:<br>$200 < V \leq 250$ km/h: 100 mm<br>$250 < V \leq 300$ km/h: 80 mm<br>Approved allowable value:<br>$200 < V \leq 250$ km/h: 150 mm<br>$250 < V \leq 300$ km/h: 130 mm | Self-determined limit value: 110 mm<br>Approved allowable value: 130 mm |
| Germany | Self-determined limit value:<br>160 mm (Ballasted)<br>170 mm (Ballastless)<br>Approved allowable value:<br>180 mm | Self-determined limit value: 130–150 mm | |
| France | 180 mm | $V \leq 300$ km/h: 153 mm<br>$V > 300$ km/h: 100 mm | 110 mm |
| Japan | 180 mm<br>(Only Tokaido Shinkansen 200 mm) | 60–110 mm (Tokaido Shinkansen 110 mm, Sanyo Shinkansen 90 mm, 300 Series 110 mm) | |
| China high-speed railway standards | 170 mm (Ballasted)<br>175 mm (Ballastless) | 40 mm (Excellent)<br>60 mm (Good)<br>90 mm (Average) | 40 mm (Excellent)<br>60 mm (Good)<br>90 mm (Average) |
| China intercity railway standards | 150 mm | 40 mm (Excellent)<br>80 mm (Good)<br>110 mm (Average) | 40 mm (Excellent)<br>80 mm (Good)<br>110 mm (Average) |

Figures 6.2 and 6.3 show the frequency distribution and cumulative distribution of lateral acceleration for the vehicle at different curve radii. From Fig. 6.2, it can be seen that when the curve radius is 6000 m, the lateral acceleration of the vehicle in the range of 0.4–0.5 m/s$^2$ accounts for nearly 30%; when the radius exceeds 7000 m, there are no components greater than 0.3 m/s$^2$. By comparing the results for curve radii of 7000 m, 8000 m, and 10,000 m, it is clear that the larger the radius, the more the acceleration distribution shifts toward smaller values. When the curve radii are 7000 m and 8000 m, the components in the range of 0.2–0.3 m/s$^2$ account for more than 20%, while for a curve radius of 10,000 m, this component is less than 5%, which is significantly better than the acceleration distribution for smaller radii.

**Table 6.2** The calculation of the minimum curve radius

| Design speed of the road section (km/h) | | Speed (km/h) | | | |
|---|---|---|---|---|---|
| | | 350 | 300 | 250 | 200 |
| $H_{max}$ (mm) | | 175 | 175 | 175 | 175 |
| $[h + h_q]$ (mm) | General | 210 | 210 | 210 | 210 |
| | Difficult | 265 | 265 | 265 | 265 |
| Minimum curve radius | General | 6883.3 | 5057.1 | 3511.9 | 2247.6 |
| | Difficult | 5454.7 | 4007.5 | 2783.0 | 1781.1 |
| Recommended value | General | 7000 | 5000 | 3200 | 2200 |
| | Difficult | 5500 | 4000 | 2800 | 2000 |

**Table 6.3** The impact of circular curve radius on ride comfort indices of railway line schemes

| $L_y$ (m) | $a_{ymax}$ (m/s$^2$) | $a_{zmax}$ (m/s$^2$) | $a_{yrms}$ (m/s$^2$) | $a_{zrms}$ (m/s$^2$) | $W_y$ | $W_z$ | $N_{mvy}$ | $N_{mvz}$ | $RCL_y$ | $RCL_z$ |
|---|---|---|---|---|---|---|---|---|---|---|
| 4000 | 1.414 | 0.091 | 0.845 | 0.027 | 1.269 | 1.202 | 0.283 | 0.16 | 83.23 | 71.58 |
| 4500 | 1.118 | 0.09 | 0.662 | 0.027 | 1.246 | 1.202 | 0.256 | 0.159 | 81.26 | 71.57 |
| 5000 | 0.883 | 0.089 | 0.398 | 0.027 | 1.226 | 1.201 | 0.225 | 0.158 | 77.37 | 71.55 |
| 5500 | 0.69 | 0.088 | 0.398 | 0.027 | 1.226 | 1.201 | 0.225 | 0.158 | 77.37 | 71.55 |
| 6000 | 0.529 | 0.087 | 0.298 | 0.027 | 1.222 | 1.201 | 0.217 | 0.158 | 75.41 | 71.54 |
| 7000 | 0.277 | 0.086 | 0.143 | 0.027 | 1.219 | 1.201 | 0.21 | 0.157 | 71.51 | 71.53 |
| 8000 | 0.271 | 0.085 | 0.14 | 0.027 | 1.218 | 1.201 | 0.21 | 0.157 | 71.42 | 71.52 |
| 9000 | 0.234 | 0.084 | 0.117 | 0.027 | 1.218 | 1.201 | 0.21 | 0.156 | 70.79 | 71.52 |
| 10,000 | 0.226 | 0.083 | 0.112 | 0.027 | 1.218 | 1.201 | 0.21 | 0.156 | 70.66 | 71.51 |
| 11,000 | 0.24 | 0.083 | 0.121 | 0.027 | 1.218 | 1.201 | 0.21 | 0.156 | 70.89 | 71.51 |
| 12,000 | 0.233 | 0.082 | 0.117 | 0.027 | 1.218 | 1.201 | 0.21 | 0.156 | 70.77 | 71.51 |

However, when using only the indicators in Table 6.3, the differences between the curve radii of 7000 m, 8000 m, and 10,000 m are not immediately apparent. From Fig. 6.3, it can be seen that when the curve radius is 6000 m, the component of lateral vibration acceleration below 0.3 m/s$^2$ only accounts for 60%, while when the curve radius exceeds 7000 m, it accounts for 100%. From both the frequency distribution and cumulative distribution graphs, it can be further concluded that the curve radius should be greater than 7000 m.

In conclusion, the curve radius has an important impact on the vehicle's lateral acceleration and lateral ride comfort. The calculation results show that when the curve radius exceeds 5000 m, the ride comfort indicators meet the corresponding requirements, and when the curve radius increases to 7000 m, the change in the curve radius has little effect on ride comfort indicators. Therefore, for designing a track with a running speed of 350 km/h, the minimum curve radius should be greater than 5000 m, and generally, it should be greater than 7000 m under normal conditions.

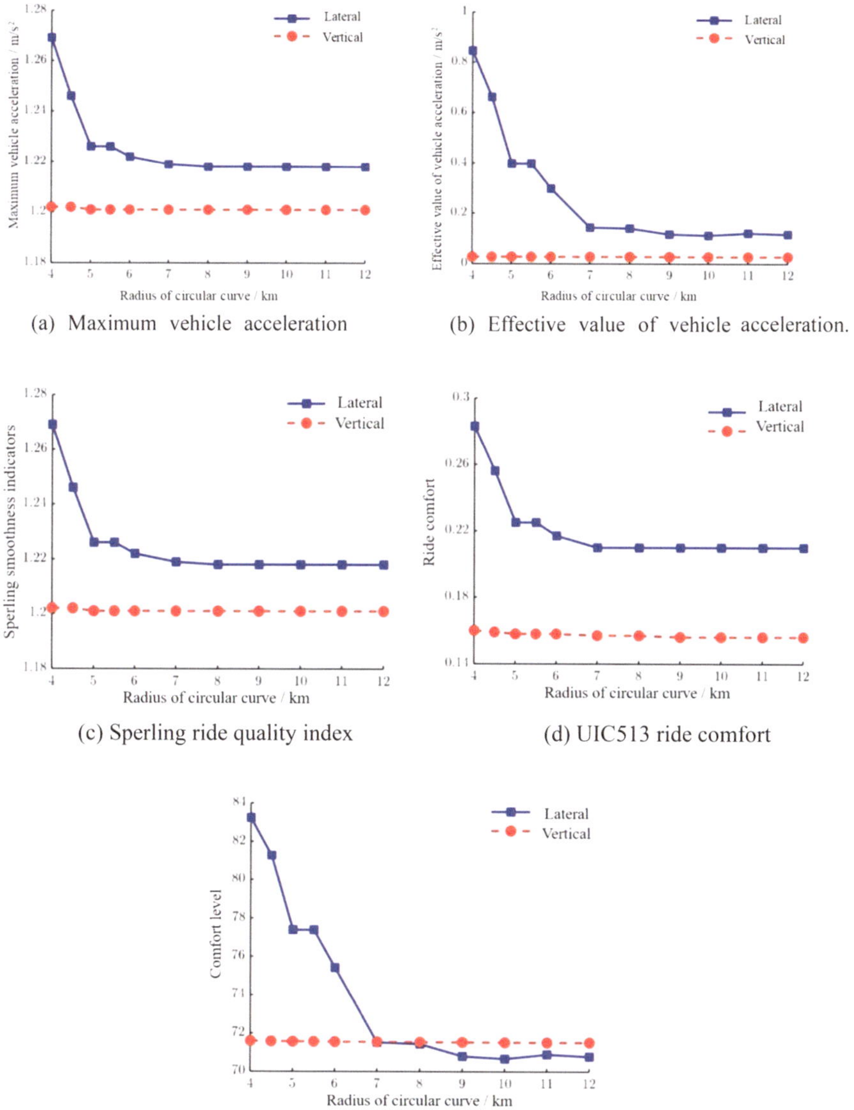


(a) Maximum vehicle acceleration (b) Effective value of vehicle acceleration.

(c) Sperling ride quality index (d) UIC513 ride comfort

(e) comfort level

**Fig. 6.1** The impact of circular curve radius on ride comfort indices of track schemes

## *6.1.2 Minimum Length of Straight Sections Analysis*

1. **Factors Affecting the Minimum Length of Straight Sections**

The design of high-speed railway lines generally prioritizes straight sections to shorten the total length of the route and improve driving conditions. However, to

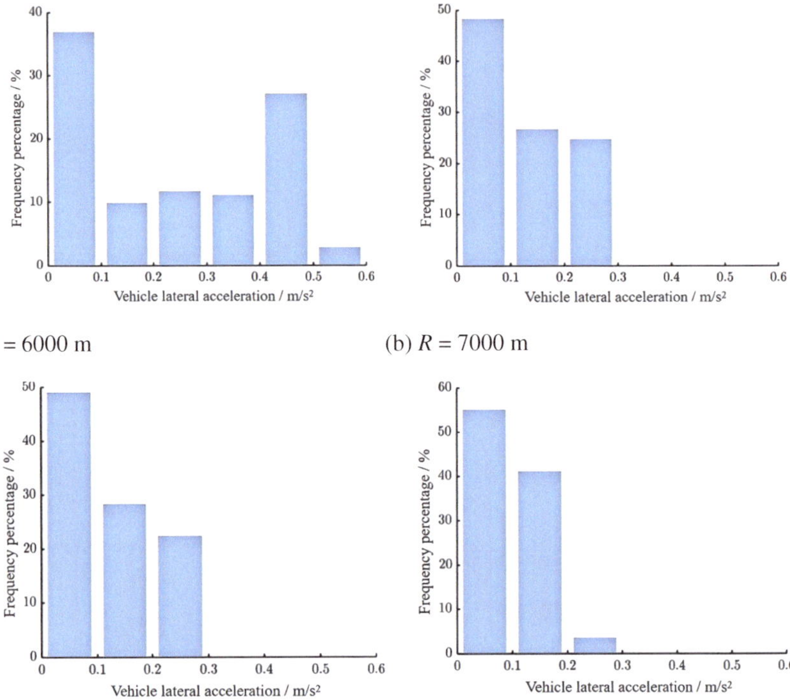


(a) $R = 6000$ m (b) $R = 7000$ m

(c) $R = 8000$ m (d) $R = 10000$ m

**Fig. 6.2** Lateral acceleration frequency distribution of the vehicle body

adapt to terrain, avoid obstacles, and reduce construction costs, circular curves and transition curves need to be incorporated. The minimum length between straight sections and circular curves is primarily controlled by train operation stability and passenger comfort conditions. Additionally, it is constrained by track inspection requirements, mechanized maintenance work, and other limitations.

Theoretically, the minimum length of a straight section or circular curve required for smooth train operation and passenger comfort is generally limited by the avoidance of vibration overlap at the entry and exit points of the transition curve (i.e., at the start and end of the straight section or circular curve). This length is related to the train's vibration and attenuation characteristics, as well as the train's operating speed. The time for the vehicle to pass through the straight section or circular curve, denoted as t, should not be less than the time for vibration attenuation, denoted as $t_z$.

$$t = \frac{L_j}{V_{max}/3.6} \geq t_z \tag{6.1.2}$$

The minimum length requirements for straight sections of high-speed railways in China and abroad are shown in Table 6.4.

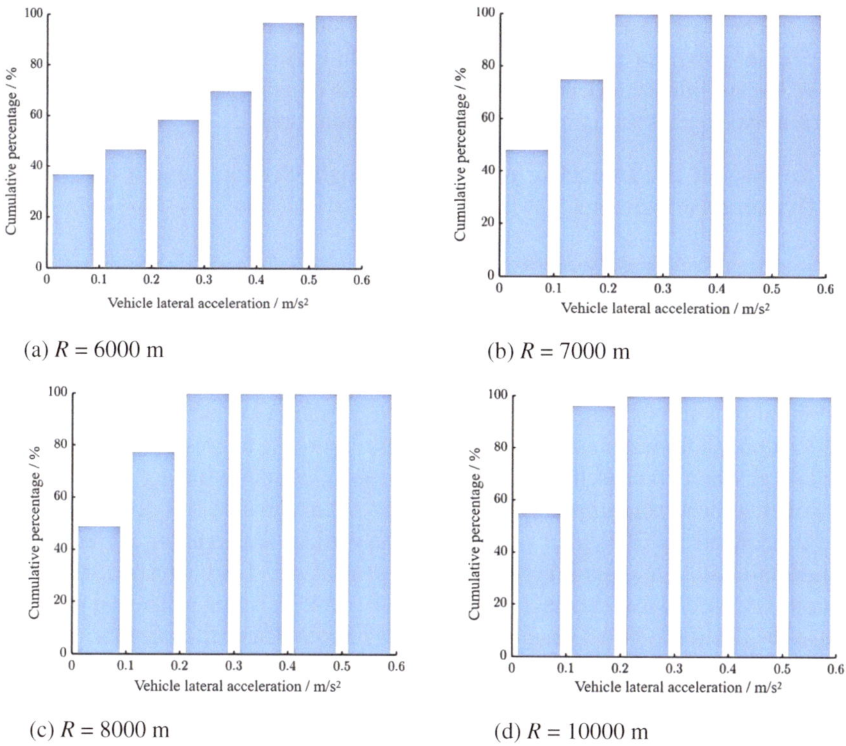


(a) $R = 6000$ m (b) $R = 7000$ m (c) $R = 8000$ m (d) $R = 10000$ m

**Fig. 6.3** Cumulative distribution of lateral acceleration of the vehicle body

**Table 6.4** The minimum length limit of the straight section for high-speed railways

| Category | Minimum length of straight section |
|---|---|
| European standard EN 13803-1 (2010) | General value: 0.67 $V$<br>Exceptional value: 0.4 $V$<br>Minimum value: 20 m |
| Germany | $\geq$ 200 km/h<br>Under normal conditions: 0.5 $V$<br>Under difficult conditions: 0.3 $V$ |
| France | Under normal conditions: 0.67 $V$<br>Under difficult conditions: 0.5 $V$ |
| Japan | 0.42 $V$ |
| China high-speed railway standards | Under normal conditions: 0.8 $V$<br>Under difficult conditions: 0.6 $V$ |
| China intercity railway standards | Under normal conditions: 0.6 $V$<br>Under difficult conditions: 0.4 $V$ |

Based on the above standards, the minimum length of the straight section and the circular curve for high-speed railways should be within the range of 0.4 *V* to 0.8 *V*. The specific value depends on factors such as vehicle vibration and attenuation characteristics, track structure, and the target running speed.

2. **Analysis of the Impact of Minimum Straight Section Length on Driving Dynamic Performance**

(1) Straight Section Between Reverse Curves

Table 6.5 presents the simulation analysis of the track matching parameters for different target speed values, where the straight section length between reverse curves is 0 m, 100 m, 0.4 *V*, and 0.6 *V*, without considering the impact of track irregularities.

Figures 6.4 and 6.5, and Table 6.6 show the vibration attenuation in the straight section under different running speed conditions. From the figures, it can be seen that the lateral vibration caused by the transition-to-straight point impact is very small. The lateral acceleration reaches its peak after 1/4 of the side roll period (0.81 s), and it attenuate by 50% after one cycle. The vertical acceleration caused by the transition-to-straight point impact reaches its peak after 1/4 of the heaving vibration period (1.45 s), and it attenuates by 98% after one cycle. It is evident that the vibration caused by the transition-to-straight point is almost completely attenuated within 1.25 vibration cycles. The lateral vibration attenuation length is 112.5 m (1.01 s), and the vertical vibration attenuation length is 200 m (1.81 s).

The sudden change in cant and curvature at the transition-to-straight point is prone to causing excessive body roll angular velocity and vertical acceleration. Eliminating the straight section and directly connecting the transition curve can instead achieve better curve smoothing effects. Therefore, eliminating the straight section results in smaller body roll angular velocity and vertical acceleration responses. From this perspective, the straight section between reverse curves can be omitted.

In actual train operations, setting a straight section between curves helps to increase the lateral resistance of reverse curves and reduce track diseases. After setting the straight section, in order to avoid excessive frequency of changes in body roll angular velocity, the length of the straight section should meet a certain minimum length requirement. From the vibration attenuation perspective, a straight section length of 0.4 *V* can achieve effective attenuation.

Considering the overall dynamic simulation results and actual track conditions, it is recommended that the straight section between reverse curves may be omitted under difficult conditions. Under general conditions, the straight section length should not be less than 0.4 *V*.

**Table 6.5** Track parameter matching

| Speed (km/h) | Curve radius (m) | Cant (mm) | Transition curve length (m) |
|---|---|---|---|
| 350 | 5500 | 175 | 610 |
| 250 | 2800 | 175 | 440 |

**Fig. 6.4** Dynamic response of the vehicle at a speed of 350 km/h

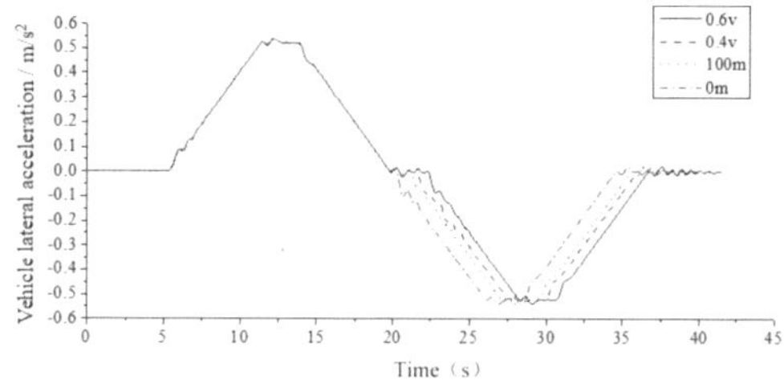


(a) Lateral acceleration of the vehicle

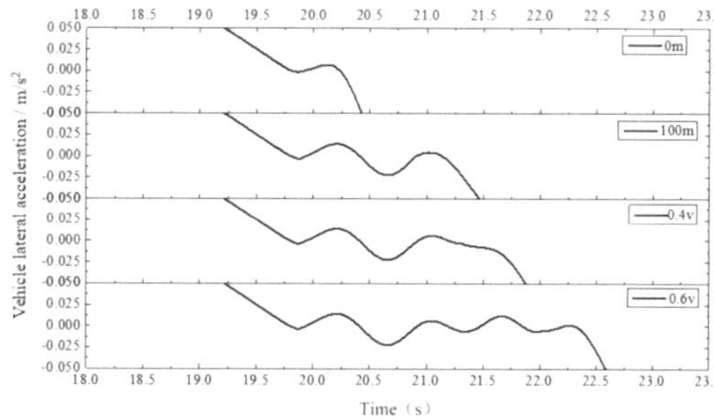


(b) Lateral acceleration attenuation of the tangent line

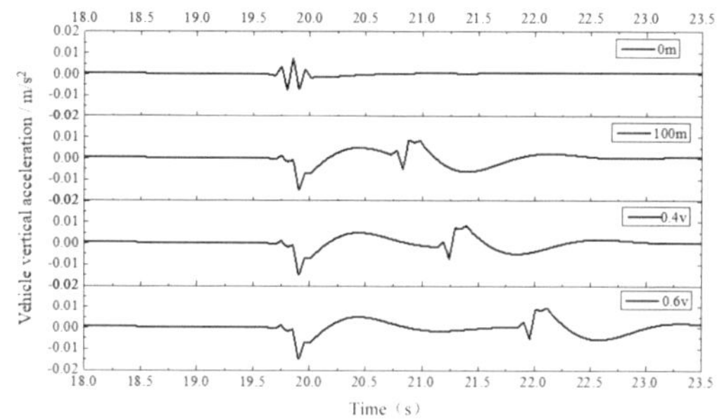


(c) Vertical acceleration attenuation of the tangent line

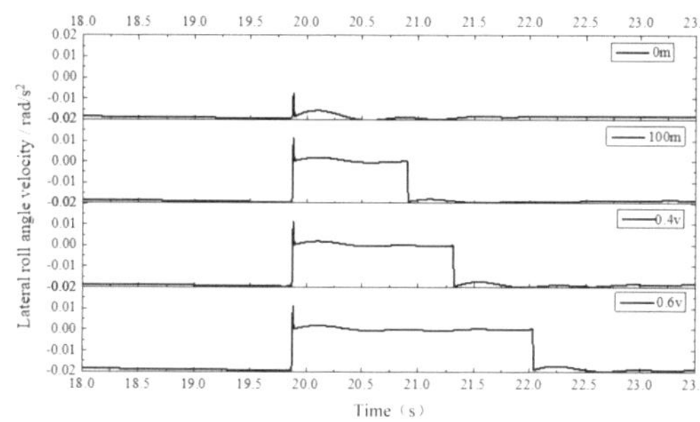


(d) Decay chart of vehicle roll angle velocity on the tangent line

**Fig. 6.5** Dynamic response of the vehicle at a speed of 250 km/h

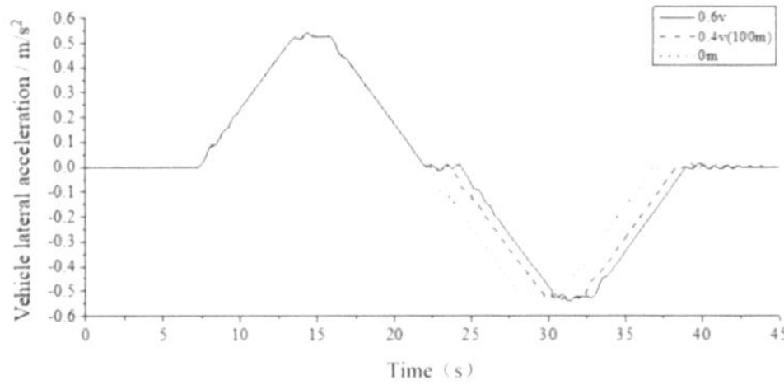


(a) Lateral acceleration of the vehicle

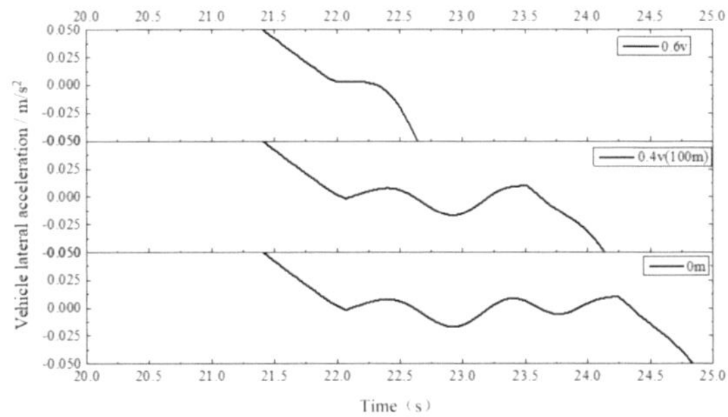


(b) Lateral acceleration attenuation of the tangent line

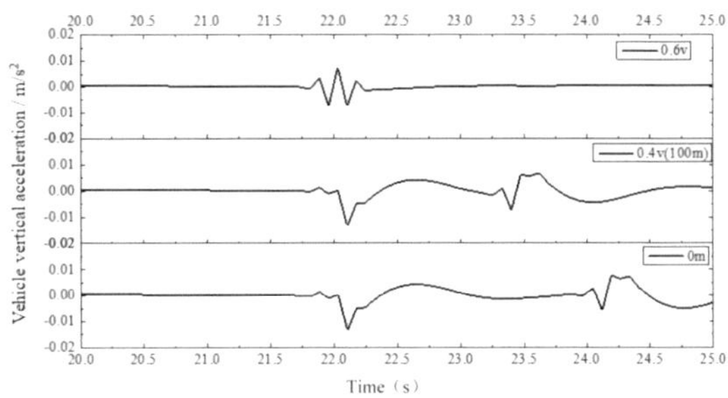


(c) Vertical acceleration attenuation of the tangent line

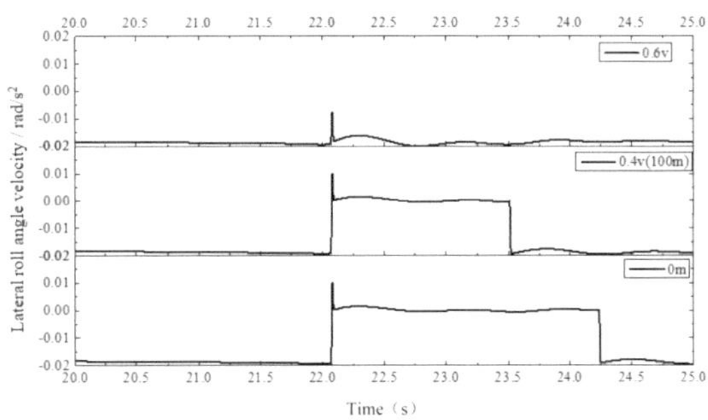


(d) Straight section body roll angular velocity attenuation diagram

**Table 6.6** Straight section vibration attenuation (350 km/h)

| | Lateral acceleration of the vehicle body | Vertical acceleration of the vehicle body |
|---|---|---|
| Main frequency (Hz) | 1.24 | 0.69 |
| Period (s) | 0.81 | 1.45 |
| Transition-to-straight point peak value ($m/s^2$) | 0.0054 | 0.01799 |
| Peak value after 1/4 period ($m/s^2$) | 0.001792 | 0.00617 |
| Decay size after one period (%) | 53 | 98 |
| Decay size after two period (%) | 39 | 100 |
| Decay size after three period (%) | 67 | |

(2) Straight Section Between Same-Direction Curves

The dynamic simulation analysis for the straight section between same-direction curves uses the track parameters in Table 6.5, with straight section lengths of 0 m, 100 m, 0.4 *V*, and 0.6 *V*, ignoring the effect of track irregularities. Figures 6.6 and 6.7 show the vibration attenuation for different speed conditions along the straight section. As seen in the figures, the magnitude of the lateral vibration caused by the transition-to-straight point impact is very small, and the attenuation characteristics are similar to those of the reverse curves. In the straight section between same-direction curves, the direction of the vehicle's roll angle velocity is opposite before and after the transition curves, which has a greater impact on passenger comfort. Setting a certain length of straight section is beneficial to improve comfort.

It is suggested that the minimum tangent line length between same-direction curves can be controlled by the effective attenuation length of 0.4 *V*. Compared to the general condition of 0.8 *V* and the difficult condition of 0.6 *V* required by high-speed railway standards, this can be somewhat relaxed.

(3) Reasonable Minimum Length of the Tangent Line

For the tangent line between reverse curves, removing the tangent line and directly connecting the transition curves can result in better circular smoothness. In actual train operations, setting a tangent line between curves is beneficial for improving lateral resistance on reverse curves and reducing track defects. To prevent frequent changes in side-roll angle velocity, the minimum tangent line length must meet certain length requirements. From the perspective of vibration attenuation, a tangent line length of 0.4 V achieves effective damping. Therefore, based on the comprehensive results of dynamic simulations and actual track conditions, it is recommended that in difficult conditions, tangent lines may not be set between reverse curves, while in general conditions, the length of the tangent line should be no less than 0.4 *V*.

**Fig. 6.6** Dynamic response of the vehicle at a speed of 350 km/h

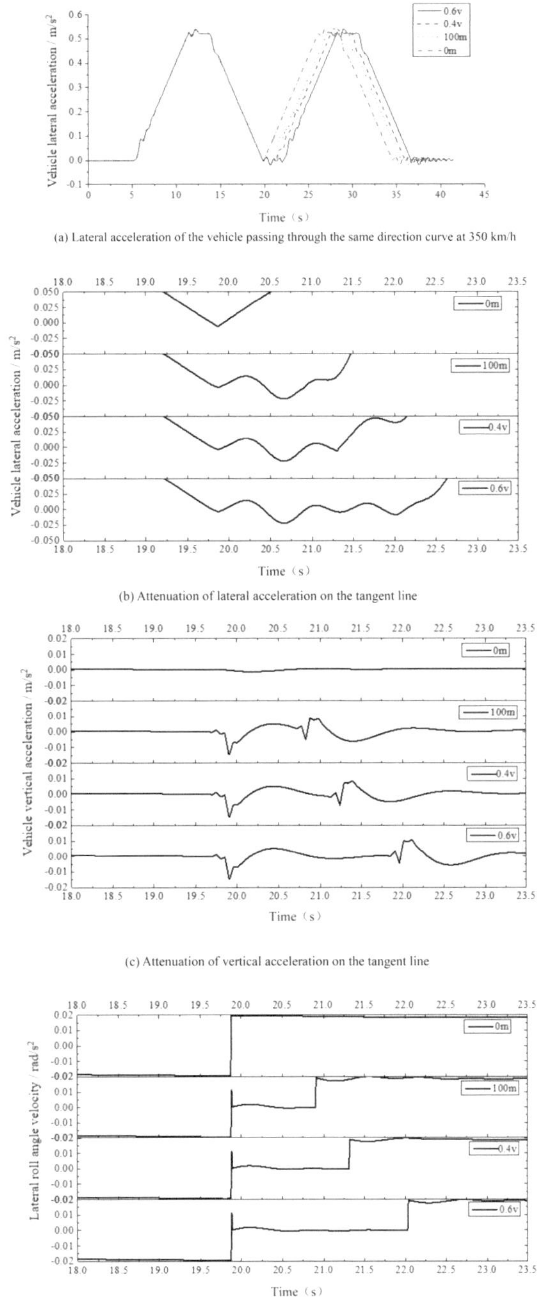


(a) Lateral acceleration of the vehicle passing through the same direction curve at 350 km/h

(b) Attenuation of lateral acceleration on the tangent line

(c) Attenuation of vertical acceleration on the tangent line

(d) Decay chart of vehicle roll angle velocity on the tangent line

**Fig. 6.7** Dynamic response of the vehicle at a speed of 250 km/h

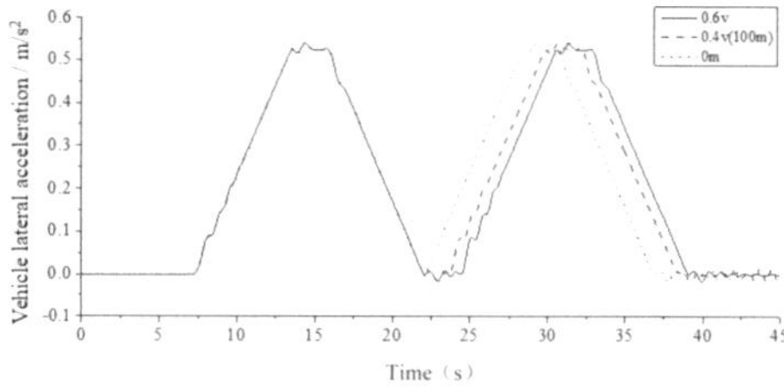


(a) Lateral acceleration of the vehicle on the same direction curve at 250 km/h

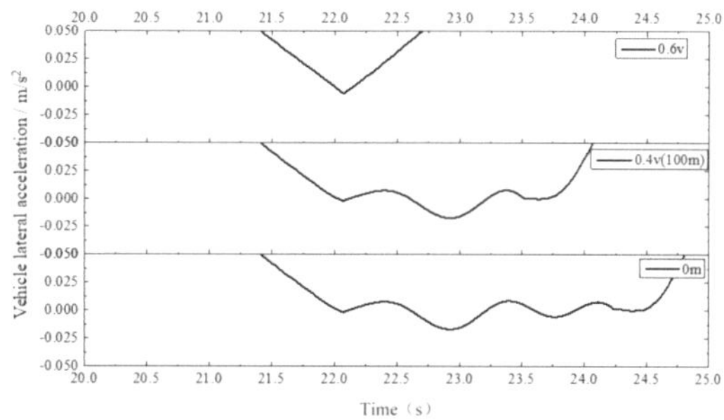


(b) Attenuation of lateral acceleration on the tangent line

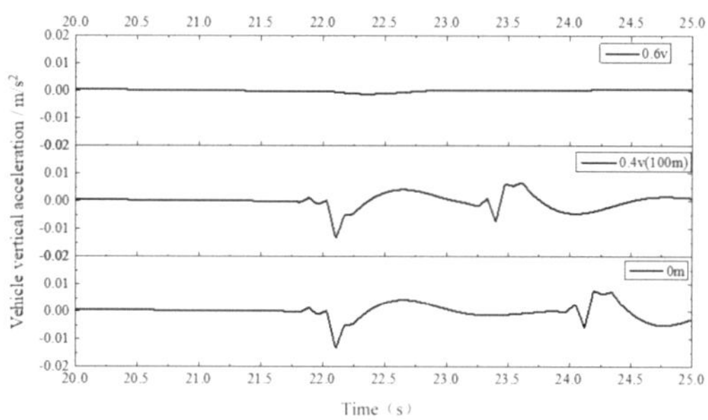


(c) Attenuation of vertical acceleration on the tangent line

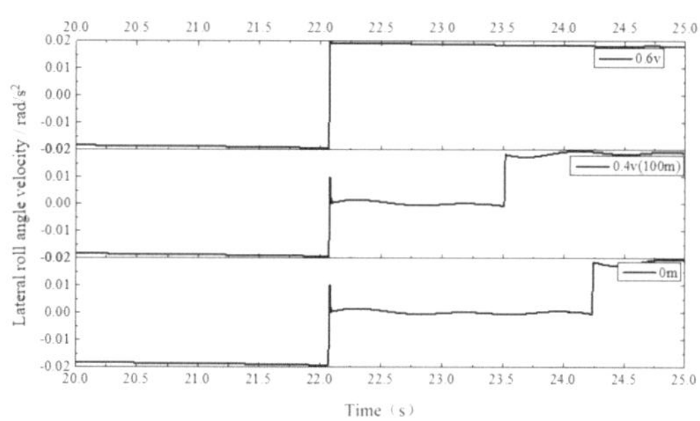


(d) Decay chart of vehicle roll angle velocity on the tangent line

For the tangent line between same-direction curves, the lateral vibration caused by easing point impacts under different speed conditions is very small, and the damping characteristics are similar to those of reverse curves. However, since the side-roll angle velocities of the vehicle before and after the transition curves in the same-direction curves are opposite, the side-roll angle velocity negatively affects passenger comfort. Setting a tangent line of a certain length can improve ride comfort. Therefore, it is recommended that the minimum length of the tangent line between same-direction curves can be controlled by 0.4 $V$.

### 6.1.3 Transition Curve Profile and Length Analysis

1. **Transition Curve Profile Requirements and Types**

In order to ensure that trains transition safely and smoothly between the curvature and cant of a straight line and a circular curve, a certain length of transition curve must be set between the straight line and the circular curve, see Fig. 6.8.

When the train runs on the transition curve, due to the changes in the horizontal curve and vertical gradient, the corresponding driving performance indicators of the train change, thereby affecting passenger comfort. The main indicators include lateral acceleration, lateral impact, yaw rate, yaw acceleration, yaw impact in the horizontal plane, vertical acceleration, vertical impact in the vertical plane, as well as roll rate, roll angle acceleration, and roll angle impact.

When determining the transition curve alignment, the following geometric and mechanical conditions should be met.

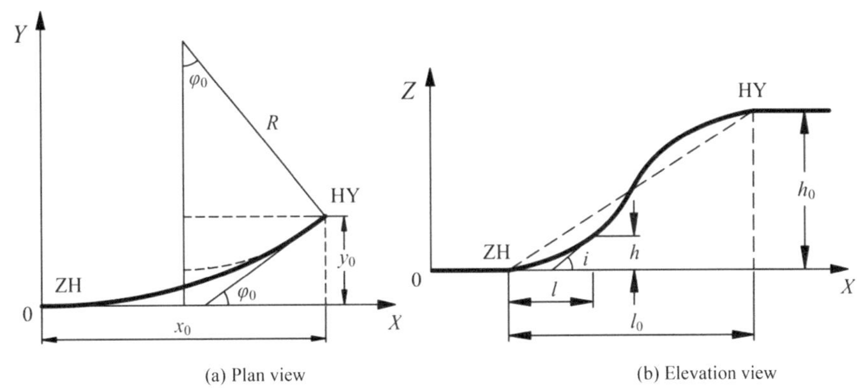


**Fig. 6.8** Schematic diagram of transition curve in plan and elevation view

(1) Coordinate requirements

At the start of the transition curve (ZH), X-coordinate $x = 0$, Y-coordinate $\mathbf{y} = 0$; At the end of the transition curve (HY), $x = x_0$, $y = y_0$, there is continuous change between the two.

(2) Deflection angle requirements

At the start of the transition curve, the deflection angle $\varphi = 0$, and at the end of the transition curve, the deflection angle $\varphi = \varphi_0$, with a continuous change between the two.

(3) Curvature requirements

At the start of the transition curve, the curvature should be $k = 0$, at the end of the transition curve, $k = 1/R$, with continuous variation between the two.

The first three conditions are the basic requirements that a transition curve must meet. Since the variation rule of the cant deficiency is the same as that of curvature, a transition curve that only meets these three conditions is called Transition curve with linear cant gradient, such as the widely used cubic parabolic transition curve. Due to the increase in train speed, in order to improve the dynamic performance of train operation, in addition to meeting the above three conditions, a transition curve generally also needs to meet the following conditions:

(4) At the start and end points of the transition curve $dk/dl = 0$; continuously varying between the start and end points.

This condition ensures that during operation, the train's lateral impact (rate of change of lateral acceleration), cant rate of change, vehicle roll angle velocity, and rotational angle acceleration do not experience sudden changes.

(5) At the start and end points of the transition curve $d^2k/dl^2 = 0$; continuously varying between the start and end points.

This condition ensures that during train operation, vertical acceleration, roll angle acceleration, and rotational angle impact do not experience sudden changes.

(6) At the start and end points of the transition curve $d^3k/dl^3 = 0$; continuously varying between the start and end points.

This condition ensures that during train operation, vertical impact and roll angle impact do not experience sudden changes.

In summary, the transition curve profile should generally meet the geometric and mechanical requirements listed in Table 6.7.

**Table 6.7** Conditions to be met by transition curves

| Serial number | Curve elements | Start point | Intermediate points | End point |
|---|---|---|---|---|
| 1 | Y-coordinate $y$ | 0 | Continuous gradual increase | $y_0$ |
| 2 | Inclination angle $\varphi = y'$ | 0 | | $\varphi_0$ |
| 3 | Curvature $k = y''$ | 0 | | $1/R$ |
| 4 | $k = y'''$ | 0 | Continuous gradual transition | 0 |
| 5 | $k'' = y^{(4)}$ | 0 | | 0 |
| 6 | $k''' = y^{(5)}$ | 0 | | 0 |

Meeting the variation conditions at the curve endpoints (start and end points) pertains to the transition curve profile, while meeting the variation conditions within the curve relates to the transition curve length. From the table, it can be seen that the six elements ($y$, $\varphi$, $k$, $dk/dl$, $d^2k/dl^2$ and $d^3k/dl^3$) of the transition curve are interdependent, and each element is related to the length of the transition curve through differentiation or integration.

The cubic parabola transition curve satisfies only the three conditions mentioned above, with both cant ramping and curvature varying linearly. Due to its simple geometric form, it offers advantages in calculation, surveying, and maintenance while also ensuring a longer effective length, making it widely used. However, as train speeds continue to increase, the cubic parabola transition curve presents angular discontinuities in the vertical profile at its start and end points, leading to sudden changes in vertical acceleration and significant vertical impact when trains pass through. To address this issue and meet the requirements of safe, smooth, and comfortable high-speed operation, transition curves with nonlinear variations in cant ramping and curvature are needed. These curve profiles can be categorized into three types. The first category involves transition curves whose curvature is the first derivative of the transition curve length, with a value of 0 at both the start and end points, that is, $dk/dl = 0$, where $d^2k/dl^2 = 0$. Common examples include the quintic parabolic curve, half-wave sine curve, and S-shaped curve, among others.

The second category involves transition curves where the curvature at the start and end points is $dk/dl = 0$, $d^2k/dl^2 = 0$, while $d^3k/dl^3 = 0$. Common examples include the septic parabolic curve and the full-wave sine curve.

The second category involves transition curves where the curvature at the start and end points is $dk/dl = 0$, $d^2k/dl^2 = 0$, $d^3k/dl^3 = 0$, for example, the nonic parabola, etc.

Several common types of transition curve line shapes are listed in Table 6.8.

Where: $l$ is the distance from any point on the transition curve to the straight-to-curve transition point (m), $l_0$ is the total length of the transition curve (m), $k_0$ is the curvature of the circular curve (1/m).

The curvature variation curves of these transition curves are shown in Fig. 6.9. Evidently, the higher the order of the derivative equal to zero at the start and end points of the transition curve, the smoother the curvature and cant variations at these

**Table 6.8** Types of transition curve line shapes

| Pattern | Combination | Transition curve line types | Curvature function | Satisfying conditions |
|---|---|---|---|---|
| Algebraic | A section | Cubic parabolic type | $k = k_0 \cdot \frac{l}{l_0}$ | First 3 items |
| Algebraic | A section | Quintic curve | $k = k_0 \cdot \left[3\left(\frac{l}{l_0}\right)^2 - 2\left(\frac{l}{l_0}\right)^3\right]$ | First 4 items |
| Algebraic | A section | Septic curve | $k = k_0 \cdot \left[10\left(\frac{l}{l_0}\right)^3 - 15\left(\frac{l}{l_0}\right)^4 + 6\left(\frac{l}{l_0}\right)^5\right]$ | First 5 items |
| Algebraic | A section | Nonic curve | $k = k_0 \cdot \left[35\left(\frac{l}{l_0}\right)^4 - 84\left(\frac{l}{l_0}\right)^5 + 70\left(\frac{l}{l_0}\right)^6 - 20\left(\frac{l}{l_0}\right)^7\right]$ | First 6 items |
| Trigonometric | A section | Half-wave sine | $k = k_0 \cdot \left[\frac{1}{2} - \frac{1}{2}\cos\left(\frac{\pi l}{l_0}\right)\right]$ | First 4 items |
| Trigonometric | A section | Full-wave sine | $k = k_0 \cdot \left[\frac{l}{l_0} - \frac{1}{2\pi}\sin\left(\frac{2\pi l}{l_0}\right)\right]$ | First 5 items |
| Trigonometric | Two section | S-shaped curve | First half: $k = 2k_0 \cdot \left(\frac{l}{l_0}\right)^2$<br>Second half: $k = k_0 \cdot \left[1 - 2\left(1 - \frac{l}{l_0}\right)^2\right]$ | First 4 items |

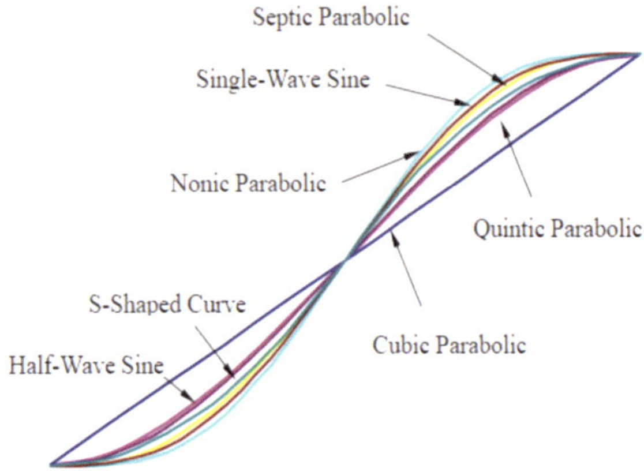


**Fig. 6.9** Schematic diagram of curvature variations for common transition curves

points, resulting in better ride stability and higher passenger comfort. At the same time, for transition curves with the same length, those with curvature and cant varying in a nonlinear manner exhibit greater changes in curvature and cant at the midpoint of the transition curve. Additionally, the higher the order of the derivative equal to zero at the start and end points, the greater the variation.

Figure 6.10 illustrates the variation patterns of different transition curve profiles. It can be observed that for transition curves where curvature and cant vary nonlinearly, the maximum rate of change occurs at the midpoint of the transition curve. Among these, the ninth-order curve exhibits the highest maximum rate, followed in descending order by the full-wave sine curve, S-shaped curve, septic curve, half-wave sine curve, quintic curve, and cubic parabola. In terms of peak values, the cubic parabola shows a significantly lower maximum rate than the other curves, while the quintic curve and half-wave sine curve exhibit similar values. The S-shaped curve and full-wave sine curve have equal maximum rates. Using the cubic parabola as a reference, Table 6.9 presents the relative ratios of the maximum curvature (cant) variation rates for different transition curves. It is evident that under the condition of limiting the curvature and cant variation rates within permissible bounds, a greater maximum rate of change necessitates a longer transition curve. Table 6.9 provides the length ratios of transition curves required to maintain the same maximum variation rate, using the cubic parabola as a benchmark.

Figure 6.11 illustrates the variation patterns of different transition curve profiles. It can be observed that the second derivative of the S-shaped curve exhibits discontinuities at both the midpoint and endpoints of the transition curve, forming singular points. For the quintic curve and half-wave sine curve, discontinuities occur only at the endpoints of the transition curve. Only the full-wave sine curve, septic curve, and ninth-order curve maintain continuity in their second derivatives throughout the transition curve, including at its endpoints.

Figure 6.12 illustrates the variation patterns of different transition curve $k'''$ profiles. It can be observed that only the half-wave sine curve and the ninth-order curve maintain continuity in their third derivatives throughout the transition curve,

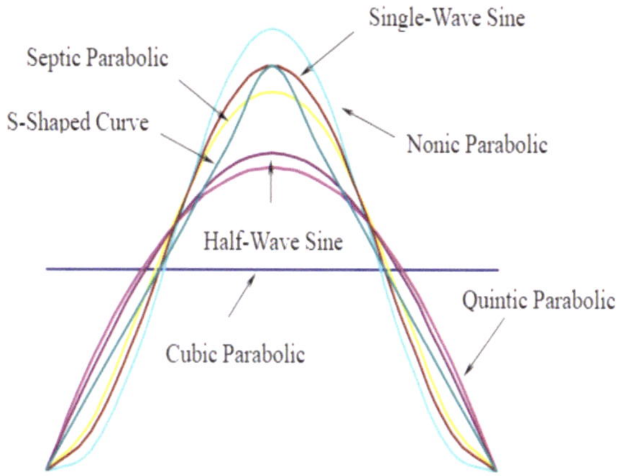


**Fig. 6.10** Schematic diagram of transition curve variation patterns

**Table 6.9** Curvature change rates of various transition curves

| Line type | Cubic | Quintic | Half-wave | S-curve | Septic | Full-wave | Nonic |
|---|---|---|---|---|---|---|---|
| Ratio of maximum curvature change rate | 1 | 1.5 | 1.57 | 2 | 1.875 | 2 | 2.1875 |
| Ratio of transition curve lengths for the same maximum curvature change rate | 1 | 1.5 | 1.57 | 2 | 1.875 | 2 | 2.1875 |

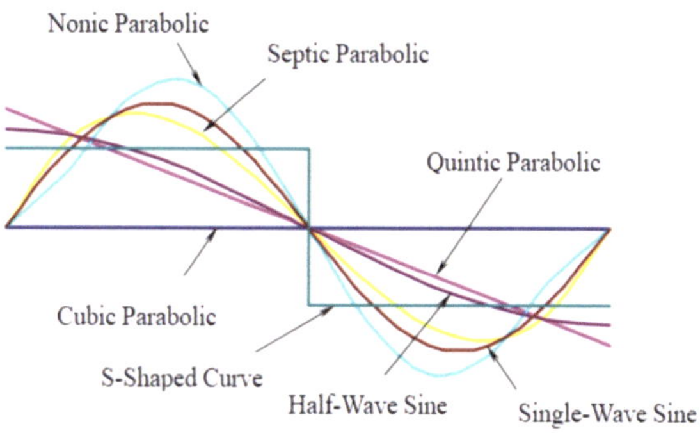


**Fig. 6.11** Schematic diagram of variation patterns of curvature $k''$ for transition curves

including at its endpoints. However, the second derivative of curvature for the half-wave sine curve exhibits discontinuities at the endpoints of the transition curve.

At lower operating speeds, it is generally sufficient to meet only the first three conditions. However, as speed increases, the impact forces at the start and end points of the transition curve escalate sharply. In transition curves with a linear cant ramp, the discontinuity at both ends of the ramp not only reduces passenger comfort but also poses a potential track irregularity hazard. As a result, many high-speed railways worldwide have adopted transition curves with curved cant ramps. For instance,

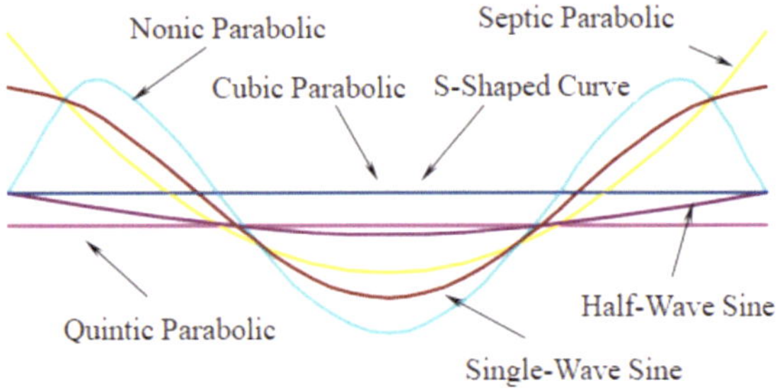


**Fig. 6.12** Schematic diagram of variation patterns of transition curves

Japan uses a half-wave sine profile, while the European Union Railway Agency (ERA) recommends a sine function or modifications to the traditional linear cant ramp, improving both ends to ensure continuity. In France, the ends of a linear cant ramp are adjusted to follow a cosine function while keeping the middle section linear, whereas in the UK, circular curves are applied at the ends to improve the transition while retaining the linear ramp in the middle. The transition curve profiles used in various high-speed rail systems are listed in Table 6.10.

2. **Analysis of the Influence of Transition Curve Profile on Vehicle Dynamic Performance**

(1) Time History Curve Analysis of Transition Curve Profiles

IIn the dynamic simulation, six representative transition curve profiles—cubic, quintic, septic, nonic, half-wave sine, and one-wave sine curves—were selected for analysis. The key parameters are as follows: train speed is 350 km/h, the radius of the circular curve is 8000 m, the length of the circular curve is 280 m, the length of the transition curve is 590 m, and the cant is 150 mm.

Figure 6.13 presents the time history curves of vehicle dynamic responses when passing through these six representative transition curve profiles. From the time history curves of vehicle body acceleration in Fig. 6.13a, b, it is evident that the cubic parabolic transition curve, with its linear changes in cant and curvature, exhibits

**Table 6.10** Transition curve line types used in foreign high-speed rail transit systems

| Line type | System type | Transition curve line types |
|---|---|---|
| Shinkansen (Japan) | Wheel-rail | Half-wave sine |
| ICE (Germany) | Wheel-rail | Cubic parabolic, s-curve, etc. |
| TGV (France) | Wheel-rail | Planar: cubic parabolic<br>Vertical: cosine + straight line + cosine |
| High speed rail (UK) | Wheel-rail | Planar: cubic parabolic<br>Vertical: circular curve + straight line + circular curve |
| Transrapid (Germany) | Eddy current magnetic levitation | Full-wave sine |

abrupt changes at both ends, causing noticeable spikes in vehicle acceleration. For quintic parabolic and half-wave sine curves, the first derivative at the start and end is zero, while the second derivative is non-zero. As the vehicle's vertical acceleration is proportional to the second derivative of curvature, this results in significant impacts on vehicle acceleration. On the other hand, septics, nonics, and one-wave sine curves, where the second derivative at the start and end is zero, do not exhibit spikes in the vehicle vibration acceleration, resulting in a smoother acceleration variation.

From Fig. 6.13c–f, it can be observed that under all transition curve conditions, the wheel-rail lateral forces and derailment coefficients change gradually at the start and end of the transition curve. However, under the quintic parabolic and half-wave sine profiles, the wheel-rail vertical forces and wheel load reduction rates show large peak values at the start and end of the transition curve. The primary reason for this is that their curvature's second derivative has a discontinuity at the start and end, resulting in vertical impact forces.

Furthermore, it can be observed from all the response curves that the influence of the transition curve profile on the system's dynamic response is primarily dependent on the order of the derivative of curvature at the start and end points, with the specific profile type (whether algebraic or trigonometric) having less impact. For example, the dynamic response of the system under quintic parabolic and half-wave sine curves, or septics and one-wave sine curves, shows similar trends and values in both the response variation and magnitude.

(2) Impact of Transition Curve Length on Profile

Figure 6.14 illustrates the variation of dynamic response indicators with transition curve length for different transition curve profiles. The key parameters for the calculation are as follows: train speed is 350 km/h, the radius of the circular curve is 8000 m, the length of the circular curve is 280 m, and the cant is 150 mm.

From Fig. 6.14a, b, which show the time history curves of vehicle lateral and vertical acceleration, it can be seen that when the transition curve length is shorter, the cubic parabolic profile has the optimal dynamic characteristics. However, as the length increases, the response values of higher-order curves decrease rapidly, with the higher-order curves exhibiting a greater rate of change, while the reduction rate for the cubic parabolic curve becomes gradually smaller.

Under the cubic parabolic profile, the variation in lateral acceleration is relatively smooth. In contrast, for higher-order curves, when the transition curve is short, the larger curvature variation at the midpoint of the transition curve results in higher lateral acceleration. When the length reaches approximately 700 m, the response values of the seventh-order, ninth-order parabolic, and one-wave sine curves fall below those of the cubic parabolic curve, while the quintic parabolic and half-wave sine curves remain higher. Therefore, in terms of the horizontal plane, the cubic parabolic curve takes full advantage of the longer effective length, which is more effective in controlling vehicle lateral vibration.

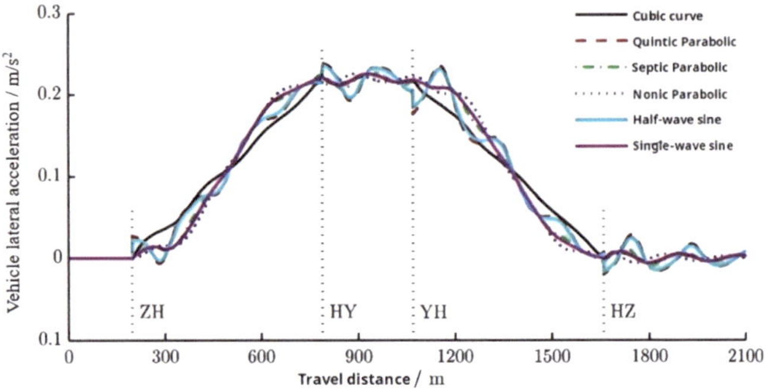


(a) Vehicle lateral acceleration

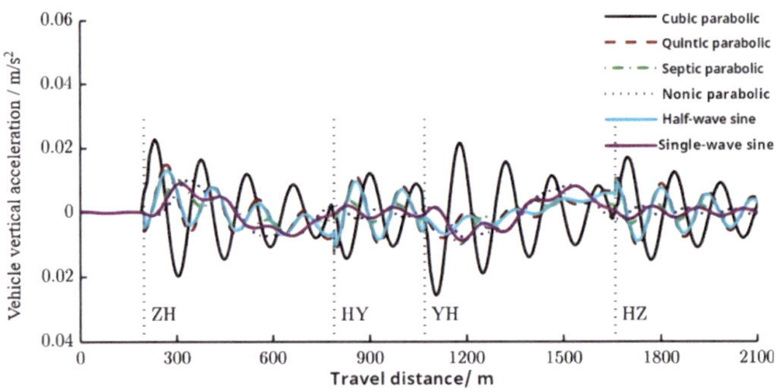


(b) Vehicle vertical acceleration

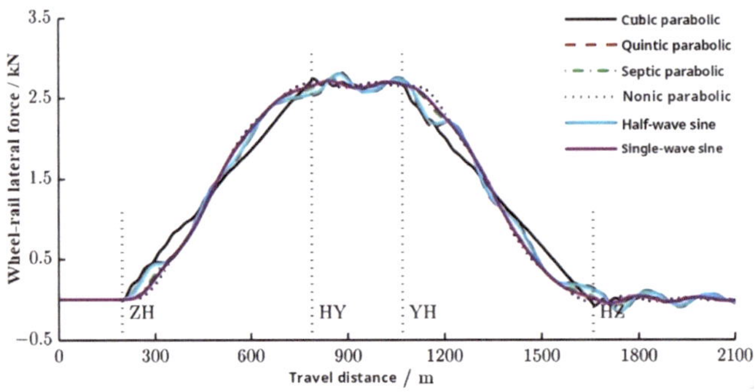


(c) Wheel-rail lateral force

**Fig. 6.13** Impact of transition curve line types on the time-history curves of key train dynamic performance indices

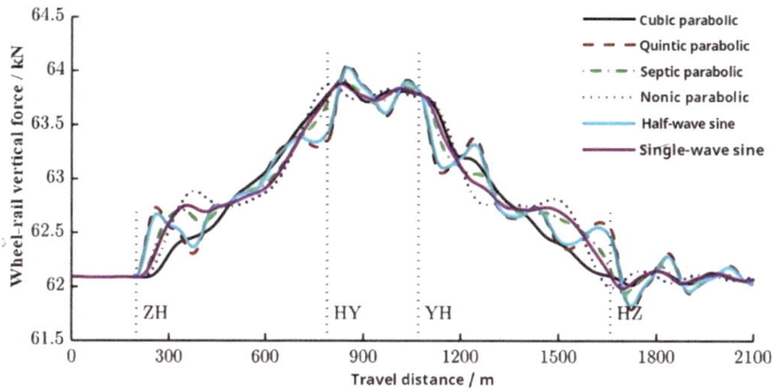


(d) Wheel-rail vertical force

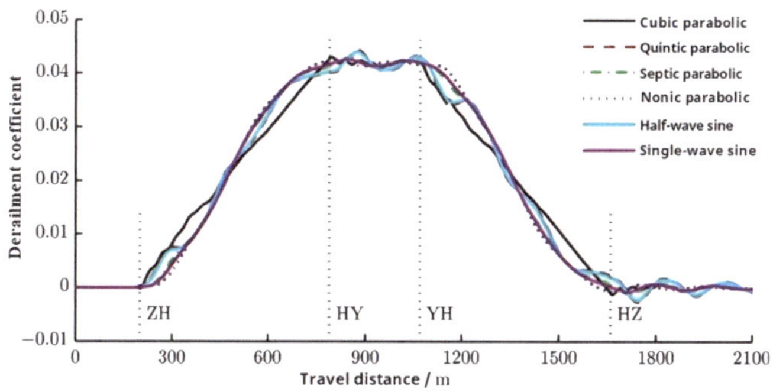


(e) Derailment coefficient

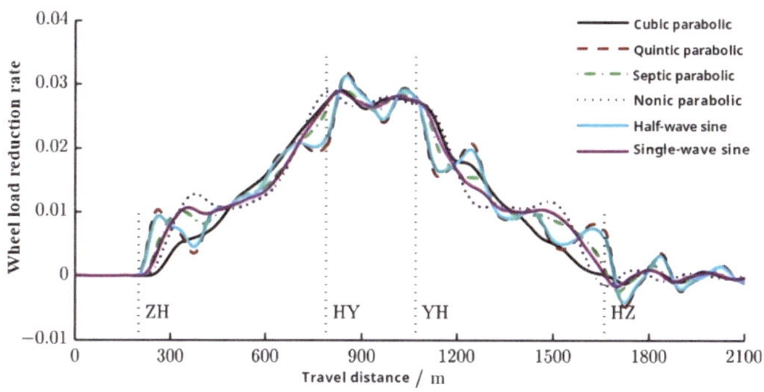


(f) Wheel load reduction rate

**Fig. 6.13** (continued)

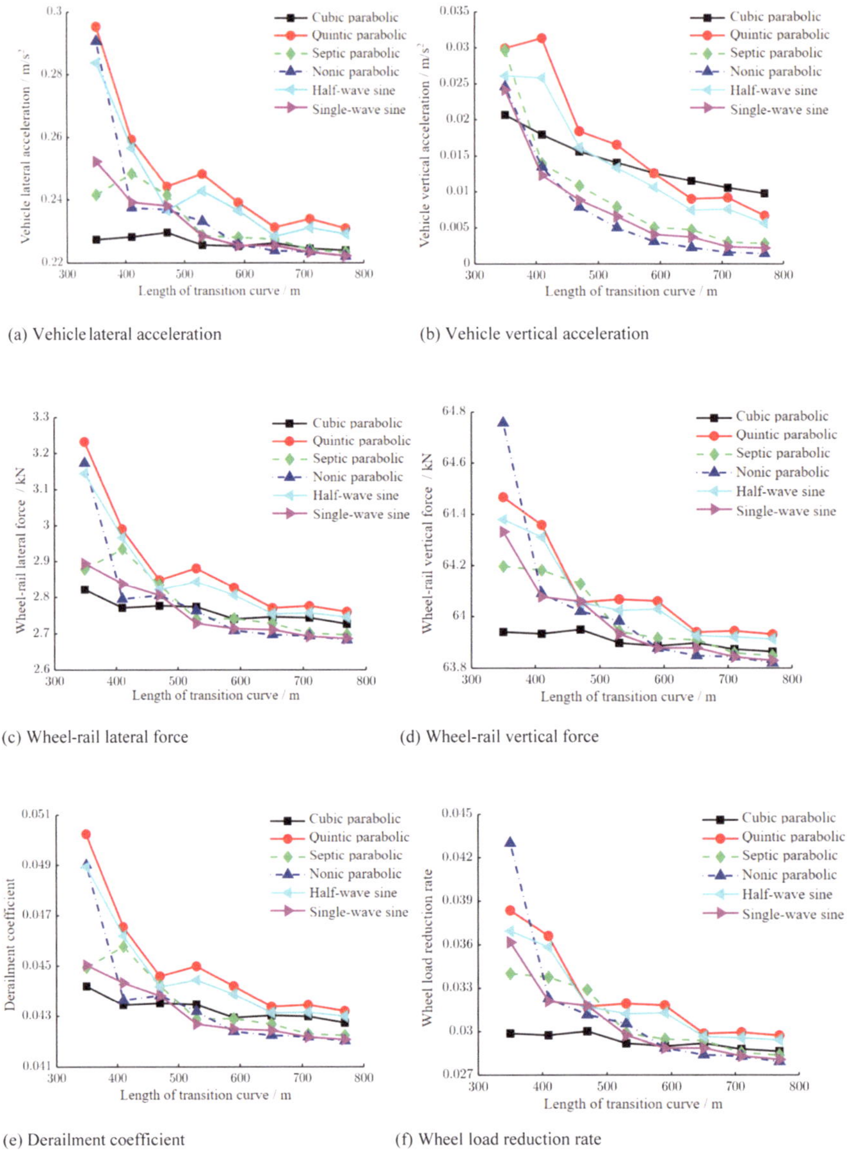


**Fig. 6.14** The impact of line type under different transition curve lengths (V = 350 km/h)

Regarding vertical vibration, when the transition curve length is shorter, the vertical acceleration for the cubic parabolic profile is lower than that for higher-order curves. However, as the length increases, the vertical acceleration for higher-order curves decreases rapidly, while the reduction rate for the cubic parabolic curve becomes smaller. When the length exceeds 600 m, the vertical acceleration for all higher-order profiles becomes smaller than the vertical acceleration for the cubic parabolic profile. Therefore, in terms of the vertical plane, extending the transition curve under the cubic parabolic profile has limited efficiency in improving vertical vibration, and there are certain constraints. When there are higher requirements for vertical comfort, selecting an appropriate transition curve length and using higher-order profiles can effectively reduce vertical vibration, thereby improving passenger comfort. On the other hand, the response reduction for linear transition curves does not significantly change with an increase in length. Under high-speed conditions, even extending the length may not meet comfort requirements.

In practice, high-speed rail systems in France and the UK use cubic parabolic curves for the horizontal plane, while they use cosine or circular curves in the vertical plane to smooth the cant transitions at the start and end of the transition curve. This aligns with the above analysis, indicating that the vertical profile, compared to the horizontal profile, has stricter requirements for the smoothness of transition curve start and end points.

Figure 6.15 shows the variation of dynamic response indicators with transition curve length under different transition curve profiles when the train speed is 400 km/h. The key parameters for the calculation are as follows: train speed is 400 km/h, the radius of the circular curve is 8000 m, the length of the circular curve is 320 m, and the cant is 150 mm.

From the figure, it can be seen that the trend of changes in dynamic performance indicators under the conditions of 400 km/h and 350 km/h is similar. However, as the speed increases, the advantages of higher-order curves become more evident. Under the condition of 400 km/h, the linear cant transitions cause greater impact on the train. When the transition curve length exceeds 500 m, the dynamic performance indicators for all higher-order curve profiles are superior to those for the cubic parabolic transition curve profile. When using a transition curve profile where the second derivative of curvature is zero at the start and end points, the vertical vibration peaks of the vehicle can be significantly improved, thereby enhancing passenger comfort, as shown in Fig. 6.15b.

The transition curve design for high-speed railways in different countries typically adopts three approaches. First, the cubic parabolic transition curve is used, with an increase in the transition curve length, as seen in China's high-speed railways. However, the improvement effect diminishes with increasing length, and excessively long transition curves can reduce the flexibility of the track alignment, leading to an increase in construction work. Second, a cubic parabolic curve is used in the horizontal plane, while a cosine or circular curve is employed at the start and end points of the transition curve in the vertical plane to smooth the transition. This method ensures that the transition curve length is not too long, while maintaining good dynamic performance in the vertical plane, as used in France and the United Kingdom. Third, higher-order transition curve profiles are used, as seen in Japan and Germany. This approach delivers optimal dynamic performance but increases the difficulty of construction and maintenance.

From the above analysis, when the train speed is 350 km/h, the cubic parabolic transition curve, due to its cant ramp angles, causes relatively low lateral and vertical

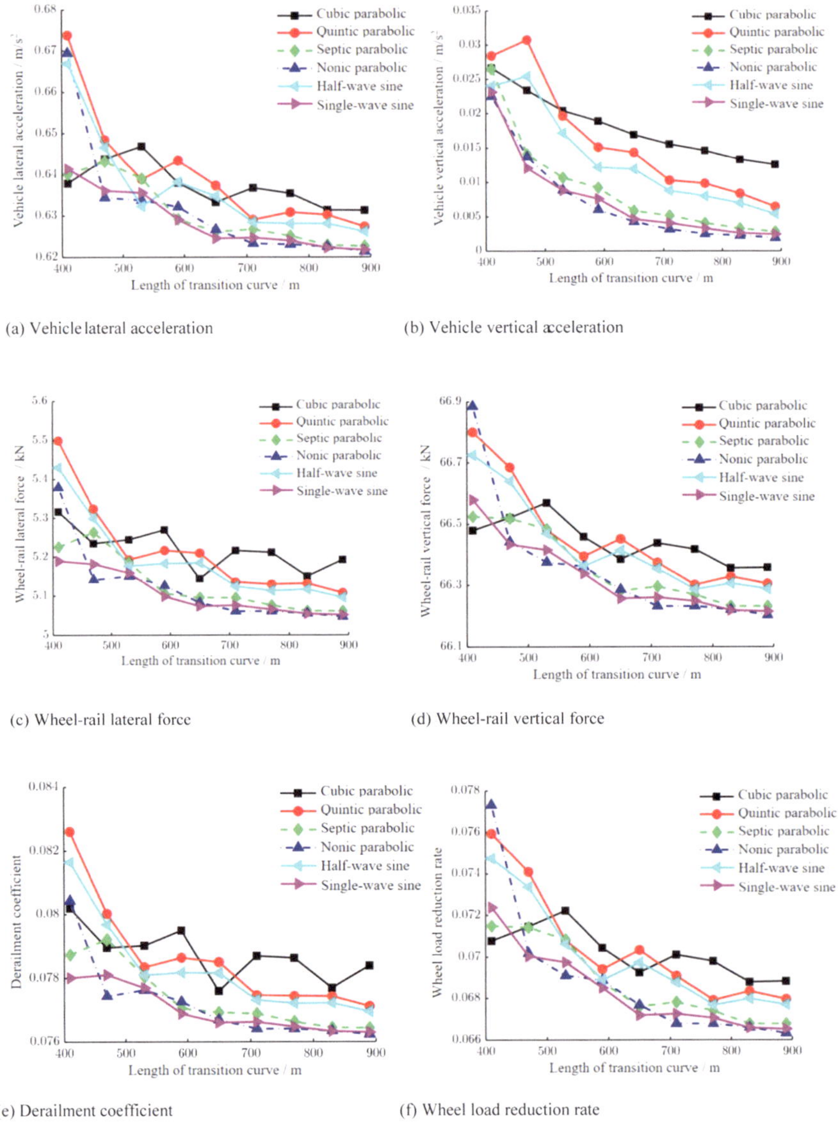


**Fig. 6.15** The impact of line type under different transition curve lengths (V = 400 km/h)

vibrations, allowing it to leverage the advantage of longer effective length. High-speed railways currently in operation with a design speed of 350 km/h also demonstrate that the cubic parabolic transition curve used can ensure good dynamic performance and passenger comfort. The current transition curve length standards effectively reduce lateral and vertical impacts at the start and end of the transition curve. Therefore, for lines designed for 350 km/h, the cubic parabolic transition curve is a suitable choice. Given the existing high-speed railway track structure, construction, measurement, and maintenance methods, to further enhance passenger comfort and improve the flexibility of the track design, higher-order curves for cant profiles should be allowed.

When the train speed reaches 400 km/h, higher-order transition curves significantly outperform cubic parabolic curves in terms of dynamic performance, especially in vertical acceleration. Therefore, for lines designed for 400 km/h and higher speeds, transition curves with a curve-shaped cant ramp should be used. As shown in Fig. 6.15, a curve profile where the second derivative of curvature is continuous at the start and end points of the transition curve should be employed. Considering the complexity of the curve profiles, a sine wave or seventh-order parabolic curve is ideal for the transition curve profile.

### 3. **Minimum Length of Transition Curves**

The minimum length of railway transition curves is primarily controlled by requirements such as cant rate of change, cant deficiency rate of change, and cant ramp rate. For high-speed railways, the transition curve length is typically controlled by the cant rate of change. The control criteria for transition curve length for high-speed railways in different countries, are shown in Table 6.11.

In the simulation analysis, the $P_{CT}$ indices, suitable for evaluating passenger comfort on transition curves, was used to analyze the length of the transition curve. The calculation parameters were: operating speed of 350 km/h, circular curve radius of 7000 m, and cant of 175 mm.

Figure 6.16 shows the variation of the $P_{CT}$ comfort indices with the length of the transition curve. From the figure, it can be seen that when the length of the transition curve exceeds 550 m (equivalent to a cant rate of change of 31 mm/s), the impact

**Table 6.11** Control indices for the length of transition curves on high-speed railways

| Category | Cant rate of change | Insufficiency of cant rate of Change | Cant transition rate |
|---|---|---|---|
| European standard EN 13803-1 (2010) | General value: 50 mm/s<br>Exceptional value: 60 mm/s | General value: 55 mm/s<br>Exceptional value: 75 mm/s | General value: 2.25 mm/m<br>Exceptional value: 2.5 mm/m |
| Germany | Standard: 50 mm/s<br>Maximum: 60 mm/s | | |
| France | Standard: 50 mm/s<br>Limitation: 60 mm /s | Standard: 30 mm/s<br>Limitation: 75 mm/s | |
| Japan | 45 mm/s | 37 mm/s | |
| China high-speed railway standards | Excellent: 25 mm/s<br>Good: 28 mm/s<br>General: 31 mm/s | Good: 23 mm/s<br>Difficult: 38 mm/s | 0.25–2‰ |
| China intercity railway standards | General: 28 mm/s<br>Difficult: 35 mm/s | General: 23 mm/s<br>Difficult: 38 mm/s | |

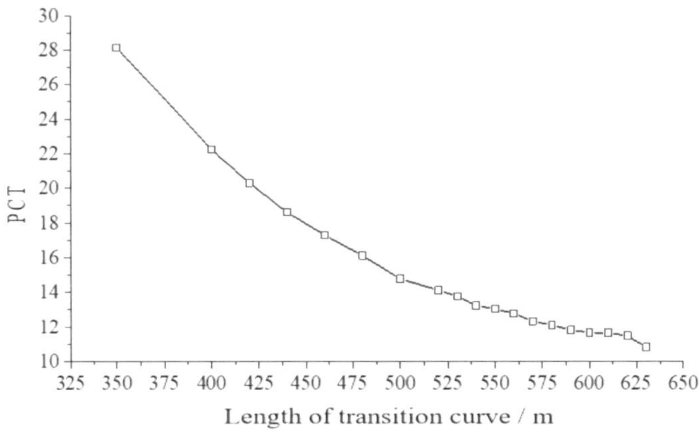


**Fig. 6.16** Relationship between the length of transition curve and $P_{CT}$

of the transition curve length on passenger comfort tends to stabilize. Therefore, the current high-speed railway transition curve length standards can be appropriately relaxed. It is recommended that the cant rate of change indices be generally relaxed to 28, 31 mm/s under normal conditions, and 35 mm/s under difficult conditions.

### *6.1.4 Analysis of the Impact of Cant Configuration*

There are generally two methods for setting up superelevation in railways. One method involves lowering half of the superelevation on the inner rail while raising the other half on the outer rail (hereafter referred to as "inner lowering and outer raising"). This method is used by the Shinkansen in Japan, as well as in countries like Switzerland and the United States. In China, this method is also commonly used in metro design when clearance restrictions are in place. The other method places the entire superelevation on the outer rail (hereafter referred to as "outer raising"). This method is used by other European countries and by China's mainline railways.

Figures 6.17, 6.18 and 6.19 show a comparison of the impact of both superelevation methods on system dynamic response under conditions of cubic, quintic, and septenic parabolic transition curves. The other line conditions are as follows: train speed of 120 km/h, circular curve radius of 800 m, curve length of 120 m, transition curve length of 80 m, and a superelevation of 150 mm with no track irregularity.

From Fig. 6.17, it can be seen that for the vehicle's vertical acceleration, the "inner lowering and outer raising" superelevation method performs significantly better than the "outer raising" method. The same result is observed for the vertical wheel-rail force, where the "inner lowering and outer raising" method also outperforms the "outer raising" method. This is because, with the "inner lowering and outer raising" configuration, the height of the vehicle's center of gravity remains nearly unchanged at the start and end of the transition curve, effectively reducing the impact caused by the abrupt change in superelevation at the beginning and end of the cubic parabolic transition curve.

Figures 6.18 and 6.19 further demonstrate that for both vertical acceleration and vertical wheel-rail force, the "inner lowering and outer raising" superelevation method still performs better than the "outer raising" method, though the difference is not as significant. This is because the higher-order transition curves already significantly mitigate the impact at the start and end of the transition curve, reducing the effect on the system's dynamic response.

It is also clear that superelevation method has a minimal effect on the vehicle's lateral acceleration and the lateral wheel-rail force. In fact, the difference between the two superelevation methods is primarily in the vertical plane, while the horizontal plane curve remains the same, meaning that the lateral dynamic response is almost identical for both methods.

In summary, the "inner lowering and outer raising" superelevation method offers better vertical dynamic performance compared to the "outer raising" method. However, if higher-order transition curves or longer transition curves are used to improve the impact at the start and end of the transition curve, the advantage of the "inner lowering and outer raising" method becomes less pronounced. Furthermore, the "outer raising" superelevation method can simplify the treatment of the subgrade and trackbed, making maintenance tasks easier.

## 6.2 Track Longitudinal Profile Parameter Analysis

### *6.2.1 Analysis of Minimum Slope Length*

Shorter slope sections can better adapt to the terrain, reduce construction quantities, and lower investment costs. However, from the perspective of train operational stability, the minimum slope length should not only ensure that two vertical curves do not overlap but also ensure that there is a certain length of transition between the two vertical curves. This is to avoid the superposition of vibrations generated by the previous vertical curve with those generated by the next vertical curve.

This study simulates and analyzes the impact of slope section length on the dynamic response of the train-track system under the condition of a travel speed of 350 km/h. Figure 6.20 shows the time history curves of the vertical acceleration of the vehicle under different algebraic differences of slope and different slope lengths. Figure 6.21 presents the corresponding peak vertical acceleration of the vehicle as a function of the transition slope length.

**Fig. 6.17** Impact of different cant methods on the dynamic performance of vehicle motion with cubic parabolic curves

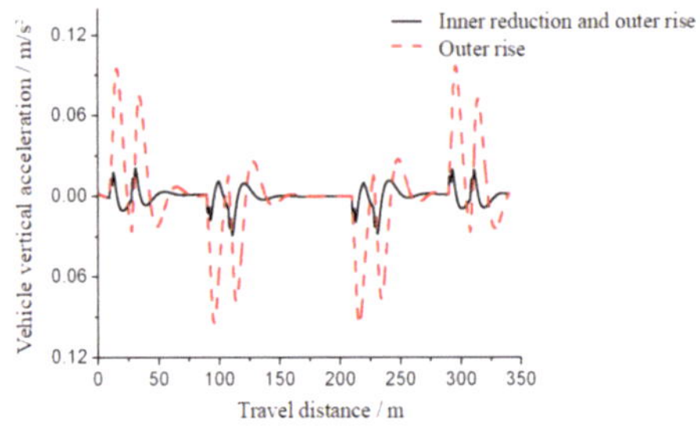


(a) Vertical acceleration of the vehicle

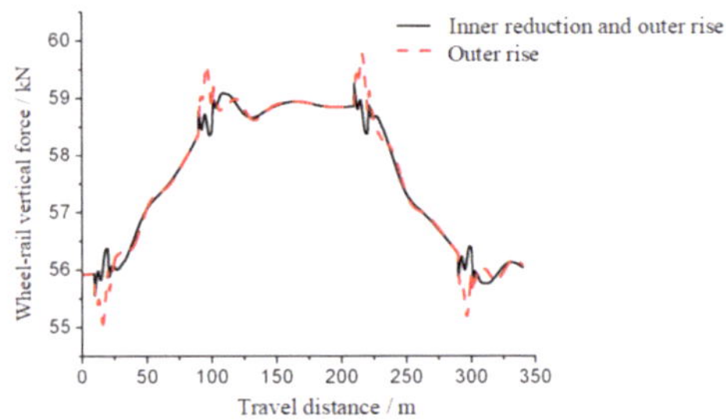


(b) Wheel-rail vertical force

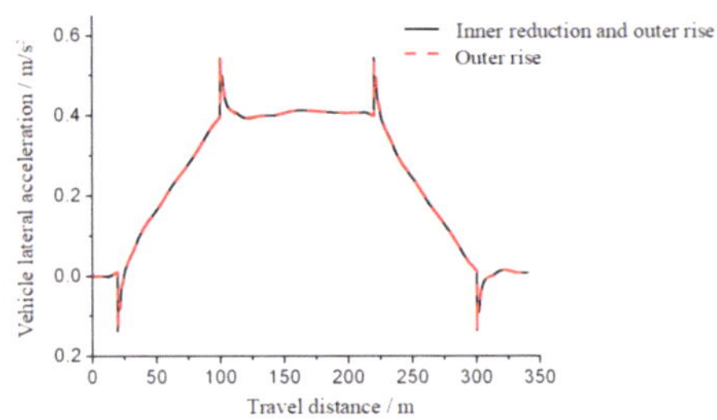


(c) Vehicle lateral acceleration

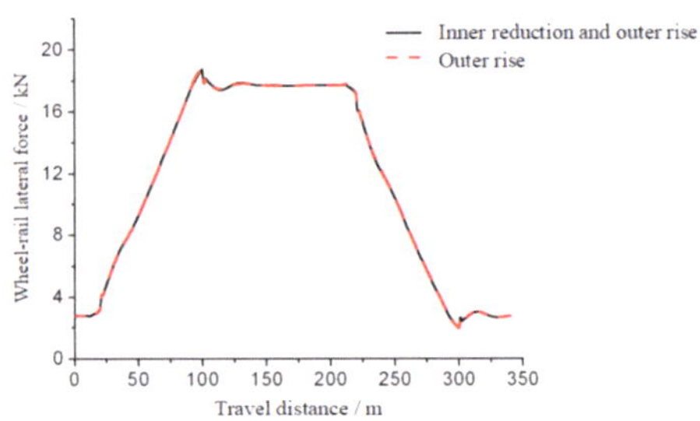


(d) Wheel-rail lateral force

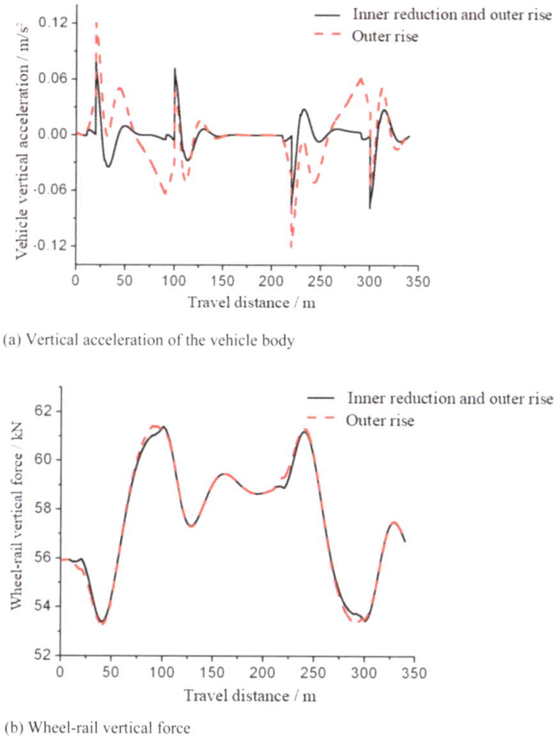


**Fig. 6.18** Impact of different cant methods on the dynamic performance of vehicle motion with quintic parabolic curves

From the figures, it can be seen that the peak vertical acceleration of the vehicle decreases gradually as the transition slope length increases, but the rate of decrease also becomes slower. When the transition slope length is within 600 m, the increase in slope length significantly reduces the peak acceleration. However, when the transition slope length exceeds 900 m, the increase in length has almost no effect on the peak acceleration.

Tables 6.12 and 6.13 show the variation curves of the peak vertical acceleration of the CRH3 train under different operating speeds, vertical curve radii, and gradient differences, with respect to the length of the transitional slope section. From the tables, it can be concluded that:

1. Under different gradient difference conditions, the influence of the length of the transitional slope section on the peak vertical acceleration follows a similar pattern. Increasing the length of the transitional slope section helps to reduce the combined vibration from the previous and subsequent vertical curves, thus reducing the peak vertical acceleration on the following vertical curve. However, as the length of the transitional slope section increases, the effect of peak acceleration reduction decreases.
2. In general, under the same transitional slope section length, the smaller the gradient difference, the smaller the peak vertical acceleration of the train body.

**Fig. 6.19** Impact of different cant methods on the dynamic performance of vehicle motion with septic parabolic curves

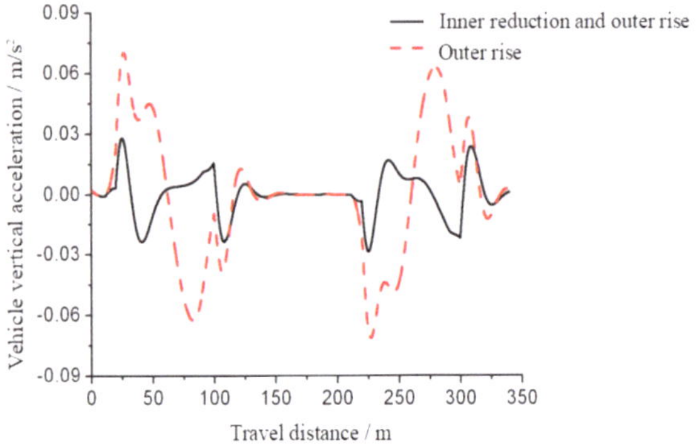


(a) Vehicle vertical acceleration

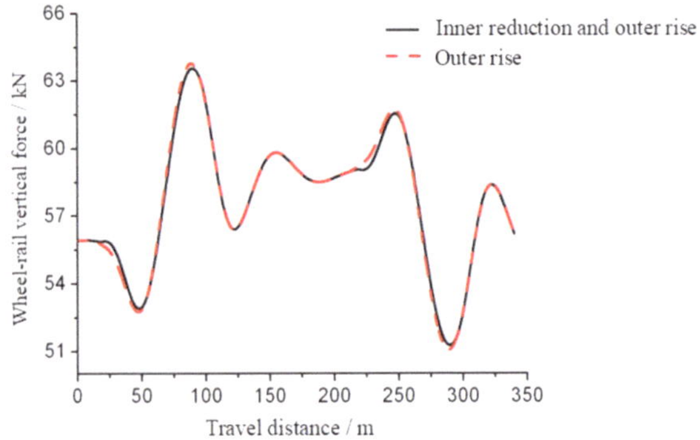


(b) Wheel-rail vertical force

The larger the gradient difference, the more significant the effect of increasing the transitional slope section length on reducing the peak acceleration.

3. Under the given calculation conditions, even when the length of the transitional slope section is 0 m, the peak vertical acceleration of the train body is still less than 1.0 m/s$^2$, which meets the relevant standards.
4. When the length of the transitional slope section is 0.4 V and 0.8 V, the difference in the impact on the peak vertical acceleration is within 10%. When the transitional slope section length reaches 1.0 V, there is a significant improvement, with the peak value reduced by approximately 40%.

In conclusion, when the line conditions are not constrained, the transitional slope section length should be greater than 1.0 V. Under more challenging conditions, a length greater than 0.4 V is sufficient, and no strict limit is necessary, as long as the vertical curves do not overlap.

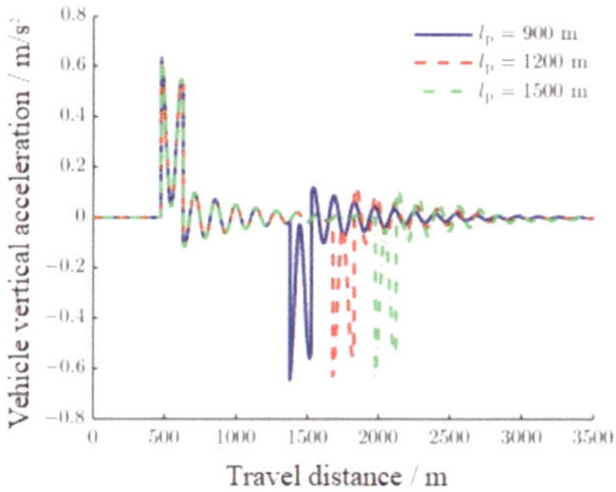


(a) Δ*i*=5‰

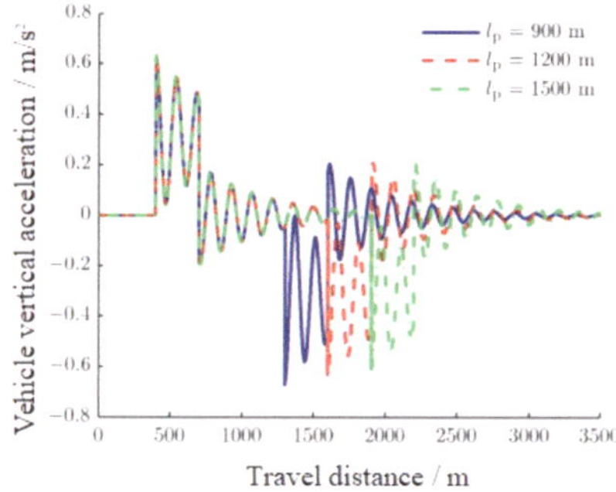


(b) Δ*i*=10‰

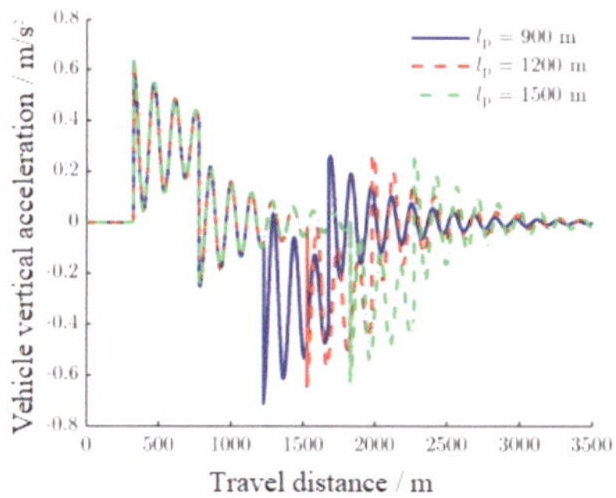


(c) Δ*i*=15‰

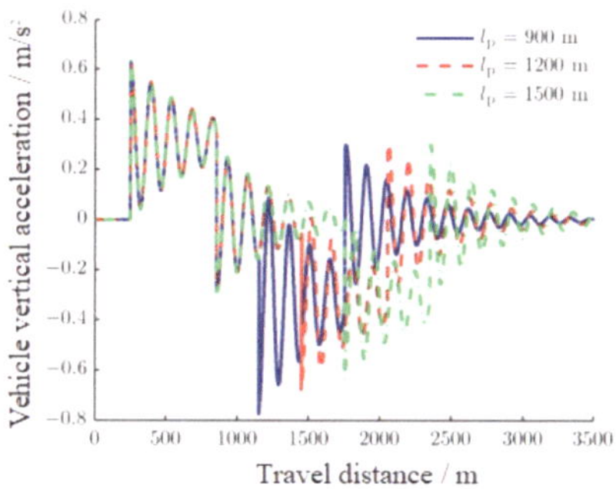


(d) Δ*i*=20‰

**Fig. 6.20** The impact of slope segment length on vertical acceleration of the vehicle

### 6.2.2 Minimum Radius of Vertical Curves Analysis

In China's high-speed railways, the allowed value of quasi-static vertical acceleration $a_v$ is generally 0.4 m/s$^2$, and it is 0.5 m/s$^2$ under difficult conditions. When using the general value of 0.4 m/s$^2$, the approximate formula for calculating the radius of a vertical curve is:

$$R_v \geq 0.193 \times V^2 \tag{6.2.1}$$

Table 6.14 presents the minimum vertical curve radius values derived from the formula for different speed conditions.

Figure 6.22 shows the dynamic performance time history curves under different vertical curve radii. The main calculation parameters are: train speed of 350 km/h and an algebraic difference in gradients of 15‰.

The radius of the vertical curve significantly affects the dynamic performance of train motion, including vertical acceleration of the vehicle body, vertical force between the wheel and rail, and wheel load reduction rate. From the time history

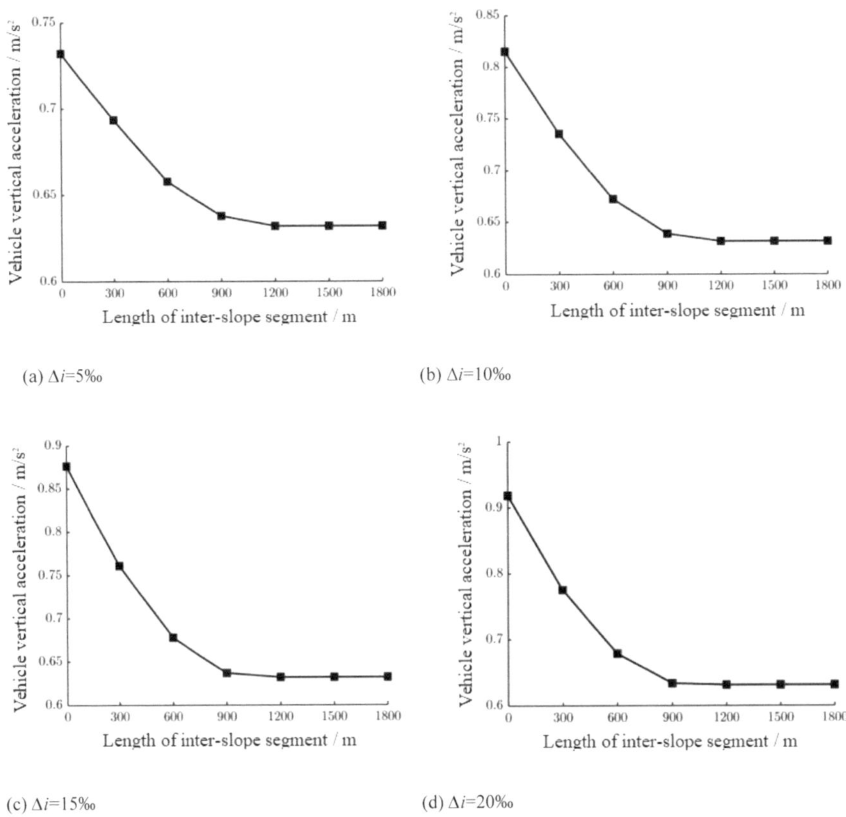


**Fig. 6.21** Impact of transition slope length on peak vertical acceleration of the vehicle

**Table 6.12** Influence of the transitional slope section length on the peak vertical acceleration of the vehicle body (Δi = 20‰)

| Inter-slope segment length (m) | The peak vertical acceleration of the vehicle (m/s$^2$) | | | |
|---|---|---|---|---|
| | Train speed (km/h)/Vertical curve radius (m) | | | |
| | 200/15,000 | 250/20,000 | 300/25,000 | 350/30,000 |
| 0 | 0.650 | 0.663 | 0.765 | 0.918 |
| 0.4 V | 0.594 | 0.621 | 0.708 | 0.835 |
| 0.8 V | 0.548 | 0.585 | 0.662 | 0.772 |
| 1.0 V | 0.296 | 0.396 | 0.469 | 0.517 |

curves, it can be observed that without the transition of transition curves, there will be a significant vertical impact on the vehicle at the start and end points of the vertical curve. At the start of the vertical curve, the smaller the radius, the greater the impact. At the end of the vertical curve, the peak value of vibration is the largest when the

**Table 6.13** Influence of the transitional slope section length on the peak vertical acceleration of the vehicle. body ($\Delta i = 40‰$)

| Inter-slope segment length (m) | The peak vertical acceleration of the vehicle (m/s$^2$) | | | |
|---|---|---|---|---|
| | Train speed (km/h)/Vertical curve radius (m) | | | |
| | 200/15,000 | 250/20,000 | 300/25,000 | 350/30,000 |
| 0 | 0.629 | 0.716 | 0.822 | 0.965 |
| 0.4 V | 0.570 | 0.657 | 0.751 | 0.876 |
| 0.8 V | 0.524 | 0.610 | 0.695 | 0.806 |
| 1.0 V | 0.319 | 0.376 | 0.441 | 0.486 |

**Table 6.14** Minimum vertical curve radius calculation table

| $V$ (km/h) | 350 | 300 | 250 | 200 |
|---|---|---|---|---|
| $[a_v]$ (m/s$^2$) | 0.4 | 0.4 | 0.4 | 0.4 |
| Calculated value (m) | 23,643 | 17,361 | 12,056 | 7720 |
| Regulatory limit value (m) | 25,000 | 25,000 | 20,000 | 10,000 |

radius is 25,000 m, and the vibration peak value at a radius of 35,000 m is greater than that at a radius of 30,000 m. The vibration at the end of the vertical curve is influenced by the vibration caused at the start point. The radius of the vertical curve affects its length, and the length of the vertical curve affects the value of the vibration decay at the start point. This value, when superimposed with the vibration at the end point, determines the size of the peak value of vibration at the end point.

Figure 6.23 presents the variation curves of the peak vertical acceleration of the vehicle body with the radius of the vertical curve under different algebraic differences in gradients at a train speed of 350 km/h.

From the figure, it can be observed that at the start of the vertical curve, the size of the algebraic difference in gradients has little effect, and the peak vertical acceleration of the vehicle body continuously decreases as the radius of the vertical curve increases. When the radius of the vertical curve is small, increasing the radius has a significant effect on reducing the peak acceleration. However, when the radius increases to a certain extent, the trend of reducing the peak acceleration slows down. Therefore, considering the dynamic performance of the train, the difficulty of construction, and maintenance, the radius of the vertical curve should not only have a minimum limit but also should not be too large.

At the end of the vertical curve, the peak vertical acceleration of the vehicle body shows a certain fluctuation with the change of the vertical curve radius, and the size of the algebraic difference in gradients has a significant impact. This is because the length of the vertical curve is determined by the algebraic difference in gradients and the radius of the vertical curve, and the decay value of the vibration at the start point affects the peak value of the vibration at the end point. However, the overall trend of the peak acceleration at the end point still decreases as the radius increases.

**Fig. 6.22** The impact of vertical curve radius on dynamic performance of train operation

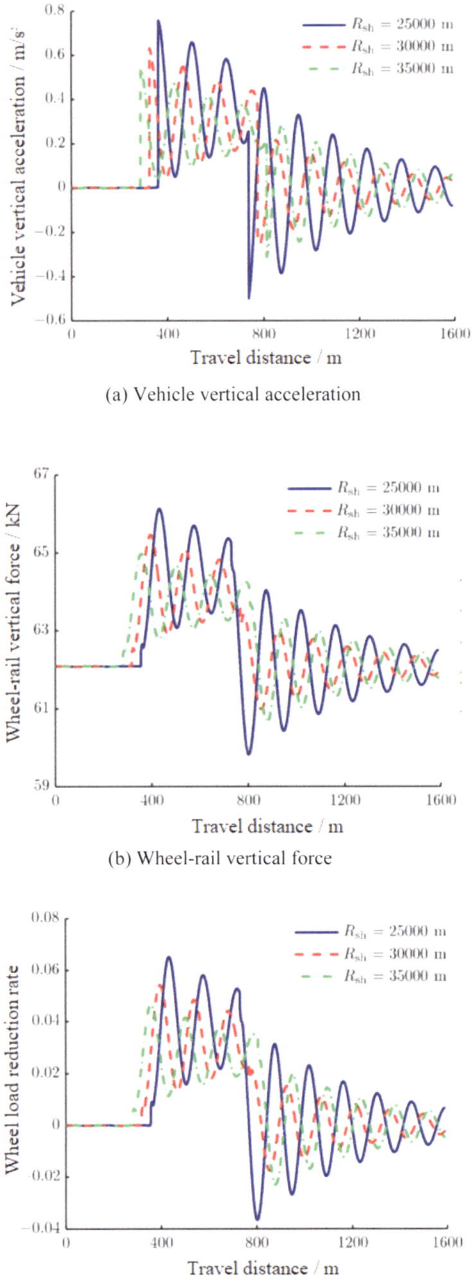


(a) Vehicle vertical acceleration



(b) Wheel-rail vertical force



(c) Wheel load reduction rate

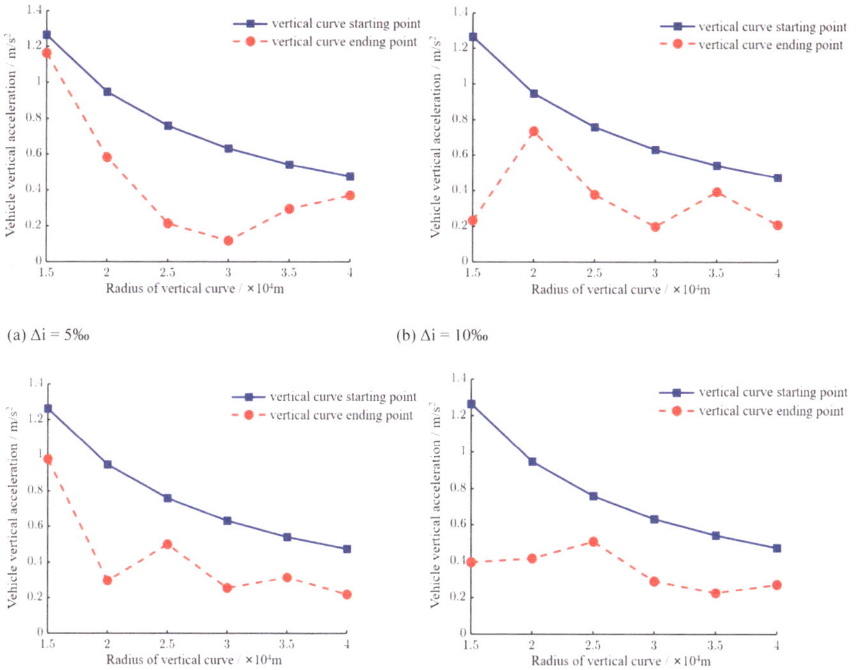


(a) Δi = 5‰ (b) Δi = 10‰

(c) Δi = 15‰ (d) Δi = 20‰

**Fig. 6.23** The impact of vertical curve radius on the peak vertical acceleration of the vehicle

## 6.3 Analysis of the Overlap Between Horizontal and Vertical Curves

### *6.3.1 Minimum Radius of Circular and Vertical Curves in Overlapping Settings*

In high-speed railway lines, the radius of the horizontal curve is large, and the curve length can be several kilometers long. It is unrealistic to completely avoid the overlap between the horizontal and vertical curves. Therefore, under the premise of ensuring the safety of train operations and passenger comfort, the "High-speed Railway Design Code" allows for the overlapping setting of vertical and circular curves. This section will analyze the impact of the radii of horizontal and vertical curves on passenger comfort when they overlap, and discuss the principles for setting horizontal and vertical curve overlaps.

1. **Analysis of the Horizontal and Vertical Curve Overlap with Circular Curve Radius**

Figure 6.24 shows the time history curve of the vehicle vibration acceleration under different circular curve radii conditions when the vertical and circular curves overlap. Main calculation parameters: train speed is 350 km/h; the vertical curve radius is 30,000 m, with a grade difference of 10‰; the circular curve length is 500 m, and the transition curve length is 590 m. From the figure, it can be seen that under different line conditions, the passenger comfort criteria are met. At the horizontal and vertical curve overlap point, the vibration acceleration that passengers experience is the resultant acceleration of both horizontal and vertical body vibrations. The resultant acceleration is greater than the individual horizontal and vertical vibration accelerations of the body.

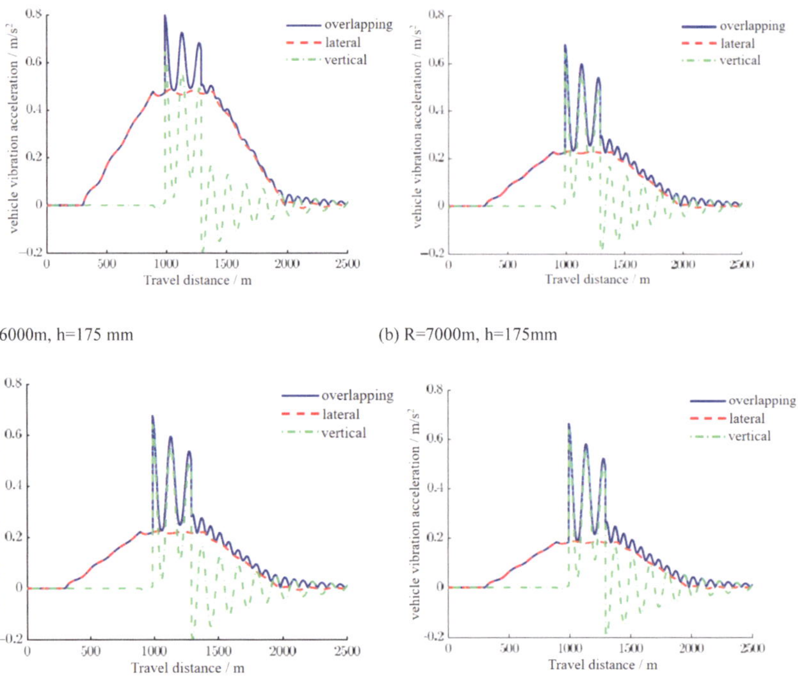


(a) R=6000m, h=175 mm (b) R=7000m, h=175mm

(c) R=8000m, h=150mm (d) R=9000m, h=135mm

**Fig. 6.24** Time history curves of vehicle vibration acceleration with overlapping lateral and vertical curves at different circular curve radii

Under normal conditions, due to the effect of the transition curve, the horizontal vibration shock caused at the transition curve point is not large. Therefore, the distance between the start of the vertical curve and the transition curve point has little effect on the resultant vibration peak value. However, when the horizontal and vertical curves overlap, the horizontal acceleration reaches its peak at the transition curve point, and the vertical acceleration peaks at the start of the vertical curve. If these two points are too close, passengers will experience rapid changes in both the magnitude and direction of the vibration acceleration in a short time, severely deteriorating ride comfort. Therefore, a certain distance should be maintained between the transition curve point and the start of the vertical curve. This distance should allow for one complete cycle of horizontal vibration buffering. The impact of the distance between the start of the vertical curve and the transition curve on comfort is similar to that between the transition curve and the vertical curve. Maintaining a certain distance helps avoid the sudden changes in magnitude and direction of the vibration acceleration that passengers would otherwise experience in a short time.

Figure 6.25 shows the variation curve of the increment of composite acceleration relative to vertical acceleration with the radius of the circular curve when the longitudinal profile parameters are constant. The main calculation parameters are consistent with Fig. 6.24. From the figure, it can be seen that, first, under different line conditions, the comfort index requirements for riding can be met. Secondly, when the curve radius is greater than 7000 m, the vehicle vibration acceleration is significantly reduced. Therefore, when conditions permit, the radius of the circular curve at the lateral and vertical overlap should be greater than 7000 m.

In summary, for sections where the horizontal and vertical curves overlap, even when both the circular curve radius and the vertical curve radius are set to the minimum values allowed by the code, passenger comfort requirements can still be met. Therefore, under difficult conditions, there is no need for special requirements regarding the radii of the horizontal and vertical curves in overlapping sections.

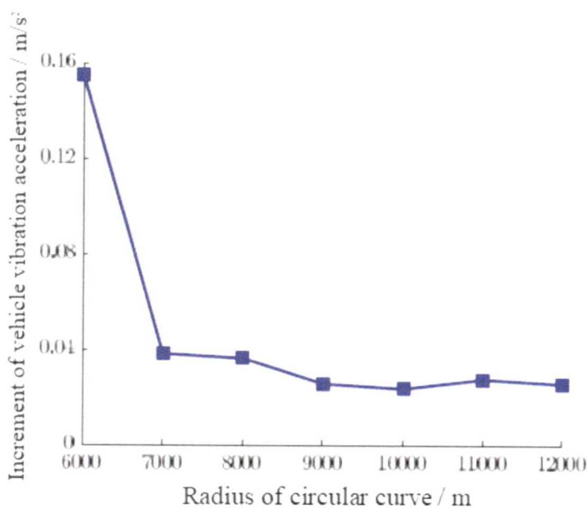


**Fig. 6.25** Variation curve of composite acceleration increment relative to vertical acceleration with circular curve radius

However, to further improve passenger comfort, when conditions permit, the radius of the circular curve should not be less than 7000 m.

2. **Vertical Curve Radius Analysis in Overlapping Horizontal and Vertical Curves**

Figure 6.26 shows the time history curve of vehicle vibration acceleration under different vertical curve radius conditions when the vertical curve overlaps with the circular curve. Figure 6.27 shows the change in the increment of the resultant acceleration relative to the lateral acceleration as the vertical curve radius varies, with the horizontal parameters fixed. The main calculation parameters are: running speed of 350 km/h; grade difference of 10‰; circular curve radius of 8000 m, length of 500 m, cant of 150 mm, and transition curve length of 590 m.

From the figures, it can be seen that under different vertical curve radius conditions, the vehicle acceleration satisfies the comfort requirements. Additionally, increasing the vertical curve radius can significantly improve the vertical vibration at the start of the vertical curve, thus reducing the peak of the resultant acceleration on the circular curve. However, when the grade difference is fixed, increasing the vertical curve radius will lead to an increase in the length of the vertical curve, which

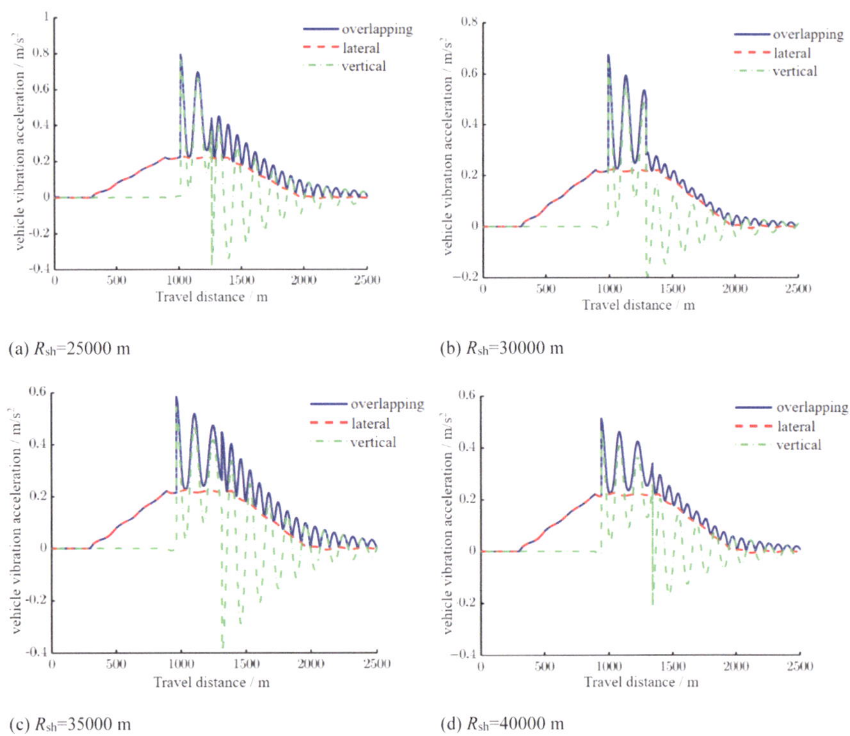


(a) $R_{sh}$=25000 m (b) $R_{sh}$=30000 m (c) $R_{sh}$=35000 m (d) $R_{sh}$=40000 m

**Fig. 6.26** Time history curves of vehicle vibration acceleration with overlapping lateral and vertical curves at different vertical curve radii

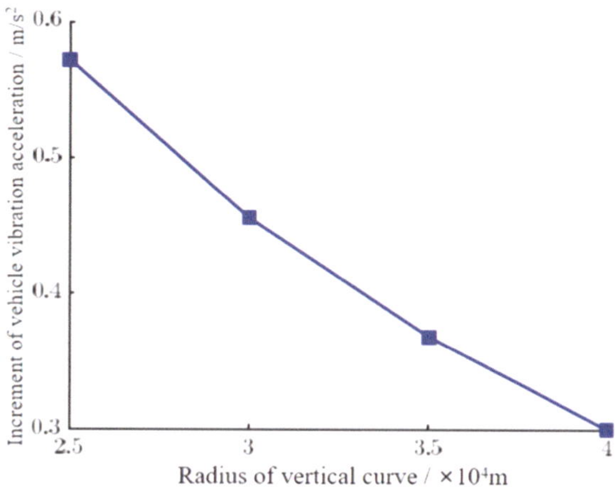


**Fig. 6.27** Variation curve of composite acceleration increment relative to lateral acceleration with vertical curve radius

will, to some extent, raise the overall level of the resultant acceleration on the circular curve. Furthermore, the distance between the transition curve and the circular curve, as well as between the circular curve and the vertical curve, will shorten, which is not conducive to passenger comfort. Therefore, the effect of the vertical curve radius on comfort is not monotonically changing when the horizontal and vertical curves overlap. The selection of the vertical curve radius should also consider factors such as the circular curve radius, length, and grade difference. When the grade difference is small and the circular curve length is long, the vertical curve radius should be appropriately increased. However, when the grade difference is large and the circular curve length is short, the vertical curve radius should not be too large.

3. **Summary of Overlapping Circular and Vertical Curve Settings**

In summary, even when both the circular curve radius and the vertical curve radius are at the minimum allowed values according to the specifications, the comfort requirements are still met in overlapping horizontal and vertical curve sections. Therefore, under difficult conditions, there is no need for special requirements for the radii of circular and vertical curves in overlapping sections. To further improve passenger comfort, when conditions allow, the circular curve radius should not be smaller than 7000 m.

Increasing the vertical curve radius can reduce the peak of the resultant acceleration. However, increasing the vertical curve radius also leads to a longer vertical curve, which raises the overall level of resultant acceleration on the circular curve. Therefore, the effect of the vertical curve radius on comfort is not monotonically changing in overlapping horizontal and vertical curve settings. When the grade difference is small and the circular curve length is long, the vertical curve radius should be appropriately increased. However, when the grade difference is large and the circular curve length is short, the vertical curve radius should not be too large.

### 6.3.2 Analysis of the Distance Between Horizontal and Vertical Curves

The "Design Code for High-Speed Railways" (*TB 10621-2014*) specifies that vertical curves should not overlap with transition curves, and the minimum distance between the start and end points of the vertical curve and the horizontal curve should not be less than 20 m. This section focuses on analyzing the effect of the distance between the horizontal curve and the vertical curve on passenger comfort under the condition that the curves do not overlap.

This section analyzes the effect of the distance between the horizontal curve and the vertical curve on vehicle dynamic performance under a train speed of 350 km/h. Other parameters for the line's vertical and horizontal profile: circular curve radius of 7000 m, transition curve length of 670 m, cant of 170 mm, circular curve length of 1000 m; vertical curve radius of 25000 m, vertical curve length of 1000 m, grade of 20‰, with both convex and concave cases. The distances between the horizontal and vertical curves are 0 m, 10 m, 20 m, 100 m, and 0.4 V. Figure 6.28 shows the schematic diagram of the vertical and horizontal curve overlap section of the railway.

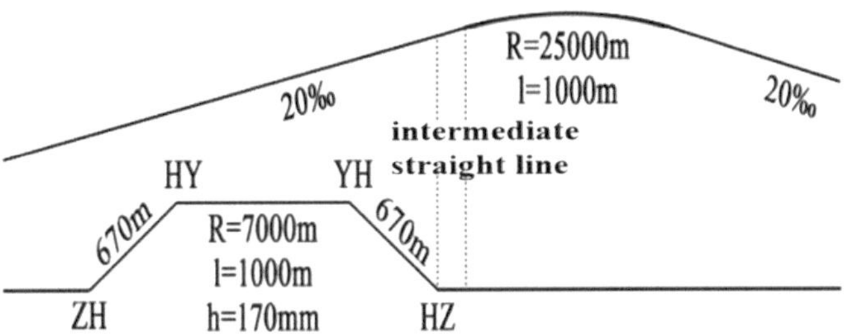


(a) Coincidence of vertical and horizontal curves

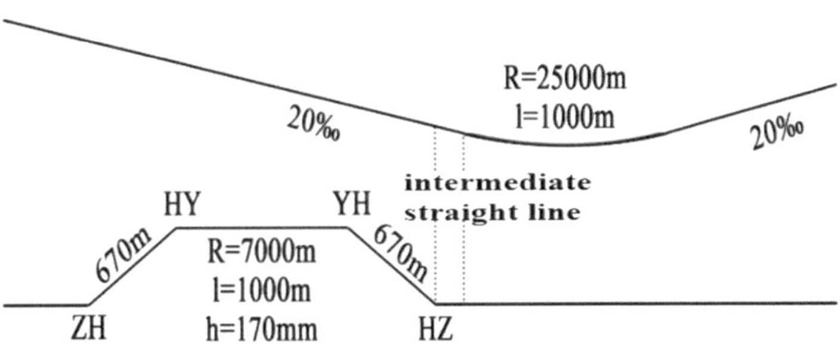


(b) Coincidence of vertical and horizontal curves

**Fig. 6.28** Schematic diagram of a track with coincident vertical and horizontal curves

Figures 6.29 and 6.30 show the impact of the distance between the horizontal and vertical curves on various dynamic performance indicators under convex and concave vertical curves, respectively. From the figures, it can be seen that under all operating conditions, all dynamic indicators are within the corresponding limits. The distance between the horizontal and vertical curves has little impact on the vehicle's lateral roll rate, derailment coefficient, and wheel-rail lateral force. As the distance increases, the peaks of the vehicle vertical acceleration and wheel load reduction rate are delayed, but the magnitude of the peaks is little affected.

The distance has a small effect on the peaks of the lateral and vertical acceleration of the vehicle. From the perspective of passenger comfort, to avoid continuous exposure to lateral and vertical vibrations, when conditions allow, the distance should be sufficient to ensure effective attenuation of lateral and vertical vibrations, typically no less than 0.4–0.6 V.

Figure 6.31 further provides the time history curves of the vehicle's lateral acceleration response under convex and concave vertical curves with different spacings (at 65 s when the train reaches the transition point). It can be observed that different spacings have a minor impact on the peak values, affecting only the time at which the peak values occur.

In summary, regarding the spacing between planar and vertical curves, under difficult conditions, there is no need for special requirements. Considering the maintenance and preservation of the line profile, setting a spacing of 20 m is appropriate. To further enhance riding comfort, when conditions permit, the spacing between planar and vertical curves should meet the requirements for effective decay of lateral and vertical vibrations, generally not less than 0.4–0.6 V.

### *6.3.3 Analysis of Vertical Curve and Transition Curve Overlap Settings*

In the design of railway lines in China, it has been stipulated that vertical curves and transition curves should not overlap. This is mainly due to considerations of train safety, passenger comfort, and the increased difficulty of line measurement and maintenance. However, for high-speed railways with ballastless track, the geometric stability of the line's vertical and horizontal profile is easier to maintain, and the difficulty of line measurement and maintenance in vertical and transition curve overlap settings is greatly reduced. Even for ballasted tracks, with current measurement and maintenance methods, the difficulty of measuring and maintaining line shape in vertical and transition curve overlap settings has also been significantly reduced.

From the practical experience of line design in engineering, under difficult conditions, allowing the overlap of the horizontal and vertical curves can greatly increase the flexibility of the line's vertical and horizontal profile design and reduce construction investment. This is especially important for high-speed railway connections

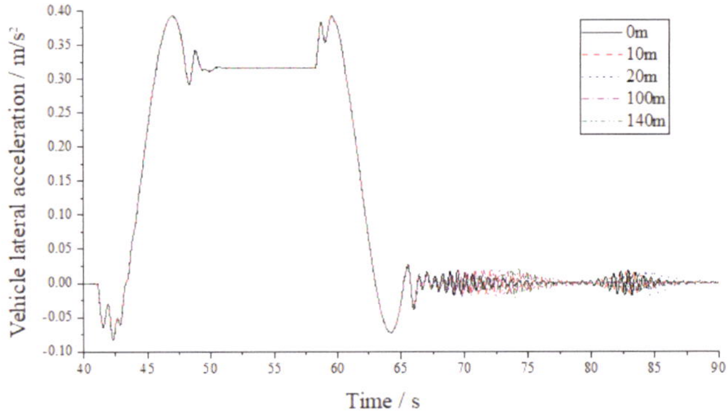


(a) Lateral acceleration of vehicle on a convex vertical curve

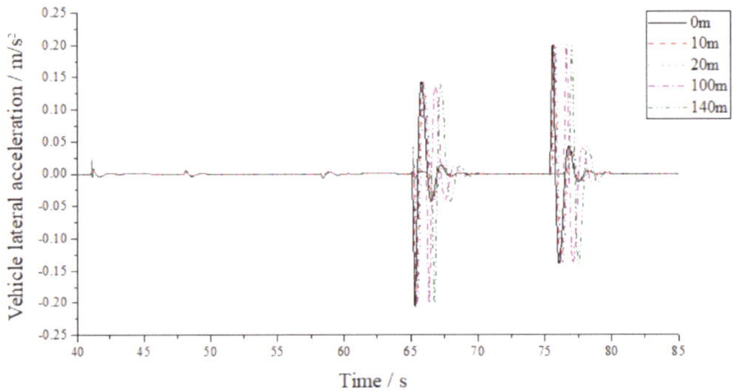


(b) Vertical acceleration of vehicle body on a convex vertical curve

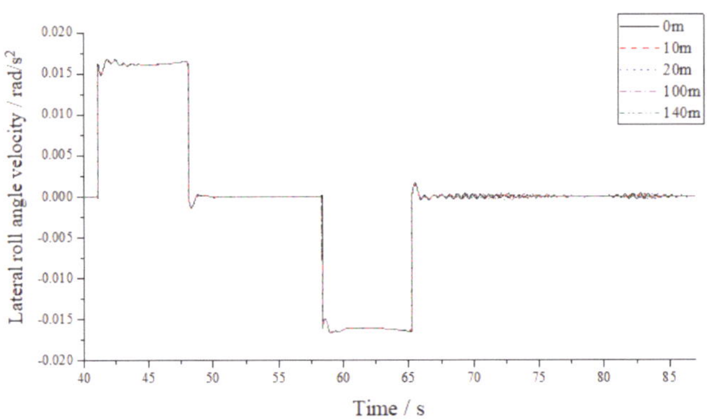


(c) Roll angular velocity of vehicle side on a convex vertical curve

**Fig. 6.29** Dynamic performance indices of convex vertical curves

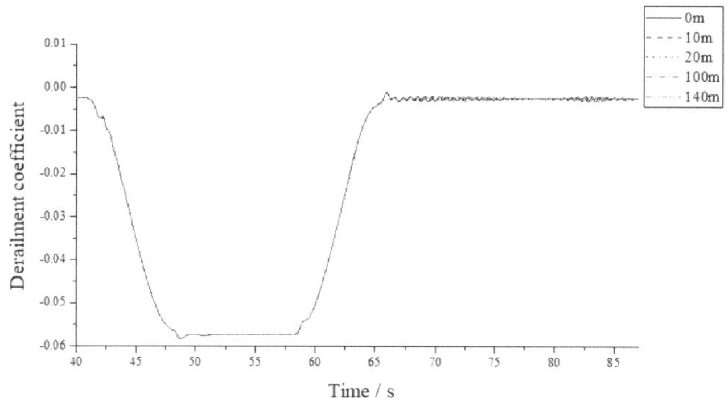


(d) Derailment coefficient on convex vertical curve

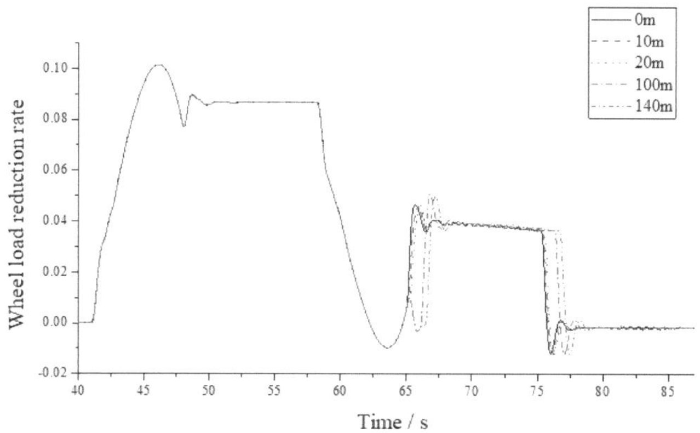


(e) Convex vertical curve wheel load reduction rate

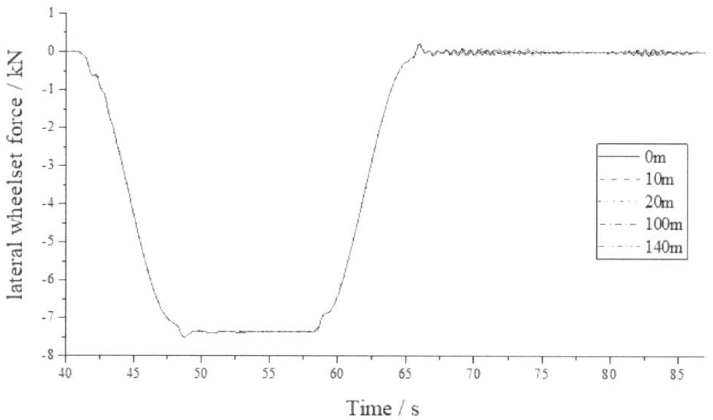


(f) Convex vertical curve lateral wheelset force

**Fig. 6.29** (continued)

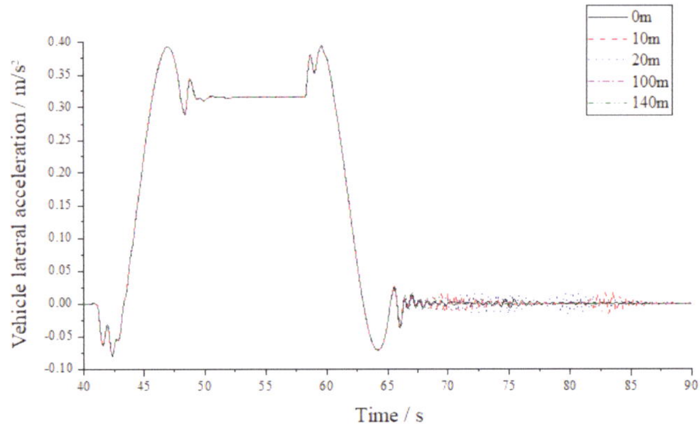


(a) Concave vertical curve lateral acceleration

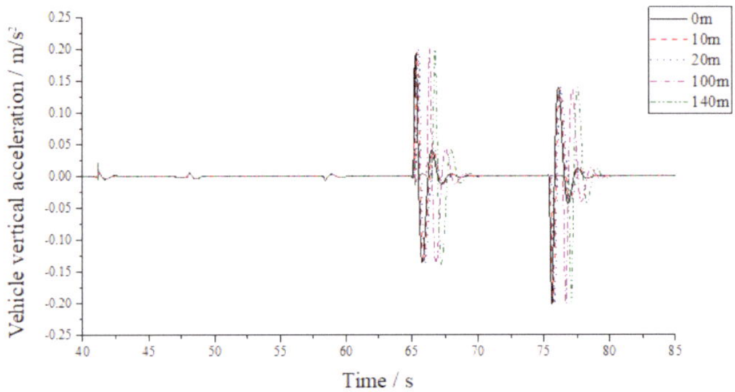


(b) Concave vertical curve vertical vibration acceleration

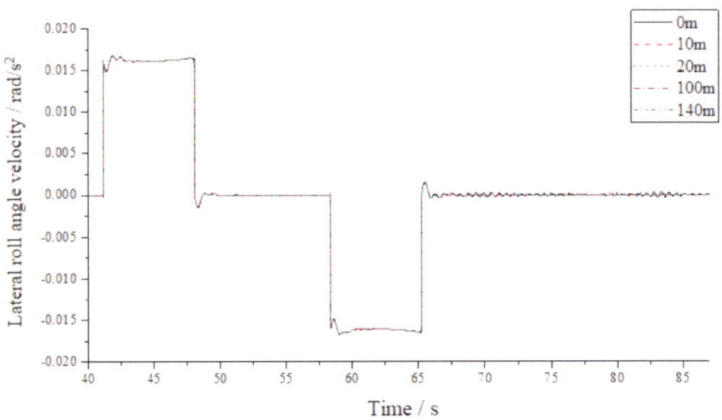


(c) Concave vertical curve vehicle roll angular velocity

**Fig. 6.30** Dynamic performance indices of concave vertical curves

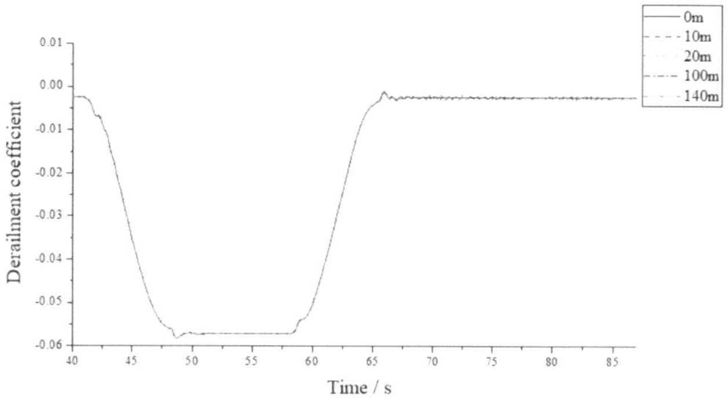


(d) Concave vertical curve derailment coefficient

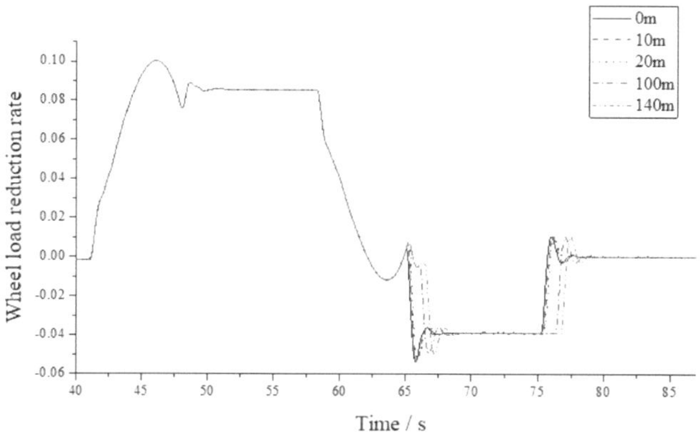


(e) Concave vertical curve wheel load reduction rate

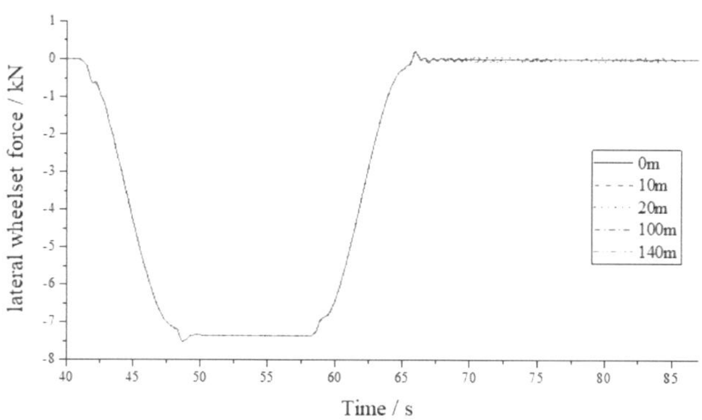


(f) Concave vertical curve lateral wheelset force

**Fig. 6.30** (continued)

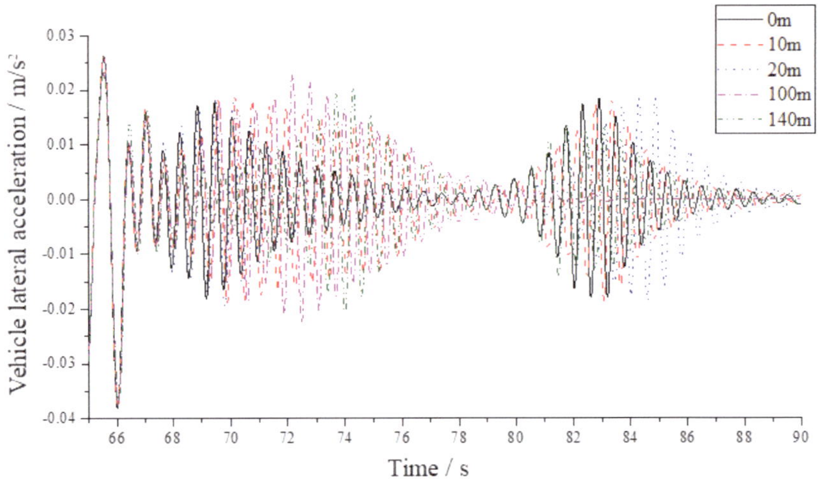


(a) Convex vertical curve vehicle lateral acceleration

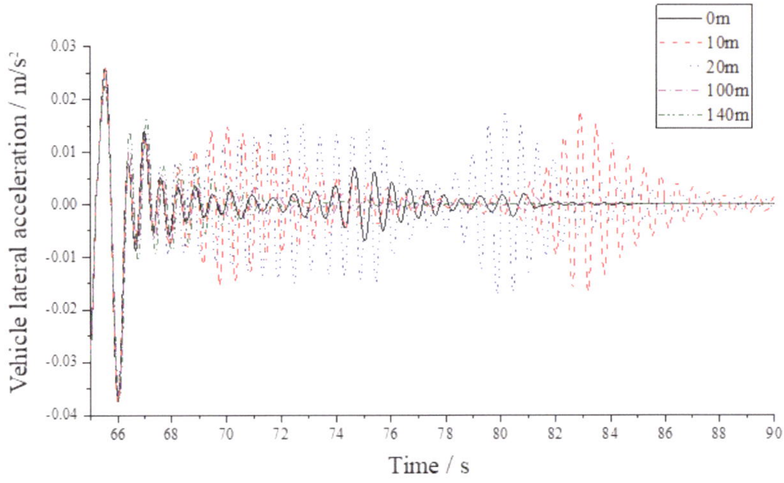


(b) Concave vertical curve segment vehicle body lateral acceleration

**Fig. 6.31** Lateral acceleration of vehicle on convex and concave vertical curve segments

within hubs, where cross-line relationships are complex, and there are many horizontal control points. When using steep gradients, the vertical curve length is relatively long, and often due to the inability to overlap the vertical and transition curves, valuable schemes are abandoned or the line is extended considerably.

Looking at relevant standards and design experience from abroad, European railway standards do not prohibit the overlap of horizontal transition curves and vertical curves, but instead provide requirements for lateral and vertical comfort indicators. The Japanese Shinkansen, however, stipulates that when the radius of the vertical curve is less than 15,000 m, it should not overlap with the transition curve; when the radius of the vertical curve is greater than 15,000 m, it can overlap with the transition curve.

Considering the current development status of railway technology in different countries, it is necessary to re-evaluate the long-standing requirement in China that vertical and transition curves cannot overlap. This section will conduct a dynamic simulation analysis of the effects of the overlap of vertical and transition curves on train safety and passenger comfort.

The dynamic simulation includes three typical conditions: non-overlapping vertical and transition curves, overlapping vertical and transition curves (where the vertical curve is longer than the transition curve, i.e., vertical curve wrapping transition curve), and overlapping vertical and transition curves (where the transition curve is longer than the vertical curve, i.e., transition curve wrapping vertical curve). The specific line parameters are shown in Table 6.15.

**Table 6.15** Line condition design and parameters

| Working condition | Speed km/h | Circular curve radius/m | Vertical curve radius/m | Length of transition curve/m | Length of circular curve/m | Cant/mm | Algebraic difference of grades/‰ | Length of Vertical Curve/m | Notes |
|---|---|---|---|---|---|---|---|---|---|
| Working condition 1 | 350 | 7000 | 0 | 670 | 500 | 170 | 0 | 0 | Planar curve only |
| Working condition 2 | 350 | 7000 | 25,000 | 670 | 500 | 170 | 10 | 250 | Horizontal curve enclosing certical curve |
| Working condition 3 | 350 | 7000 | 25,000 | 670 | 500 | 170 | 40 | 1000 | Vertical curve enclosing horizontal curve |
| Working condition 4 | 350 | 7000 | 30,000 | 670 | 500 | 170 | 10 | 300 | Horizontal curve enclosing vertical curve |
| Working condition 5 | 350 | 7000 | 30,000 | 670 | 500 | 170 | 40 | 1200 | Vertical curve enclosing horizontal curve |

Figure 6.32 shows the time history curves of key train dynamic performance indicators under various conditions of vertical and transition curve overlap. Table 6.16 provides the peak values of each indicator. The following conclusions can be drawn:

(1) Under all conditions, the safety and passenger comfort indicators meet the corresponding requirements, and the peak value differences are small, generally less than the effect of track irregularities on train safety. In other words, the overlap of the vertical and transition curves has little effect on train safety and passenger comfort.
(2) In conditions 2 and 4 (transition curve wrapping vertical curve), the derailment coefficient peak value is 0.058. In conditions 3 and 5 (vertical curve wrapping transition curve), the peak values are 0.065 and 0.064, respectively. Regarding the derailment coefficient, the condition where the vertical curve wraps the transition curve is more unfavorable for train operations.
(3) Since the vertical curve radius in conditions 2 and 3 is smaller than in conditions 4 and 5, it can be seen that increasing the vertical curve radius can reduce the vertical acceleration peak value in the overlap section, thereby improving passenger comfort.

In summary, due to the relatively high standards for horizontal and vertical curve parameters in current high-speed railways, after implementing the vertical and transition curve overlap, all dynamic performance indicators meet the corresponding requirements, with only small differences compared to the non-overlapping conditions. Therefore, from the perspective of train safety and passenger comfort, the feasibility of vertical and transition curve overlap should not be denied.

The overlap of the vertical curve and transition curve can greatly improve the design flexibility of the line in difficult sections, reduce construction investment, and enhance the overall project benefits. Therefore, from a theoretical standpoint, relaxing the restrictions on the overlap of vertical and transition curves is feasible. It is recommended to further conduct research, evaluations, and corresponding tests and experiments, considering construction, maintenance, and technical requirements.

## 6.4 Key Line Parameter Analysis for Upgrading Existing 350 km/h High-Speed Rail

### *6.4.1 Minimum Horizontal Curve Radius Analysis*

This section analyzes the impact of horizontal curve radius on vehicle dynamic response under different speed conditions, with cant (superelevation) set to 175 mm and considering track irregularity. For speeds of 300 km/h, 350 km/h, 400 km/h, and 450 km/h, the corresponding curve radii matching the actual cant are 6068 m, 8260 m, 10,788 m, and 13,654 m, respectively.

**Fig. 6.32** Dynamic simulation analysis results of vertical and transition curve overlap setup

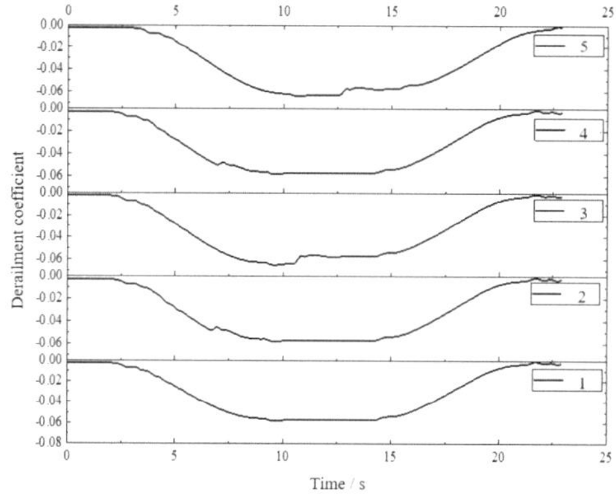


(a) Derailment coefficient

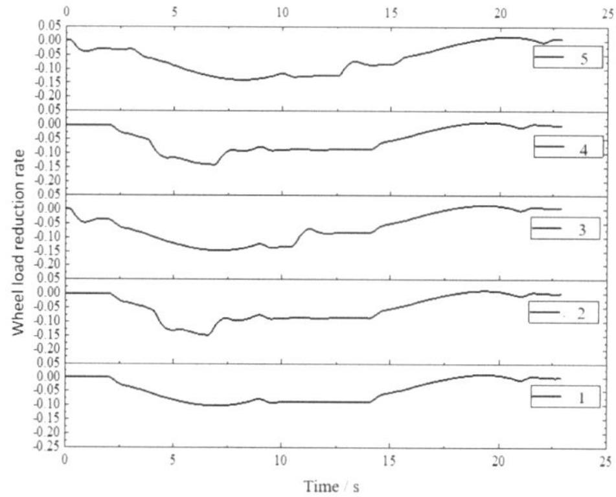


(b) Wheel load reduction rate

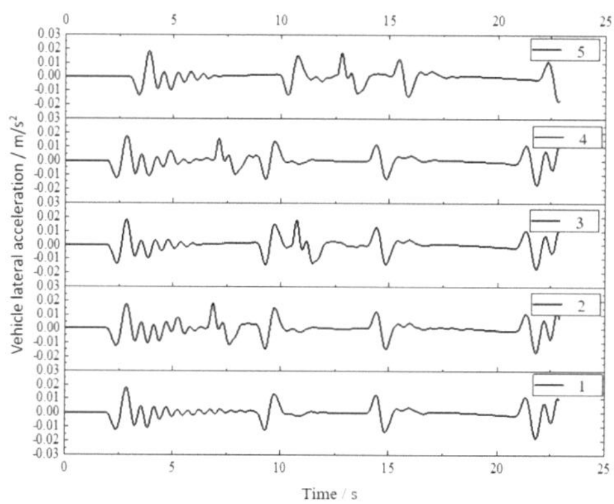


(c) Lateral acceleration of vehicle

**Fig. 6.32** (continued)

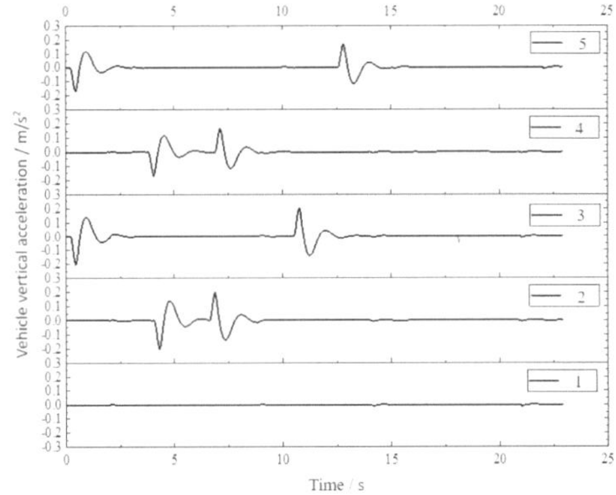


(d) Vertical acceleration of vehicle

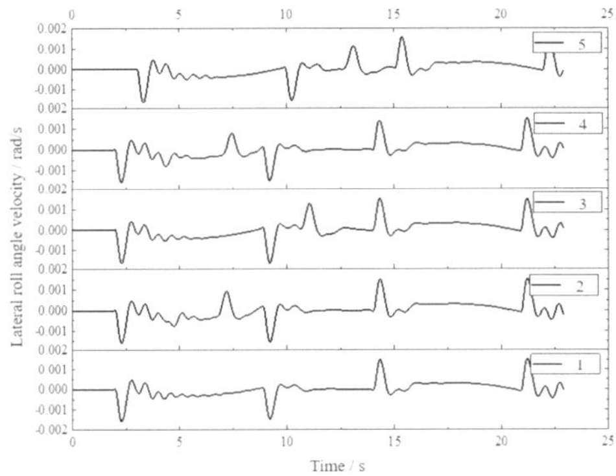


(e) Vehicle side roll angular velocity

**Table 6.16** Peak value statistics of dynamic performance indices for vehicle operation with vertical curve and transition curve overlapping settings

| Working condition | Derailment coefficient | Wheel load reduction rate | Lateral acceleration of vehicle m/s$^2$ | Vertical acceleration of vehicle m/s$^2$ | Vehicle side roll angular velocity rad/s |
|---|---|---|---|---|---|
| Working condition 1 | 0.05829 | 0.10158 | 0.01798 | 0.01172 | 0.00158 |
| Working condition 2 | 0.05826 | 0.14733 | 0.01825 | 0.20486 | 0.00158 |
| Working condition 3 | 0.06491 | 0.14708 | 0.01793 | 0.20387 | 0.00162 |
| Working condition 4 | 0.05822 | 0.14022 | 0.01780 | 0.17021 | 0.00158 |
| Working condition 5 | 0.06379 | 0.13932 | 0.01807 | 0.16956 | 0.00162 |

Figure 6.33 shows the simulation results of safety indicators under different speeds and curve radii conditions. Under the 300 km/h condition, the curve radius of 6068 m matches the equilibrium cant of 175 mm. As the curve radius increases, the cant also increases, resulting in a progressive increase in the peak values of all three safety indicators. Under the 350 km/h condition, as the curve radius increases, the safety indicator peak values first decrease and then increase, with the minimum value occurring near a radius of 8500 m. Under the 400 km/h and 450 km/h conditions, when the curve radius increases from 6000 m to 9000 m, the under-cant decreases, and all three safety indicators show a decreasing trend, effectively improving the interaction between the wheel and rail. Under the 400 km/h condition, the derailment coefficient and wheel-axle lateral force are far below the limit values. At a curve radius of 6000 m, the wheel load reduction rate reaches 0.79, close to 0.8 but not exceeding the limit. Under the 450 km/h condition, when the curve radius is smaller than 8000 m, the wheel load reduction rate exceeds 0.8. Therefore, it is recommended that, under difficult conditions, a curve radius of 7500 m be used to meet the safety requirements for operating trains at 400 km/h.

Figures 6.34 and 6.35 show the simulation results of the vehicle body acceleration and Sperling ride quality index under different speed and curve radius conditions. The vehicle's lateral acceleration is judged using the 10 Hz low-pass filtered values based on Class III and IV standards, while the vertical acceleration is judged using the 20 Hz low-pass filtered values. Under the 300 km/h and 350 km/h conditions, the variation pattern of the vehicle's lateral acceleration peak is the same as that of the safety indicators peak, both reaching a minimum near the curve radius matching the actual cant. The Sperling lateral ride quality index shows minimal variation, rated as "Excellent" and "Good" respectively. Under the 400 km/h and 450 km/h conditions, as the curve radius increases, the under-cant decreases, and both the vehicle lateral acceleration peak and Sperling lateral ride quality index show a decreasing trend, with a significant decline. Under the 400 km/h condition, when the curve radius is smaller than 6500 m, the vehicle lateral acceleration exceeds the Class III standard. When the curve radius is smaller than 7000 m, the Sperling lateral ride quality index is rated as "Unqualified". With a curve radius of 7500 m, the vehicle lateral acceleration reaches 0.78 $m/s^2$, which is well below the Class III and IV standards, meeting comfort requirements, and the Sperling lateral ride quality index reaches 2.66, rated as "Good". Therefore, it is recommended that, under difficult conditions, a curve radius of 7500 m be used to meet the comfort requirements for high-speed trains operating at 400 km/h. Under the 450 km/h condition, when the curve radius is smaller than 8000 m, the vehicle lateral acceleration exceeds the Class III standard; when the curve radius is smaller than 8500 m, the Sperling lateral ride quality index is rated as "Unqualified". Under all speed conditions, as the horizontal curve radius increases, the vehicle's vertical acceleration decreases, and the change is more gradual. The Sperling vertical ride quality index shows minimal variation, indicating that the change in curve radius has little effect on the vehicle's vertical acceleration.

**Fig. 6.33** Safety indexes

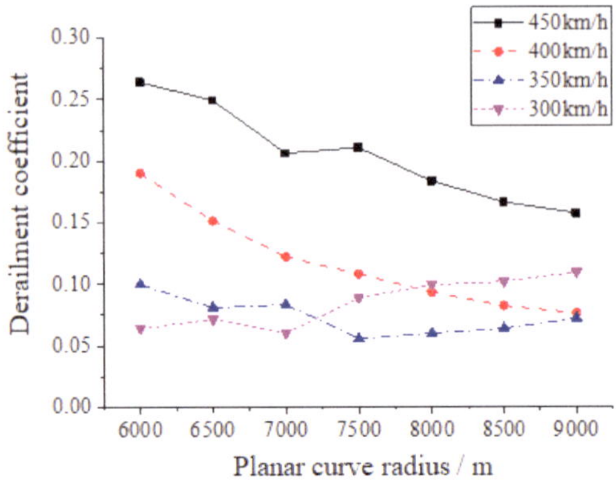


(a) Derailment coefficient

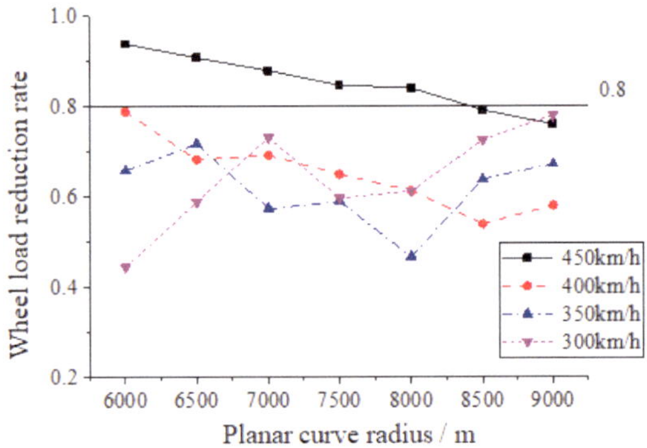


(b) Wheel load reduction rate

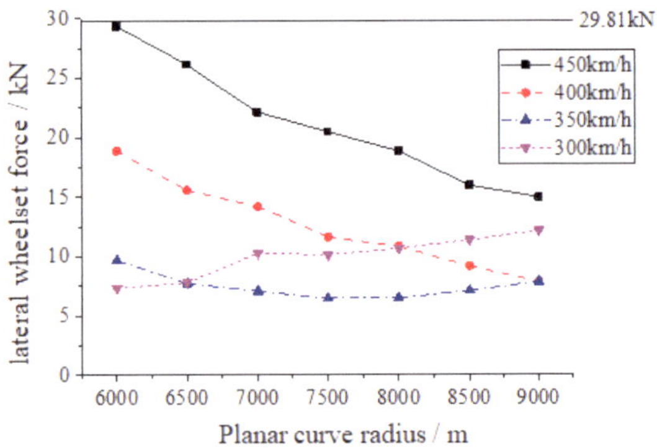


(c) Lateral wheelset force

**Fig. 6.34** Vehicle acceleration

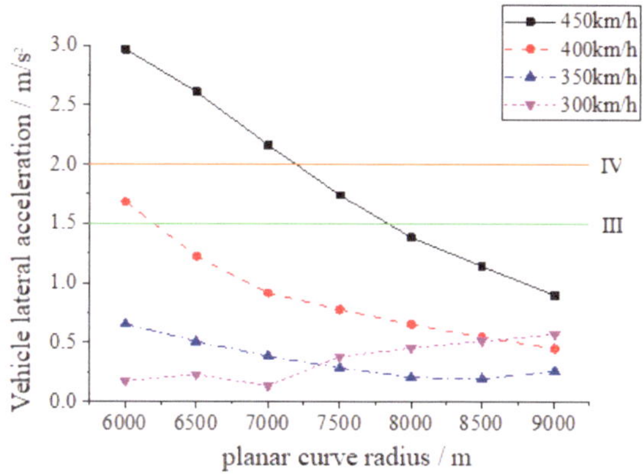


(a) Vehicle lateral acceleration

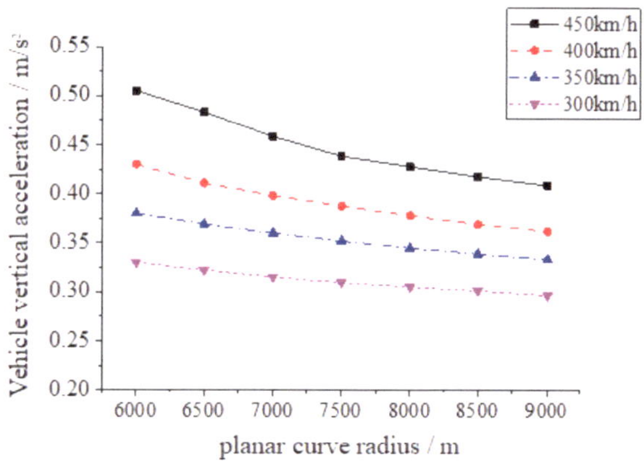


(b) Vehicle vertical acceleration

## *6.4.2 Cant Analysis*

This section analyzes the impact of cant on vehicle dynamic responses under different speed conditions, with the horizontal curve radius set to 8000 m and track irregularities considered. Under the 350 km/h and 400 km/h conditions, the corresponding balanced cant values are 181 mm and 236 mm, respectively.

As shown in Fig. 6.36, when the cant increases from 80 to 260 mm, the peak values of the three safety indicators initially decrease and then increase. When the designed cant is 190 mm, the high-speed trains running at 350 km/h and 400 km/h on the 8000 m radius curve exhibit similar driving performance. Under the 350 km/h condition, the safety indicator peaks reach their lowest values when the cant is between 140 and 160 mm. At this point, the actual cant is 20 mm less than the balanced cant. Under the 400 km/h condition, the safety indicator peaks reach their

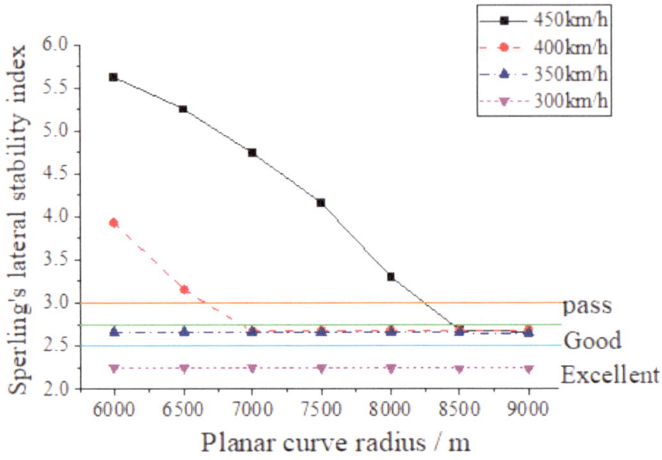


(a) Sperling lateral ride quality index

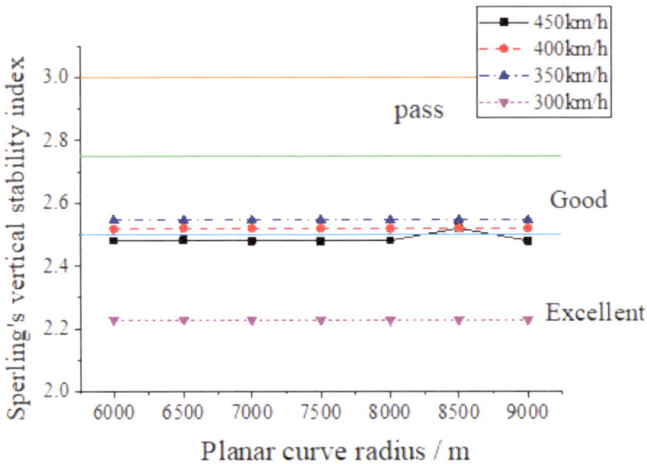


(b) Sperling vertical ride quality index

**Fig. 6.35** Sperling ride quality index

lowest values when the cant is between 200 and 220 mm. At this point, the actual cant is 20 mm less than the balanced cant. In summary, a 20 mm under-cant results in a better safety evaluation compared to the balanced cant condition.

As shown in Fig. 6.37, the variation pattern of the peak vehicle lateral acceleration is the same as that of the safety indicator peaks under the 350 km/h and 400 km/h conditions. With a horizontal curve radius of 8000 m and a speed of 400 km/h, when the cant increases from 80 to 140 mm, the vehicle lateral acceleration rapidly

**Fig. 6.36** Safety indexes

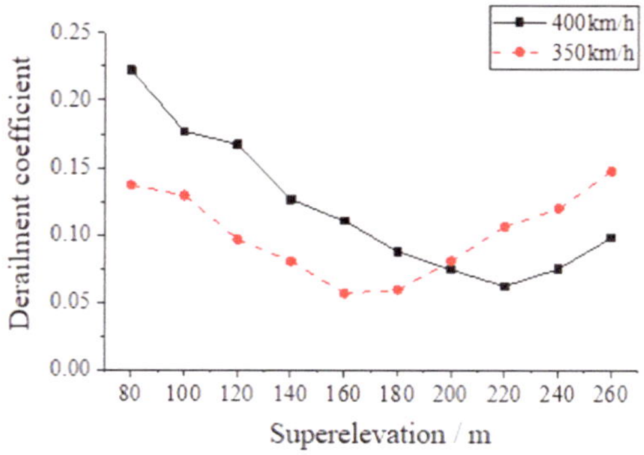


(a) Derailment coefficient

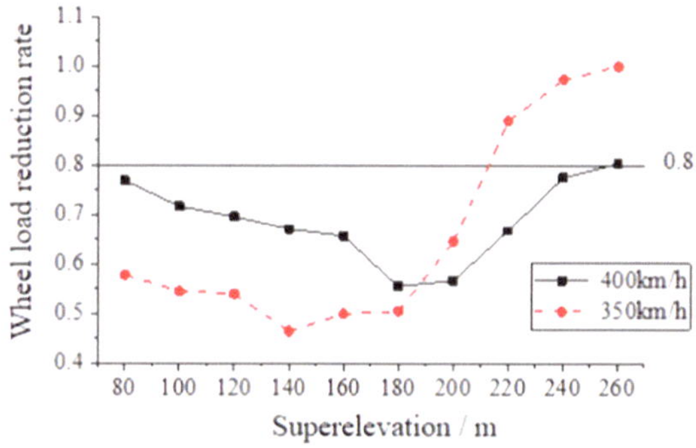


(b) Wheel load reduction rate

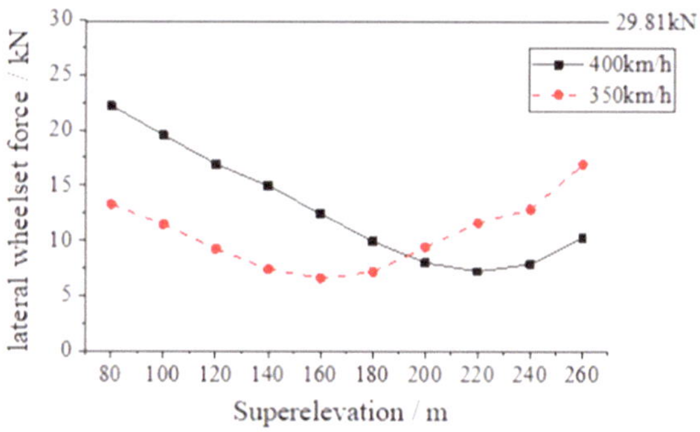


(c) Lateral wheelset force

decreases from 2.02 to 0.93 m/s$^2$. This is because the under-cant has exceeded the maximum allowed under-cant value of 90 mm, and the change in cant significantly impacts comfort. The vehicle's vertical acceleration increases with the increase in cant, and all comfort requirements are met. The increase in cant causes the cant-induced slope change to intensify, which leads to a rise in vertical acceleration. Thus, the variation in cant has a considerable impact on vertical comfort.

This section, based on the vehicle-track system dynamics theory, calculates the recommended design parameters for a 400 km/h high-speed railway key section according to current high-speed rail design standards. Through statistical analysis of simulation results, the matching relationship between the horizontal curve radius, cant, and speed under higher speeds is studied from the perspectives of safety and comfort. Additionally, a dynamic evaluation of a certain high-speed rail section is conducted to assess the adaptability of operating higher-speed trains after matching various design parameters. The research conclusions are as follows:

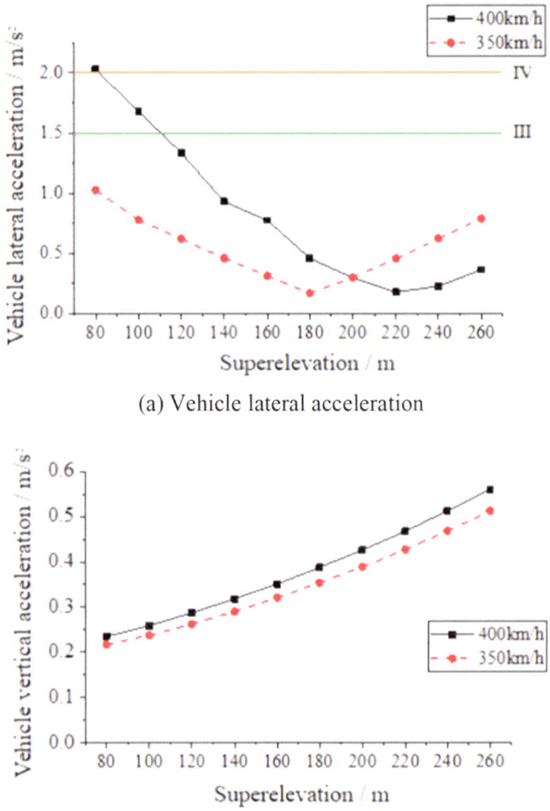


(a) Vehicle lateral acceleration

(b) Vehicle vertical acceleration

**Fig. 6.37** Vehicle acceleration

(1) It is recommended to design the minimum horizontal curve radius for a 400 km/h high-speed railway to be 9000 m under optimal conditions. Under general conditions, the radius can be 8500 m, and under difficult conditions, it can be relaxed to 7500 m. The design cant should be 175 mm.

(2) A vehicle's dynamic response is optimal with an under-cant of around 20 mm. Therefore, a slight under-cant is more beneficial to both the safety and comfort of the train operation.

# Chapter 7
# Dynamic Evaluation of Line Schemes

The alignment of the track and its compatibility with the subgrade structures largely determine the dynamic response of the vehicle-track system and directly impact the operational quality. Conducting dynamic evaluation and optimization during the line scheme research phase is of great significance for improving the train's dynamic performance after the high-speed railway begins operation and reducing the maintenance and repair costs of the subgrade structures. Furthermore, in overseas high-speed railway projects, a dynamic performance evaluation report of the line scheme is often required. This chapter introduces the application of track system dynamics theory in the dynamic evaluation and optimization of high-speed railway line schemes, based on the dynamic model and the line scheme evaluation indicators and methods established earlier, using actual engineering cases such as the Beijing-Shenyang High-Speed Railway (Jinshen HSR) and the Jakarta-Bandung High-Speed Railway of Indonesia.

## 7.1 Dynamic Evaluation of the Beijing-Shenyang High-Speed Railway Trial Section Line Scheme

The Beijing-Shenyang High-Speed Railway is located in Beijing, Hebei Province, and Liaoning Province. The line starts from Beijing Chaoyang Station and extends northeast along the Beijing-Chengde Expressway through Chengde City in Hebei Province, Chaoyang City, Fuxin City, Jinzhou City in Liaoning Province, and ends in Shenyang City at Shenyang Station. The total operational length of the line is 696.764 km, with a design speed of 350 km/h.

The Beijing-Shenyang High-Speed Railway sets up a pilot section in Liaoning Province, which serves as part of the Beijing-Shenyang High-Speed Railway while also conducting comprehensive high-speed railway tests in this section. The tests are planned to focus on the development needs of China's high-speed rail and the

© China Railway Publishing House Co., Ltd. 2026

X. Long et al., *Vehicle–Track–Bridge System Subjected to Track Alignment*,
https://doi.org/10.1007/978-981-95-1649-0_7

climate characteristics of the northeast region, involving research and testing on self-developed high-speed railway signaling systems, the next-generation LTE-R broadband mobile communication system, infrastructure state prediction and health management (PHM) based on big data systems, key materials, structures and technologies in civil engineering, key technologies and applications of the next-generation traction dynamic supply system, and high-speed testing for trains running at speeds over 350 km/h, among other systems and different professional fields.

### 7.1.1 Line Conditions

1. **Overview of Trial Section**

The high-speed comprehensive trial section is located between Chaoyang North Station (DK417 + 000) and Guchengzi Station (DK676 + 000). The total length of this newly constructed section is 255.025 km. It includes 7 stations: Chaoyang North, Beipiao East, Ulanmutu, Fuxin North, Heishan North, Xinmin North, and Guchengzi. The subgrade length is 74.761 km. There are 78 bridges, with a total length of 157.56 km, accounting for 61.78% of the section's total length. The longest bridge is the Xinglongdian Extra Large Bridge, which is 23.6 km long. There are 15 tunnels, with a total length of 22.704 km, accounting for 8.9%. The longest tunnel is the Koumo Mingdong Tunnel, which is 2.47 km long. The track type is the CRTS III slab ballastless track.

The trial section of the Beijing-Shenyang High-Speed Railway from DK417 + 000 to DK676 + 000 includes 23 designed curves, with a total curve length of 65.191 km, accounting for 25.56% of the total line length. The smallest curve radius is 8000 m, with 13 locations; the largest curve radius is 12,000 m, with 1 location; there are 5 curves with a radius of 9000 m and 3 curves with a radius of 10,000 m. The longest straight section is 27.330 km, located between DK609 + 657.743 and DK640 + 987.694. This straight section includes bridges, subgrade, and stations, and has a maximum gradient of 5.6‰ (slope length 2100 m). Another long straight section is located between DK575 + 849.241 and DK595 + 145.230, with a length of 19.296 km. This section includes subgrade, bridges, tunnels, and stations, and has a maximum gradient of 10.3‰ (slope length 2000 m). These two long straight sections provide suitable conditions for conducting high-speed comprehensive tests.

2. **Dynamic Analysis Section Description**

To simulate the dynamic performance of the test section, the typical section from DK473 + 238 to DK533 + 228 is selected as the input for the dynamic performance analysis of the train operation. The longitudinal and horizontal profile parameters of this section are shown in Fig. 7.1. The calculation model uses the CRH380 train. The track irregularity spectrum of China's high-speed railway ballastless track is inverse-transformed, with a wavelength range of 2–120 m, to obtain the left and right rail track vertical irregularity and track alignment irregularity. This section has a total

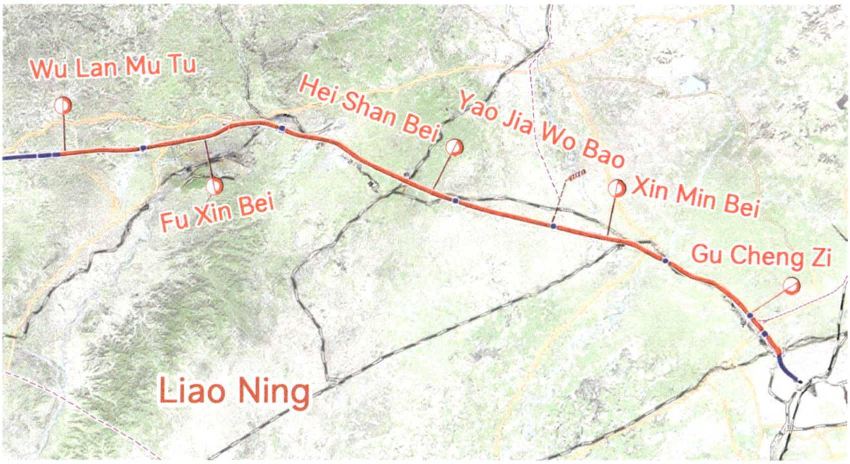


**Fig. 7.1** Schematic diagram of the horizontal alignment of the pilot section of the Beijing-Shenyang railway passenger dedicated line

**Table 7.1** Typical section plan curve statistical table

| Curve radius (m) | Actual cant (mm) | Length of transition curve (m) |
|---|---|---|
| 8000 | 125 | 590 |
| 9000 | 120 | 530 |
| 10,000 | 110 | 470 |
| | 100 | |
| 12,000 | 90 | 370 |

length of 60 km, with 6 curves and 24 longitudinal slope segments. The parameters of the plan curves for this section are listed in Table 7.1.

## 7.1.2 Dynamic Analysis and Evaluation

1. **System Dynamic Response Time History Analysis**

Figure 7.2 shows the dynamic response time history curves at operating speeds of 350 and 400 km/h through the test section. From the figure, the following observations can be made:

① The vertical acceleration of the vehicle ranges between − 0.03 and 0.03 g, meeting the comfort control limit of 0.13 g.

② At a speed of 350 km/h, the lateral acceleration of the vehicle ranges between − 0.03 and 0.03 g, which meets the comfort control limit of 0.1 g. At 400 km/h, the lateral acceleration response is larger on curves with a radius of 8000 m, ranging between – 0.15 and 0.15 g, with some local sections exceeding the 0.1 g comfort control limit.

③ At 350 km/h, the derailment coefficient is very small, ranging between – 0.15 and 0.1, which is well below the safety limit of 0.8. At 400 km/h, the derailment coefficient is slightly larger than at 350 km/h, ranging between – 0.3 and 0.3, but still smaller than the safety limit of 0.8.

④ At a running speed of 350 km/h, the wheel load reduction rate ranges approximately from – 0.35 to 0.4, which is below the safety threshold of 0.8 for dynamic wheel load reduction. At 400 km/h, the ratio ranges approximately from – 0.55 to 0.55, which also remains within the safe limit of 0.8.

⑤ According to the definition of the wheelset lateral force limit, the safety threshold is determined to be 56.28 kN. At an operating speed of 350 km/h, the wheelset lateral force ranges between – 17 and 15 kN, which is within the safety limit. When the train operates at 400 km/h and passes through a curve with a radius of 8000 m, the local maximum lateral wheel–rail force reaches approximately 40 kN, which is approaching the safety threshold of 56.28 kN.

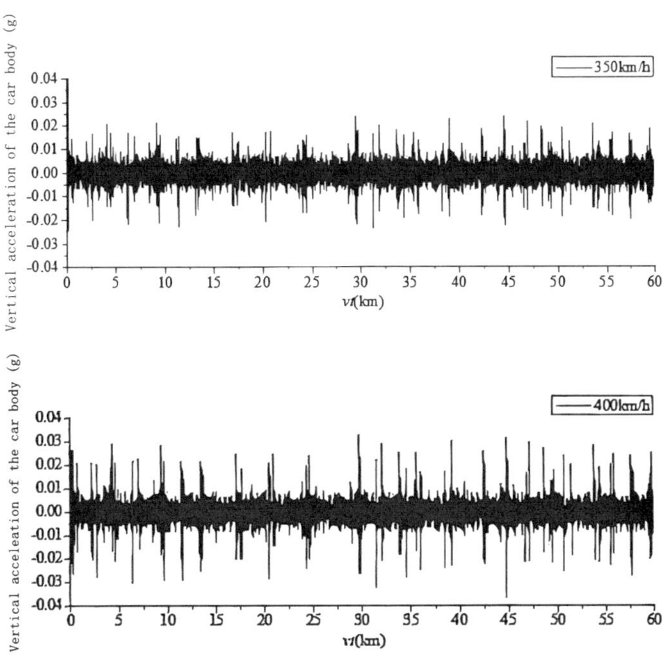


(a) The vertical acceleration of the car body

**Fig. 7.2** Time-dependent curves of the dynamic response of the test section under 350 km/h and 400 km/h operating conditions

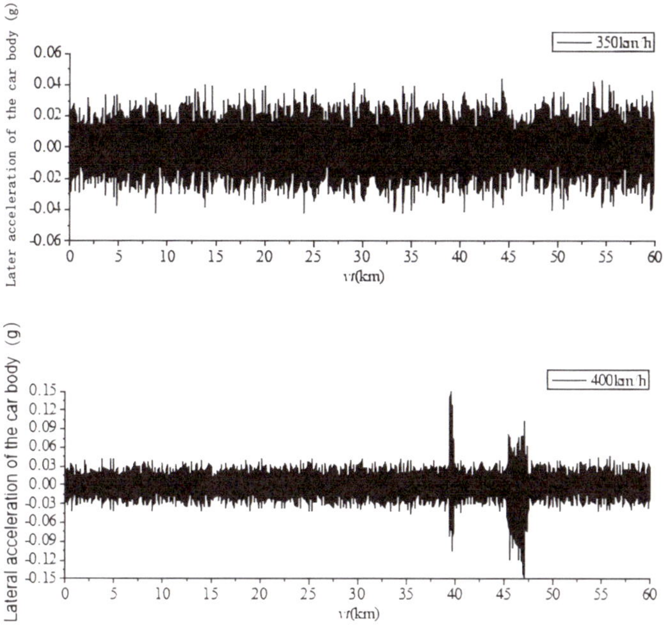


(b) The lateral acceleration of the car body

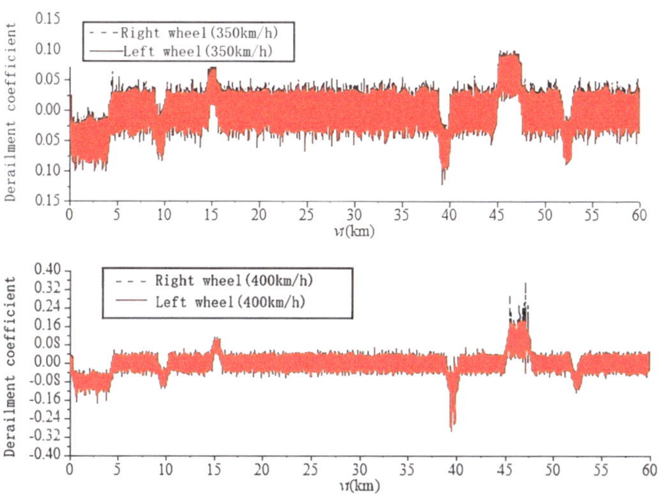


(c) Derailment coefficient

**Fig. 7.2** (continued)

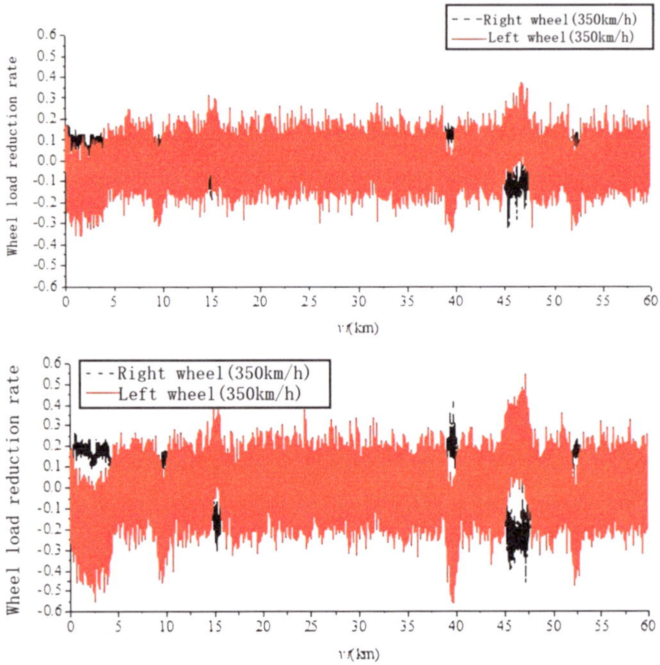


(d) Wheel load reduction rate

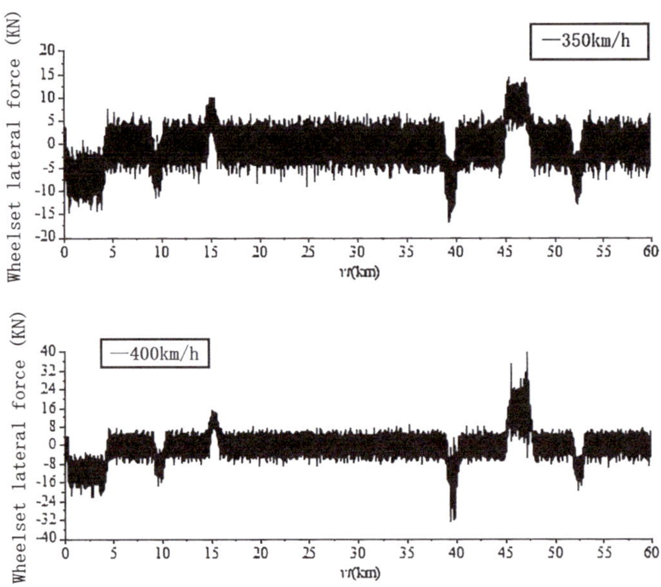


(e) Wheelset lateral force

**Fig. 7.2** (continued)

2. **System Dynamic Response Statistical Analysis**

(1) Dynamic performance analysis of the overall track alignment scheme

Figure 7.3 shows the frequency statistics of the dynamic response indicators for each line section. From the overall statistical results of the line scheme, it can be observed that when the operating speed increases to 400 km/h, the frequency of occurrence of the higher values of various dynamic indicators increases. Among these indicators, the wheel load reduction rate shows a particularly significant change.

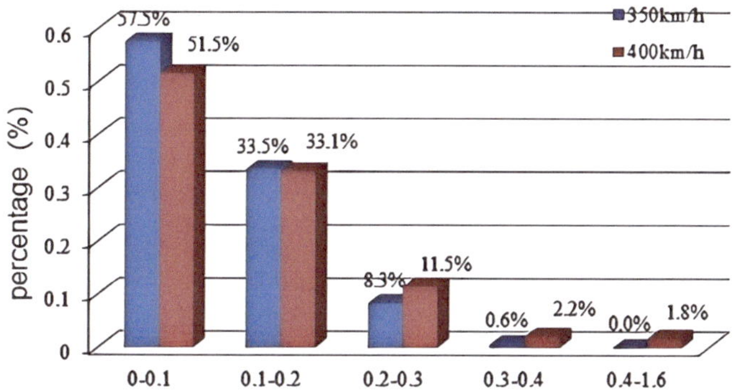


(a) Lateral acceleration of car body (m/s $^2$)

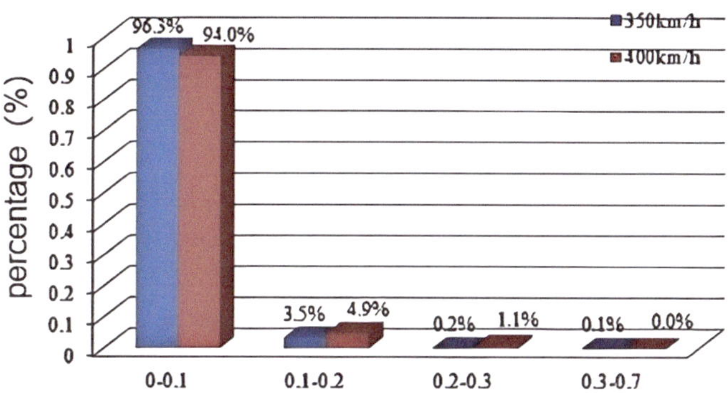


(b) Vertical acceleration of car body (m/s $^2$)

**Fig. 7.3** Line program dynamics indicator statistics

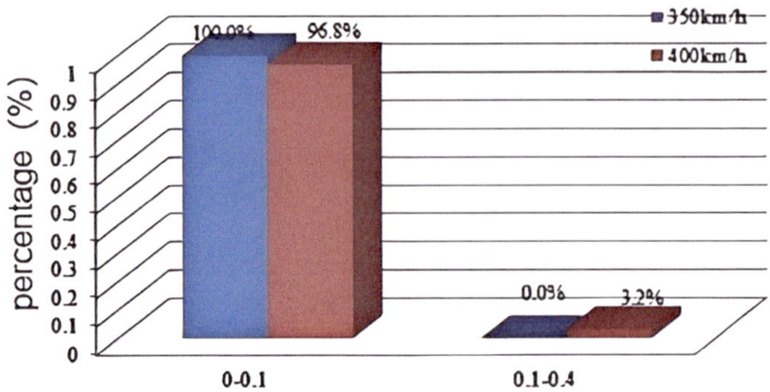


(c) Derailment coefficient

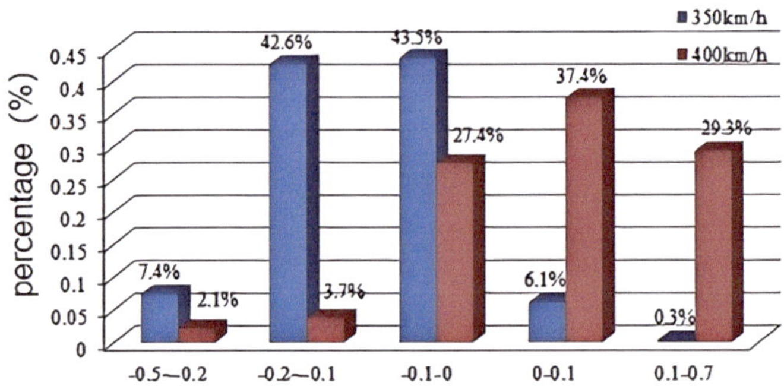


(d) Wheel load reduction rate

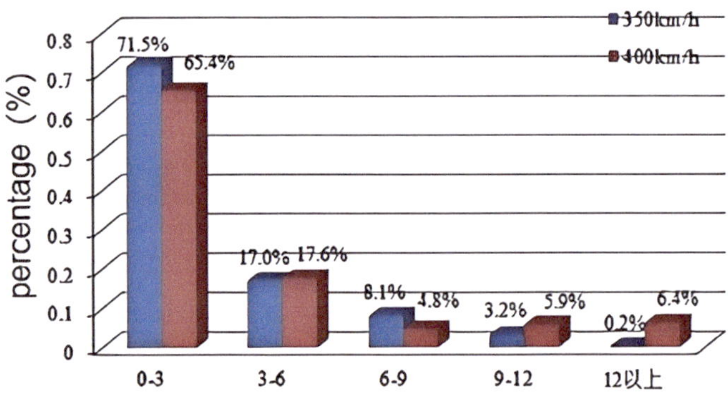


(e) Wheelset lateral force (kN)

**Fig. 7.3** (continued)

(2) Dynamic indicators under different radius conditions

Figure 7.4 shows the statistical results of the dynamic indicators under different curve radius conditions. From the figure, the following can be observed:

**Fig. 7.4** Dynamic indicators for different radius conditions

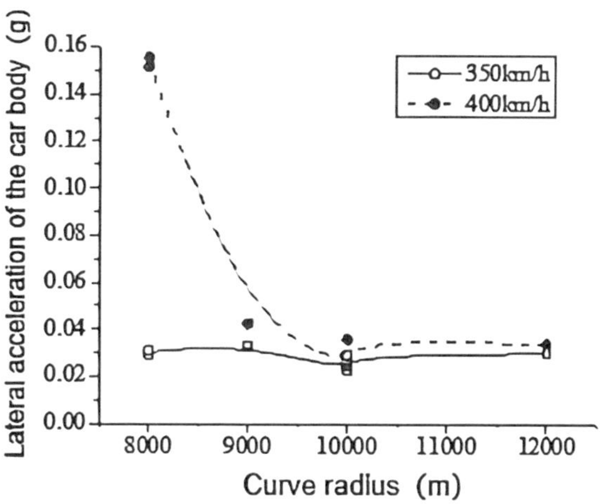


(a) Maximum value

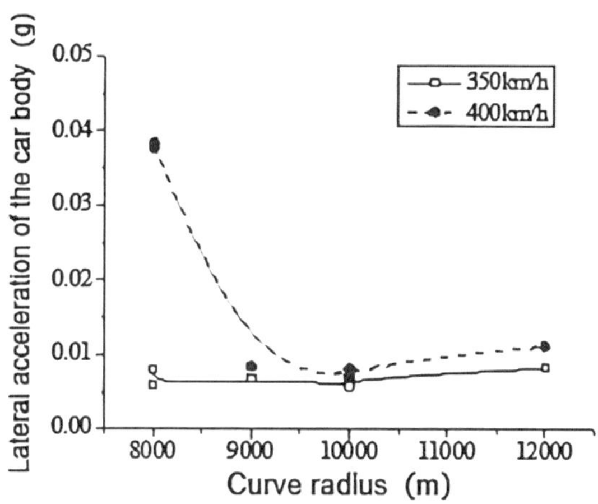


(b) Average value

**Fig. 7.4** (continued)

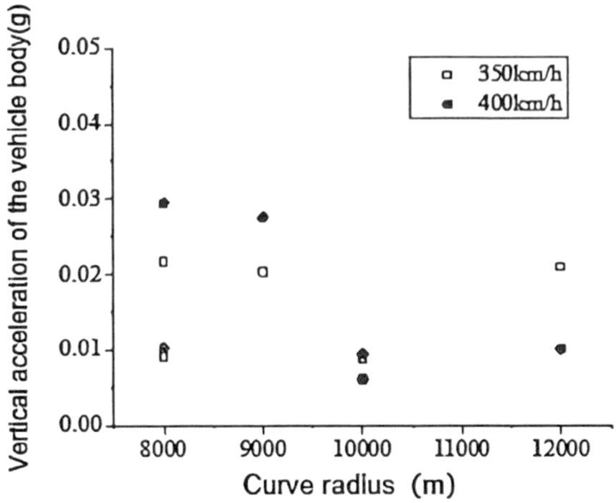


(c) Maximum value

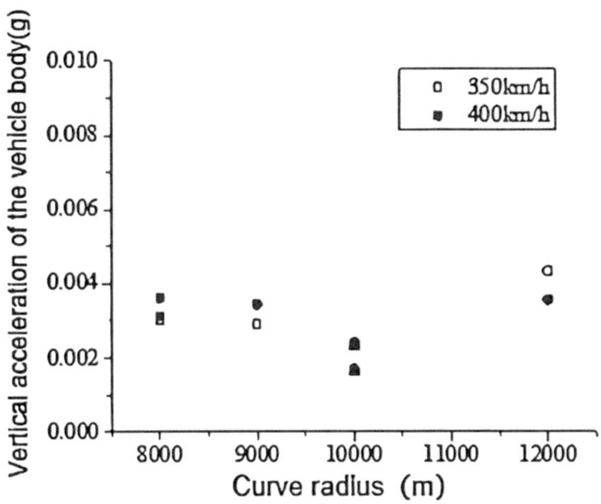


(d) Average value

**Fig. 7.4** (continued)

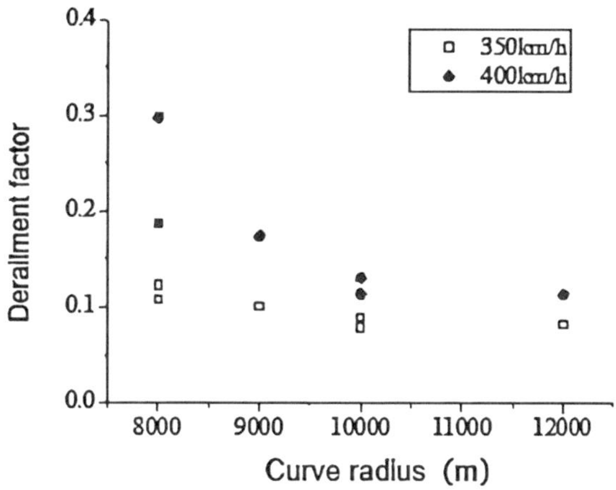


(e) Maximum value

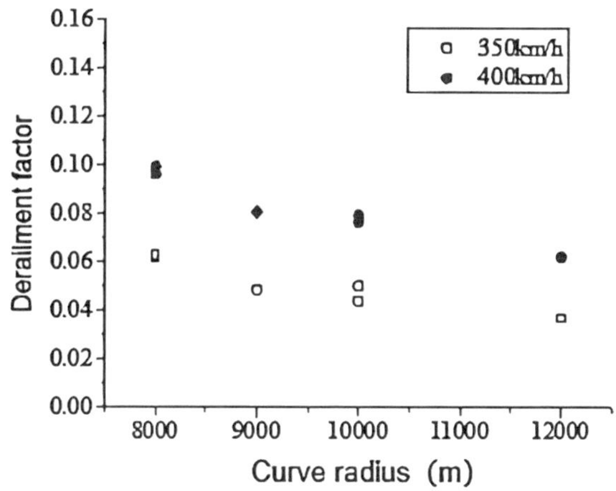


(f) Average value

**Fig. 7.4** (continued)

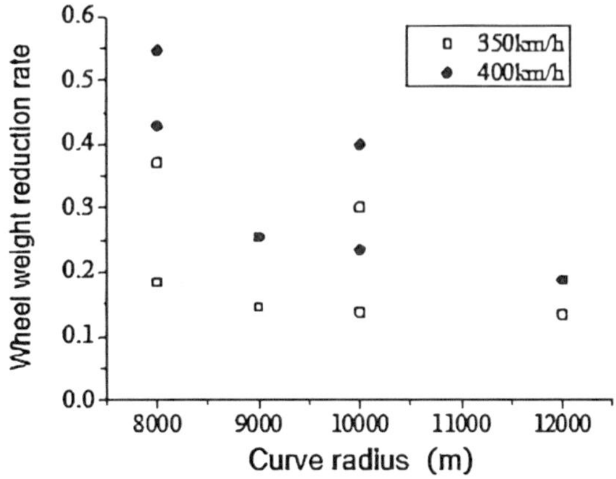


(g) Maximum value

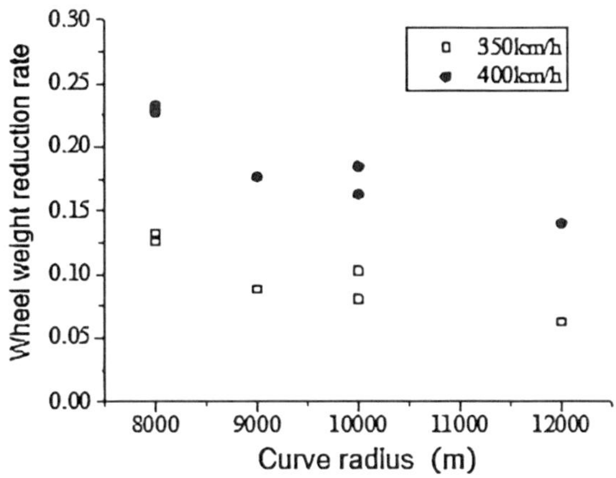


(h) Average value

**Fig. 7.4** (continued)

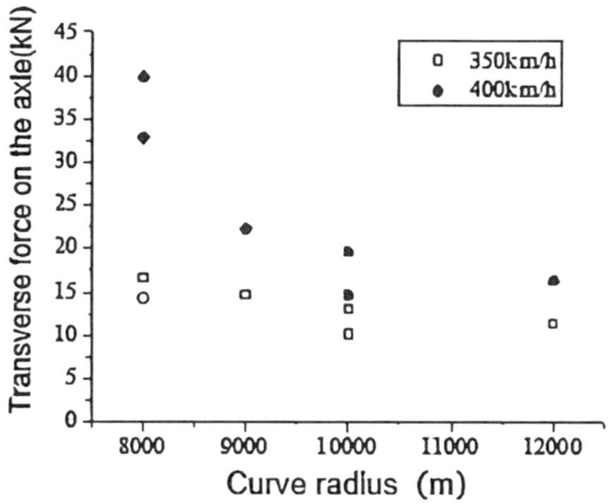


(i) Maximum value

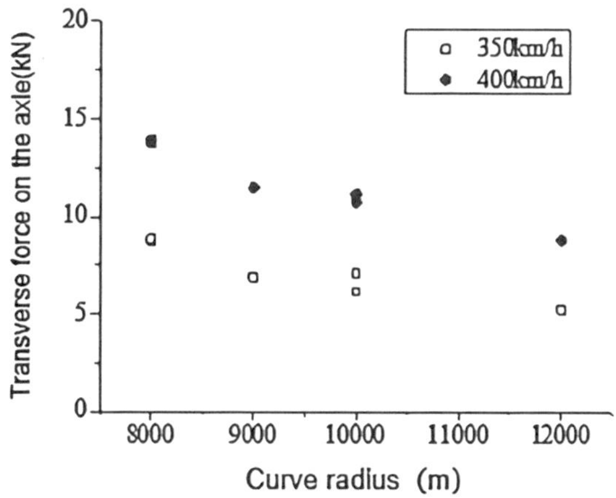


(j) Average value

① At a speed of 350 km/h, there is little difference in the lateral acceleration of the vehicle under different curve radii conditions. The maximum value is generally stable around 0.03 g, and the average value remains stable around 0.008 g. At a speed of 400 km/h, the maximum and average values of the lateral acceleration on the 8000 m radius curve are higher than those on other curve radii, with the maximum value reaching approximately 0.15 g, and certain sections exceeding the 0.1 g comfort limit.

② At both 350 and 400 km/h, the vertical acceleration of the vehicle shows little difference and meets the comfort limit of 0.13 g.

③ At a speed of 400 km/h, the derailment coefficient on the 8000 m radius curve is significantly higher than on other radius curves, but it does not exceed the safety limit of 0.8 for derailment.

④ At both 350 and 400 km/h, the wheel load reduction rate on the 8000 m radius curve is greater than on other radius curves. Specifically, at 400 km/h, the wheel load reduction rate reaches a maximum value of approximately 0.55, which is still below the dynamic wheel load reduction rate safety limit of 0.8.

⑤ At both 350 and 400 km/h, the Wheelset lateral force on the 8000 m radius curve is greater than on other radius curves. Under the 400 km/h condition, the maximum wheelset lateral force is approximately 40 kN, which is close to the limit of 56.28 kN.

(3) Dynamics index on 8000 m radius curve

Figure 7.5 shows the results of the frequency distribution of the dynamic response indicators on 8000 m curve radius section.

① At 350 km/h, the transverse acceleration is mainly concentrated in the range of 0–0.2 g, accounting for 79.3% of the total response; at 400 km/h, the frequency of transverse acceleration in the range of 0.4–0.8 g is significantly increased, with an increase of 55.9%.

② For the vertical acceleration of the car body, there is little difference in the frequency statistics at different train speeds.

③ At 350 km/h, the derailment coefficient is mainly concentrated in the range of 0.05–0.1, accounting for 80.1% of the total response; at 400 km/h, the frequency of derailment coefficients in the range of 0.15–0.3 increases significantly, with an increase of 23.4%.

④ At 350 km/h, the wheel load reduction rate is mainly concentrated in the range of 0.1–0.3, accounting for 90.7% of the total response; at 400 km/h, the frequency of wheel load reduction rate in the range of 0.3–0.6 increases significantly by 33.2%.

⑤ At 350 km/h, the wheelset lateral force is mainly concentrated in the range of 10–15 kN, accounting for 75.5% of the total response; at 400 km/h, the frequency of wheelset lateral force in the range of 15–40 kN increases significantly, with an increase of 55.6%.

(4) Dynamic performance indicators for a curve with a radius of 9000 m

Figure 7.6 shows the results of the frequency statistics of dynamic response indexes in the R9000 m curve section.

① At 350 km/h, the transverse acceleration is mainly concentrated in 0–0.2 g, accounting for 87.6% of the total response; at 400 km/h, the frequency statistics of the transverse acceleration is not much different from that at 350 km/h.

② For the vertical acceleration of the car body, there is little difference in the frequency statistics at different train speeds.
③ At 350 km/h, the derailment coefficient is mainly concentrated in the range of 0.05–0.1, accounting for 89.7% of the total response; at 400 km/h, the frequency of derailment coefficient in the range of 0.1–0.15 is significantly increased, with an increase of 58.4%.
④ At 350 km/h, the wheel load reduction rate is mainly concentrated in the range of 0.1–0.15, accounting for 74.0% of the total response; at 400 km/h, the frequency of wheel load reduction rate in the range of 0.2–0.3 increases significantly, with an increase of 59.4%.
⑤ At 350 km/h, the wheelset lateral force is mainly concentrated in the range of 6–12 kN, accounting for 82.6% of the total response; at 400 km/h, the frequency of wheelset lateral force in the range of 12–24 kN increases significantly by 61.3%.

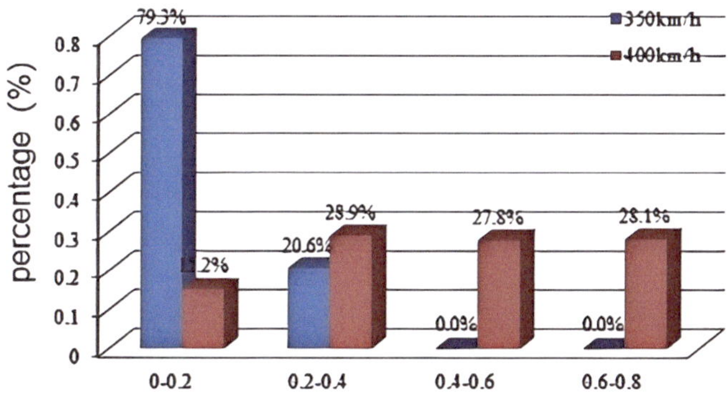


(a) Lateral acceleration of vehicle (m/s $^2$)

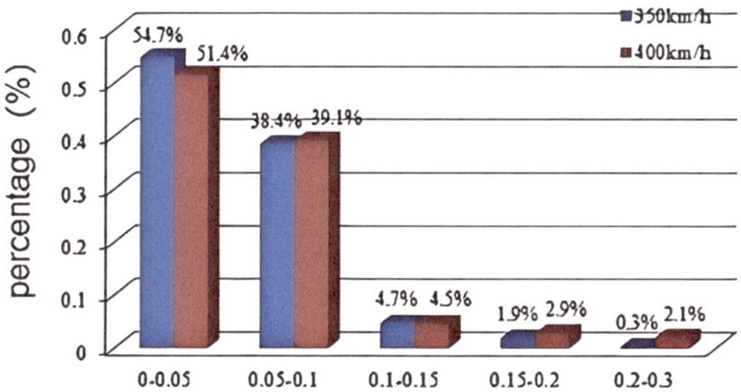


(b) Vertical acceleration of car body (m/s $^2$)

**Fig. 7.5** Dynamic index R = 8000 m curve dynamics index

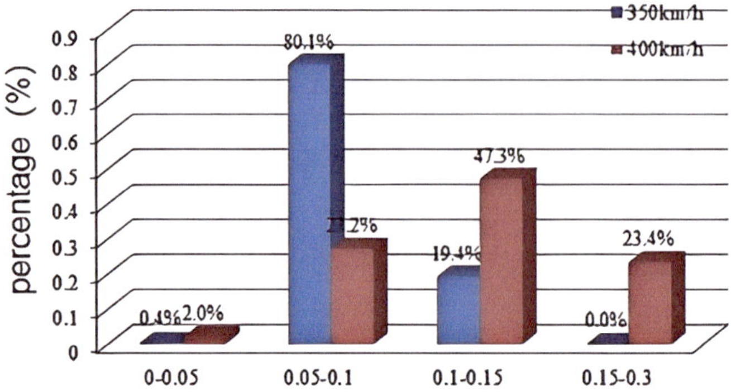


(c) Derailment coefficient

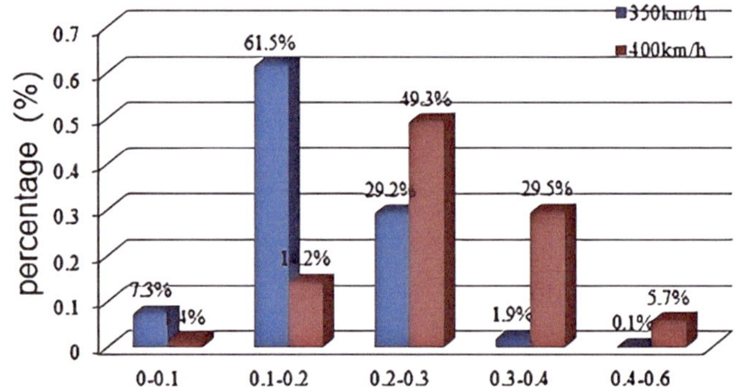


(d) Wheel load reduction rate

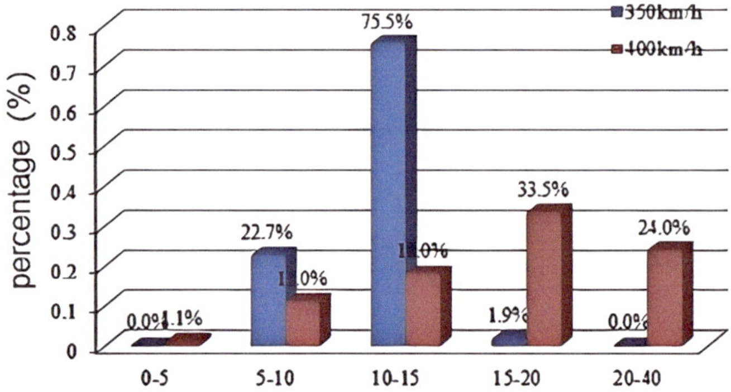


(e) Wheelset lateral force (kN)

**Fig. 7.5** (continued)

3. **Improvement Effect of R8000 m Radius**

From the previous analysis, it can be seen that on the 8000 m radius curve at a speed of 400 km/h, the lateral acceleration of the vehicle exceeds the 0.1 g limit in certain sections. To reduce the lateral acceleration peak, the superelevation in this area was adjusted to 170 mm. The calculated dynamic response time-history curve at a speed of 400 km/h is shown in Fig. 7.7. From the figure, it can be observed that after the improvement of the line conditions, the dynamic response is significantly reduced. The lateral and vertical accelerations of the vehicle are now between − 0.03 and 0.03 g, respectively, meeting the 0.1 g and 0.13 g limits. Additionally, the derailment

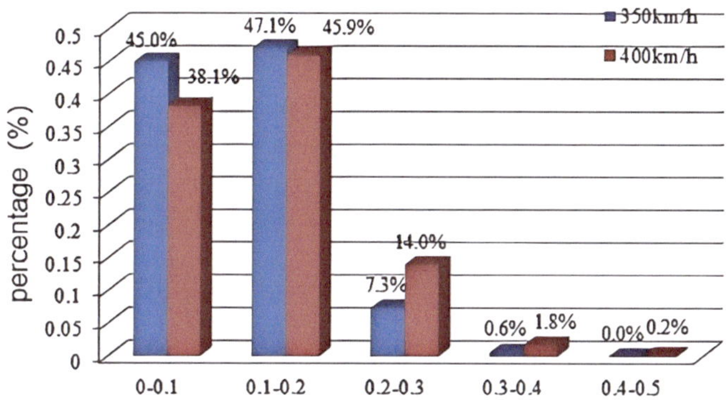


(a) Lateral acceleration of car body (m/s $^2$)

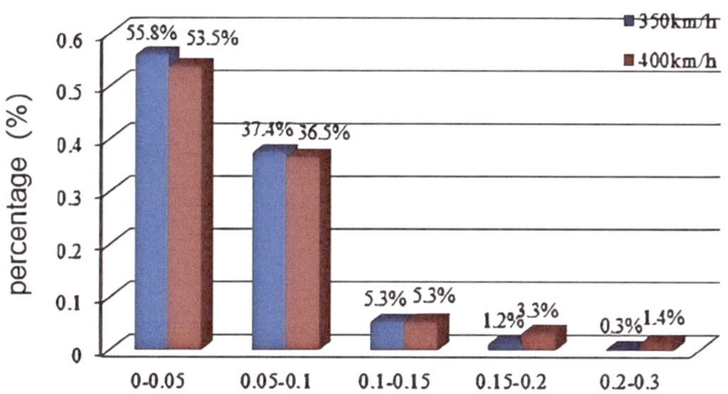


(b) Vertical acceleration of car body (m/s $^2$)

**Fig. 7.6** R9000 m radius curve dynamics index

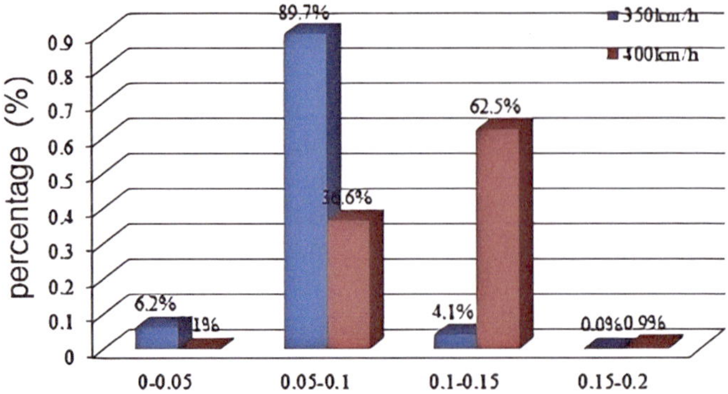


(c) Derailment coefficient

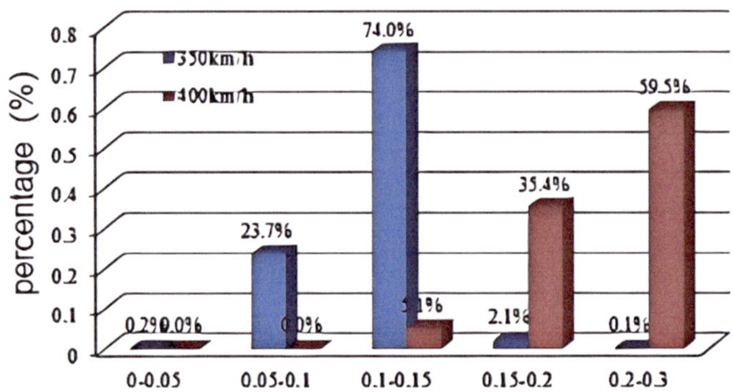


(d) Wheel load reduction rate

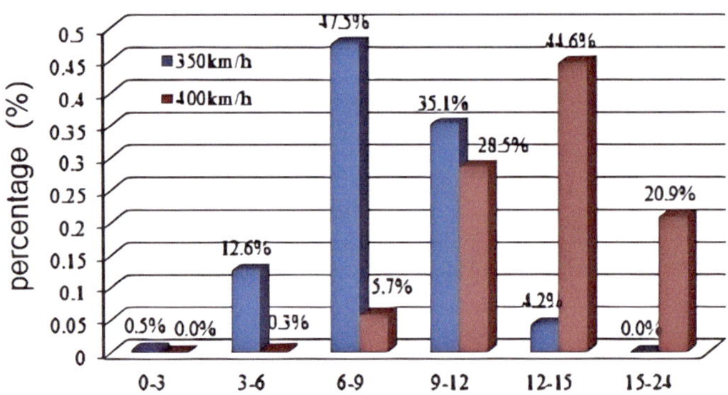


(e) Wheelset lateral force (kN)

**Fig. 7.6** (continued)

coefficient, wheel load reduciton rate, and wheelset lateral force have all significantly decreased compared to the previous line conditions.

A further statistical analysis of the calculation results before and after the improvement of line conditions is presented in Fig. 7.8. As shown in the figure, after the improvement, the frequency of lateral acceleration responses exceeding 0.04 g is significantly reduced. The derailment coefficient is primarily concentrated within the range of 0.05–0.15, while the wheel load reduction rate is mainly distributed between 0.1 and 0.3. The frequency of derailment coefficients exceeding 0.15 and wheel load reduction rates exceeding 0.3 is notably decreased. Additionally, the occurrence frequency of wheelset lateral forces exceeding 15 kN is reduced. Following the improvement of line conditions, the dynamic response indices of the train-track system at a speed of 400 km/h are generally equivalent to those before improvement at a speed of 350 km/h.

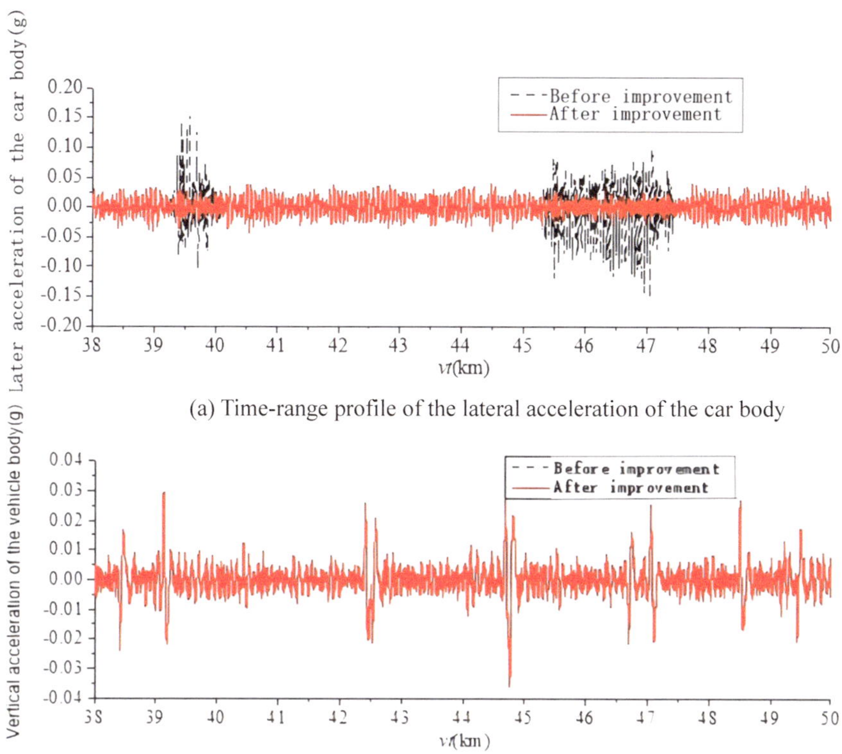


(a) Time-range profile of the lateral acceleration of the car body

(b) Vertical acceleration of car body time-range profile

**Fig. 7.7** Comparison of dynamic indicators before and after line condition improvement at 400 km/h

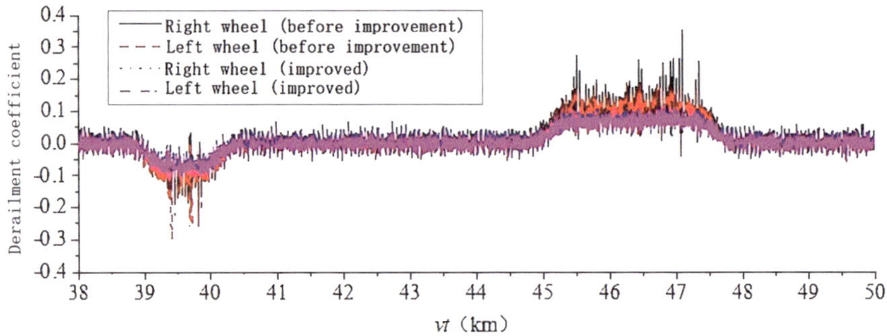


(c) Patterns of change in derailment coefficient

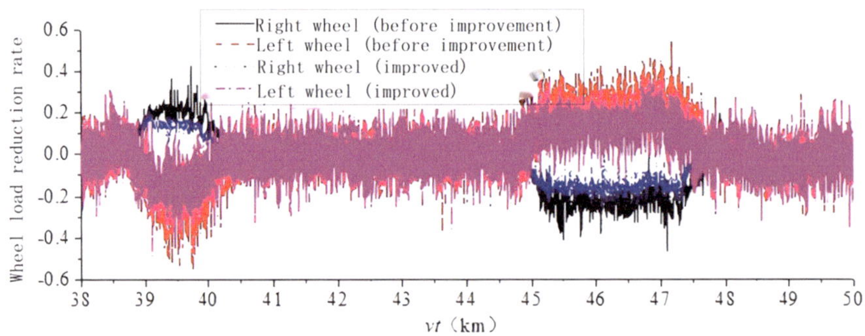


(d) Patterns of change in wheel load reduciton rate

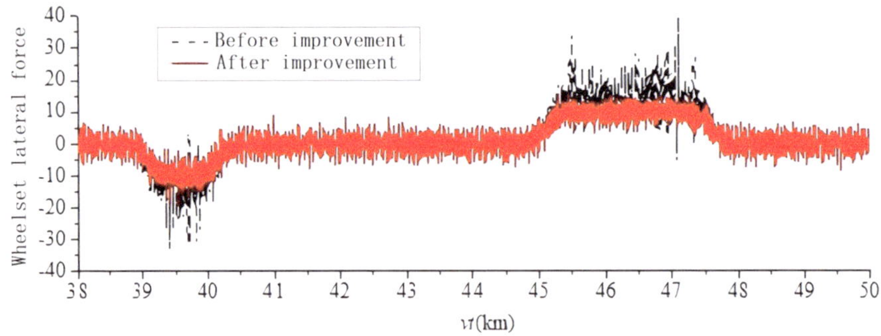


(e) The pattern of change in the wheelset lateral force

**Fig. 7.7** (continued)

### 7.1.3 Summary

The dynamic simulation of speed increase tests conducted on the test section of the Beijing–Shenyang High-Speed Railway provides a theoretical basis for superelevation setting and safe test operation. The main research conclusions are as follows:

① When the CRH380 high-speed train passes through this section at speeds of 350 km/h and 400 km/h, the vehicle safety indices meet the regulatory requirements. Except for the 8000 m radius curve section, where the maximum lateral acceleration of the car body reaches 0.15 g at 400 km/h, exceeding the 0.1 g limit requirement, all other sections meet the ride comfort index requirements at both 350 and 400 km/h.

② The setting of the horizontal curve radius and superelevation significantly affects train passing performance, particularly the necessity of matching superelevation with the target speed.

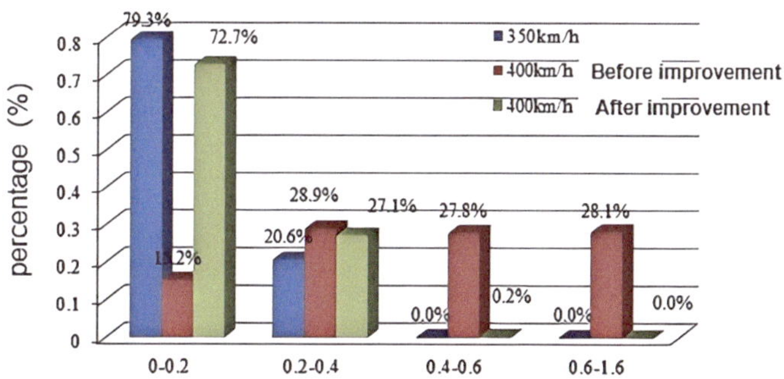


(a) Lateral acceleration of car body (m/s $^2$)

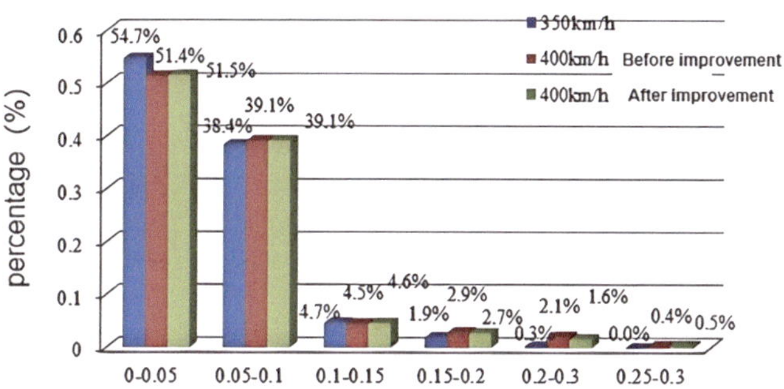


(b) Vertical acceleration of car body (m/s $^2$)

**Fig. 7.8** Comparison of the frequency statistics of dynamic response indicators before and after line condition improvement

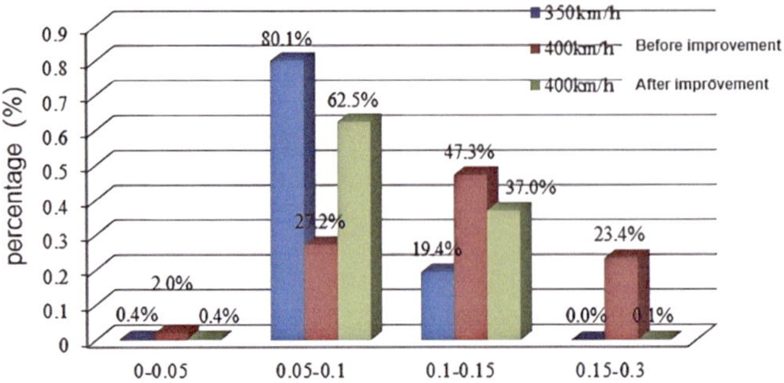


(c) Derailment coefficient

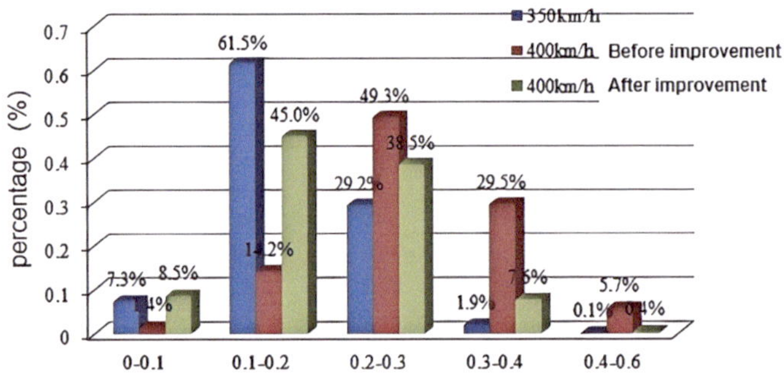


(d) Wheel load reduction rate

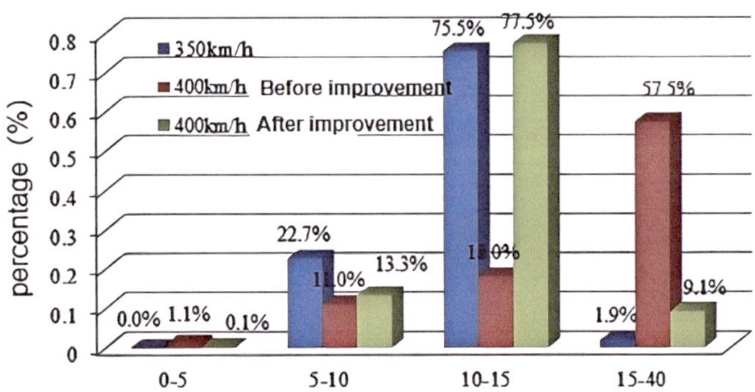


(e) Wheelset lateral force (kN)

**Fig. 7.8** (continued)

③ Vertical alignment parameters, such as gradient, gradient length, and vertical curve radius, have minimal sensitivity to the target speed.

④ Compared with the speed of 350 km/h, when the CRH380 high-speed train runs at 400 km/h on this section, the frequency of extreme values in various dynamic indices increases.

⑤ By increasing the design superelevation of the 8000 m radius curve from 120 to 170 mm, the dynamic indices of the CRH380 high-speed train at 400 km/h improve significantly. After the improvement of track conditions, the vehicle-track dynamic response indices at 400 km/h are generally equivalent to those before the improvement at 350 km/h.

## 7.2 Dynamic Analysis of the Jakarta–Bandung High-Speed Railway Alignment

The Indonesia Jakarta–Bandung High-Speed Railway serves as a crucial link between Jakarta, the capital city of Indonesia, and Bandung, the capital of West Java Province. It functions as a rapid passenger transport corridor connecting major cities along the economic belt of Java Island. The railway has a total length of 142.3 km, with a bridge-to-tunnel ratio of approximately 73%. Along the entire line, four stations are established: Halim, Karawang, Walini, and Tegal Luar, as shown in Fig. 7.9.

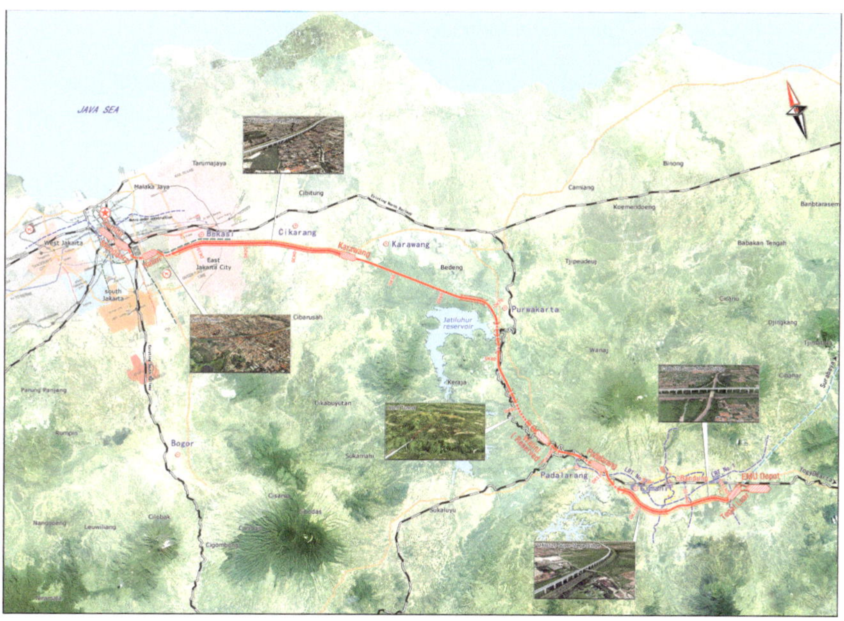


Fig. 7.9 Diagram of the route of the Javan high speed railway in Indonesia

### 7.2.1 *Line Difficult Section Dynamics Simulation Analysis*

1. **Overview of Line Difficult Section Alignment**

In the feasibility study of the project, the 14 km section from CK17 + 000 to CK31 + 000 is identified as a typical line difficult section. This section is designed for a speed of 350 km/h. The horizontal alignment includes three curves with radii of 6000 m, 7000 m, and 9000 m, respectively. Notably, the 6000 m radius curve is smaller than the general condition limit of 7000 m specified in China's "High-Speed Railway Design Specification". The straight section between the 6000 and 7000 m radius curves is 162.3 m, which is shorter than the minimum straight section length under difficult conditions of 210 m (0.6 V) as stipulated in the same specification. The vertical alignment comprises nine gradient sections, with the longest being 3600 m and the shortest 900 m, averaging 1577 m in length. The maximum algebraic difference in gradient reaches 39‰. Specific line conditions are shown in Fig. 7.10. This section's horizontal and vertical alignment design, tailored to meet the economic applicability requirements of foreign projects, adopts certain standards that exceed the corresponding limits specified in China. It represents a typical section in the design of the Jakarta-Bandung High-Speed Railway in Indonesia.

2. **Time History Curve Analysis**

Figure 7.11 presents the time history curves of key operational dynamic performance indicators for the CRH380 EMU traveling at 350 km/h through the CK17 + 000 to CK31 + 000 section. The results indicate the following:

① The vertical acceleration of the car body ranges from − 0.03 to 0.03 g, meeting the comfort control limit of 0.13 g. Peak values mainly occur at transition points between the 7000 and 6000 m curve radii and at vertical profile variation points.
② The lateral acceleration of the car body ranges from − 0.04 to 0.04 g, satisfying the comfort control limit of 0.1 g. Peak values appear in sections with challenging horizontal and vertical alignment conditions.
③ At a speed of 350 km/h, the derailment coefficient remains low, within the range of − 0.13 to 0.13, far below the safety limit of 0.8. Peak values are primarily concentrated in sections with curve radii of 7000 and 6000 m.
④ At a speed of 350 km/h, the wheel load reduction ratio ranges from − 0.4 to 0.4, which is below the dynamic wheel load reduction safety limit of 0.8. Peak values are primarily found in sections with curve radii of 7000 and 6000 m.
⑤ Regarding the wheelset lateral force, the safety limit is calculated as 56.28 kN based on the defined limit formula. At 350 km/h, the wheelset lateral force remains within the range of − 18 kN to 18 kN, meeting the safety limit. Peak values are mainly concentrated in sections with curve radii of 7000 and 6000 m.

These results indicate that train operation through this section meets the requirements for both ride comfort and operational safety.

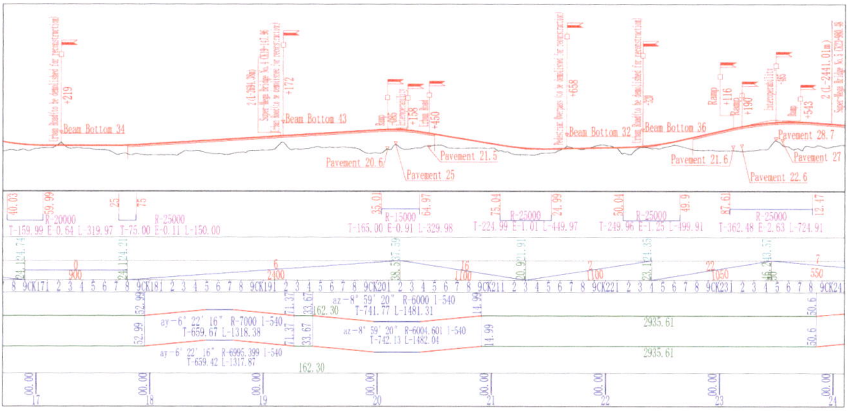

(a) Longitudinal section of the line from CK17+000 to CK24+000

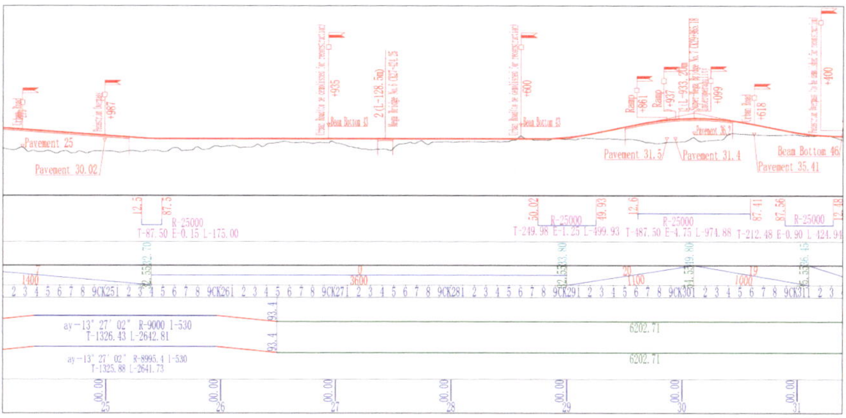

(b) Longitudinal section of the line from CK24+000 to CK31+000

**Fig. 7.10** Longitudinal section of the line from CK17 + 000 to CK31 + 000 of the Yavan high speed railway in Indonesia

3. **Statistical Analysis**

Figure 7.12 presents the frequency statistics of dynamic response indicators for this section of the track. The analysis reveals the following:

① At an operating speed of 350 km/h, the lateral acceleration of the car body is primarily concentrated in the range of 0–0.01 g, accounting for 61.1% of the total responses.

② At an operating speed of 350 km/h, the vertical acceleration of the car body is mainly within 0–0.01 g, making up 95.1% of the total responses.
③ At an operating speed of 350 km/h, the derailment coefficient is mostly below 0.05, constituting 82.4% of the total responses.
④ At an operating speed of 350 km/h, the wheel load reduction rate is predominantly below 0.1, accounting for 77.6% of the total responses, while values between 0.1 and 0.2 account for 18.7%.
⑤ At an operating speed of 350 km/h, the wheelset lateral force is mainly concentrated within 6 kN, representing 78.2% of the total responses.

From the statistical results, it is evident that all major dynamic performance indicators remain at a favorable level, with considerable safety margins compared to the corresponding limit values.

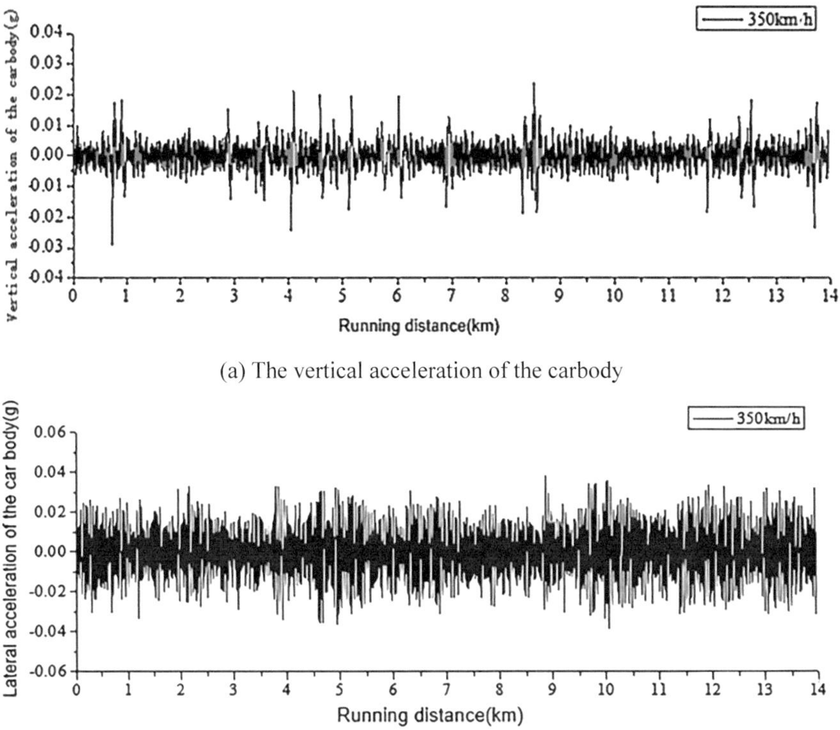


(a) The vertical acceleration of the carbody

(b) The lateral acceleration of the car body

**Fig. 7.11** Time history curve of dynamic response under 350 km/h operating conditions

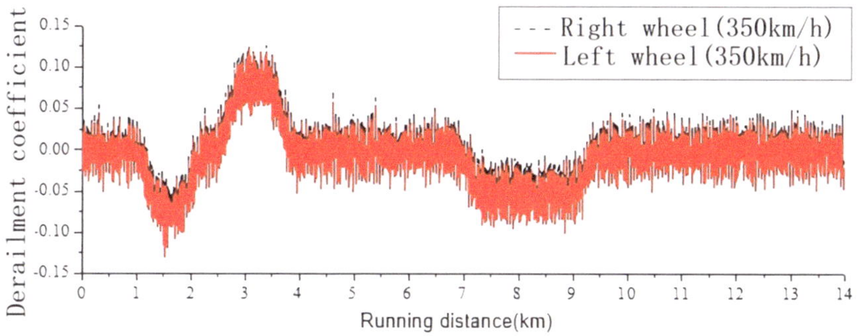


(c) Derailment coefficient

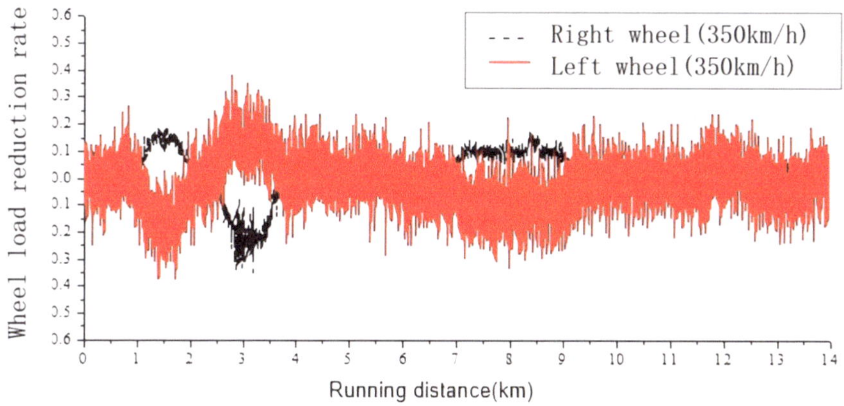


(d) Wheel load reduction rate

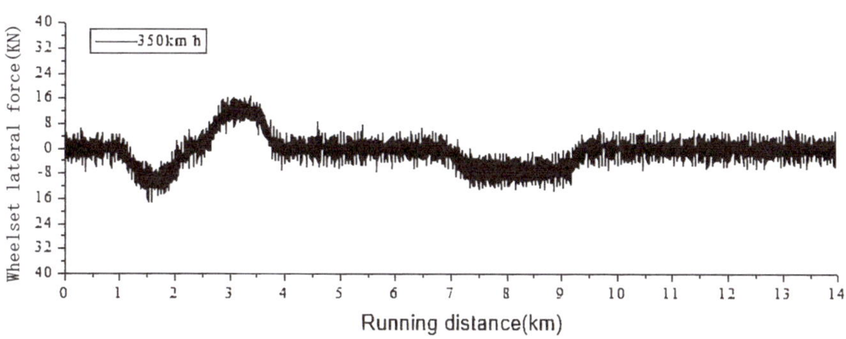


(e) Wheelset lateral force

**Fig. 7.11** (continued)

4. **Summary**

To fulfill China's strategic goal of high-speed railway expansion abroad and enhance the competitiveness of Chinese high-speed rail, this project systematically studies key parameters that significantly impact economic feasibility, in the context of the Jakarta-Bandung High-Speed Railway alignment design. Compared with the Code for Design of High-Speed Railway in China, parameters such as the minimum radius of circular curves and the minimum length of transition straights have been relaxed. In track alignment design, the relaxation of these standards has played a crucial role in adapting to terrain conditions, avoiding control points, reducing relocations, and meeting local planning requirements. This approach has substantially reduced construction costs and decreased implementation difficulties. Through

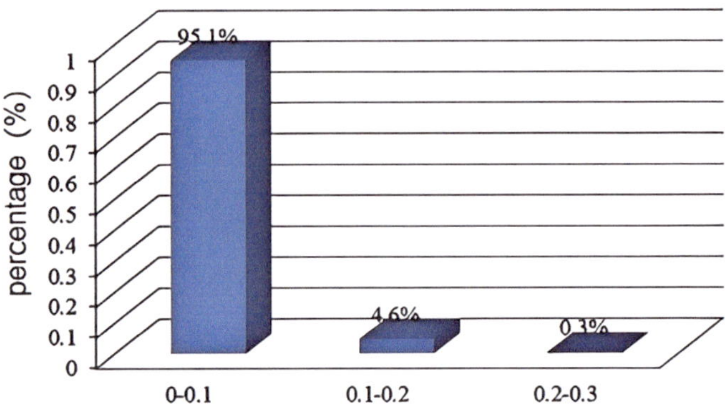


(a) Lateral acceleration of car body (m/s $^2$)

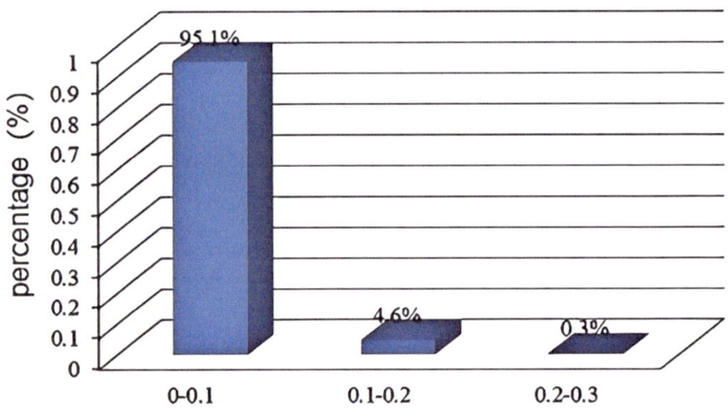


(b) Vertical acceleration of car body (m/s $^2$)

**Fig. 7.12** Statistical analysis of dynamic performance indicators for the track scheme

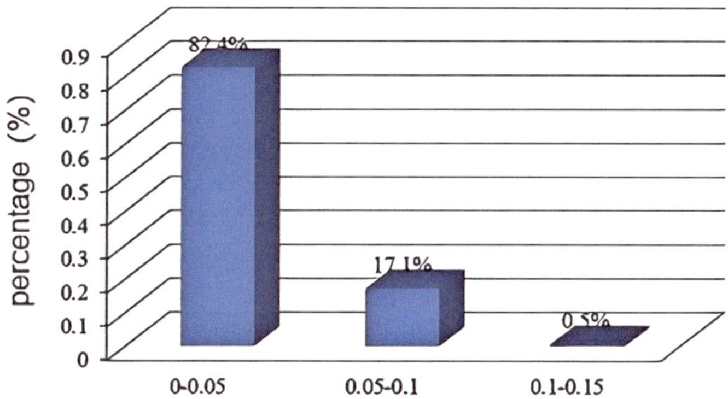


(c) Derailment coefficient

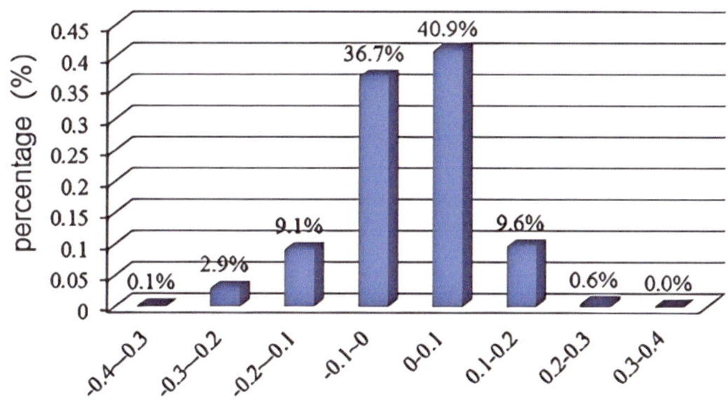


(d) Wheel load reduction rate

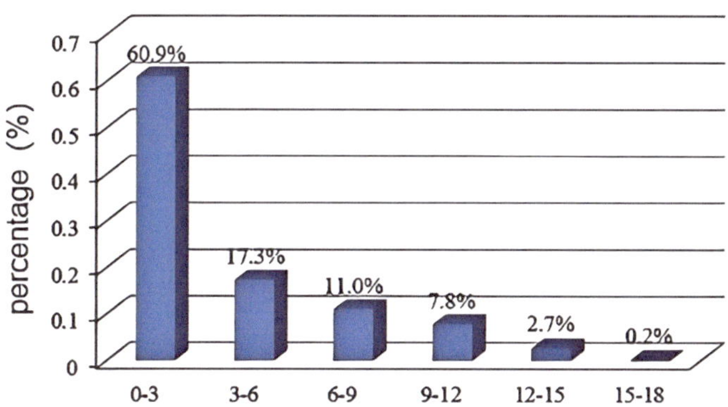


(e) Wheelset lateral force (kN)

**Fig. 7.12** (continued)

dynamic simulation analysis of the track scheme, it is demonstrated that in sections where alignment parameters exceed the current Chinese standards, both ride comfort and operational safety criteria are met, with a certain margin of safety.

## 7.2.2 Evaluation and Comparison of the Dynamic Performance of Track Alignment Schemes

1. **Overview of the Track Alignment Scheme**

In the feasibility study of the Jakarta-Bandung High-speed Railway, two main alignment options for the high-speed rail introduction to the Bandung area were studied, in accordance with the urban planning of Bandung city and the rail transit planning. These alignment options are shown in Fig. 7.13.

Option 1 follows the existing railway alignment, which is closer to the city center. The total length of this segment is approximately 24 km, with 9 curves, 15 slope sections, a minimum curve radius of 1600 m, the shortest tangent length of 126 m, and a maximum slope of 20‰. The maximum algebraic difference of the slope reaches 20‰, and there are sections where horizontal and vertical curves overlap. The designed running speed is 200 km/h. A local profile of the horizontal and vertical alignment is shown in Fig. 7.14a, with the specific layout details in Tables 7.2 and 7.3.

Option 2 follows the highway alignment, located on the southern edge of the city, with less demolition required on both sides. The total length of this segment is approximately 33 km, with 16 curves, 14 slope sections, a minimum curve radius

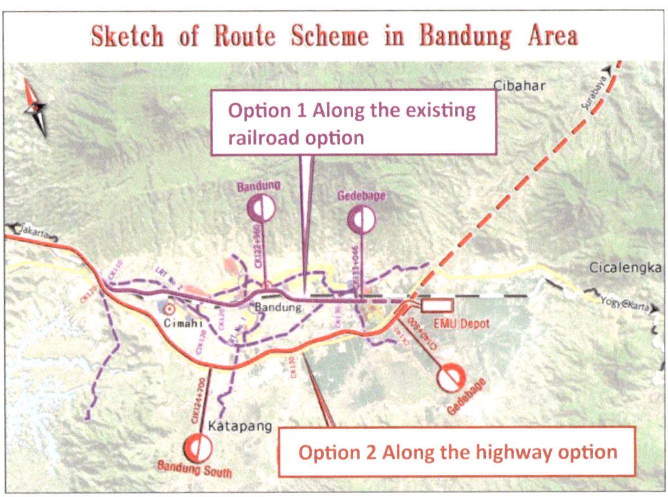


**Fig. 7.13** Schematic diagram of the route scheme of the Javan high speed railway in Indonesia

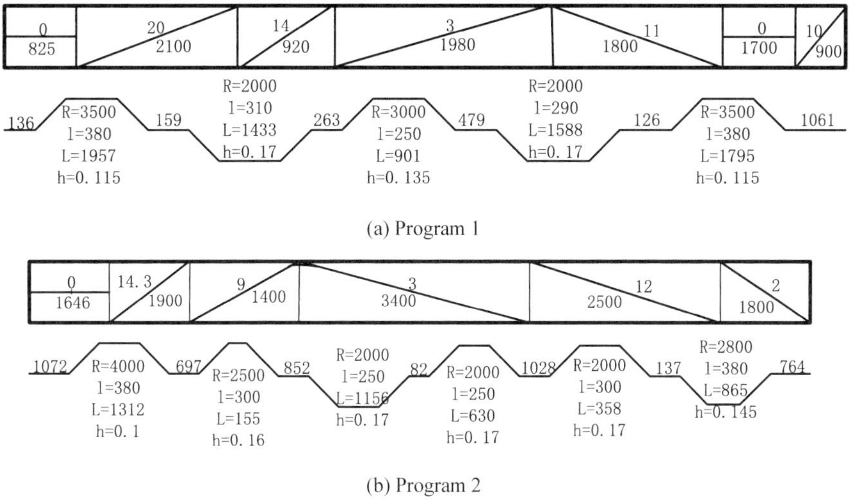


**Fig. 7.14** Schematic of the local alignment plan's horizontal and longitudinal sections for the Indonesia Jakarta-Bandung high-speed railway

**Table 7.2** Indonesia Jakarta-Bandung high-speed railway local alignment plan: plan view parameter statistics

| Curve radius (m) | Under 2000 | 2000–3000 | 3000–4000 | 4000–5000 | 5000 or more |
|---|---|---|---|---|---|
| Option 1 | 1 | 3 | 3 | 1 | 1 |
| Option 2 | 3 | 10 | 2 | 1 | 0 |

**Table 7.3** Statistical analysis of the longitudinal gradient sections of the local alignment plan for the Indonesia Jakarta-Bandung high-speed railway

| Slope (‰) | 0–5 | 5–10 | 10–15 | 15–20 | 20 or more |
|---|---|---|---|---|---|
| Option 1 | 7 | 3 | 3 | 2 | 0 |
| Option 2 | 10 | 3 | 1 | 0 | 0 |

of 1600 m, the shortest tangent length of 82 m, a maximum slope of 14.3‰, and a maximum algebraic difference of the slope of 12‰. The designed running speed is 200 km/h. A local profile of the horizontal and vertical alignment is shown in Fig. 7.14b, with the specific layout details in Tables 7.2 and 7.3.

2. **Time History Analysisime-Course Analysis**

The dynamic simulation for the comparison section uses the CRH380 train, with the Chinese high-speed railway ballastless track spectrum as the track irregularity disturbance.

Figure 7.15 presents the time history curves of the dynamic response for the two alignment options as the train passes through. Table 7.4 shows the peak values of the dynamic response for both alignment plans. From the table, it is clear that the safety and comfort indices are well below the control limits, with the safety indices and body vibration accelerations of Plan 2 being slightly higher than those of Plan 1. The Sperling index and UIC513 evaluation results for both plans are very close. The vibration response quantities used in the Sperling index and UIC513 evaluation methods involve frequencies above 0.4 Hz, which do not capture the low-frequency vibrations caused by the track alignment. The peak values of all dynamic responses are mainly caused by track irregularities, and it is difficult to make a comprehensive evaluation of the alignment plans based solely on peak values.

3. **Statistical Analysis**

To further analyze the composition of each dynamic response component, the frequency of each indicator was statistically analyzed, as shown in Fig. 7.16. From the figure, it can be observed that the car body vibration acceleration is mainly concentrated between 0 and 0.05 m/s$^2$, with a very low frequency of occurrence above 0.1 m/s$^2$. The derailment coefficient for Scheme 1 appears less frequently in the range of 0 to 0.05 compared to Scheme 2, while the frequency of derailment coefficient greater than 0.05 is significantly higher in Scheme 1 than in Scheme 2.

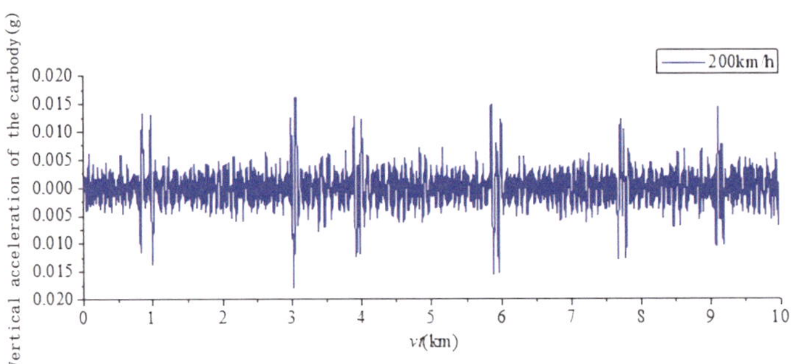


(I) Program 1

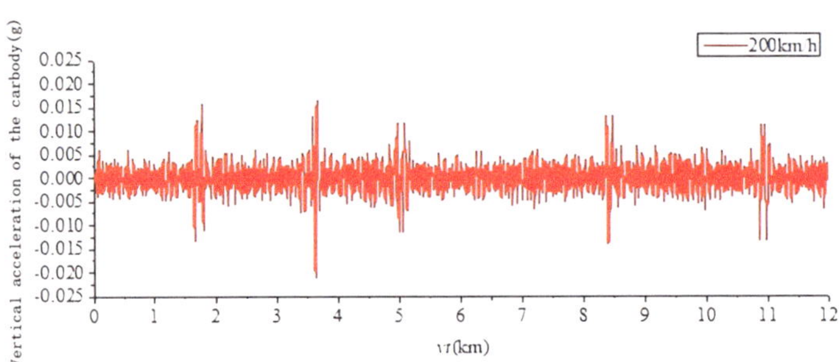


(II) Program 2

(a) The vertical acceleration of the car body

**Fig. 7.15** Dynamic response time history curve under 200 km/h operating conditions

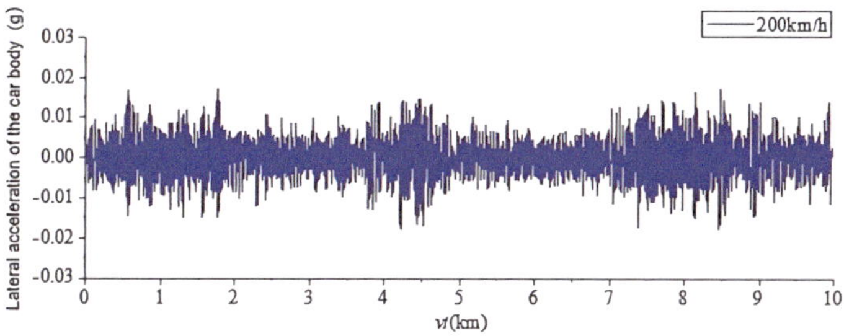


(I) Program 1

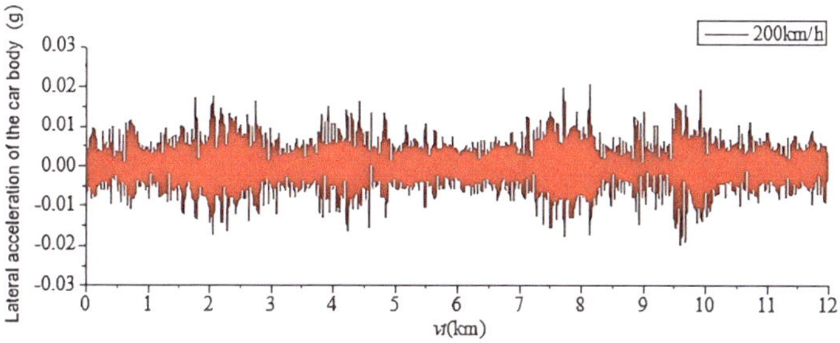


(II) Program 2

(b) The lateral acceleration of the car body

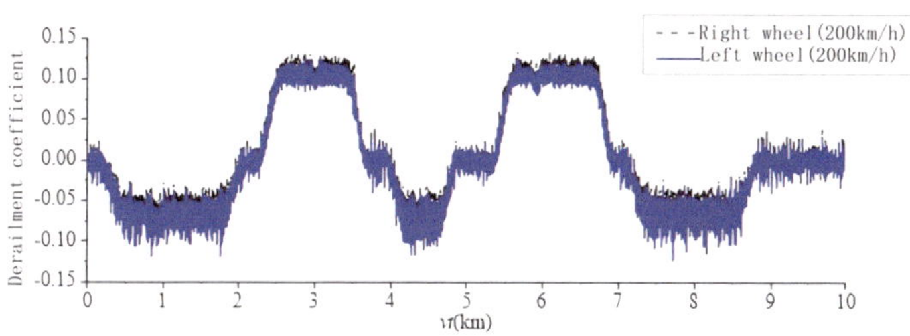


(I) Program 1

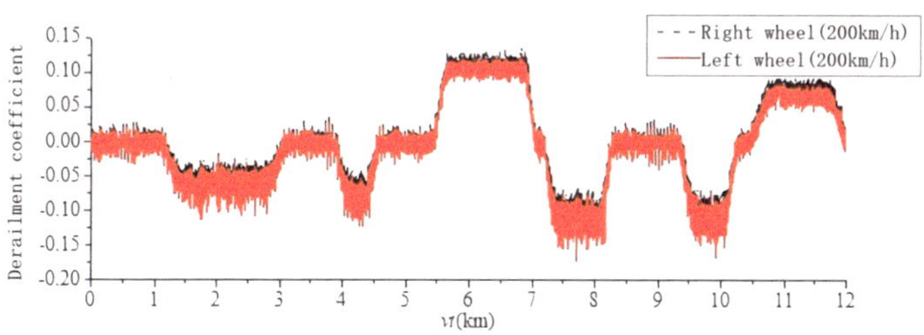


(II) Program 2

(c) Derailment coefficient

**Fig. 7.15** (continued)

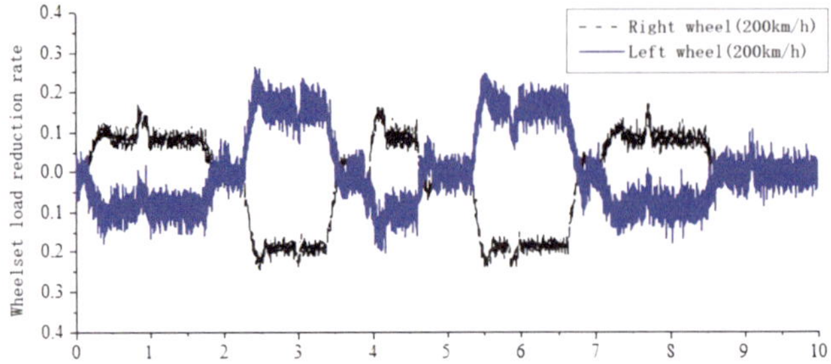


(I) Program 1

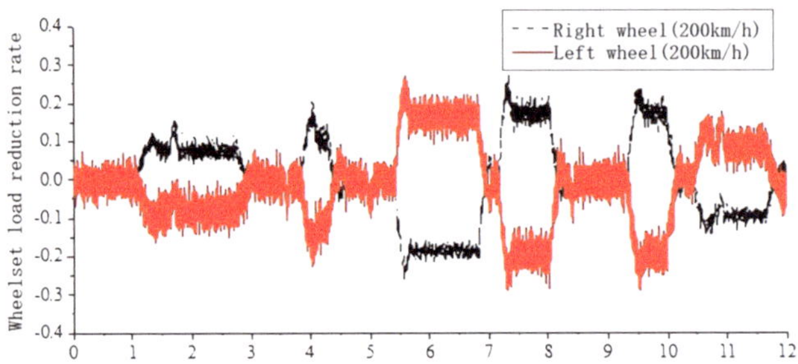


(II) Program 2

(d) Wheel load reduction rate

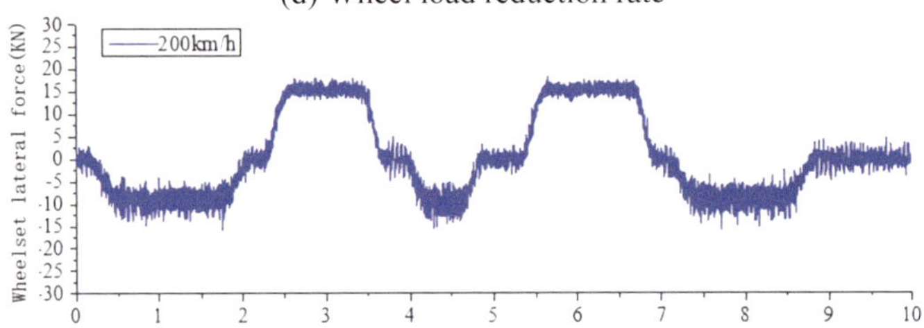


(I) Program 1

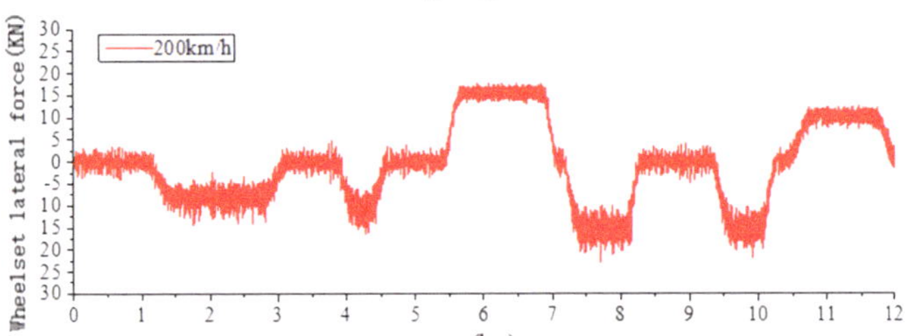


(II) Program 2

(e) Wheelset lateral force

**Fig. 7.15** (continued)

**Table 7.4** Peak values of dynamic response for the alignment plans

| Evaluation of indicators | | Calculations | The conclusions of the evaluation |
|---|---|---|---|
| Derailment coefficients | Option 1 | 0.14 < [0.8] | Meeting the requirements |
| | Option 2 | 0.17 < [0.8] | Meeting the requirements |
| Wheel load reduction rate | Option 1 | 0.26 < [0.6] | Meeting the requirements |
| | Option 2 | 0.28 < [0.8] | Meeting the requirements |
| Wheelset lateral force/kN | Option 1 | 18.32 < [57.84] | Meeting the requirements |
| | Option 2 | 22.94 < [57.84] | Meeting the requirements |
| Lateral acceleration/g | Option 1 | 0.018 < [0.1] | Meeting the requirements |
| | Option 2 | 0.020 < [0.1] | Meeting the requirements |
| Vertical acceleration/g | Option 1 | 0.018 < [0.13] | Meeting the requirements |
| | Option 2 | 0.021 < [0.13] | Meeting the requirements |
| Sperling's index | Option 1 | Vertical: 1.66; horizontal: 2.09 | Excellent |
| | Option 2 | Vertical: 1.68; horizontal: 2.13 | Excellent |
| UIC513 | Option 1 | 0.22 | Very comfortable |
| | Option 2 | 0.24 | Very comfortable |

The wheel load reduction rate for Scheme 1 occurs less frequently in the range of − 0.1 to 0.1 compared to Scheme 2 (negative values represent increased load), whereas in the range of 0.1 to 0.2 (− 0.2 to − 0.1), the frequency of occurrence is noticeably higher in Scheme 1 than in Scheme 2. The wheelset lateral force for Scheme 1 appears less frequently in the 0 to 3 kN range compared to Scheme 2, but the frequency of occurrence for wheelset lateral force between 3 and 18 kN is noticeably higher in Scheme 1 than in Scheme 2. From this, it can be concluded that although the dynamic response peak values for Scheme 1 are slightly smaller than those of Scheme 2, the occurrence frequency of larger wheel-rail forces is higher under the conditions of Scheme 1, indicating that under the conditions of Scheme 1, the train-line system dynamic response is larger.

From the above analysis, it can be observed that although there are differences between the two schemes in terms of driving safety indicators, they are both far below the control limits. The body vibration acceleration and overall comfort evaluation results show that the two schemes have very little difference in terms of comfort. It is worth noting that the comfort evaluation method used above does not fully account for the low-frequency vibrations caused by the track alignment. Therefore, a further evaluation using the curve $P_{CT}$ method was employed to assess the comfort of the two schemes.

Figure 7.17 shows the curve comfort $P_{CT}$ and other indicators for the two line schemes. This figure includes the $P_{CT}$ value, lateral acceleration of the car body, and the lateral roll angle velocity of the car body. From the figure, it can be seen that Scheme 2, due to the presence of short tangents, results in greater lateral acceleration and lateral roll angle velocity compared to Scheme 1. In Scheme 1, there are two locations where the $P_{CT}$ is about 3%, while in Scheme 2, three locations exceed 3.5%. Therefore, the passenger comfort in Scheme 1 is better than in Scheme 2.

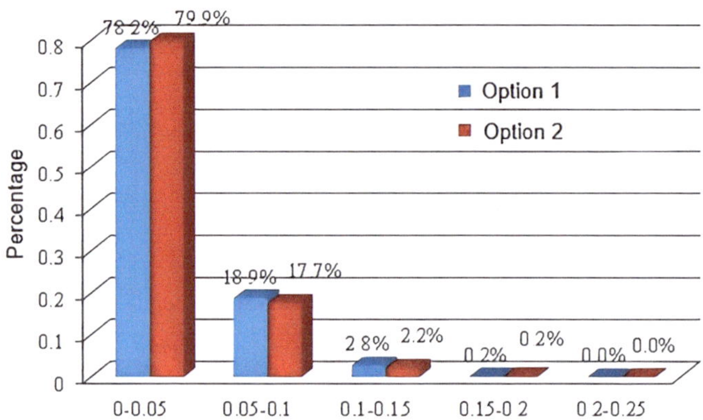


(a) Lateral acceleration of car body (m/s$^2$)

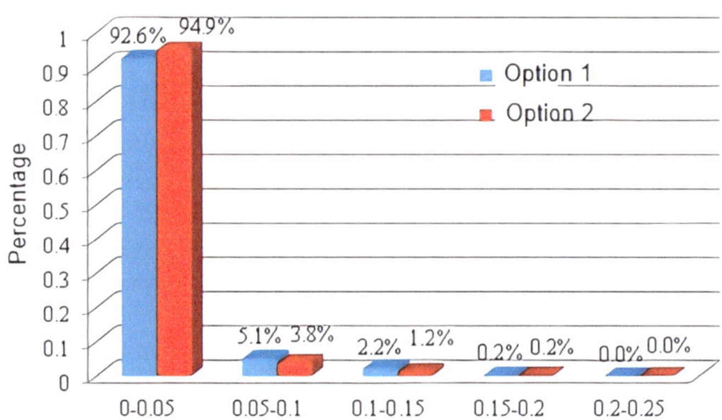


(b) Vertical acceleration of car body (m/s$^2$)

**Fig. 7.16** Overall dynamic indicators of the line scheme

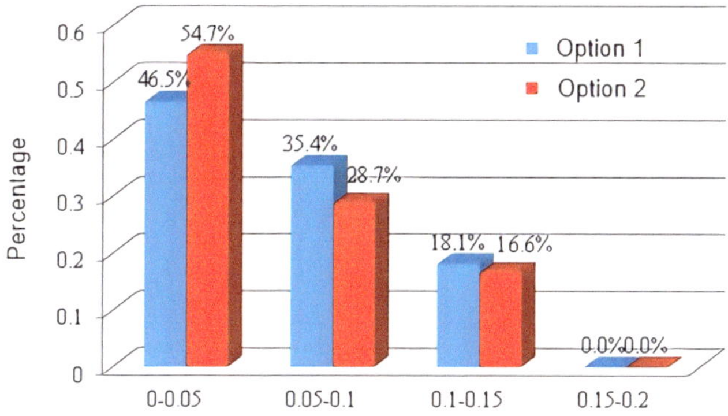


(c) Derailment coefficient

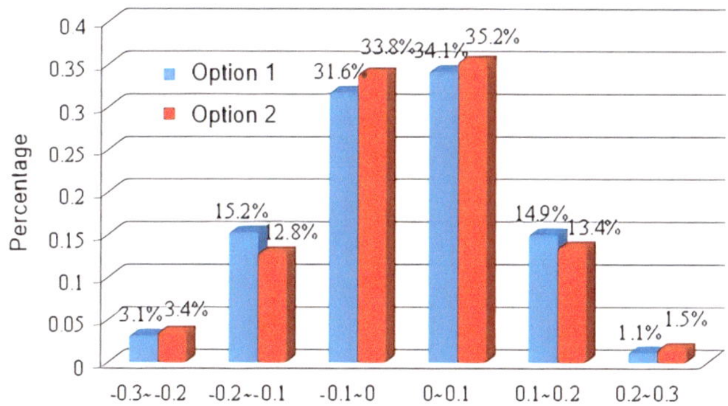


(d) Wheel load reduction rate

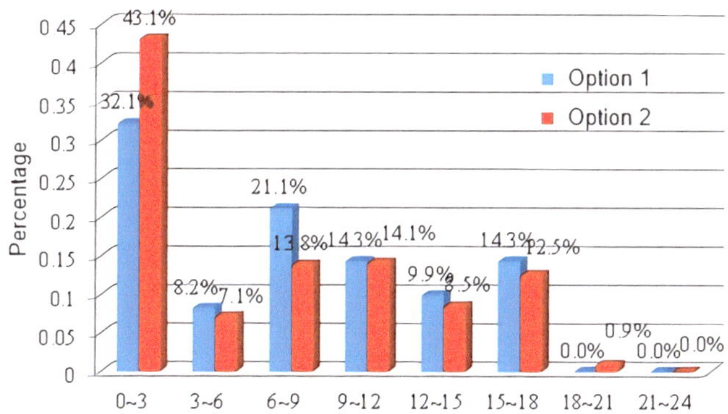


(e) Wheelset lateral force (kN)

**Fig. 7.16** (continued)

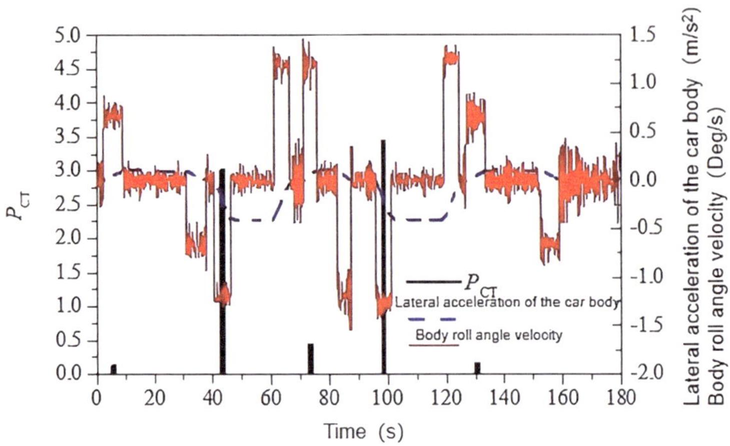


(a) Program 1 P CT Evaluation

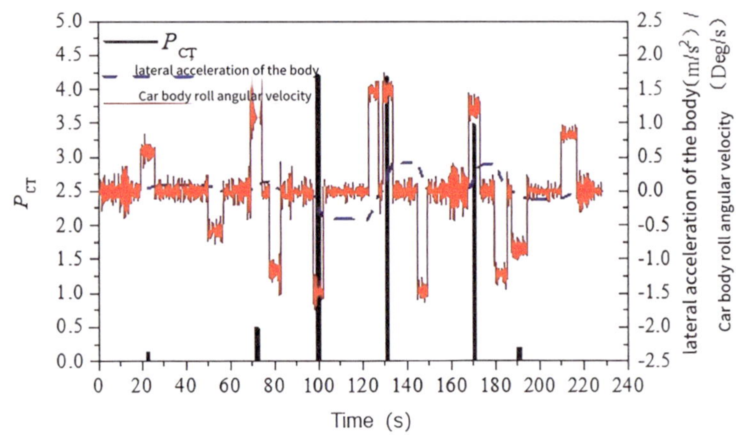


(b) Program 2 P CT Evaluation

**Fig. 7.17** Evaluation of line program $P_{CT}$

4. **Summary**

This section evaluates the track alignment options in the feasibility study of the Jakarta-Bandung High-Speed Railway project in Indonesia, considering operational safety and ride comfort. The key findings are as follows:

① The $P_{CT}$-based dynamic evaluation method accounts for the asynchronous response of the train body to the track plane during operation, providing a more objective assessment of passenger comfort.

② Traditional comfort evaluation methods, such as the Sperling Index and UIC 513, do not fully capture the low-frequency vibrations induced by track geometry, making it difficult to distinguish between alignment options. In contrast, the $P_{CT}$ method emphasizes vibration components caused by track geometry, making it more effective in assessing the impact of different track alignments on passenger comfort.

③ The frequency statistics of operational safety indicators indirectly reflect the intensity of the train-track dynamic interaction under different alignment conditions, aiding in the evaluation of the track design's performance during operation.

④ The dynamic evaluation method proposed in this study can be applied to similar projects for assessing track alignment options.

## 7.3 Evaluation and Comparison of an Overseas High-Speed Railway Alignment

### *7.3.1 Research Background*

Track geometry has a significant impact on the operational safety and ride comfort of high-speed railways, influencing not only train performance but also maintenance requirements. As China continues its "High-Speed Rail Going Global" strategy, international railway projects often require justifications for track parameter selection and dynamic performance assessments. Therefore, applying railway system dynamics to evaluate track alignment options is of great engineering and theoretical value.

This section analyzes a high-speed railway project overseas, using railway system dynamics to assess and compare track alignment alternatives. A novel approach is proposed that combines peak value statistics of dynamic indicators to comprehensively evaluate track alignment performance.

A holistic evaluation of track alignment should consider both operational safety and ride comfort. Common comfort evaluation indicators include the Sperling ride index, UIC 513, and ISO 2631–1. However, these indicators primarily focus on frequency components between 0.4 and 80 Hz. Since train vibrations caused by track geometry are mainly in the low-frequency range and occur primarily at horizontal and vertical curves, these traditional indicators may not effectively differentiate between alignment options for long track sections.

Instantaneous indicators, such as vibration acceleration and wheel-rail interaction forces, effectively reflect the impact of track geometry on system dynamics. For long track segments, statistical analysis methods can further enhance the evaluation. To better distinguish the effects of different track geometries on train dynamics, this study proposes an evaluation approach that supplements traditional comfort indices with root-mean-square values, frequency distributions, and cumulative characteristics of train vibration acceleration and wheel-rail interaction forces.

### 7.3.2 *Overview of Track Alignment Options*

A certain high-speed railway has a total length of approximately 240 km and adopts a segmented design speed scheme. The maximum design speed for high-speed sections is 350 km/h for passenger trains and 100 km/h for freight trains, while for conventional-speed sections, the maximum design speed is 220 km/h for passenger trains and 70 km/h for freight trains. The PK171 + 000 to PK210 + 500 section traverses mountainous terrain with steep topography and complex geological conditions, primarily consisting of tunnels, with a maximum design speed of 220 km/h. Based on factors such as embankment height, tunnel length, geological conditions, and track geometry, two alignment options were studied. The original design has a total length of 39.5 km with 11 curves, while the optimized design has a total length of 39.285 km with 9 curves, featuring two major optimizations.

PK181 + 500 − PK182 + 600 Section: The original alignment involved deep excavation, with a maximum cut depth of 60 m, requiring extensive earthworks and slope protection, posing high construction and operational risks. The optimized alignment was shifted eastward, closer to the highway, reducing the maximum excavation depth to 31 m, thereby mitigating construction risks.

PK185 + 800 − PK202 + 460 Section: The original alignment included a 16.6 km tunnel with four curves inside, complicating drainage and ventilation. The optimized alignment straightened the track, reducing tunnel length to 12.2 km, decreasing total tunnel length by 330 m. The number of curves inside the tunnel was reduced from four to one, enhancing train dynamics and ventilation conditions. Additionally, the optimized alignment shortened the overall track length by 210 m, reducing construction costs.

A schematic diagram of the track geometry is shown in Fig. 7.18.

1. **Time History Analysis of Train Dynamic Performance Indicators**

Figure 7.19 presents a comparison of the time history curves of key train dynamic performance indicators before and after the optimization of the 39.5 km track alignment from PK171 + 000 to PK210 + 500. Overall, track sections with horizontal and vertical curves are critical areas for train dynamic performance. Optimizing the relevant parameters of horizontal and vertical curves plays a significant role in improving train dynamic performance, enhancing operational safety, and increasing passenger comfort. As shown in Fig. 7.19, the peak values of various indicators in the optimized track alignment are reduced to varying degrees compared to the original design, indicating an improvement in train dynamic performance.

By filtering and applying Fourier transformation to the lateral and vertical accelerations of the car body, the Sperling ride comfort indices for the original design are calculated as 1.61 and 1.56, respectively, while the optimized design yields 1.58 and 1.55, both of which are below the "excellent" comfort threshold ($< 2.5$). According to the UIC 513 standard, the ride comfort index N for the original and optimized designs are 1.68 and 1.66, respectively, corresponding to comfort level 2 ("comfortable," $1 \leq N < 2$). This also confirms that the optimization enhances train dynamic

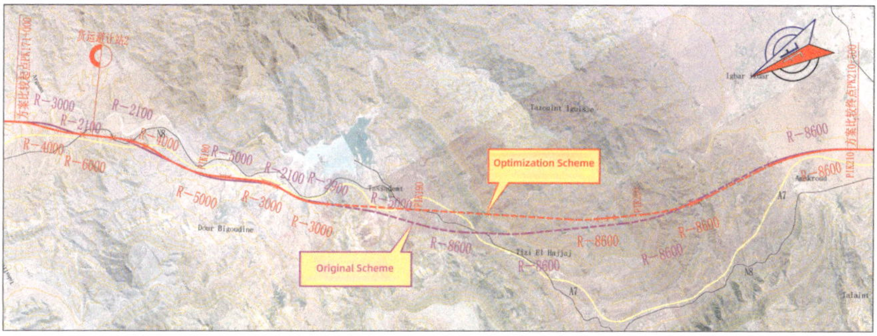


(a) A schematic plan of the horizontal alignment

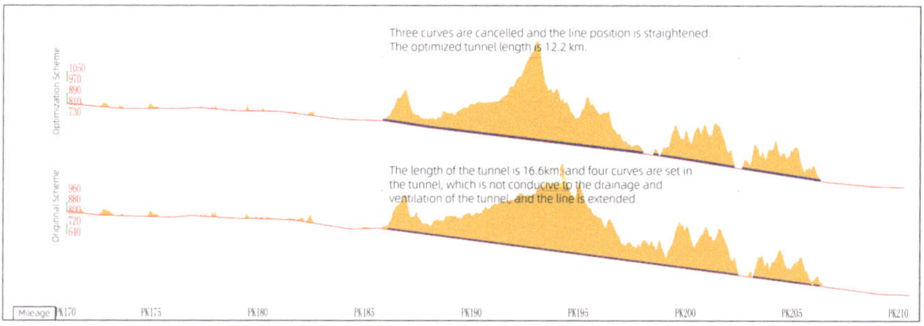


(b) A schematic longitudinal section of the line

**Fig. 7.18** Track alignment plan and profilediagram & dynamic evaluation of track alignment

performance, though the absolute differences between the two schemes are relatively small.

To further analyze the impact of increasing the minimum curve radius on train dynamic performance, a comparative study was conducted on the PK178 + 000 to PK187 + 000 section. This segment includes three curves in the original design with radii of 5000 m, 2100 m, and 2900 m, respectively, whereas the optimized design adopts radii of 5000 m, 3000 m, and 3000 m at a design speed of 220 km/h. The results indicate that increasing the curve radius significantly improves train dynamic performance, with a noticeable reduction in dynamic indicators. The average lateral acceleration of the car body in the original design reaches 1.5 m/s$^2$ in the 2100 m radius curve section, whereas in the optimized alignment, it is reduced to below 1.0 m/s$^2$, markedly improving passenger ride comfort. Additionally, the wheel load reduction rate and derailment coefficient show significant improvements, enhancing operational safety.

2. **Statistical Analysis of Train Dynamic Performance Indicators**

Figure 7.20 presents the statistical analysis of the main train dynamic performance indicators before and after the optimization of the PK171 + 000–PK210 + 500 section. From the overall statistical results of this section, the optimized track alignment features an increased minimum curve radius and a well-matched horizontal and vertical alignment, leading to a noticeable reduction in the peak values of various dynamic indicators and a significant improvement in train dynamic performance.

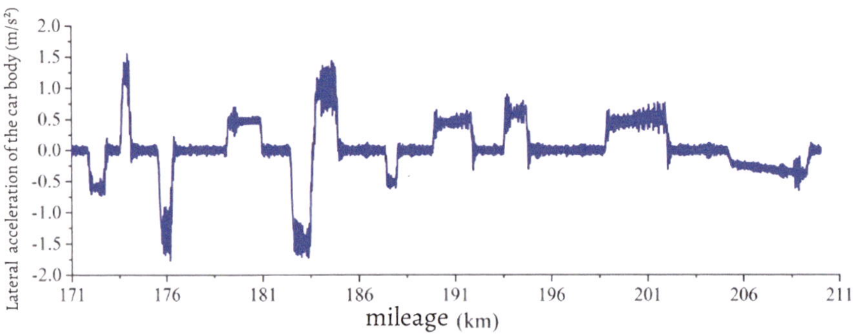


(a) Time history curve of car body lateral acceleration in the original design scheme

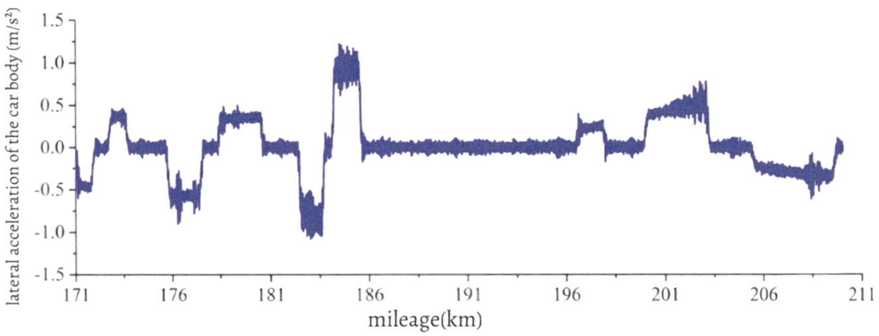


(b) Time history curve of car body lateral acceleration in the optimized scheme

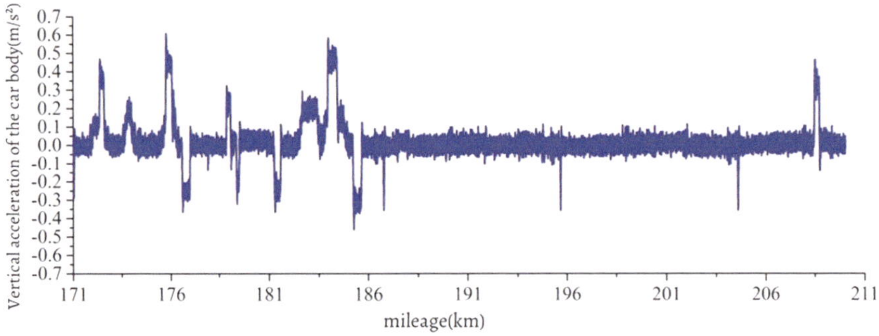


(c) Time history curve of car body vertical acceleration in the original design scheme

**Fig. 7.19** Time history analysis of train dynamic performance indicators

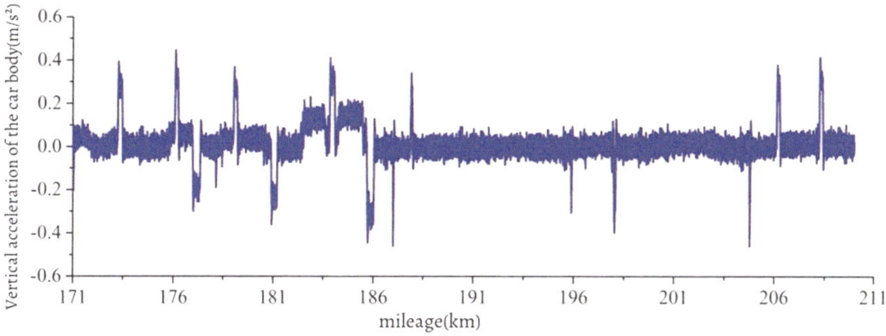


(d) Time history curve of car body vertical acceleration in the optimized scheme

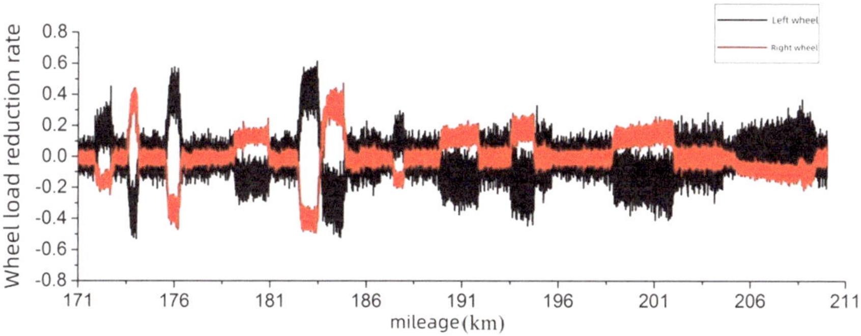


(e) Time history curve of derailment coefficient in the optimized scheme

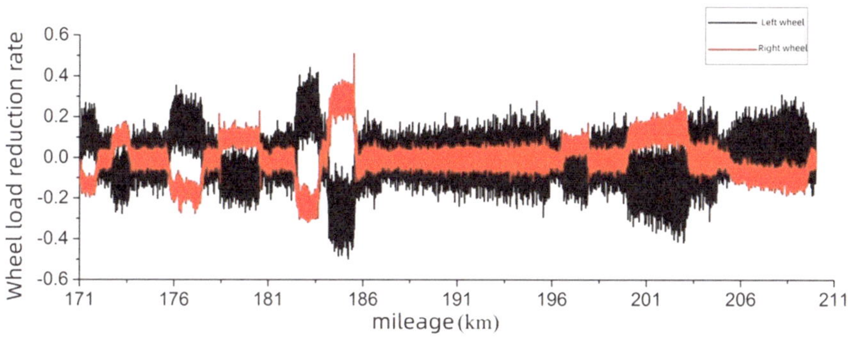


(f) Time history curve of wheel load reduction rate in the original scheme

**Fig. 7.19** (continued)

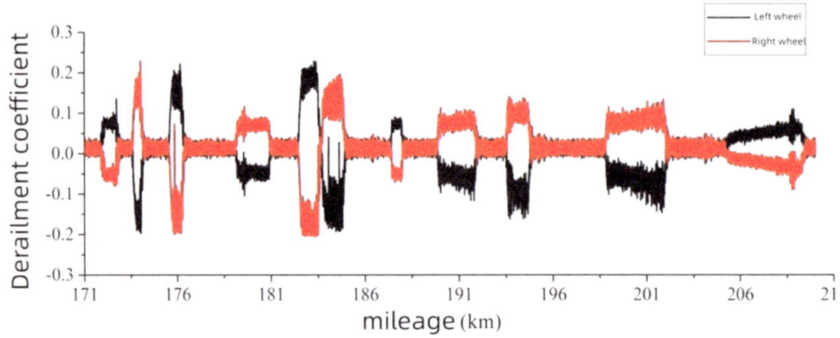


(g) Time history curve of wheel load reduction rate in the optimized scheme

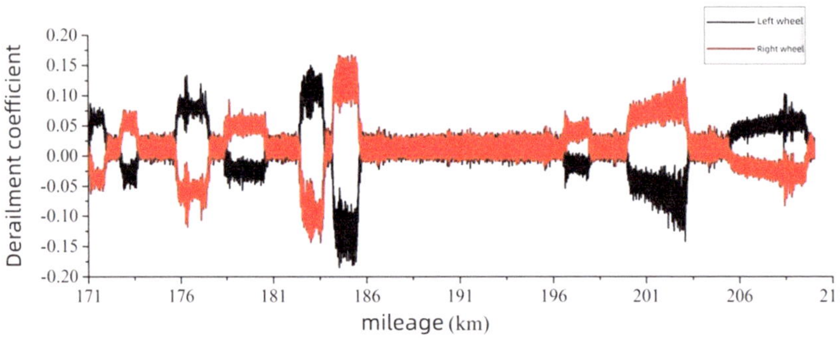


(h) Time history curve of derailment coefficient in the original scheme

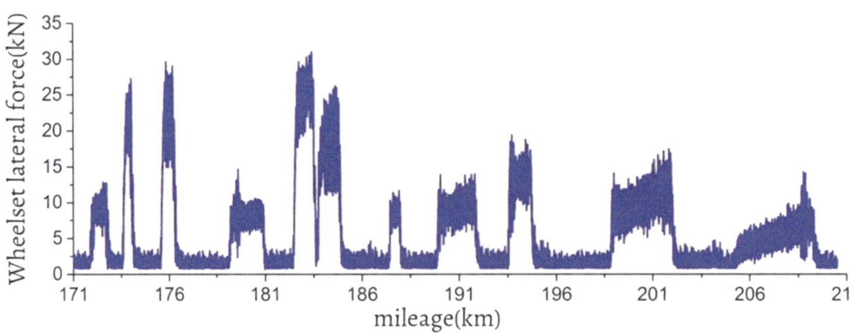


(i) Time history curve of wheelset lateral force in the original scheme

**Fig. 7.19** (continued)

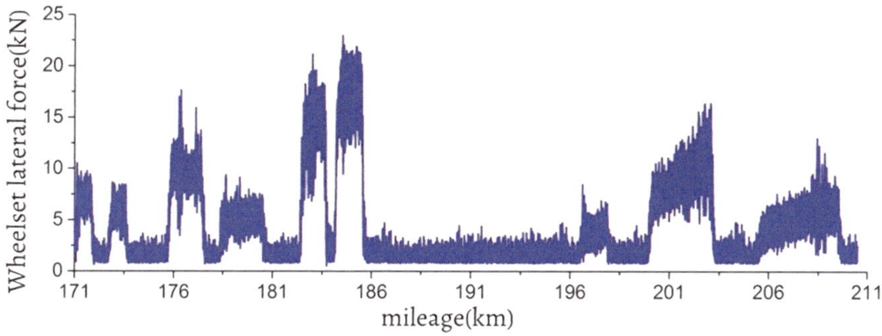


(j) Time history curve of wheelset lateral force in the optimized scheme

**Fig. 7.19** (continued)

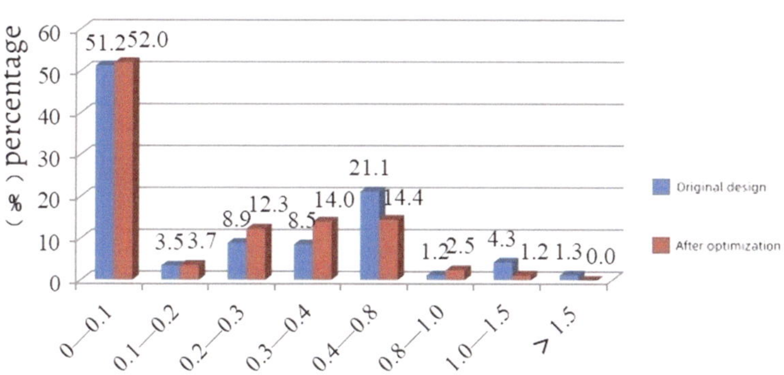


(a) The lateral acceleration of the car body (m/s$^2$)

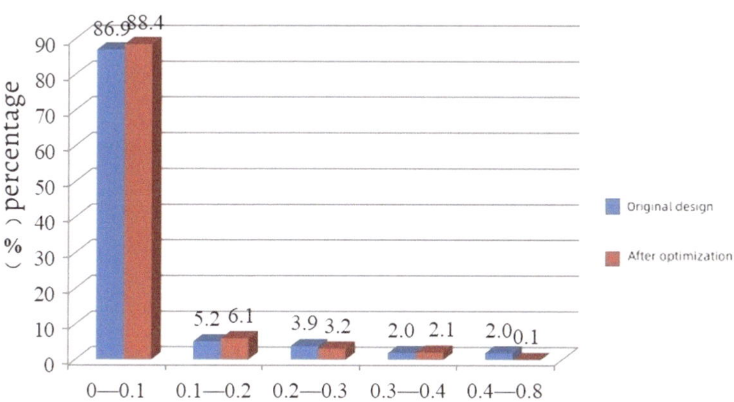


(b) The vertical acceleration of the car body (m/s$^2$)

**Fig. 7.20** Statistical analysis of train dynamic performance indicators for track alignment schemes

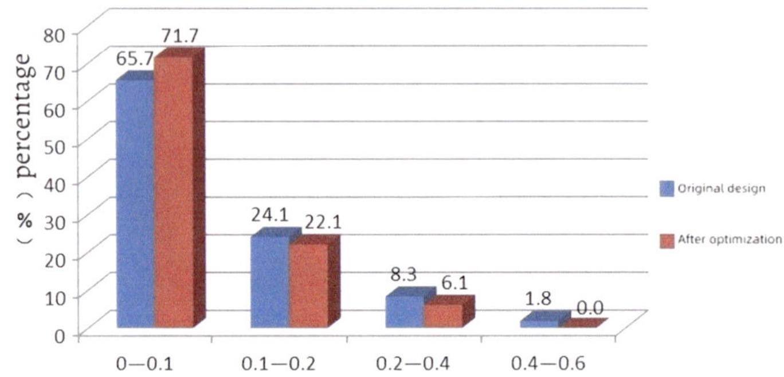


(c) Wheel load reduction rate

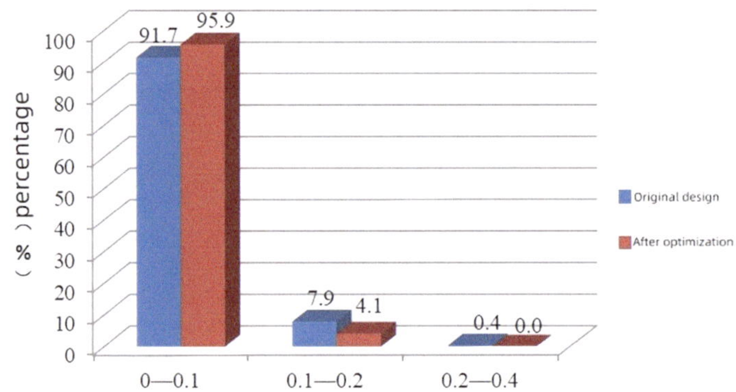


(d) Derailment coefficient

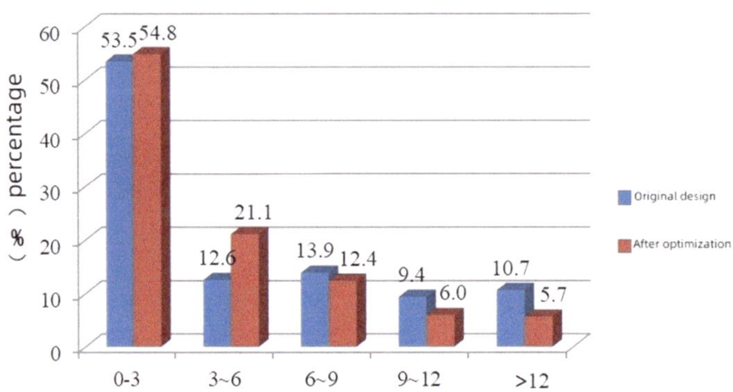


(e) Wheelset lateral force (kN)

**Fig. 7.20** (continued)

In the original design scheme, 73.1% of the track segments exhibited a lateral car body acceleration of less than 0.4 $m/s^2$, with 1.3% of the peak values exceeding 1.5 $m/s^2$. After optimization, 82% of the track segments had a lateral car body acceleration below 0.4 $m/s^2$, and all peak values remained under 1.5 $m/s^2$.

For vertical car body acceleration, 2% of the peak values in the original design fell within the range of 0.4–0.8 $m/s^2$, while in the optimized scheme, nearly all peak values were below 0.4 $m/s^2$. The optimization of the horizontal and vertical alignments significantly improved passenger ride comfort.

Following the optimization, the peak values of wheelset lateral force, wheel load reduction rate, and derailment coefficient were all reduced to some extent, contributing to an improvement in operational safety.

The statistical analysis of dynamic indicators suggests that frequency-based statistical evaluation of dynamic performance can effectively distinguish the differences between various track alignment schemes. This method is suitable for application in the dynamic evaluation and comparison of track alignment schemes.

### *7.3.3 Summary*

This section focuses on the track alignment schemes in the preliminary research of a certain overseas high-speed rail project. It adopts a railway system dynamics approach to evaluate and compare track alignment schemes from the perspective of system dynamic responses and train dynamic performance. The applicable models for track alignment scheme dynamic evaluation are introduced, and an evaluation method based on the conventional dynamic evaluation indicators, with the addition of peak value frequency statistics, is proposed. Through this dynamic evaluation and comparison of track alignment schemes, the following conclusions can be drawn:

① The dynamic evaluation and comparison can provide reference for decision-making in selecting track alignment schemes and guidance for optimizing track alignment.

② The frequency statistics of dynamic indicators can effectively reflect the differences in the impact of different track alignment schemes on system dynamic responses, making it suitable for the dynamic evaluation and comparison of track alignment schemes. The dynamic evaluation and comparison method proposed in this paper can be applied to other similar projects.

# Chapter 8
# Dynamic Analysis of the Train-Track-Bridge System in Curved Sections

High-speed railways have numerous bridges located in curved sections. When trains navigate through curves, the centrifugal forces further amplify the spatial dynamic responses of the train-track system, which may adversely affect passenger comfort and operational safety. This chapter first analyzes the dynamic behavior characteristics of the train-track-bridge system in curved sections. Secondly, it introduces a case study on the performance verification of the train-track-bridge dynamic system under the condition of reverse curves and short tangents on the Taiyuan-Jiaozuo Railway.

## 8.1 Dynamic Behavior Characteristics of the Train-Track-Bridge System in Curved Sections

### *8.1.1 Calculation Conditions*

1. **Curve Configuration Conditions**

The speed, cant, and radius of the curves in high-speed railways must be mutually compatible. To align with engineering practices, the design specifications for China's high-speed railways and the actual radius sequences were referenced. The matching relationships of speed, cant, and radius are shown in Table 8.1.

The gravitational component due to cant in curved sections can partially offset the centrifugal force. The centrifugal acceleration not countered by gravity can be calculated using the following formula:

$$a_u = \frac{v^2}{R} - g \sin\theta_s \tag{8.1.1}$$

© China Railway Publishing House Co., Ltd. 2026

X. Long et al., *Vehicle–Track–Bridge System Subjected to Track Alignment*,
https://doi.org/10.1007/978-981-95-1649-0_8

**Table 8.1** High-speed railway curve design parameters

| Speed (km/h) | Superelevation (mm) | Radius (m) |
|---|---|---|
| 25–80 | 30 | 2000 |
| 80–120 | 80 | 2000 |
| 120–140 | 100 | 2000 |
| 140–160 | 150 | 2000 |
| 160–180 | 150 | 2000 |
| 180–200 | 150 | 2500 |
| 200–250 | 150 | 3500 |
| 250–300 | 150 | 5000 |
| 300–350 | 150 | 7000 |
| 350–400 | 150 | 8000 |
| 400–450 | 150 | 10,000 |
| 450–500 | 175 | 10,000 |

The unbalanced centrifugal acceleration not only reflects the impact of curves on passenger comfort but also indicates the centrifugal forces actually exerted on the bridge in curved sections. This is a crucial parameter for assessing the performance of curves in railway design. In railway standards, the unbalanced centrifugal acceleration is expressed in terms of insufficient cant ($h_q$), and the specific conversion formula is as follows:

$$h_q = \frac{S_1}{g} a_u \tag{8.1.2}$$

Let $S_1$ be the track gauge, which is taken as 1500 mm. Thus:

$$h_q = 153 a_u \tag{8.1.3}$$

Through extensive practice on high-speed railways, the allowable value for insufficient cant can be taken as 90–110 mm. Figure 8.1 shows the variation of insufficient cant under different speed and curve radius conditions, where a negative $h_q$ indicates a situation of excessive cant. For comparative analysis in subsequent sections, the absolute value of insufficient cant $|h_q|$ is also presented.

As observed in Fig. 8.1, with increasing speed, the insufficient cant exhibits multiple peaks and troughs. This is due to the actual cant increasing with speed, ensuring that the insufficient cant remains within the allowable range. When the insufficient cant is a positive value, the resultant lateral force on the track acting on the bridge directs outward, creating a clockwise torque. Conversely, when the insufficient cant is negative, indicating a state of excessive cant, the resultant lateral force on the track produces a counterclockwise torque on the bridge.

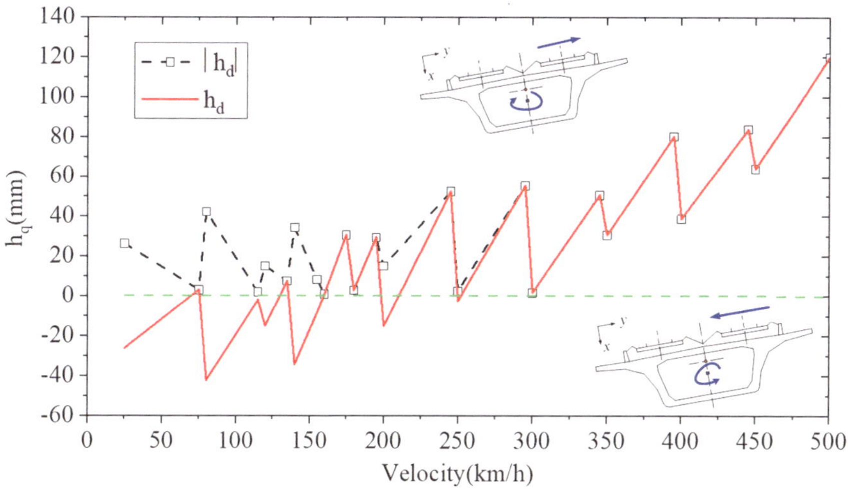


**Fig. 8.1** Variation of insufficient cant under different speed and curve radius conditions

2. **Model Overview**

Based on the structural diagram of the bridge, a double-track simply supported box girder model was established using finite element software. This allows for the determination of the vertical, lateral, and torsional frequencies and mode shapes of the bridge. Figure 8.2 illustrates a 32-m simply supported girder for high-speed railways operating at 350 km/h. As shown in Table 8.2, the calculation results indicate that the lateral and torsional mode shapes frequently occur in conjunction, with the lateral-torsional vibration frequency being relatively high.

### *8.1.2 Train-Bridge Resonance and Damping Conditions*

Referencing Fig. 8.3, consider a torsional beam of length $L$ subjected to a series of moments $T_b$ that move at a speed $v$ with a constant interval $d$. The torsional vibration equation of the beam under the influence of these moving moments is given by:

$$-\rho J \frac{\partial^2 \theta_x}{\partial t^2} + GJ \frac{\partial^2 \theta_x}{\partial x^2} = \sum_{k=1}^{N} T_b \delta[x - v(t - t_k)] \times [H(t - t_k) + H(t - t_k - L/v)] \tag{8.1.4}$$

In the equation, $\rho$ is the unit volume mass, $G$ is the shear modulus, $J$ is the polar moment of inertia, $T_b$ is the moving moment, and $N$ is the total number of moving moments that have passed through the left endpoint of the bridge. $\delta$ is the Dirac

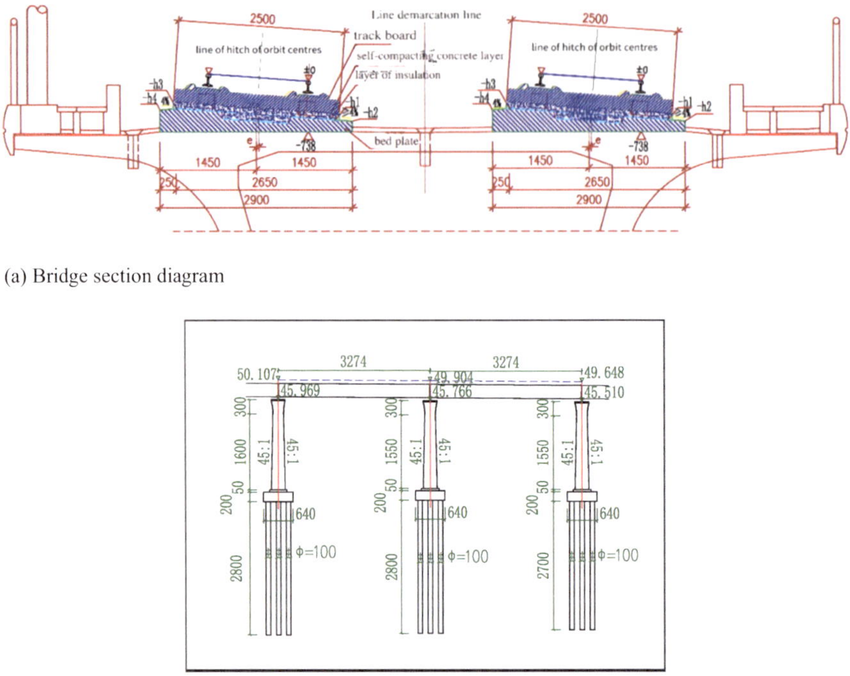


(a) Bridge section diagram

(b) Bridge longitudinal section diagram

**Fig. 8.2** Schematic diagram of a 32 m simply supported girder for high-speed railways at 350 km/h

delta function, $t_k$ represents the time at which the $k$ moving moment enters the left endpoint of the bridge, and $H(t - t_k)$ is the unit step function.

For a simply supported beam, the torsion angle $\theta_x(x,\ t)$ can be expressed in terms of the vibration mode shapes as follows:

$$\theta_x(x, t) = \sum_{i=1}^{\infty} \phi_i(x) \cdot q_i(t) \tag{8.1.5}$$

where $q_i(t)$ represents the generalized coordinates associated with the $n$-th vibration mode of the beam. By substituting the angle $\theta_x(x,\ t)$ from Eq. (8.1.5) into Eq. (8.1.4), the torsion equation of the beam can be expressed as:

$$\begin{aligned} & -\rho J \sum_{i=1}^{\infty} \phi_i(x) \frac{d^2 q_i(t)}{dt^2} + GJ \sum_{i=1}^{\infty} q_i(t) \frac{d^2 \phi_i(x)}{dx^2} \\ & = \sum_{k=1}^{N} T_b \delta[x - v(t - t_k)] \times [H(t - t_k) + H(t - t_k - L/v)] \end{aligned} \tag{8.1.6}$$

**Table 8.2** Self-vibration characteristics of the double-track simply supported box girder

| Order | (Hz) | Vibration mode description | Vibration mode |
|---|---|---|---|
| 1 | 4.9179 | Perpendicular |  |
| 2 | 14.023 | Transverse-torsion | 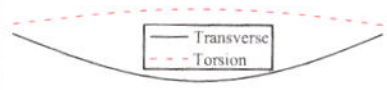  |
| 3 | 16.522 | Transverse-torsion | 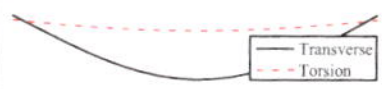  |
| 4 | 15.682 | Perpendicular |  |
| 5 | 21.015 | Perpendicular |  |
| 6 | 35.361 | Perpendicular |  |
| 7 | 29.462 | Transverse-torsion |   |
| 8 | 44.219 | Transverse |  |
| 9 | 53.427 | Perpendicular |  |
| 10 | 44.657 | Torsion |  |

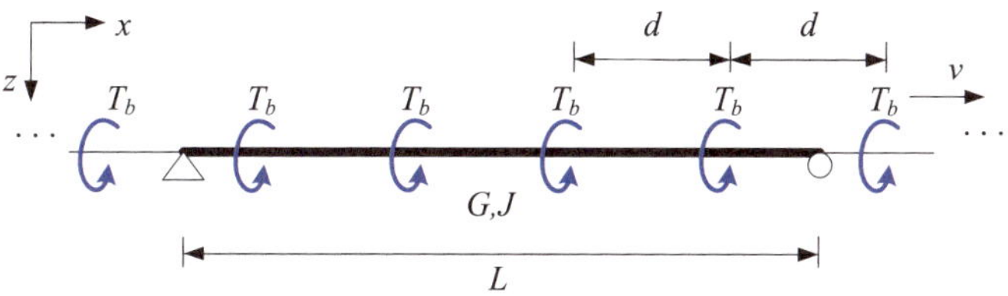


**Fig. 8.3** Simply supported beam subjected to moving bending moment

Integrating along the beam axis x over the length L, and then multiplying both sides of the equation by the $n$-th shape function while utilizing the orthogonality of the mode shapes, the above equation can be transformed into:

$$-\rho J \frac{d^2 q_n(L)}{dt^2} \int_0^L \frac{d^2 \phi_n(x)}{dx^2} \phi_n(x) dx$$

$$= \int_0^L \sum_{k=1}^{N} T_b \phi_n(x) \delta[x - v(t - t_k)] \times [H(t - t_k) + H(t - t_k - L/v)] dx \quad (8.1.7)$$

The free torsional vibration of the beam can be expressed by a sine function, that is, $\phi_n(x) = \sin(n\pi x/L)$, so the above equation can be rewritten as

$$\ddot{q}_n + \omega_{tn}^2 q_n = \frac{2T_b}{\rho JL} \left\{ \sum_{k=1}^{N} \sin\left( \frac{\pi v(t - t_k)}{L} \right) \times [H(t - t_k) + H(t - t_k - L/v)] \right\} \quad (8.1.8)$$

In the above formula, $N$ is the number of torques that have been moved, and $\omega_{tn}$ is the $n$-th torsional vibration frequency of the beam.

$$\omega_{tn} = \frac{n\pi}{L} \sqrt{\frac{G}{\rho}} \quad (8.1.9)$$

By Duhamel integral, the response $\theta_x(x, t)$ in the generalized coordinate system can be derived directly from Eq. (8.1.10).

$$\theta_x(x, t) = \frac{2T_b L}{\pi^2 GJ} \frac{1}{1 - S_{tn}^2} \sin \frac{\pi x}{L} \tau_n(v, t) \quad (8.1.10)$$

$$\begin{aligned} \tau_n(v, t) &= \sum_{k=1}^{N} \{\sin \Omega_n (t - t_k) - S_{tn} \sin \omega_{tn}(t - t_k)\} H(t - t_k) \\ &+ \sum_{k=1}^{N} \left\{ \sin \Omega_n \left( t - t_k - \frac{L}{v} \right) - S_{tn} \sin \omega_{tn} \left( t - t_k - \frac{L}{v} \right) \right\} H\left( t - t_k - \frac{L}{v} \right) \end{aligned} \quad (8.1.11)$$

where $\Omega_n$ is the excitation frequency ($\Omega_n = n\pi v/L$) and $S_{tn}$ is the velocity parameter ($S_{tn} = \Omega_n/\omega_{tn}$).

According to the reference, the above formula is further integrated by the triangular transformation formula to obtain the first $N - 1$ moving torque leaving the beam, and when the $N$-th moving torque enters the beam, Formula (8.1.11) can be organized as follows:

$$\begin{aligned} \tau_n(v, t) &= [\sin \Omega_n (t - t_N) - S_{tn} \sin \omega_{tn}(t - t_N)] H(t - t_N) \\ &- 2S_{tn} \cos \frac{\omega_{tn} L}{2v} \times \left[ \sin \omega_{tn} \left( t - \frac{L}{2v} \right) \right. \\ &\left. + \sin \omega_{tn} \left( t - \frac{L}{2v} - \frac{t_N}{2} \right) \frac{\sin (N-2) \frac{\omega_{tn} d}{2v}}{\sin \frac{\omega_{tn} d}{2v}} \right] H\left( t - t_{N-1} - \frac{L}{v} \right) \end{aligned} \quad (8.1.12)$$

In the above formula, $H(t - t_N)$ represents that the moving torque is located on the beam at time t, and $H\left(t - t_{N-1} - \frac{L}{v}\right)$ represents the residual vibration effect of the moving torque on the beam after the $N - 1$ moving torque leaves the beam.

According to the above formula, when $\sin \frac{\omega_{tn} d}{2v} = 0$, in the formula, it is obtained by L'Hospital rule:

$$\begin{aligned}\tau_n(v,t) &= [\sin \Omega_n(t - t_N) - S_{tn} \sin \omega_{tn}(t - t_N)]H(t - t_N) \\ &\quad - 2(N-1)S_{tn} \cos \frac{\omega_{tn} L}{2v} \times \sin \omega_{tn}\left(t - \frac{L}{2v}\right)H\left(t - t_{N-1} - \frac{L}{v}\right)\end{aligned} \tag{8.1.13}$$

The above formula shows that the torsional angular displacement of the beam is continuously amplified with the increase of the number of torques $\boldsymbol{N}$, forming a resonance phenomenon. The torsional resonance speed can be expressed as:

$$v_{res}^{bt} = 3.6 \frac{f_{tn}^b d_v}{i} \quad (i = 1, 2, 3 \ldots) \tag{8.1.14}$$

In the formula, $f_{tn}^b$ is the torsion frequency of the beam, and $d_v$ is the moving torque spacing (generally for the length of the train).

When $\cos \frac{\omega_{tn} L}{2v} = 0$ in the formula, there is only one item left, namely:

$$\tau_n(v,t) = [\sin \Omega_n(t - t_N) - S_{tn} \sin \omega_{tn}(t - t_N)]H(t - t_N) \tag{8.1.15}$$

This shows that the sum of the free vibration of the bridge caused by the $N - 1$ concentrated force is zero, and the vertical displacement is only determined by the $N$-th concentrated force acting on the beam.

$$v_{can}^{bt} = 7.2 \frac{f_{tn}^b L}{2i - 1} \quad (i = 1, 2, 3 \ldots) \tag{8.1.16}$$

Considering that there are several different spacings in the actual axle load arrangement of the train, this spacing can be the length $L_v$ of the vehicle, the fixed distance $L_c$ of the vehicle and the wheelbase $L_t$. According to the literature derivation, the torsional vibration reduction speed of the beam can be uniformly expressed as:

$$v_{can}^{bt} = \frac{7.2\alpha f_{tn}^b L}{2i - 1} \quad (i = 1, 2, 3 \ldots) \tag{8.1.17}$$

In the formula, $\alpha$ is a dimensionless parameter, which can be taken as 1, /, $L_v/L$, $L_c/L$, $L_t/L$.

According to Yang et al. [15, 20], the critical speed of a simply supported beam bridge under bending resonance is:

$$v_{res}^{bv(h)} = 3.6\frac{f_{v(h)n}^{b} L_v}{i} \quad (i = 1, 2, 3 \ldots) \tag{8.1.18}$$

where $f_{v(h)n}^{b}$ is the vertical and horizontal bending frequency of the bridge in the nth mode, in Hz. The critical speed of bridge resonance is not directly affected by the bridge span.

The critical speed of bending vibration reduction of simply supported beam bridge is similar to that of Eq. (8.1.18), which can be expressed as:

$$v_{can}^{bv(h)} = \frac{7.2\alpha f_{v(h)n}^{b} L}{2i - 1} \quad (i = 1, 2, 3 \ldots) \tag{8.1.19}$$

For the train structure in this paper, the train length, fixed distance and wheelbase are 24.65 m, 17.5 m and 2.5 m respectively, and the dimensionless coefficients are 1,0.77,0.55 and 0.08 respectively. According to the formula, the fundamental frequency resonance and vibration elimination speed of 32 m double-line simply supported box girder can be obtained, as shown in Table 8.3. According to the analysis of Sect. 8.1.1, it can be seen that the transverse vibration frequency and the torsional fundamental frequency of the simply supported box girder appear together, and the transverse displacement and torsional angle resonance of the beam are consistent with the vibration elimination speed.

**Table 8.3** Resonance and damping speed of fundamental frequency of 32.1 m double-line simply supported box girder

| I | | 1 | 2 | 3 | 4 | 5 | 6 | 7 | 8 | 9 | 10 |
|---|---|---|---|---|---|---|---|---|---|---|---|
| $v_{res}^{bv}$ (km/h) | | 436 | 218 | 146 | 109 | 87 | 73 | 62 | 55 | – | – |
| $v_{res}^{bh}$ (km/h) | | – | – | 415 | 311 | 249 | 207 | 178 | 156 | 138 | 124 |
| $v_{can}^{bv}$ (km/h) | $\alpha = 1$ | – | 379 | 227 | 162 | 126 | 103 | 87 | 76 | 67 | – |
| | $\alpha = 0.77$ | – | 291 | 175 | 125 | 97 | 79 | 67 | 58 | 51 | – |
| | $\alpha = 0.55$ | – | 207 | 124 | 89 | 69 | 56 | – | – | – | – |
| | $\alpha = 0.08$ | 89 | – | – | – | – | – | – | – | – | – |
| $v_{can}^{bh(t)}$ (km/h) | $\alpha = 1$ | – | – | – | 463 | 360 | 295 | 249 | 216 | 191 | 171 |
| | $\alpha = 0.77$ | – | – | 498 | 356 | 277 | 226 | 191 | 166 | 146 | 131 |
| | $\alpha = 0.55$ | – | – | 354 | 253 | 196 | 161 | 136 | 118 | 104 | 93 |
| | $\alpha = 0.08$ | 253 | 84 | 51 | – | – | – | – | – | – | – |

### 8.1.3 *Dynamic Interaction Analysis of Wheel-Rail on Curved Sections*

The interaction between the wheels and rails on curved sections is crucial. This paper's dynamic analysis model takes the wheel-rail relationship into account in detail, allowing for a thorough analysis of the interaction on curved sections.

Figure 8.4 illustrates the changes in wheel load reduction rates and speed under straight and curved conditions. It can be observed that the wheel load reduction rate on curved sections is significantly higher than that on straight sections, exhibiting several peaks and troughs. Under the condition of cant (superelevation), the outer wheel experiences a reduction in load, while under the condition of negative cant (cant deficiency), the inner wheel experiences a reduction in load. Consequently, the patterns of peaks and troughs in the load reduction rate as a function of speed resemble those shown in the $|h_q|$ graph.

Figure 8.5 shows the derailment coefficient of the outer wheel as a function of speed under straight and curved conditions. It can be seen that the derailment coefficient on curved sections is significantly higher than that on straight sections, exhibiting several peaks and troughs. The derailment coefficient of the outer wheel is positively correlated with cant deficiency, which means that the patterns of peaks and troughs in the derailment coefficient as a function of speed resemble those shown in the $h_q$ graph.

When the vehicle passes through the curve, due to the centrifugal force, the wheel is easy to deviate to one side and cause wear to the wheel and rail. Here, the wear index is used to represent the strength of wheel-rail wear, as follows:

$$W = T_x\xi_x + T_y\xi_y \tag{8.1.20}$$

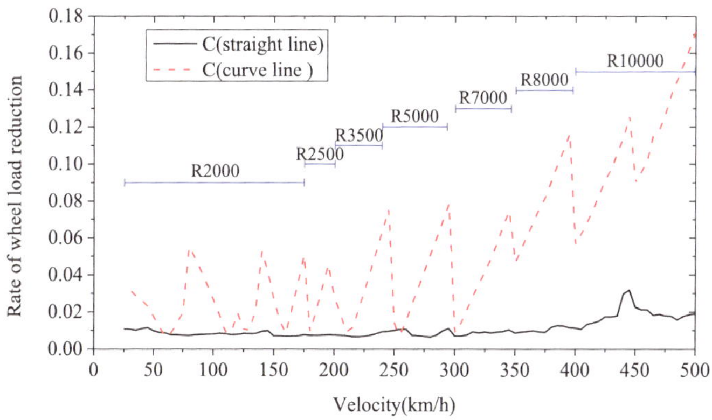


**Fig. 8.4** Changes in wheel load reduction rate and speed under straight and curved conditions

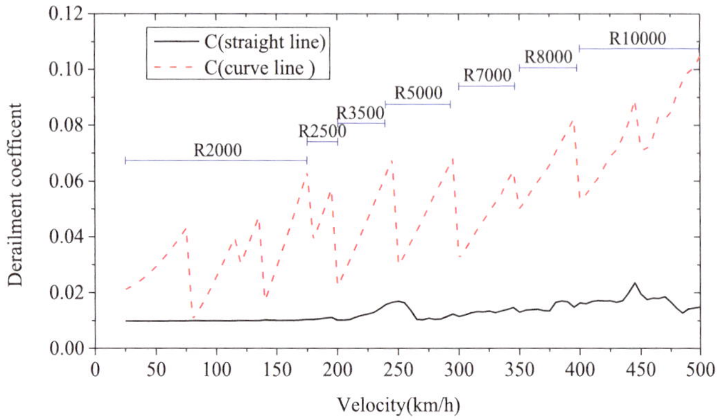


**Fig. 8.5** Curves of derailment coefficient and speed of outer wheel under straight section and curve conditions

In the formula, $T_x$ and $T_y$ are wheel-rail longitudinal creep force and lateral creep force respectively, $\xi_x$ and $\xi_y$ are wheel-rail longitudinal creep rate and lateral creep rate respectively.

The surface fatigue index proposed by Ekberg et al. is used to evaluate the wheel-rail rolling contact fatigue characteristics. The fatigue index is defined as follows:

$$FI_{\text{surf}} = \mu - \frac{2\pi abk}{3F_z} \tag{8.1.21}$$

In the formula, $a$ and $b$ are the short half-axis and the long half-axis of the elliptical contact spot respectively, $k$ is the pure shear yield strength of the material, take 303 MPa, $F_z$ is the wheel-rail normal force, $\mu$ is the traction coefficient, and its value is determined by the following formula:

$$\mu = \frac{\sqrt{T_x^2 + T_y^2}}{F_z} \tag{8.1.22}$$

In the above expression, if the fatigue index $FI_{\text{surf}}$ is positive, it is believed that the wheel-rail material will experience a ratchet effect, and plastic deformation will accumulate until the material loses its toughness and develops fatigue cracks. As seen from the equation, poor guiding performance of the wheelset or inadequate wheel-rail geometric relationships will lead to an increase in the fatigue index, accelerating fatigue failure.

Figures 8.6 and 8.7 show the wear index and surface fatigue index of the outer wheel on straight and curved sections as functions of speed, derived from the above expression. It can be observed that both the wear index and fatigue index on curved sections are significantly higher than those on straight sections, with both indices decreasing progressively as speed increases. When the train speed exceeds

200 km/h on curved sections, both indices decline rapidly; after exceeding 300 km/h, the decline becomes more gradual. This indicates that the key factors influencing wheel-rail wear and fatigue failure on curved sections are the radius of curvature, with radii greater than 7000 m significantly improving wear and fatigue issues.

Figure 8.8 further illustrates the contact points of the inner and outer wheelsets and the rails on straight and curved sections as functions of speed. It can be seen that on straight sections, the contact points of the wheelsets and the rails are located near the nominal rolling circle and the centerline of the rails, with little influence from increasing speed. On curved sections, the contact point on the tread of the inner wheelset shifts toward the flange, while the contact point on the inner rail moves inward toward the centerline. As speed increases and the radius of curvature enlarges, the contact points approach the nominal rolling circle and the centerline of the rails. For the outer wheelset on curved sections, the contact points on the tread and the rails are less affected by speed and radius, and the positions are relatively close to

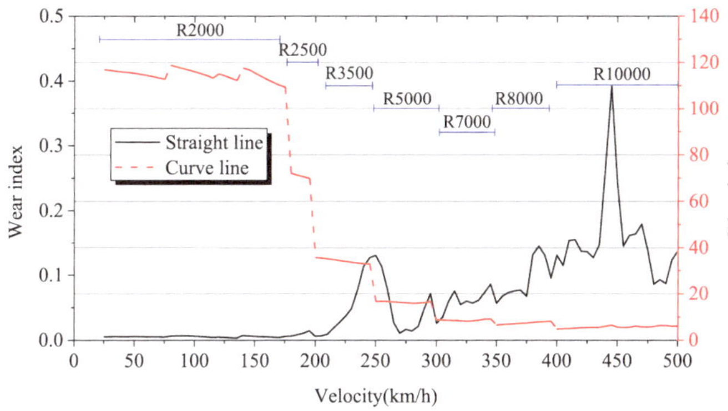


**Fig. 8.6** The curve of the wear index of the outer wheel with the speed in the straight section and the curve section

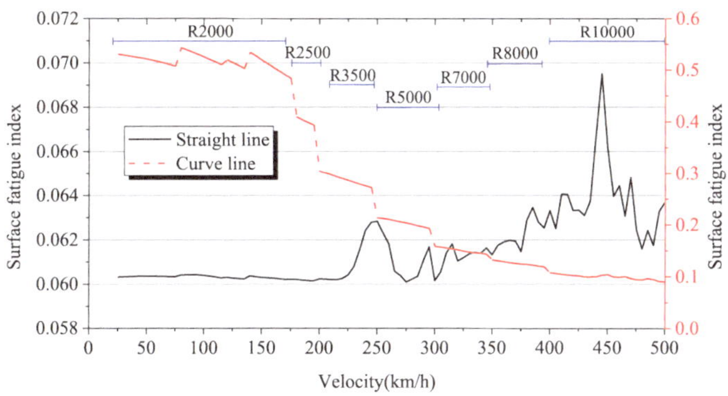


**Fig. 8.7** The curve of the fatigue index of the outer wheel surface with the speed in the straight section and the curve section

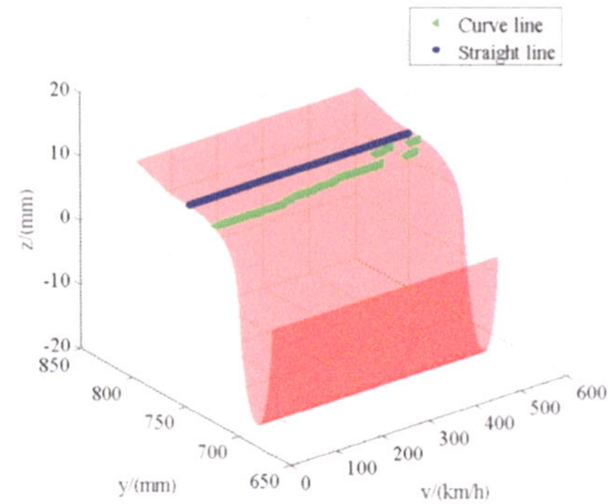


(a) Inner wheel tread

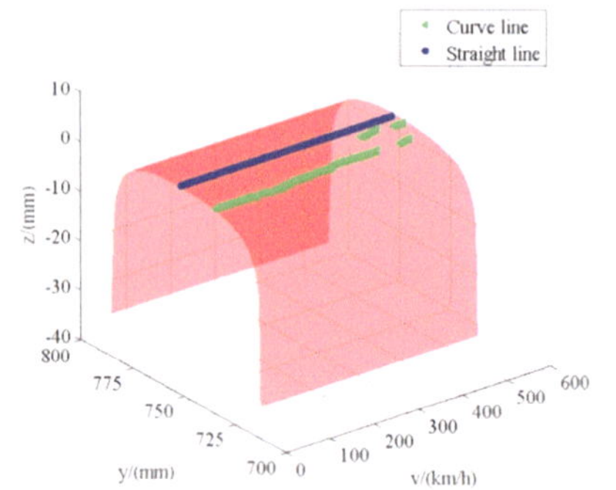


(b) Inner rail

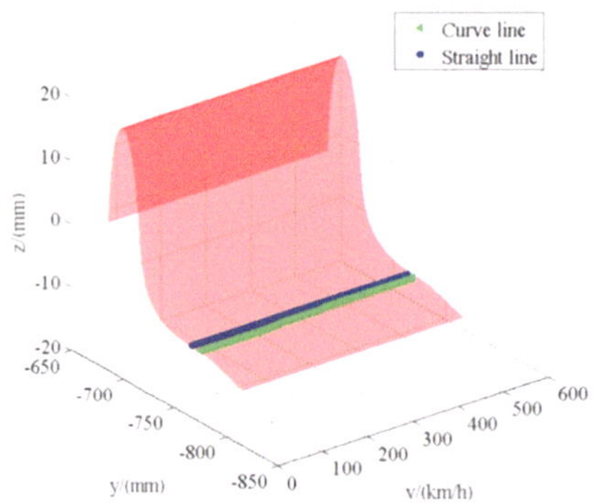


(c) Outer wheelset tread

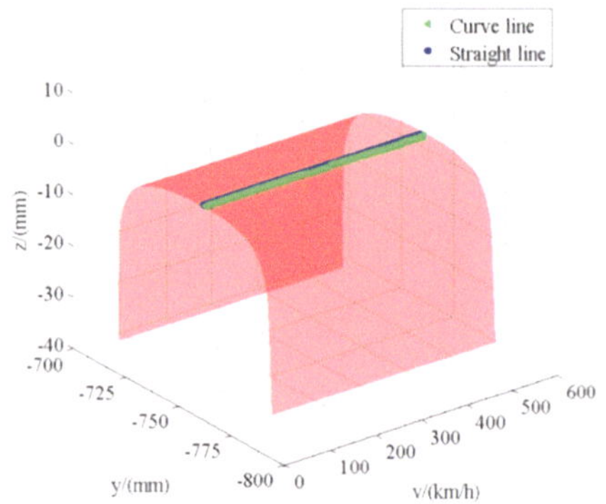


(d) Outer rail

**Fig. 8.8** Variation trend of wheel-rail contact points with speed under straight section and curve conditions

those on straight sections. However, the contact positions of the inner wheelset and the rails on curved sections are influenced by both radius and speed, which should be a focal point in practical wheel-rail friction management.

### 8.1.4 *Dynamic Response Analysis of Bridges on Curved Sections*

Based on the theoretical analysis results in the table, the first-order resonance speed of the beam is selected at 436 km/h (R = 10,000 m) and the second-order damping speed at 379 km/h (R = 8000 m). Using the vehicle-bridge coupling model, the time history curve of the vertical displacement at point A in the middle of the span on

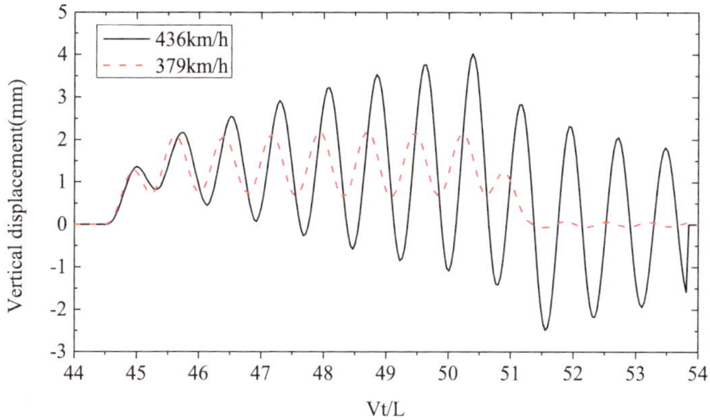


**Fig. 8.9** A vertical displacement time history curve of point 89 a in mid-span

the curved section is obtained, as shown in Fig. 8.9. It can be observed that under resonance speed, the vertical displacement response is gradually amplified, while under damping speed conditions, the residual vibration caused by the moving load is suppressed, with the vertical displacement of the beam determined by the forced vibration induced by the load acting on it.

According to the theoretical analysis results in the table, the third-order resonance speed for the beam in lateral-torsional motion is selected at 415 km/h (R = 10,000 m) and the fourth-order damping speed at 463 km/h (R = 10,000 m). Using the vehicle-bridge coupling model, the time history curves of the lateral displacement and torsional angle at point C in the middle of the span on the curved section are obtained, as illustrated in Figs. 8.10 and 8.11. It can be seen that the lateral displacement and torsional angle at resonance speed are significantly greater than at damping speed. Under resonance conditions, the residual vibrations caused by centrifugal force and moving torque are continuously amplified, while under damping speed conditions, these residual vibrations are suppressed. The lateral displacement and torsional angle of the beam are determined by the forced vibrations induced by the loads and torques acting on the beam.

Figure 8.12 shows the relationship between the vertical displacement at the midpoint of the beam on straight and curved sections as a function of vehicle speed. For easier comparison, the resonance and damping speeds listed in the table are marked on the curves. From the figure, it can be observed that as vehicle speed increases, the displacement curve does not monotonically increase but instead exhibits several peaks and troughs. When the resonance speed is at 436 km/h, the vertical displacement is significantly high, indicating that resonance occurs at this speed. At other resonance speeds of 218, 109, 87, 73, and 62 km/h, corresponding damping speeds of 227, 103, 87, 76, 56, and 55 km/h are observed. Consequently, there are no significant vertical resonance peaks at these resonance speeds, suggesting a phenomenon of resonance disappearance. Additionally, the radius of curvature has a minimal impact on the vertical displacement response of the bridge. Due to the

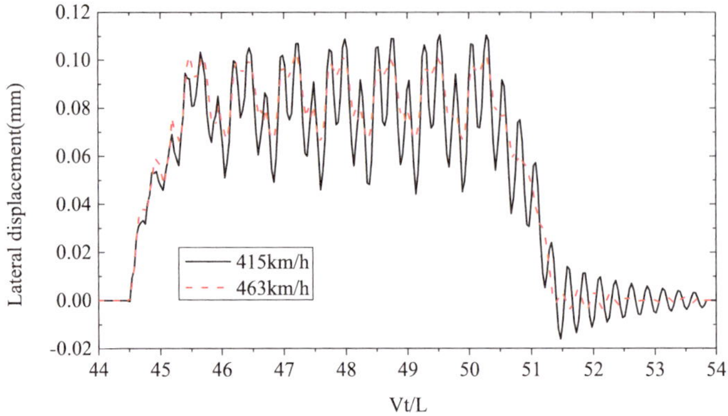


**Fig. 8.10** C Point mid-span lateral displacement time history curve

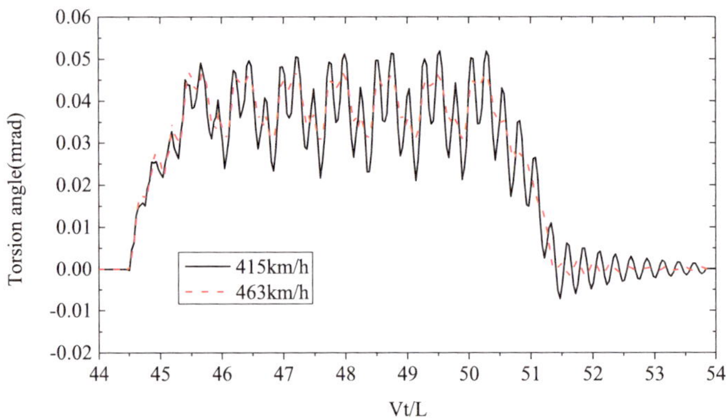


**Fig. 8.11** C Point mid-span torsion angle time history curve

torsional effects of the double-track box girder, the dynamic response at the edge position (B) of the beam cross-section is greater than that at the centerline position (A).

Figures 8.13 and 8.14 show the relationship between lateral and torsional angle displacements at the midpoint of the beam on straight and curved sections as functions of vehicle speed. For comparison, the resonance and damping speeds listed in the table are marked on the curves. From the figures, it can be seen that as vehicle speed increases, the curves for lateral and torsional angle displacements exhibit similar patterns, with several peaks and troughs. On straight sections, there is no centrifugal force from curves; instead, the moving torque caused by the train running eccentrically on the double-track box girder plays a decisive role in lateral and torsional angle displacements. Notably, when the resonance speeds are 415, 178, and 124 km/h, both lateral and torsional angle displacements are significantly high, indicating a pronounced torsional resonance phenomenon. For the straight section, resonance

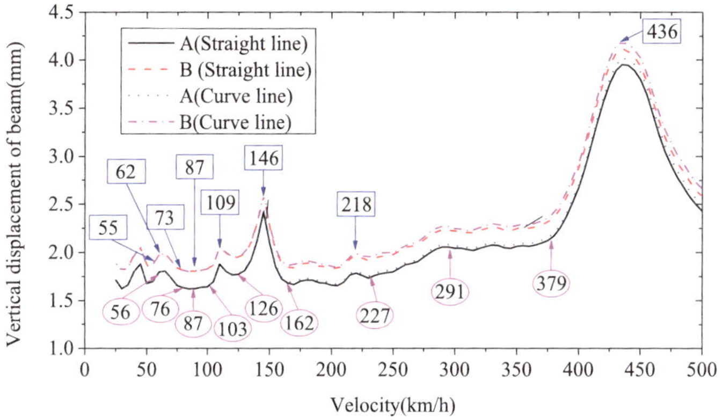


**Fig. 8.12** Relationship between vertical displacement at beam midpoint and vehicle speed

speeds of 311, 249, 207, 156, and 138 km/h correspond to nearby damping speeds of 295, 249, 201, 161, and 136 km/h. Thus, there are no significant resonance peaks at these resonance speeds. On curved sections, the bridge's lateral and torsional displacements are influenced by both the train's centrifugal force and the eccentric torque. In addition to the peaks occurring at the previously mentioned resonance speeds on straight sections, the lateral and torsional angle displacements on curved sections maintain the same pattern of peaks and troughs as shown in Fig. 8.1, which reflects the magnitude of the actual centrifugal force distributed along the bridge. This indicates that the effect of centrifugal force on lateral and torsional angle displacement responses is significant. In conditions of excessive cant on curved sections, the torque generated by centrifugal force opposes the torque generated by the eccentric load of the train. Conversely, under cant deficiency conditions, the directions of the centrifugal torque and the eccentric torque are aligned. Therefore, the displacement response under cant deficiency conditions is lower than that on straight sections, while the response under excessive cant conditions is higher than that on straight sections. This shows that adjusting the actual cant on curves can reduce the peaks of torsional resonance responses.

### 8.1.5 *Summary*

(1) Impact of Eccentric Torque and Cant Deficiency: In curved sections, both lateral and torsional displacements of the bridge are significantly influenced by the train's centrifugal force and eccentric torque. Cant deficiency plays a crucial role in the responses of lateral and torsional displacements. Adjusting the actual cant on curves can effectively reduce the peaks of torsional resonance responses.

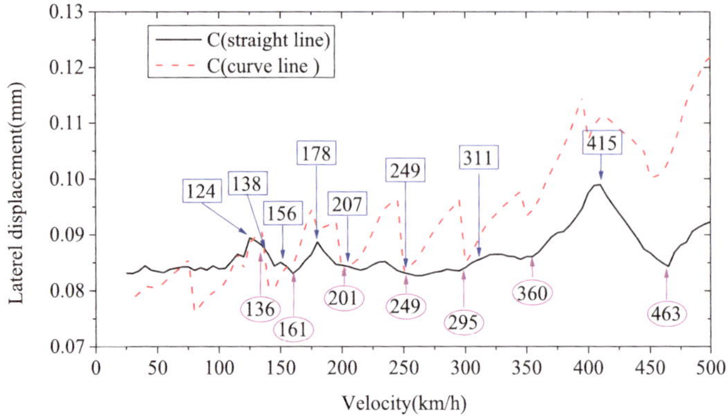


**Fig. 8.13** Relationship between lateral displacement at beam midpoint and vehicle speed

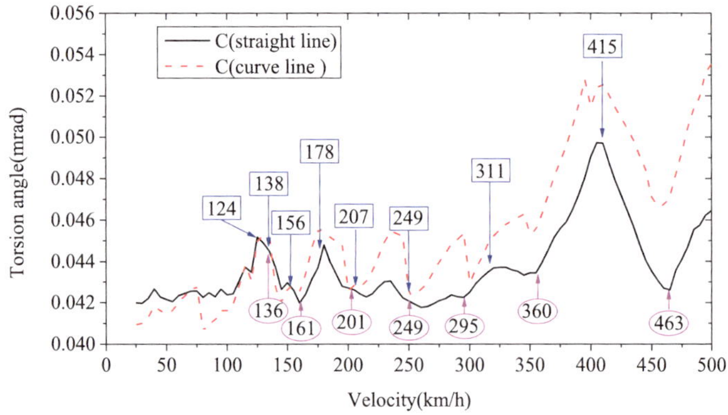


**Fig. 8.14** Relationship between torsional angle displacement at beam midpoint and vehicle speed

(2) Correlation with Wheel Load Reduction and Track Quality: The reduction rate of wheel load and the derailment coefficient in curved sections are positively correlated with cant deficiency. The radius of curvature is a key factor affecting wheel-rail wear and fatigue. Curvature radii greater than 7000 m can notably improve wear and fatigue issues. Additionally, operating trains on curved bridges in a cant-deficient state can also mitigate wear and fatigue effects. As speed and curvature radius increase, the contact points between the wheel and rail move closer to the nominal rolling circle of the wheelset and the centerline of the rail. The contact position of the inner wheelset is significantly affected by the radius and speed, warranting careful attention in actual wheel-rail friction management.

(3) Limitations of Current Methods: Due to mechanical constraints of vehicle components and the interactions between the vehicle and bridge, balanced cant is not the most optimal position for load distribution in the vehicle-bridge system.

Existing quasi-static calculation methods fail to accurately reflect the wheel-rail interactions in curved sections.

## 8.2 Vehicle-Track-Bridge Dynamic System Performance Verification Case

### *8.2.1 Conditions on the Reverse Curve Section of the Taiyuan-Jiaozuo Railway*

The Taiyuan to Jiaozuo railway passenger dedicated line begins at Taiyuan South Station in the Taiyuan hub, traversing through Taiyuan City, Jinzhong City, and Changzhi City, and ends at the Jiaozuo station of the Zheng-Jiao Intercity Railway. The total length of the line is 358.8 km, with 325.4 km located in Shanxi Province and 33.4 km in Henan Province. The design operating speed is 250 km/h, as illustrated in Fig. 8.15.

The Taiyuan-Jiaozuo Railway Bridge spans approximately 8 km and includes four continuous curves. The length of the straight sections between reverse curves is less than the length of a train. When the straight sections between reverse curves are shorter than the length of the train, it may adversely affect operational safety and the structural integrity of the bridge. According to the revised provisions of the *Design Code for Railway Bridges and Culverts* (Iron General Construction [2013] No. 92), it is stated: "Under difficult conditions, when the length of the straight section between reverse curves on a bridge does not meet the length of a train, sufficient technical and economic justification must be provided, along with necessary technical measures, to ensure safe train operation." Given this, a dynamic analysis of the continuous reverse curves on the Taiyuan to Jiaozuo Railway Bridge is conducted, incorporating structural performance and safety assessments for the curve design.

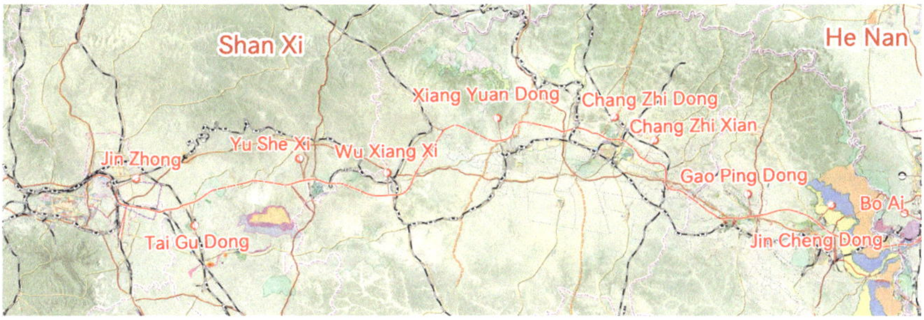


**Fig. 8.15** The horizontal and vertical section diagram of Taijiao railway line

### 8.2.2 Dynamic Behavior Characteristics of the Vehicle-Track-Bridge System in R3000m Curve Sections

1. **Calculation of Operating Conditions**

The spatial model of the vehicle-track-bridge system consists of vehicle, track, and bridge models. The vehicle model considers the spatial vibration characteristics of vehicle components (see Sect. 3.4). The track model uses a ballasted track model (see Sect. 4.5.1). The bridge is modeled as a common double-track single-chamber box girder used in high-speed railways. The finite element method is employed to establish the bridge model, and the equations of motion are derived using modal synthesis (for bridge modes and modeling methods, see Sect. 4.5.2). The contact geometry nonlinearity and contact force relationship between the wheel and rail are taken into account (see Sects. 3.1 and 3.2 for the wheel-rail relationship model).

The CRH5 high-speed train model is selected for the calculations, consisting of 8 carriages with a running speed of 250 km/h. The radius of curvature is set at 3000 m, with a transition curve length of 380 m, a circular curve length of 1383 m, and a cant of 150 mm. The train runs on one side of the double-track bridge, as illustrated in Fig. 8.16. To highlight the structural vibration characteristics under curvilinear conditions, track irregularities are not considered in this case.

2. **Time-History Analysis**

Figure 8.17 illustrates the time history curves for the vertical displacement of the rail, lateral displacement of the rail, mid-span acceleration of the bridge,

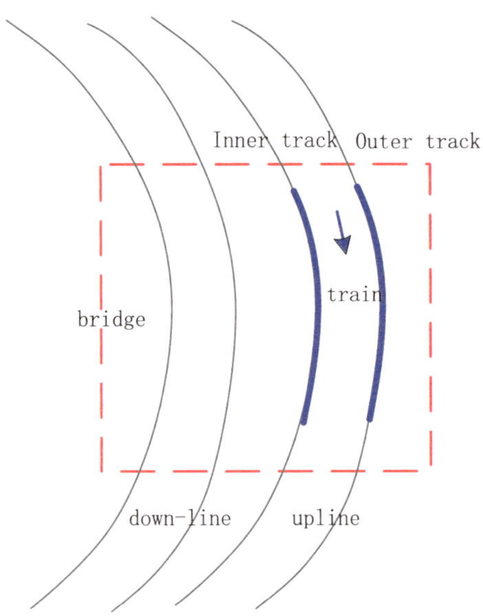


**Fig. 8.16** Train running position diagram

and fastening forces in the curved section as a train passes through the double-track curved bridge at a speed of 250 km/h. Key observations from the figure include: due to the influence of cant deficiency, the vertical deformation of the outer rail and the fastening forces are higher than those of the inner rail. The maximum vertical rail displacement is 1.8 mm, the maximum lateral rail displacement is 0.4 mm, the maximum vertical acceleration at mid-span is 0.09 m/s$^2$, and the maximum lateral acceleration at mid-span is 0.02 m/s$^2$. All observed metrics comply with the relevant standards and specifications, indicating that the structural performance under these conditions is satisfactory.

3. **Statistical Analysis**

Figure 8.18 presents the statistical curves for the maximum values of track gauge widening, sleeper lateral displacement, and fastening forces as the train passes sequentially through straight, transition, and circular curves at a speed of 250 km/h. The figure indicates that the track gauge widening gradually increases from the straight section to the circular curve, with the gauge widening on the circular curve being approximately twice that of the straight section. The lateral displacement of the sleepers also shows a slow increase from the straight section to the circular curve, with an increment of about 0.1–0.2 mm. As the train enters the curve and operates in a cant-deficient state, the vertical force on the outer rail's fastenings gradually increases, while the vertical force on the inner rail's fastenings gradually decreases.

### *8.2.3 Dynamic Analysis of Reverse Curve Setting on Bridge*

1. **Calculation Conditions**

(1) Vehicle model

All passenger trains are operated on Taijiao Railway, and CRH series EMUs with speed of 250 km/h and above are used for intercity passenger trains and cross-line passenger trains. The design size and parameters of EMUs are selected according to the actual situation.

(2) Line parameter

The starting mileage of the calculation section is DK263 + 064.32, and the end mileage is DK274 + 164.52. The section of the line plane contains a total of four curve segments, the radius of the curve is 3500 m, 11,000 m, 3500 m, 3000 m; a total of 6 slope sections and 6 vertical curves are set on the vertical section, and the radius of the vertical curve is 20,000 m. The line parameters are shown in Fig. 8.19 and Table 8.4.

Referring to the 'Code for Design of High-speed Railway' (*TB10621-2014*), the superelevation of ballast track is specified, and the actual superelevation of the curve is shown in Table 8.5.

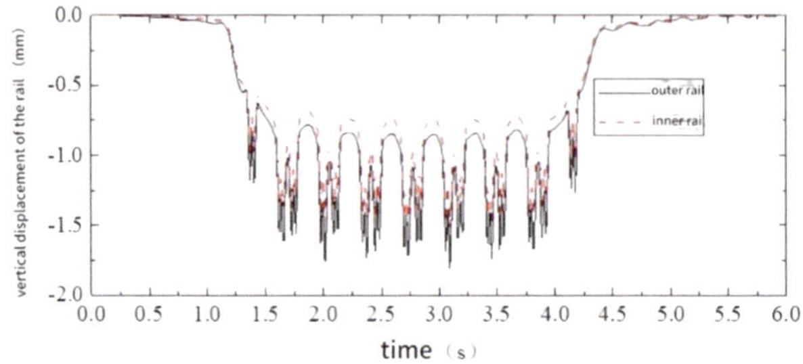


(a) Vertical displacement of the rail

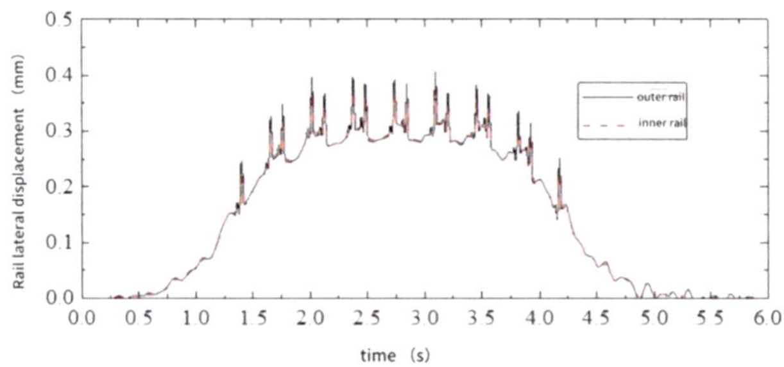


(b) Rail lateral displacement

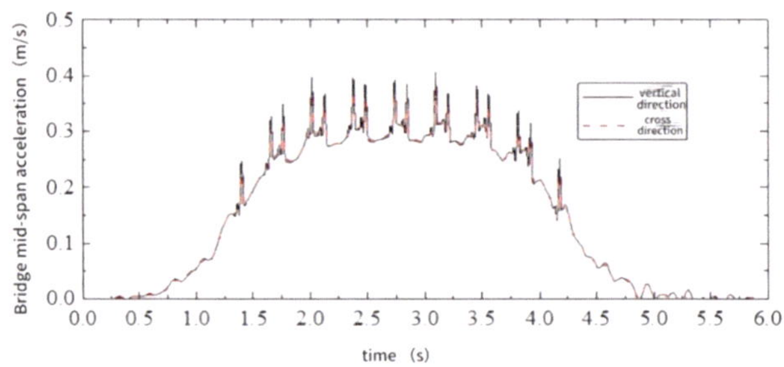


(c) Bridge mid-span acceleration

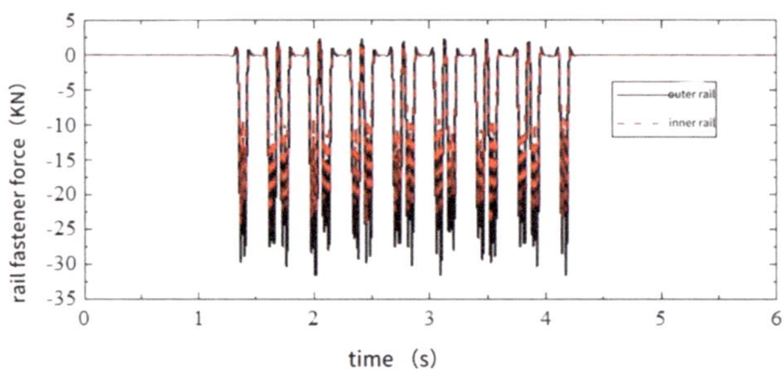


(d) Vertical fastener force in curve section

**Fig. 8.17** Time history curves of track-bridge dynamic response

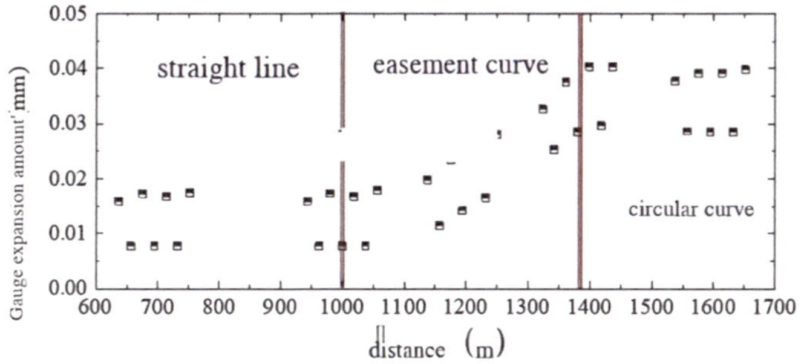


(a) The extent of gauge expansion

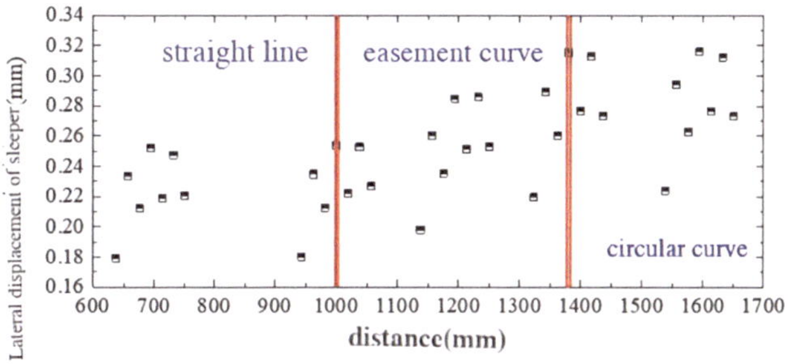


(b) The amount of sleeper transverse displacement

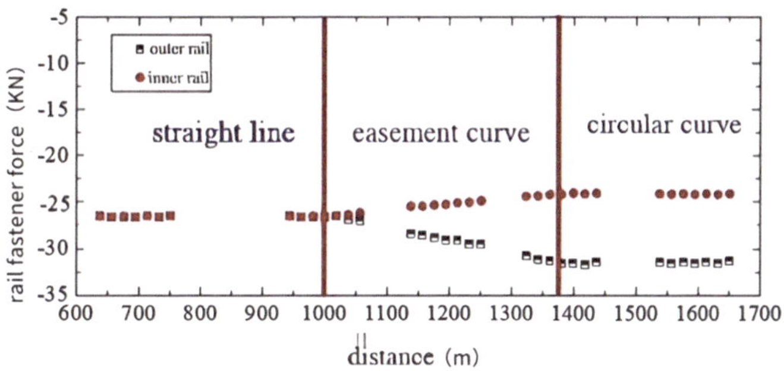


(c) The vertical force of fastener

**Fig. 8.18** Peak values of dynamic response of track structure at different positions

In the simulation calculation, the train running direction and line position are shown in Fig. 8.20, and the continuous reverse curves are numbered as Curve 1, Curve 2, Curve 3 and Curve 4.

(3) Track irregularity

Track irregularity is modeled using the irregularity spectrum for Chinese high-speed railway ballasted tracks.

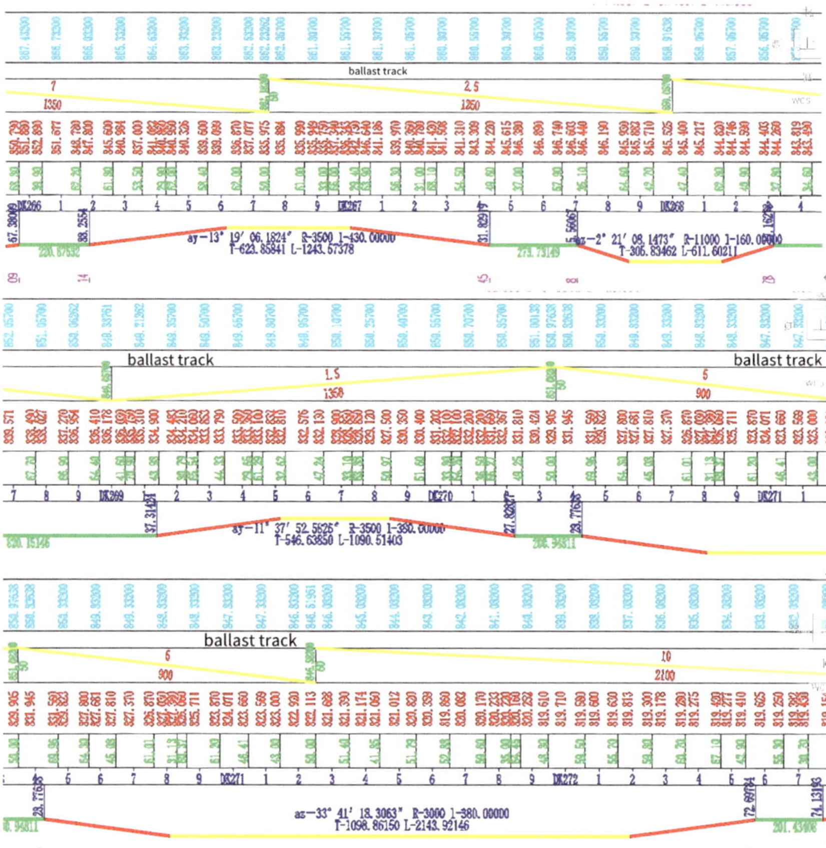


**Fig. 8.19** Setting of horizontal and vertical sections in reverse curve section of Taiyuan-Jiaozuo railway bridge

**Table 8.4** Table of curve parameters of reverse curve section on bridge of Taiyuan-Jiaozuo railway

| Radius of curve R (m) | The length of the front transition curve l (m) | Length of the trailing transition curve l (m) | Length of curve L (m) | Length of straight section (m) |
|---|---|---|---|---|
| 3500 | 430 | 430 | 383 | |
| | | | | 273 |
| 11,000 | 160 | 160 | 291 | |
| | | | | 820 |
| 3500 | 380 | 380 | 330 | |
| | | | | 200 |
| 3000 | 380 | 380 | 1383 | |

**Table 8.5** Curve superelevation setting in reverse curve section of Taijiao railway bridge

| Radius of curve (m) | Operating speed (km/h) | Equilibrium super high (mm) | Actual elevation (mm) |
|---|---|---|---|
| 3000 | 250 | 245 | 150 |
| | 275 | 297 | 150 |
| | 300 | 354 | 150 |
| 3500 | 250 | 210 | 150 |
| | 275 | 255 | 150 |
| | 300 | 303 | 150 |
| 11,000 | 250 | 67 | 105 |
| | 275 | 81 | 105 |
| | 300 | 96 | 105 |

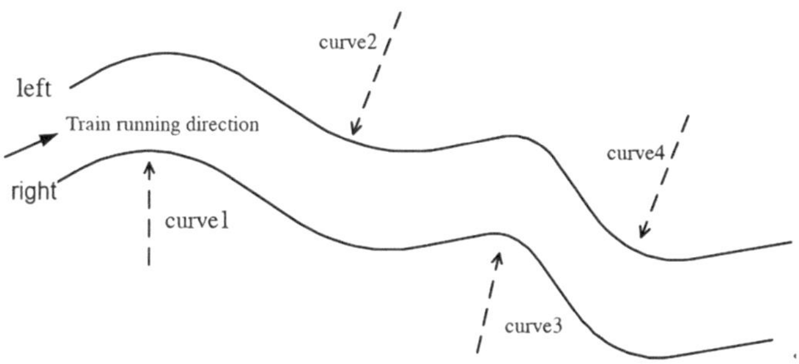


**Fig. 8.20** Schematic diagram of reverse curve section on the bridge of Taijiao railway

2. **Dynamic Analysis of Reverse Curves on the Bridge**

(1) Time History Analysis

The dynamic responses of a 16-car high-speed train passing through continuous reverse curves on the bridge at a speed of 250 km/h were calculated. The resulting time history curves are shown in Figs. 8.21, 8.22, 8.23, 8.24, 8.25, 8.26 and 8.27. The main findings are as follows:

(1) As the train passes through the continuous reverse curves, the vertical acceleration of the train body is minimally affected by the alignment, with a maximum vertical acceleration of 0.112 $m/s^2$. The lateral acceleration shows a significant amplification on Curve 4, with a maximum lateral acceleration of 0.117 $m/s^2$ and an unbalanced maximum lateral acceleration of 0.7 $m/s^2$. On the straight section, the maximum lateral acceleration is 0.075 $m/s^2$. Overall, the vibration accelerations of the train body meet the high-speed railway design specification limits of less than 0.13 g (vertical), 0.10 g (lateral), and 0.784 $m/s^2$ (unbalanced lateral).

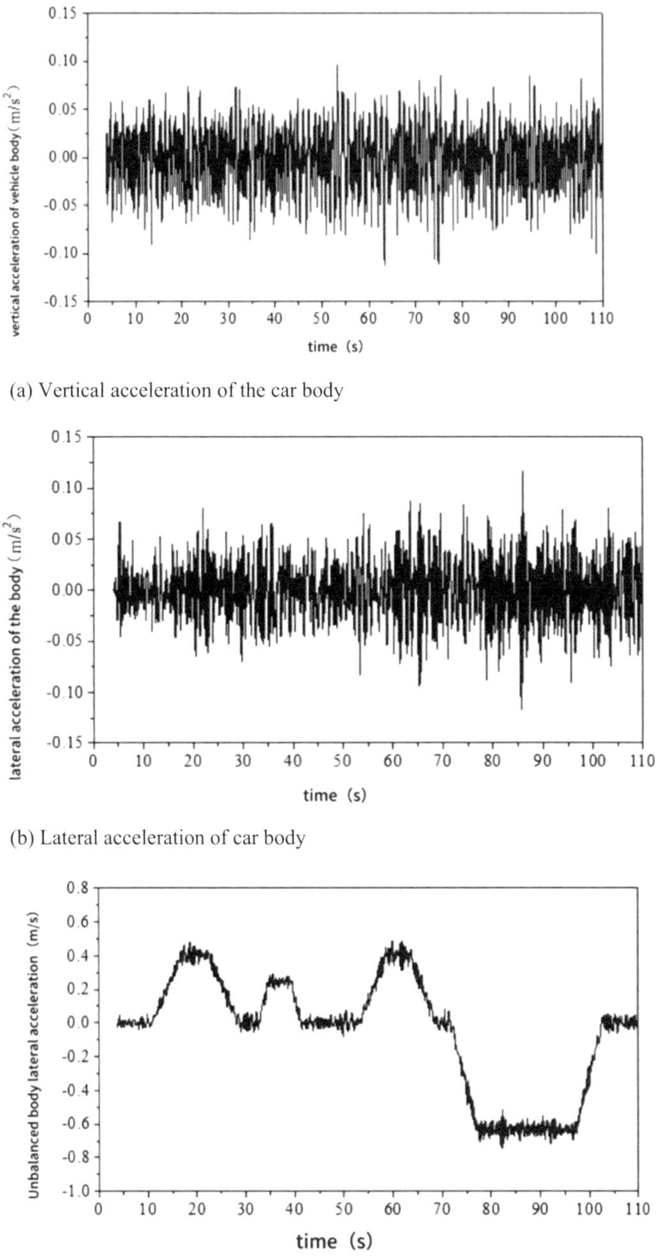


(a) Vertical acceleration of the car body



(b) Lateral acceleration of car body



(c) Unbalanced lateral acceleration

**Fig. 8.21** Vehicle dynamic response time history curve

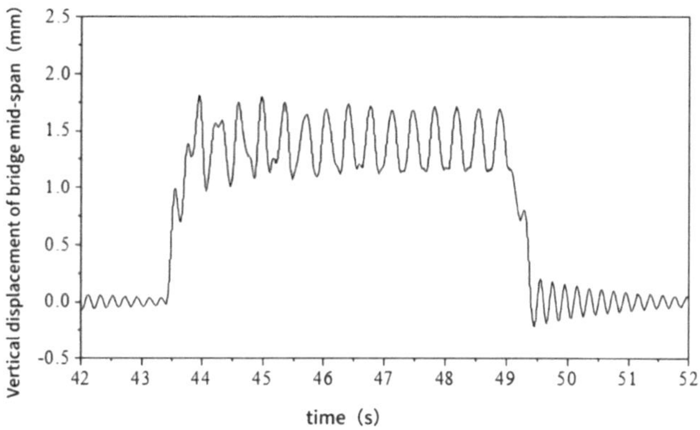


(a) Bridge mid-span vertical displacement

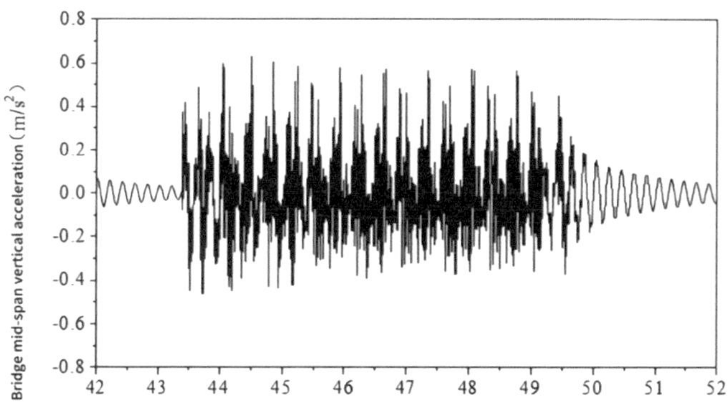


(b) Bridge mid-span vertical acceleration

**Fig. 8.22** Mid-span vertical dynamic response of the bridge

(2) The mid-span vertical dynamic displacement and acceleration of the bridge are also minimally influenced by the alignment, with a maximum dynamic displacement of 1.834 mm and a maximum acceleration of 0.743 m/s$^2$. On Curve 4, the lateral dynamic displacement and acceleration of the bridge show significant amplification, with a maximum dynamic displacement of 0.087 mm and a maximum lateral acceleration of 0.307 m/s$^2$. On the straight section, the maximum lateral dynamic displacement is 0.066 mm, and the maximum lateral acceleration is 0.290 m/s$^2$. These dynamic performance indicators for the bridge meet the control standards of 0.35 g vertical acceleration, 0.14 g lateral acceleration, and 8 mm lateral amplitude for ballasted track bridges.
(3) The lateral forces on the axles are relatively small on the straight section, with a maximum value of 2.003 kN. In the curves, the lateral forces increase significantly, with the maximum axle lateral force on Curve 4 reaching 9.55 kN, which is below the allowable value of 57.84 kN.

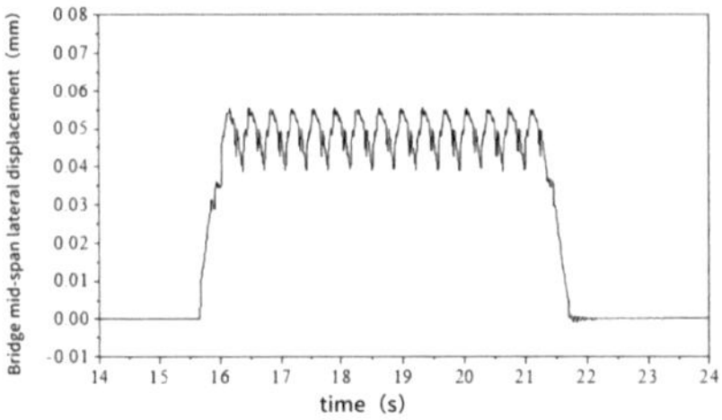


(a) Curve 1

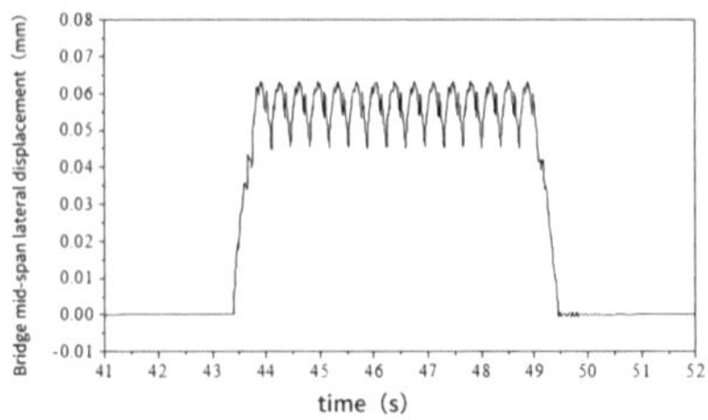


(d) Straight section 2

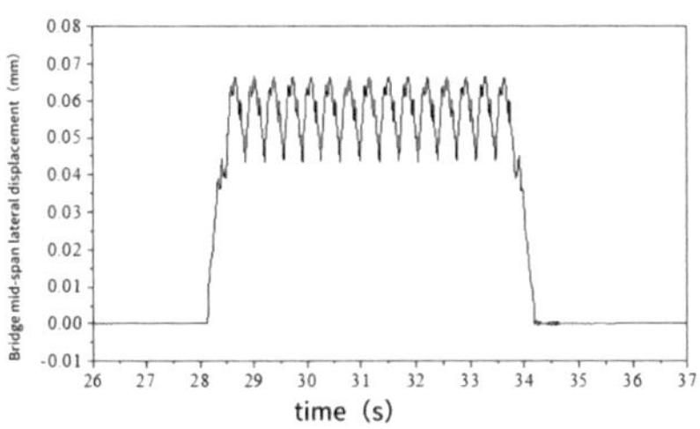


(b) Intermediate straight section 1

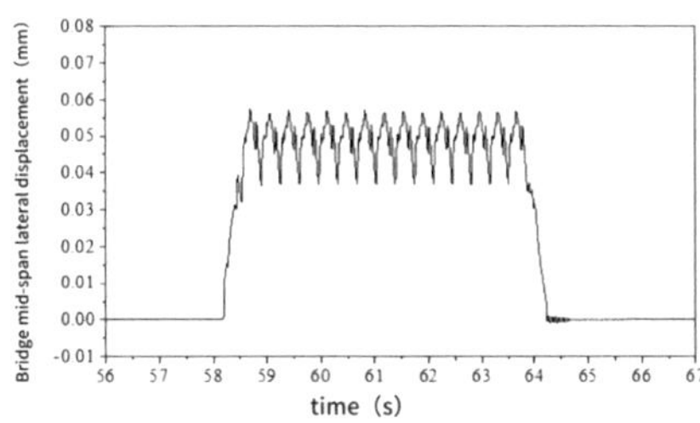


(e) Curve 3

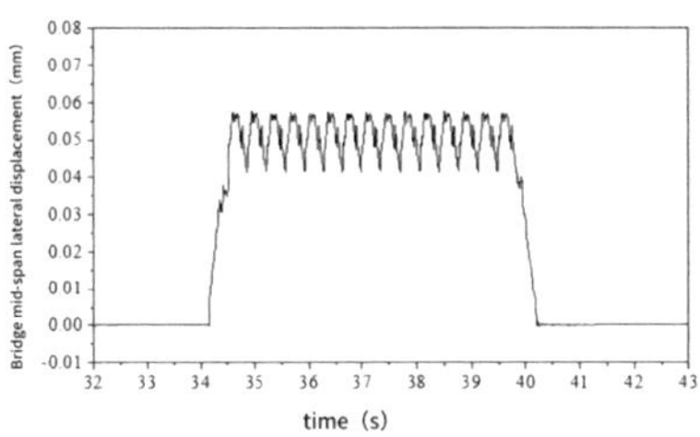


(c) Curve 2

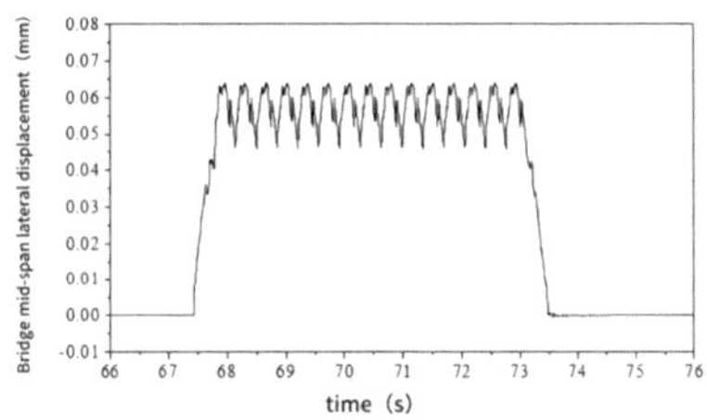


(f) Intermediate straight section 3

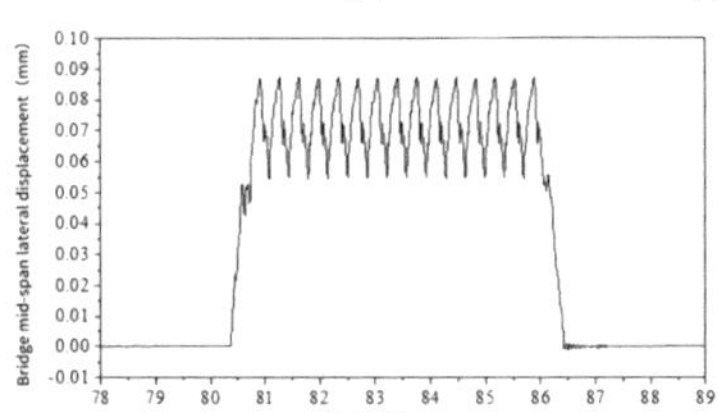


(g) Curve 4

**Fig. 8.23** Mid-span lateral displacement of bridge

(4) The safety indicators are lower on the straight section, with a maximum wheel load reduction rate of 0.035 and a maximum derailment coefficient of 0.049. On the curves, one side of the wheelset experiences an increase in load while the other side experiences a decrease, leading to larger lateral forces. The derailment coefficients and wheel load reduction rates on Curves 1, 2, and 4 show

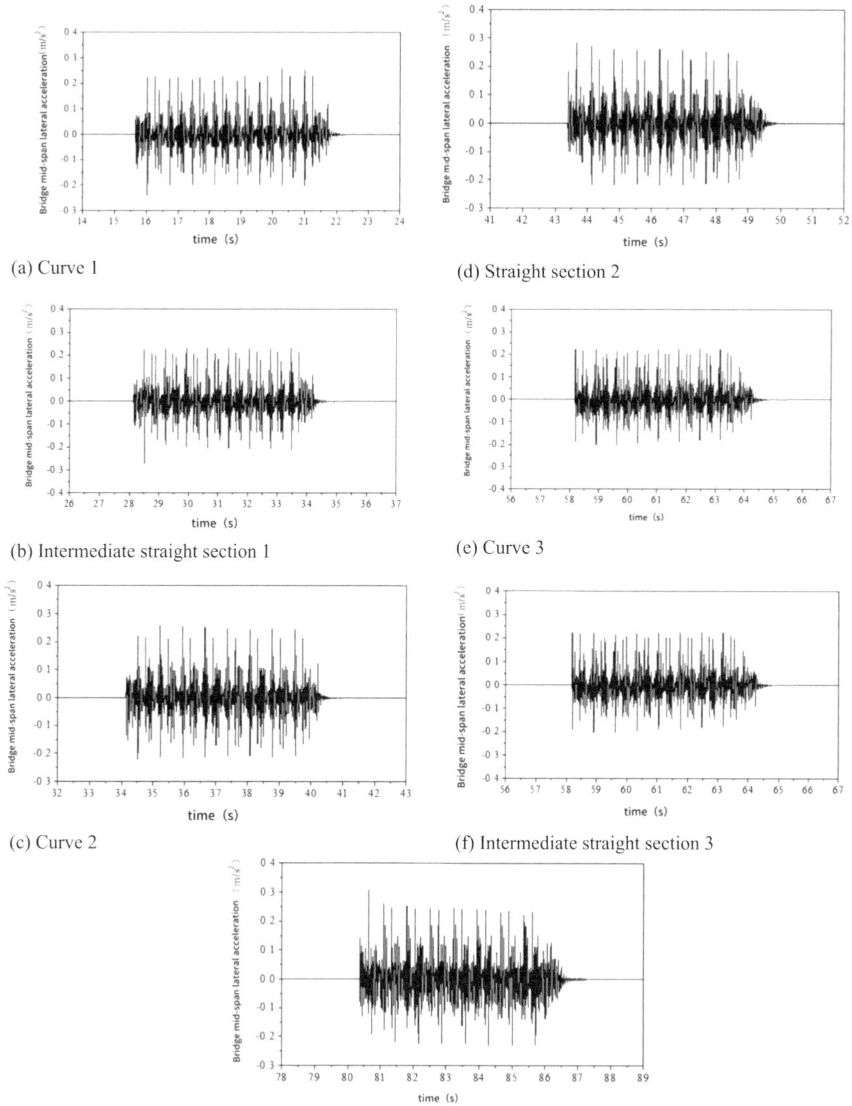


(a) Curve 1

(d) Straight section 2

(b) Intermediate straight section 1

(e) Curve 3

(c) Curve 2

(f) Intermediate straight section 3

(g) Curve 4

**Fig. 8.24** Mid-span lateral acceleration of bridge

higher values, with maximum wheel load reduction rates of 0.234 and derailment coefficients of 0.188. Both meet the safety control standards of a wheel load reduction rate below 0.8 and a derailment coefficient below 0.8.

In summary, the train passing through the continuous reverse curves on this segment at a speed of 250 km/h meets comfort, safety, and structural dynamic performance

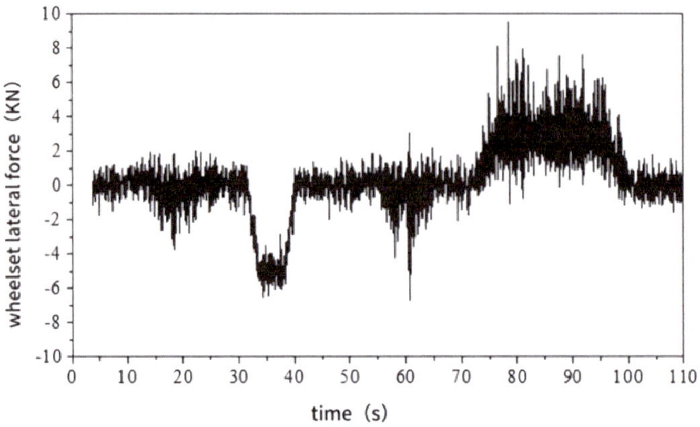


**Fig. 8.25** Transverse force of axle

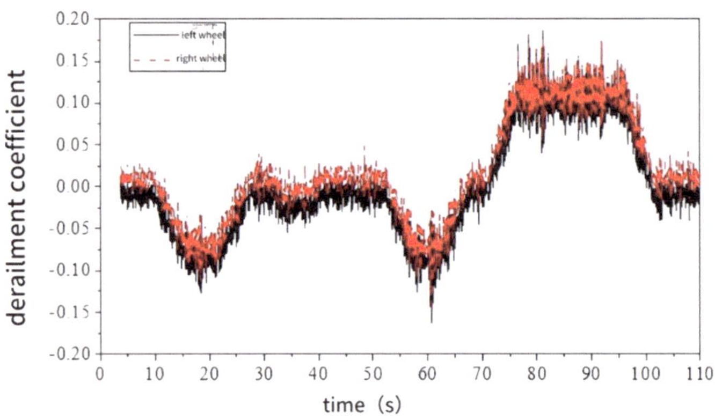


**Fig. 8.26** Derailment coefficient

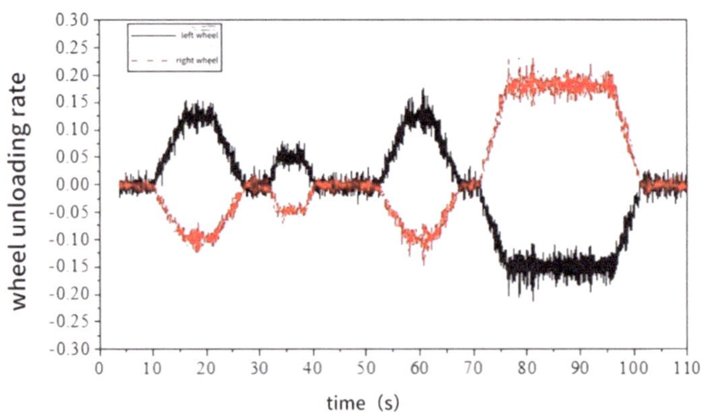


**Fig. 8.27** Wheel load reduction rate

control standards, indicating that the alignment and structural design comply with relevant requirements.

(2) **Statistical Analysis**

Further analysis of the maximum value of the dynamic response when the 250 km/h speed passes through the reverse curve on the bridge, the statistical results are shown in Figs. 8.28, 8.29, 8.30 and 8.31 and Table 8.6. The statistical results show that:

(1) The indexes of driving safety, bridge dynamic performance and operation comfort are basically the same on the intermediate straight section, and the dynamic responses are less than those on the curve section. The length of

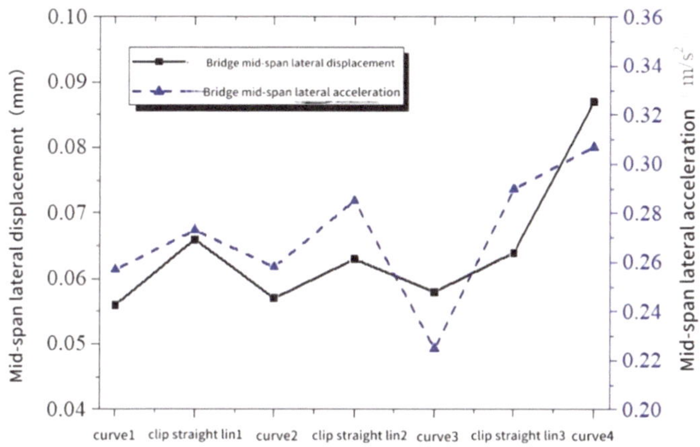


**Fig. 8.28** Peak values of mid-span transverse dynamic response of bridges at different track locations

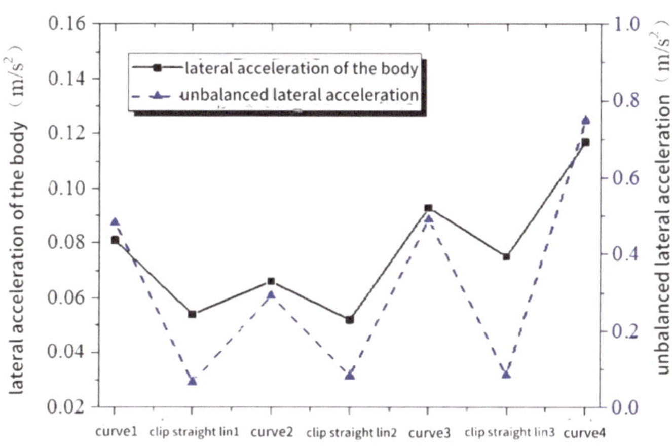


**Fig. 8.29** The peak value of lateral dynamic response of car body at different track locations

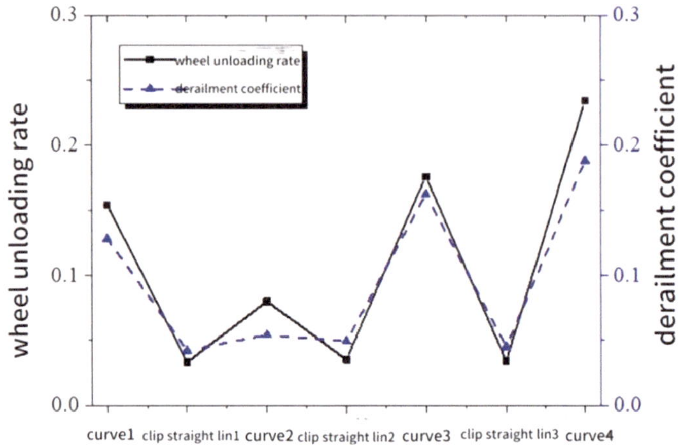


**Fig. 8.30** Peak values of traffic safety indicators at different track locations

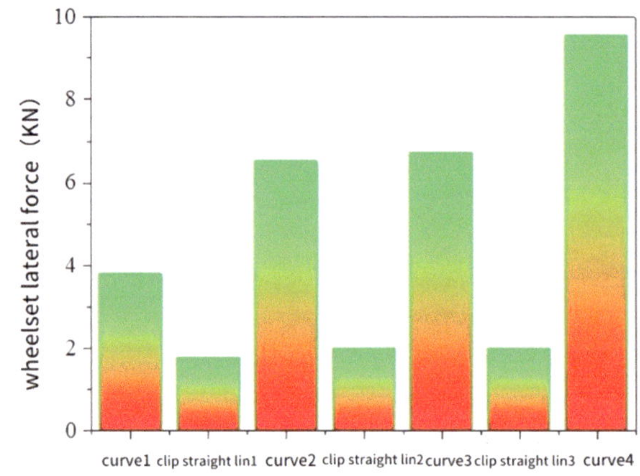


**Fig. 8.31** Peak lateral force of axle at different track locations

the intermediate straight section is not the key factor affecting the driving performance.

(2) The dynamic response indexes on the curve are higher than those on the straight section, and the maximum dynamic performance index is obtained on the 3000 m radius curve 4, but the indexes of driving safety, running comfort and structural dynamic performance on the curve all meet the requirements of the standard limits.

**Table 8.6** The maximum value of the dynamic index at the operating speed of 250 km/h

| | Dynamic displacement of bridge (mm) | | Bridge acceleration ($m/s^2$) | | Acceleration of the car body ($m/s^2$) | | |
|---|---|---|---|---|---|---|---|
| | Vertical | Lateral | Vertical | Lateral | Vertical | Lateral | Unbalanced |
| Curve 1 | 1.827 | 0.056 | 0.639 | 0.257 | 0.090 | 0.081 | 0.482 |
| Intermediate straight section 1 | 1.801 | 0.066 | 0.650 | 0.273 | 0.075 | 0.054 | 0.065 |
| Curve 2 | 1.809 | 0.057 | 0.656 | 0.258 | 0.085 | 0.066 | 0.291 |
| Intermediate straight section 2 | 1.809 | 0.063 | 0.635 | 0.285 | 0.081 | 0.052 | 0.081 |
| Curve 3 | 1.821 | 0.058 | 0.611 | 0.225 | 0.113 | 0.093 | 0.490 |
| Intermediate straight section 3 | 1.809 | 0.064 | 0.667 | 0.290 | 0.080 | 0.075 | 0.083 |
| Curve 4 | 1.834 | 0.087 | 0.743 | 0.307 | 0.112 | 0.117 | 0.749 |
| | *Wheelset lateral force (kN)* | *Load reduction rate* | *Derailment coefficient* | | | | |
| Curve 1 | 3.786 | 0.154 | 0.128 | | | | |
| Intermediate straight section 1 | 1.768 | 0.033 | 0.042 | | | | |
| Curve 2 | 6.517 | 0.080 | 0.054 | | | | |
| Intermediate straight section 2 | 2.001 | 0.035 | 0.049 | | | | |
| Curve 3 | 6.721 | 0.176 | 0.162 | | | | |
| Intermediate straight section 3 | 2.003 | 0.034 | 0.045 | | | | |
| Curve 4 | 9.552 | 0.234 | 0.188 | | | | |

### (3) The Influence of the Length of the Straight Section on the Dynamic Response of the Bridge

(1) Calculated work condition

Two reverse curves with a radius of 3000 m are set, and the lengths of the intermediate straight sections are selected as 150 m (0.6v), 200 m (0.8v), and 432 m (one train length). The calculation speed is set at 250 km/h, as shown in Table 8.7.

**Table 8.7** Curve parameter table

| Radius of curve $R$ (m) | The length of the front transition curve $l$ (m) | Length of the trailing transition curve $l$ (m) | Length of curve $L$ (m) | Length of intermediate straight section (m) |
|---|---|---|---|---|
| 3000 | 380 | 380 | 400 | |
| 3000 | 380 | 380 | 400 | 150(0.6v); 200(0.8v); 432(one train length) |

(2) Result Analysis

Figure 8.32 illustrates the variation of lateral acceleration of the vehicle body under straight section lengths of 150 m (0.6v), 200 m (0.8v), and 432 m (length of one train). It can be observed from the figure that the straight section length does not affect the impact vibrations caused at the starting point of the next curve. Furthermore, Fig. 8.33 and Table 8.8 indicate that the lateral vibrations of the vehicle body can diminish by over 80% within one cycle (0.44v).

Figure 8.34 shows the changes of wheel load reduction rate and derailment coefficient under 150 m (0.6v), 200 m (0.8v) and 432 m (a train length) intermediate straight section length. It can be seen from the diagram that the key factor affecting the driving safety index is the curve, and the shortening of the intermediate straight section length will not affect the driving safety.

(3) **Summary**

This section conducted a dynamic check on the conditions where the straight section lengths between consecutive reverse curves on the Tai-Jiao Railway do not meet

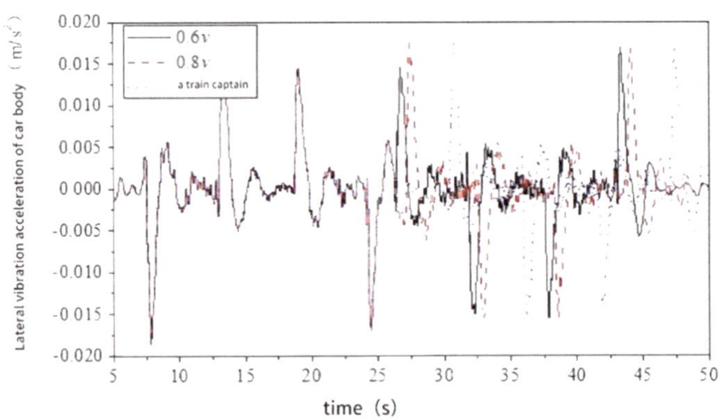


**Fig. 8.32** The change of lateral acceleration of car body under different length of intermediate straight section

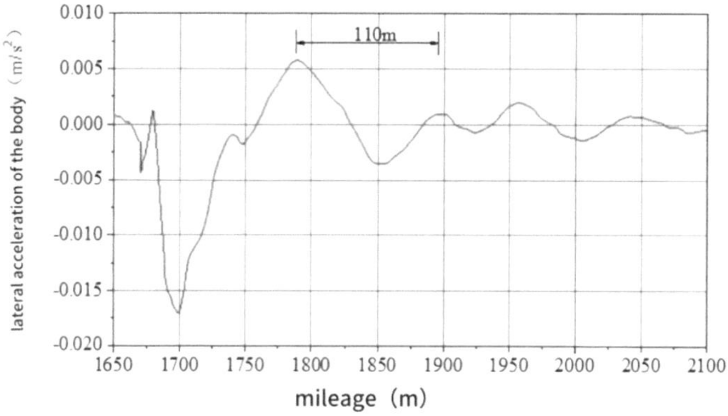


**Fig. 8.33** Vibration attenuation of intermediate straight section

**Table 8.8** Vibration attenuation of car body

| Fundamental frequency (Hz) | 0.63 |
|---|---|
| Period (s) | 1.585 |
| Wave length (m) | 110 |
| The lateral acceleration peak of starting point car body (m/s$^2$) | 0.00577 |
| Lateral acceleration attenuation ratio after a period (0.44v) | 82.4% |

the length of one train. It further analyzed the impact of straight section lengths on the safety, riding comfort, and structural dynamic performance of high-speed trains passing over reverse curves. The main conclusions are as follows:

(1) When the train passes over consecutive reverse curves at 250 km/h, various dynamic performance indicators reach their maximum values, including: maximum vertical acceleration of the vehicle body at 0.112 m/s$^2$, maximum lateral acceleration of the vehicle body at 0.117 m/s$^2$, and maximum unbalanced lateral acceleration at 0.7 m/s$^2$. The maximum vertical acceleration at the bridge mid-span is 0.743 m/s$^2$, with a maximum lateral displacement of 0.087 mm and maximum lateral acceleration of 0.307 m/s$^2$. The maximum lateral force on the axle is 9.55 kN, with a maximum wheel load reduction rate of 0.234 and a maximum derailment coefficient of 0.188. The dynamic response indicators on the straight sections are relatively small, with maximum lateral acceleration of the vehicle body at 0.075 m/s$^2$, maximum lateral dynamic displacement of the bridge at 0.066 mm, maximum lateral acceleration of the bridge at 0.290 m/s$^2$, maximum lateral force on the axle at 2.003 kN, maximum wheel load reduction rate of 0.035, and a maximum derailment coefficient of 0.049. Indicators for riding comfort, driving safety, and structural dynamics meet the requirements of the "High-Speed Railway Design Specification" (*TB10621*) and the "Testing and Evaluation Standards for Railway Locomotive Dynamic Performance" (*TB/T2360*).

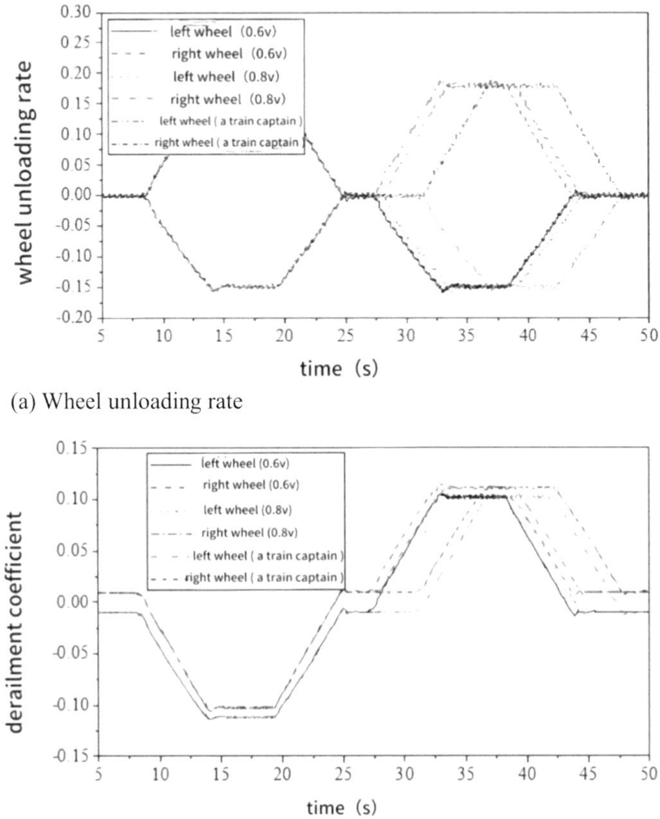


(a) Wheel unloading rate



(b) Derailment coefficient

**Fig. 8.34** Change of driving safety index under different intermediate straight section length

(2) When the train passes over consecutive reverse curves at 275 km/h, various dynamic performance indicators reach their peak values: maximum vertical acceleration of the vehicle body at 0.111 m/s$^2$, maximum lateral acceleration at 0.162 m/s$^2$, and maximum unbalanced lateral acceleration at 1.132 m/s$^2$. The maximum vertical acceleration at the bridge mid-span is 1.056 m/s$^2$, with a maximum lateral displacement of 0.102 mm and maximum lateral acceleration of 0.322 m/s$^2$. The maximum lateral force on the axle is 16.04 kN, with a maximum wheel load reduction rate of 0.362 and a maximum derailment coefficient of 0.317. The dynamic response indicators on the straight sections are also small, with maximum lateral acceleration of the vehicle body at 0.092 m/s$^2$, maximum lateral dynamic displacement of the bridge at 0.07 mm, maximum lateral acceleration of the bridge at 0.253 m/s$^2$, maximum lateral force on the axle at 2.326 kN, maximum wheel load reduction rate of 0.038, and a maximum

derailment coefficient of 0.054. Driving safety and structural dynamic performance meet relevant safety standards, although riding comfort is relatively poor on curves with radii of 3000 and 3500 m.

(3) When the train passes over consecutive reverse curves at 300 km/h, various dynamic performance indicators reach their maximum values: maximum vertical acceleration of the vehicle body at 0.114 $m/s^2$, maximum lateral acceleration at 0.261 $m/s^2$, and maximum unbalanced lateral acceleration at 1.604 $m/s^2$. The maximum vertical acceleration at the bridge mid-span is 1.472 $m/s^2$, with a maximum lateral displacement of 0.109 mm and maximum lateral acceleration of 0.429 $m/s^2$. The maximum lateral force on the axle is 26.02 kN, with a maximum wheel load reduction rate of 0.504 and a maximum derailment coefficient of 0.507. The dynamic response indicators on the straight sections are small, with maximum lateral acceleration of the vehicle body at 0.078 $m/s^2$, maximum lateral dynamic displacement of the bridge at 0.068 mm, maximum lateral acceleration of the bridge at 0.329 $m/s^2$, maximum lateral force on the axle at 2.546 kN, maximum wheel load reduction rate of 0.043, and a maximum derailment coefficient of 0.06. Driving safety and structural dynamics meet relevant safety standards, but riding comfort is also poor on curves with radii of 3000 and 3500 m.

(4) The driving safety, bridge dynamic performance, and riding comfort indicators in the straight sections between reverse curves are all lower than those in the curve sections, indicating that the reduction of straight section length has a minor impact on driving safety. The lateral vibrations of the vehicle body can diminish by over 80% within one cycle (0.44v). Considering the length of the vehicle, setting the straight section length above 0.6v can ensure sufficient damping of vibrations from the previous curve.

In conclusion, when the train passes over the Gaoping Bridge on National Highway 207 at a speed of 250 km/h through consecutive reverse curves, the riding comfort, driving safety, and structural dynamic performance all meet control standards. The alignment and structure of this section comply with relevant requirements. The dynamic simulation analysis provides a theoretical basis for the implementation of this plan.

# Bibliography


1. Liu J (2022) Structural dynamics, 2nd edn. China Machine Press, Beijing
2. Yu K, Zou J (2015) Structural dynamics, 3rd edn. Harbin Institute of Technology Press, Harbin
3. Hong J (1999) Computational dynamics of multibody systems. Higher Education Press, Beijing
4. Yuan S, (Lü/LV/LU/LYU) Z (1992) Dynamics of multi rigid body systems. Beijing Institute of Technology Press, Beijing
5. Zhang X, Wang T, Liu Y (2015) Computational dynamics, 2nd edn. Tsinghua University Press, Beijing
6. Wang F (1994) Vehicle system dynamics. China Railway Publishing House, Beijing
7. Zhang D (1996) Dynamics of rolling stock track system. China Railway Publishing House, Beijing
8. Zhai W (2015) Vehicle-track coupled dynamics, 4th edn. Science Press, Beijing
9. Ren Z (2009) Fundamentals of vehicle dynamics. China Railway Publishing House, Beijing
10. Luo R, Shi H (2019) High-speed train system dynamics. Southwest Jiaotong University Press, Chengdu
11. Yao J (2014) Dynamics of rolling stock. Science Press, Beijing
12. Shen G (2014) Railway vehicle system dynamics. China Railway Publishing House, Beijing
13. Zhai W, Xia H (2011) Train-track-bridge dynamic interaction: theory and engineering application. Science Press, Beijing
14. Zhang N, Guo W, Xia H (2018) Dynamic interaction of train-bridge systems in high-speed railway. Beijing Jiaotong University Press, Beijing
15. Xia H, Zhang N, Guo W (2014) Coupling vibrations of train-bridge system. Science Press, Beijing
16. Lei X (2021) High speed railway track dynamics. 2nd edn. Science Press, Beijing
17. Gao L (2020) Spatial refinement analysis method of ballastless track of high-speed railway and its application. Science Press, Beijing
18. Li X, Jin Z, Zhu Yan (2017) Random vibration of vehicle-bridge time-varying system:theory and engineering application. Science Press, Beijing
19. Frýba L (1999) Vibration of solids and structures under moving loads, 3rd edn. Thomas Telford, UK
20. Garg VK, Dukkipati RV (1984) Dynamics of railway vehicle systems. Elsevier, Amsterdam
21. Knothe K, Stichel S (2017) Rail vehicle dynamics. Springer Nature, Switzerland
22. Iwnicki S, Spiryagin M, Cole C, McSweeney T (2019) Handbook of railway vehicle dynamics. CRC Press, US
23. Wickens AH (2004) Fundamentals of rail vehicle dynamics: guidance and stability. Swets & Zeitlinger Publishers
24. Shabana AA, Zaazaa KE, Sugiyama H (2010) Railroad vehicle dynamics: a computational approach. CRC Press


© China Railway Publishing House Co., Ltd. 2026

X. Long et al., *Vehicle–Track–Bridge System Subjected to Track Alignment*,
https://doi.org/10.1007/978-981-95-1649-0

25. Yang YB, Yau JD, Wu YS (2004) Vehicle-bridge interaction dynamics-with applications to high-speed railways. World Scientific
26. Frýba L (1996) Dynamics of railway bridges. Thomas Telford, London
27. Sauvage G, Pascal J-P (1990) Solution of the multiple wheel and rail contact dynamic problem. Veh Syst Dyn 19(5):257–272
28. Shen ZY, Hedrick JK, Elkins JA (1983) A comparison of alternative creep force models for rail vehicle dynamic analysis. In: Proceedings 8th IAVSD symposium, MIT, Cambridge, 1983, pp 591–605
29. Nagurka ML (1983) Curving performance of rail passenger vehicles. Massachusetts Institute of Technology
30. Yang G (1993) Dynamic analysis of railway wheelsets and complete vehicle systems. Delft University of Technology
31. Xia F (2002) The dynamics of the three-piece-freight truck. Technical University of Denmark
32. Hoffmann M (2006) Dynamics of European two–axle freight wagons. Technical University of Denmark, Kongens Lyngby
33. Jin X (1999) Theory and experimental study of wheel rail creep. Southwest Jiaotong University, Chengdu
34. Yang G, Wei Y, Zhao G et al (2015) Research progress on the mechanics of high speed rails. Adv Mech 45(0):217–460
35. International Union of Railways (1994) UIC 513R guidelines for evaluating passenger comfort in relation to vibration in railway vehicles
36. International Organization for Standardization (1997) ISO 2631-1 mechanical vibration and shock-evaluation of human exposure to whole-body vibration. Part 1: general requirements
37. Suzuki H (1998) Research trends on riding comfort evaluation in Japan. Proc Inst Mech Eng, Part F: J Rail Rapid Transit 212(1):61–72
38. Kim YG, Won HB, Kim SW et al (2003) Correlation of ride comfort evaluation methods for railway vehicles. Proc Inst Mech Eng, Part F: J Rail Rapid Transit 217(2):73–88
39. Wang K (1984) The track of wheel contact points and the calculation of wheel/rail geometric contact parameters. J Southwest Jiaotong Univ 19(1):89–99
40. Wang C, Wang Y, Li H et al (2006) Comparative analysis on high-speed wheel/rail contact geometric. Railw Locomotive Car 26(4):1–5
41. Zhai W (2006) Optimum matching design between railway vehicles and tracks: principle, method and engineering practice. China Railw Sci 27(2):60–65
42. Zhai W, Cai C, Wang K (2010) Dynamics assessment method for design of high speed railway plan and profile. High Speed Railw Technol 1(1):1–5
43. Liu P, Wang K, Zhai W (2018) Simulation solution method for long and heavy-haul trains negotiating elastic curved tracks. J Southwest Jiaotong Univ 53(1):31–37
44. Zhai W, Yao L, Sun L et al (2021) Research on route plan design parameters of 400 km/h high-speed railway based on vehicle-track coupling dynamics. High Speed Railw Technol 12(2):1–10, 16
45. Wang K, Zhou W, Zhai W et al (2005) Research on reasonable matching of the level section and surface on the high or medium speed passenger-dedicated railway and high or low speed passenger and freight mixed railway based on the theory of dynamics. Railw Stand Des 49(7):1–3
46. Wang K, Zhai W (2014) Study on performance matching of wheel-rail dynamic interaction on curved track of speed-raised and high-speed railways. China Railw Sci 35(1):142–144
47. Wang K, Yuan X, Zhai W (2015) The adaptability design method and its application for high-speed electric multiple unit on the track. Eng Sci 17(4):90–97
48. Wang K, Lyu K (2016) Dynamic evaluation index system for spatial alignment of high-speed railway. J Southwest Jiaotong Univ 51(2):227–235
49. Liang C, Si D, Xu Y (2016) Research of minimum curve radius value for high speed railway at 400/380 km/H. Railw Eng 56(1):63–66
50. Yi S, Nie L, Qin F (2013) Study on minimum curve radius of high-speed railway based on dynamics analysis. J Southwest Jiaotong Univ 48(1):16–20, 35

51. Krzysztof Z (1998) Dynamical investigation of railway vehicles on a curved track. Eur J Mech, A/Solids 17(6):1001–1020
52. Kufver B, Andersson E (1998) Optimisation of length of transition curves with vehicle reaction taken into consideration. Trans Built Environ 34:33–42
53. Kufver B (1997) Optimisation of single horizontal curves in railway alignments. TRITA-FKT Report 1997(45)
54. Um JH, Choi IY, Yang SC et al (2011) Optimization of alignment considering ride comfort for superimposition of vertical and horizontal curves. Proc Inst Mech Eng, Part F: J Rail Rapid Transit 255(6):649–662
55. Long X (2008) Study on dynamic effect of alignment parameter on train running quality and its optimization for high-speed railway. Beijing Jiaotong University, Beijing
56. Long X, Shi J, Wang Y et al (2012) Dynamic simulation and ride comfort evaluation of track alignment for high speed railway. J Railw Sci Eng 9(3):26–33
57. Long XY, Wei QC, Zheng FY (2010) Dynamic analysis of railway transition curves. Proc Inst Mech Eng, Part F: J Rail Rapid Transit 224(1):1–14
58. Long X, Shi J, Wang Y (2016) Research on the key parameters of plane and profile for heavy haul railway. J Railw Eng Soc 33(1):30–35
59. Long XY, Wei QC, Shi J, Wang YJ (2009) Dynamic analysis of slab track due to high speed trains. In: Environmental vibration: prediction, monitoring and evaluation. Beijing
60. Long X, Shi J, Fang W (2015) Evaluation of ride comfort caused by track vertical alignment of highspeed railway. J Railw Eng Soc 32(4):48–53
61. Wang Y, Shi J, Long X et al (2013) Dynamic responses of high speed railway human body-vehicle system and comfort evaluation. J Central South Univ (Sci Technol) 44(7):3046–3052
62. Shi J, Long X, Wang Y (2016) Research of curve parameters and analysis of dynamic response caused by train-bridge interaction on reverse curve section of heavy haul railway bridge. J China Railw Soc 38(3):119–126
63. Gao Y, Shi J, Lu CX (2019) A two-step composite time integration scheme for vehicle-track interaction analysis considering contact separation. Shock Vib 2019(1):1212069. [LinkOut]
64. Jin S, Feng HD (2020) Reliability assessment of a curved heavy-haul railway track-bridge system. Struct Infrastruct Eng 16(3):465–480. [LinkOut]
65. Shi J, Gao Y, Long XY et al (2021) Optimizing rail profiles to improve metro vehicle-rail dynamic performance considering worn wheel profiles and curved tracks. Struct Multidiscip Optim 63(1):419–438
66. Shi J, Sun Z, Sun X et al (2020) Research on the adaptability of high-speed train running at 400 km/h under current 350 km/h high-speed railway technical conditions. J Railw Sci Eng 17(9):2171–2180
67. Shi J, Ma DK, Gao Y (2021) Interaction dynamic response of a high-speed train moving over curved bridges with deficient or surplus superelevation. Int J Struct Stab Dyn 21(8):2150103
68. Neves SGM, Montenegro PA, Azevedo AFM et al (2014) A direct method for analyzing the nonlinear vehicle–structure interaction. Eng Struct 69:83–89. [LinkOut]
69. Bell CE, Horak D, Hedrick JK (1981) Stability and curving mechanics of rail vehicles. J Dyn Syst Meas Contr 103(3):181–190
70. Kuo CM, Huang CH, Chen YY (2008) Vibration characteristics of floating slab track. J Sound Vib 317(3/4/5):1017–1034
71. Yang YB, Yau JD (1997) Vehicle-bridge interaction element for dynamic analysis. J Struct Eng 123(11):1512–1518
72. Dong RG, Sankar S, Dukkipati RV (1994) A finite element model of railway track and its application to the wheel flat problem. Proc Inst Mech Eng, Part F: J Rail Rapid Transit 208(1):61–72
73. Kortüm W (1993) Review of multibody computer, codes for vehicle system dynamics. Veh Syst Dyn 22(sup1):3–31
74. Newton SG, Clark RA (1979) An investigation into the dynamic effects on the track of wheelflats on railway vehicles[J]. ARCHIVE J Mech Eng Sci 1959–1982 (vols 1–23), 21(4):287–297

MIX
Papier aus verantwortungsvollen Quellen
Paper from responsible sources
FSC® C105338

If you have any concerns about our products,
you can contact us on
**ProductSafety@springernature.com**

In case Publisher is established outside the EU,
the EU authorized representative is:
**Springer Nature Customer Service Center GmbH**
**Europaplatz 3, 69115 Heidelberg, Germany**

Printed by Libri Plureos GmbH
in Hamburg, Germany